Toward a Whole-of-Europe Approach

Svenja Post

Toward a Whole-of-Europe Approach

Organizing the European Union's and Member States' Comprehensive Crisis Management

With a foreword by Professor Dr. Annette Jünemann

Springer VS

Svenja Post
Berlin, Germany

Helmut-Schmidt-Universität/Universität der Bundeswehr Hamburg, Fakultät für Wirtschafts- und Sozialwissenschaften, Dissertation, 2014

ISBN 978-3-658-08022-8 ISBN 978-3-658-08023-5 (eBook)
DOI 10.1007/978-3-658-08023-5

Library of Congress Control Number: 2014957856

Springer VS

Printed on acid-free paper

Springer VS is a brand of Springer Fachmedien Wiesbaden
Springer Fachmedien Wiesbaden is part of Springer Science+Business Media
(www.springer.com)

Vorwort

Spätestens die Einsätze im Irak und in Afghanistan haben deutlich vor Augen geführt, dass militärische Mittel nur sehr begrenzt zur nachhaltigen Konfliktlösung beitragen können. Vor diesem Hintergrund richtet sich das Interesse von Theoretikern und Praktikern zunehmend auf die optimale Kombination unterschiedlicher Formen der *zivilen* Konfliktbearbeitung und ihrem Zusammenspiel mit gegebenenfalls notwendigen militärischen Instrumenten. Genau das verspricht der *Comprehensive Crisis Management Approach*, für dessen Umsetzung die Europäische Union historisch und systemisch geradezu prädestiniert erscheint. Unter den zahlreichen Akteuren, die innerhalb der EU in deren auswärtige Beziehungen involviert sind, besteht ein breiter Konsens über die Notwendigkeit und den besonderen Mehrwert des *Comprehensive Crisis Management Approach*. Ausgangspunkt der vorliegenden Arbeit von Svenja Post ist jedoch die Beobachtung, dass zwischen Anspruch und Wirklichkeit eine enorme Lücke klafft. Der *Comprehensive Crisis Management Approach* steht zwar im Zentrum aller einschlägigen EU-Dokumente und Politikerreden, seine Umsetzung bleibt jedoch hinter den damit geweckten Erwartungen zurück. Von dieser Beobachtung leitet sich die Frage nach den Ursachen für die defizitäre Umsetzung des *Comprehensive Crisis Management Approach* ab. Liegen die Defizite bereits im konzeptionellen Bereich oder allein auf Ebene der Implementierung? Fehlt es an politischem Willen und wenn ja auf welcher Ebene? Oder liegen die Probleme im systemischen Bereich, konkret dem mangelhaften Akteursstatus der EU? Mit diesem Fragenkomplex ist das Thema des Buches von hoher Aktualität und von großer sowohl akademischer wie auch politischer Relevanz.

Zweifel an der Akteursqualität der EU sind stark verbreitet und werden auch in den Medien regelmäßig (re-)produziert. Diese gängige Lesart europäischer Außenpolitik mündet in der immer gleichen Klage, dass die EU noch immer nicht mit einer Stimme spräche. Auch Henry Kissingers oft zitierte Frage nach der Telefonnummer, unter der man im Krisenfall die EU erreichen könne, wird in diesem Kontext gerne zitiert. Frau Post schließt sich dieser begrenzten, in Neorealismus und liberalen Intergouvernementalismus wurzelnden Perspektive auf die EU jedoch nicht an und kritisiert den staatzentrierten *bias* beider Theoriestränge. Um das sehr viel komplexere Wechselspiel zwischen der EU und ihren Mitgliedstaaten analytisch zu durchdringen, zieht sie den institutionalistischen

‚Europäisierungsansatz' vor. Danach prägen Staaten zwar internationale Organisationen, diese haben aber auch ihr Eigenleben und prägen reziprok die Mitgliedstaaten. Um dieses Wechselspiel inklusive des institutionellen Eigenlebens auf allen Ebenen des politischen Systems noch näher unter die Lupe nehmen zu können, ergänzt Svenja Post den Europäisierungsansatz durch Ansätze der öffentliche Verwaltungslehre. Organisationstheorien, so das Ergebnis ihrer theoretischen Erörterungen, sind sehr praxisorientiert und damit hilfreich, um politische Prozesse auch auf der untersten Ebene der Implementierung erklären und interpretieren zu können. Mit dieser Herangehensweise gelingt es Frau Post, den *Comprehensive Crisis Management Approach* nicht nur auf EU-Ebene, sondern auch seine Implementierung in den EU-Mitgliedsstaaten Großbritannien, Deutschland und Schweden tiefgehend zu analysieren. Die Kombination der Europäisierungsansatzes mit Ansätzen aus der Organisationstheorien, die zur Analyse der unterschiedlich involvierten EU-Institutionen und nationalen Ministerien und Akteure herangezogen werden, ist ausgesprochen originell und hat neue Erkenntnisse über die tiefer liegenden Ursachen für die defizitäre Umsetzung des *Comprehensive Crisis Management Approach* generiert.

Das Kernproblem des *Comprehensive Crisis Management Approach* fasst Frau Post mit einem Satz zusammen *„The desire to coordinate is greater than the wish to be coordinated"*. Dieses Kernproblem, so ihre Schlussfolgerung, kann nur durch eine stärkere politische Führung gelöst werden, wobei sie ihr Postulat jedoch in einem genuin politischen Sinn verstanden wissen will. Um die konstatierten Defizite bei der Umsetzung des *Comprehensive Crisis Management Approach* zu beheben geht es nicht um ein mehr oder weniger an Hierarchie in den relevanten Entscheidungsprozessen. Es reicht auf keinen Fall aus, allein am institutionellen Design zu basteln. Viel wichtiger erscheint es, auf informeller Ebene von Werten, Normen und Wahrnehmungen anzusetzen und die Entwicklung einer gesamteuropäischen Kultur des Krisenmanagements voranzutreiben. So ist das wichtigste Ergebnis dieser Arbeit, dass der *Comprehensive Crisis Management Approach* ein holistischer Ansatz ist, dessen Defizite nicht mit einer mechanischen Problemlösung zu beheben sind, sondern allein durch einen den Realitäten angepassten Paradigmenwechsel im Verständnis eines gemeinsamen europäischen Krisenmanagements.

Ich hatte große Freude, die dieser Publikation zugrundeliegende Dissertation zu betreuen, die Svenja Post im Frühjahr 2014 an der Helmut-Schmidt Universität in Hamburg erfolgreich verteidigte. Frau Post hat mit ihrer Arbeit ein Schlüsselwerk zur Problematik europäischer Konfliktbearbeitung vorgelegt, das wichtige Forschungslücken schließt und damit sowohl die Integrationsforschung als auch die Friedens- und Konfliktforschung bereichert. Zugleich ist dieses Buch ausgesprochen praxisorientiert, so dass ich es nicht nur akademisch Interessierten, sondern auch allen EntscheidungsträgerInnen, die im komplexen Mehrebenensystem der EU und ihrer Mitgliedsstaaten in die Politik des Krisenmanagements involviert sind, zur Lektüre ans Herz legen möchte.

Annette Jünemann
Hamburg im Juli 2014

Danksagung

Ohne die vielfältige Unterstützung, motivierende aber auch kritische Worte vieler Menschen hätte ich dieses Dissertationsprojekt niemals realisieren können. Ohne kollegialen, freundschaftlichen und familiären Rückhalt wäre eine Fertigstellung undenkbar gewesen.

Meinen besonderen Dank möchte ich daher Prof. Dr. Annette Jünemann dafür aussprechen, mich in dieser spannenden Phase meiner akademischen Laufbahn so umfassend betreut und begleitet zu haben. Bedanken möchte ich mich auch für den inhaltlichen und persönlichen Zuspruch von Prof. Dr. Berthold Meyer. Auf ihren Rat und ihre Unterstützung konnte ich mich stets verlassen. Besonders bedanken möchte ich mich zudem für die Freiheit, welche sie mir gewährten, die Promotionszeit im Sinne meines angestrebten beruflichen Werdegangs und meiner eigenen Vorstellungen zu gestalten.

Für diese Forschungsarbeit wurden 48 Interviews geführt und ich bin jedem einzelnen meiner Interviewpartner dankbar, sich die Zeit genommen zu haben, ihr Wissen und ihre Erfahrungen mit mir zu teilen. Ohne die großzügige finanzielle Unterstützung im Rahmen eines Promotionsstipendiums der FAZIT-Stiftung wäre die Bearbeitung dieses Projektes nicht möglich gewesen. Für die immer schnelle und unbürokratische finanzielle Hilfe, insbesondere auch in Form von Reisekostenzuschüssen, bin ich sehr dankbar.

Hervorzuheben sind auch mehrere Einrichtungen, welche mir als institutionelle und wissenschaftliche „Heimathäfen" dienten. Die wissenschaftliche Tätigkeit am Stockholm International Peace Research Institute (SIPRI) im Anschluss an mein Studium machte die Formulierung des Dissertationsthemas erst möglich. Für die grundlegende Beratung in dieser Anfangsphase bin ich bis heute sehr dankbar. Der Deutschen Gesellschaft für Auswärtige Politik (DGAP) danke ich für die Gastfreundschaft im Rahmen ihres Gastwissenschaftlerprogramms. Insbesondere der außenpolitischen Bibliothek und Dokumentationsstelle der DGAP, in der ich mit meinen zahlreichen Literaturwünschen stets willkommen war, gilt an dieser Stelle großer Dank. Weiterhin war ein Forschungsaufenthalt am American Institute for Contemporary German Studies (AICGS) in Washington DC eine sehr bereichernde Erfahrung, welche mich motiviert zum Endspurt ansetzen ließ. Zahlreichen Damen von Women In International Security bin ich für wiederkehrenden Austausch in verschiedenen Bereichen der Außen- und Sicherheitspolitik der letzten Jahre äußerst dankbar.

Insbesondere danke ich jedoch meinen Eltern, die immer an mich, meine Ziele und meine Projekte geglaubt und mich auf vielfältige und beste Weise unterstützt haben. Stets bestärkt haben mich auch meine Schwestern und Großmütter. Mein größter Dank gilt meiner Familie und ausdrücklich auch meinen Freunden dafür, mich stets auch an andere wichtige Dinge im Leben erinnert zu haben.

Ihnen ist diese Dissertation gewidmet.

Svenja Post
Berlin, Juli 2014

Contents

Abbreviations 15

List of Figures 19

1 Introduction 21

1.1 Central Problem Statement 21

1.2 Research Questions and Purpose 27

1.3 Current Research and Literature Review 33

1.4 Research Design 41

1.4.1 Structure 42

1.4.2 Theoretical Considerations 43

1.4.3 Methodology 53

1.4.4 Challenges and Limitations 58

2 Analytical Framework 63

2.1 Crisis, Conflict and Preoccupation with Conflict 64

2.1.1 Terminological Clarifications 64

2.1.2 Characteristics of Contemporary Crisis Management in Today's International Security Environment 68

2.2 The Comprehensive Approach 77

2.2.1 Conceptual Overview and Key Dimensions 77

2.2.2 Motivations and Challenges 84

2.2.3 Understandings 89

2.2.4 Application in this Dissertation 103

2.3 Theoretical Approach 105

2.3.1 Coherence and Coordination – Semantics and Conceptualizations 106
2.3.2 Organization Theory and Public Administration 119
2.3.3 Europeanization 135

2.4 Analytical Dimensions and Parameter 148
2.4.1 Crisis Management Approach 150
2.4.2 Conceptual Dimension 152
2.4.3 Organizational Dimension 155
2.4.4 European Dimension and EU-Interface 158

2.5 Analytical Framework Synopsis 160

3 Comprehensive EU Crisis Management 163

3.1 The EU's Crisis Management Approach 164

3.2 The Comprehensive Approach in EU Crisis Management 169
3.2.1 Conceptual Dimension 169
3.2.2 Organizational Dimension 182

3.3 Synopsis: Comprehensive EU Crisis Management 198

4 Member State Comprehensive European Crisis Management 209

4.1 Case Study UK 210
4.1.1 The British Crisis Management Approach 210
4.1.2 UK Comprehensive Crisis Management 216
4.1.3 European Dimension and EU Interface in British Crisis Management 249
4.1.4 Synopsis: The UK and Comprehensive European Crisis Management 258

4.2 Case Study Germany 267
4.2.1 The German Crisis Management Approach 267
4.2.2 German Comprehensive Crisis Management 273
4.2.3 European Dimension and EU-Interface in German Crisis Management 298

4.2.4 Synopsis: Germany and Comprehensive European Crisis Management ... 306

4.3 Case Study Sweden ... 317

4.3.1 The Swedish Crisis Management Approach ... 317

4.3.2 Sweden's Comprehensive Crisis Management ... 325

4.3.3 European Dimension and EU-Interface in Swedish Crisis Management ... 342

4.3.4 Synopsis: Sweden and Comprehensive European Crisis Management ... 355

5 Conclusion ... 367

5.1 Toward a Whole-of-Europe Approach – Modelling Comprehensive European Crisis Management ... 368

5.1.1 Crisis Management Approaches and Conceptual Dimension ... 368

5.1.2 Organizational Dimension ... 377

5.1.3 European Dimension and EU Interface ... 390

5.2 Dilemmas of European Crisis Management Coherence ... 395

5.3 Further Research ... 399

5.4 Final Reflections ... 403

6 References ... 405

6.1 Primary Sources ... 405

6.1.1 Official Documents, Speeches and Press Releases ... 405

6.1.2 List of Interviewees ... 415

6.2 Secondary Sources ... 418

Abbreviations

AA	Federal Foreign Office (German: *Auswärtiges Amt*)
ACPP	Africa Conflict Prevention Pool
BMVg	Federal Ministry of Defence (German: *Bundesministerium der Verteidigung*)
BMZ	Federal Ministry for Economic Cooperation and Development (German: *Bundesministerium für wirtschaftliche Zusammenarbeit und Entwicklung*)
BSOS	Building Security Overseas Strategy
CA	Comprehensive Approach
CDU	Christian Democratic Union
CFSP	Common Foreign and Security Policy
CIMIC	Civil-Military Cooperation
CIVCOM	Committee for Civilian Aspects of Crisis Management
CMCO	Civil-Military Coordination
CMPD	Crisis Management Planning Directorate
CPCC	Civilian Planning and Conduct Capability
CPPs	Conflict Prevention Pools
CSDP	Common Security and Defence Policy
CSG	Civilian Stabilisation Group
CSU	Christian Social Union
DAC	Development Assistance Committee
DDR	Disarmament, Demobilization, and Reintegration
DFID	Department for International Development
DG	Directorate General
DG Dev	Directorate General for Development and Relations with ACP States
DG DEVCO	Directorate General for Development and Cooperation – EuropeAid
DG ECHO	Directorate General for Humanitarian Aid and Civil Protection
DG RELEX	Directorate General for External Relations
EBAO	Effects Based Approach to Operations
ECOWAS	Economic Community of West African States
EEAS	European External Action Service

ESDP	European Security and Defence Policy
EU	European Union
EUBG	European Union Battle Group
EUMC	European Union Military Committee
EUMS	European Union Military Staff
FBA	Folke Bernadotte Academy
FCO	Foreign and Commonwealth Office
FDP	Free Democratic Party
GCPP	Global Conflict Prevention Pool
GDP	Gross Domestic Product
HR/VP	High Representative of the Union for Foreign Affairs and Security Policy/Vice-President of the European Commission
ICG	International Crisis Group
IfS	Instrument for Stability
IO	International Organization
IR	International Relations
ISIS	International Security Information Service
MFA	Swedish Ministry for Foreign Affairs
MNE 5	Multinational Experiment 5
MoD	Ministry of Defence
MSB	Swedish Civil Contingencies Agency (Swedish: *Myndigheten för Samhällsskydd och Beredskap*)
MSSG	Military Stabilisation Support Group
NATO	North Atlantic Treaty Organization
NBG	Nordic Battle Group
NCW	Network-Centric Warfare
NGO	Non-Governmental Organization
NPM	New Public Management
NSC	National Security Council
NSID	Ministerial Committee on National Security, International Relations and Development
NSS	National Security Strategy
ODA	Official Development Assistance
OECD	Organization for Economic Cooperation and Development
OSCE	Organization for Security and Cooperation in Europe
PCD	Policy Coherence for Development
PCRU	Post-Conflict Reconstruction Unit
PGD	Policy for Global Development
PMSU	Prime Minister's Strategy Unit
PRT	Provincial Reconstruction Teams

PSA	Public Service Agreement
PSC	Political and Security Committee
SAF	Swedish Armed Forces
SDSR	Strategic Defence and Security Review
Sida	Swedish International Development Agency (Swedish: *Styrelsen för Internationellt Utvecklingssamarbete*)
SIPRI	Stockholm International Peace Research Institute
SRs	Special Representatives
SSR	Security Sector Reform
SU	Stabilisation Unit
SWEDINT	Swedish Armed Forces International Center
TEU	Treaty on the European Union
TFEU	Treaty on the Functioning of the European Union
UK	United Kingdom
UN	United Nations
USA/US	United States of America/Unites States
WEU	Western European Union
WoG	Whole-of-Government
ZIF	Center for International Peace Operations (German: *Zentrum für Internationale Friedenseinsätze*)
3D	Diplomacy, Defence and Development

List of Figures

Figure 1: The Double Challenge of European Crisis Management Coherence 26
Figure 2: Dimensions of CA Coherence 81
Figure 3: Dimensions of CA Coherence Analyzed in this Dissertation 104
Figure 4: Positive and Negative Coordination 115
Figure 5: Policy Coordination Scale 118
Figure 6: Ideal-type Organization Forms 128
Figure 7: Europeanization Processes 147
Figure 8: Analytical Framework and Dimensions 162

1 Introduction

1.1 Central Problem Statement

"We do crisis management the European way. With a Comprehensive Approach."[1]

This quote outlines the leitmotif of contemporary crisis management: with the objective to reach sustainable and effective conflict transformation, the international community has been committed to comprehensive crisis management for almost two decades now, be it in the Balkans, Iraq or Afghanistan. International organizations (IOs) and states alike preach the *Comprehensive Approach*[2] (CA) in international crisis management. The CA has been developed against the background of an international security environment which has changed decisively since the end of the Cold War and has produced extensive pressure for adaptation in many security related contexts. In this regard, also the nature of international crisis management has undergone significant transformation. Not only did crisis management operations increase in frequency and size, but they became, like the rest of the international security environment, increasingly complex (Rintakoski/Autti 2008, pp. 10/11). Crises today cover security, political, economic and social dimensions of conflict. Crisis management is multifaceted, and policies to prevent and resolve conflicts ought to address all relevant dimensions of security. Challenges can no longer be met by individual crisis management actors or means on their own; they rather call for traditional instruments to be integrated into a broader approach.[3]

Dealing with this complexity of crisis management is the major aim of the CA concept. As an answer to these complex crisis management challenges, both policy makers and academics have become convinced that crisis management

---

1 Statement by the European Union's High Representative for Foreign Affairs and Security Policy and Vice-President of the European Commission Catherine Ashton (Ashton 2010, p. 6).

2 Capitalization and italic typography is used to denote proper names as well as independent disciplines or conceptions of their own when first mentioned in this dissertation. Italic typography is also utilized for words in a language other than English. Please note that although this dissertation follows American English, citations and proper names are kept in British English.

3 This is elaborated in detail in chapter 2.1.2 on contemporary crisis management characteristics. For an overview on 'new' security challenges see, for instance, also Gärtner, Hyde-Price and Reiter (2001, pp. 1-8). See also Möckli (2010) as well as Van de Goor and Major (2012) for a detailed review on crisis management developments and increasing complexity.

instruments need to be realigned in a comprehensive manner. Various crisis management means are intimately linked, which places a premium on close co-ordination and cooperation among formerly separated actors and fields involved in international crisis management. Managing crises calls for the coordinated implementation of diplomatic, military, police, humanitarian and development related crisis management instruments. The expansion in terms of tasks and timelines and the increasing number of internal and external actors involved have made effective coordination an urgent priority and a condition of successful, sustainable crisis management. The basic idea behind the CA is that if different crisis management instruments are not implemented in a coordinated way, they can result in unintended consequences, undermine each other's impact and ultimately compromise the effectiveness of crisis management (Mölling 2008, p. 1).

The European Union (EU) and its member states are part of this highly interdependent, dynamic and complex world. However, they also "dispose of a uniquely wide array of political, operational, diplomatic, economic, and structural instruments for the management of crises and conflicts" (Gebhard/Norheim-Martinsen 2010). In the face of new security challenges, crisis management activity undertaken by the EU has evolved significantly in the last two decades. Just as the nature of conflict has changed, so too have the ways the EU responds. Launched in 1999, the European Security and Defence Policy (ESDP),[4] which completes the set of policy instruments available for the broader Common Foreign and Security Policy (CFSP), entered its operational phase in 2003. Since then, the EU has launched almost 30 civilian and military missions in and beyond Europe.[5] At the same time, the European Commission[6] expanded its civilian crisis management activities. The EU has declared crisis management to be a key priority of its external policy and has created a comprehensive range of instruments and mechanisms to enable it to mobilize respective capacities. It has thereby acknowledged that "complex crisis demand a comprehensive response including a mixture of instruments and actors" (Council of the European Union 2005, p. 4). In accordance with the scientific as well as its own conceptual cognition, the EU seeks a CA in crisis management which aims at implementing different instruments in an effective and coherent way

4 With the entry into force of the Treaty of Lisbon in 2009, the European Security and Defense Policy has been renamed into Common Security and Defense Policy (CSDP). Depending on the period of time dealt with, either pre- or post-Lisbon, respective terminology is adopted in this dissertation. Thus, both acronyms are used, depending on whether there is reference to the pre-Lisbon Treaty era ESDP or the post-Lisbon Treaty era CSDP.

5 An overview of ongoing and completed missions and operations of the EU can be found on the European Union's External Action Website. Also, the *International Security Information Service* (ISIS) Europe provides a monthly up-dated chart of EU mission (ISIS Europe 2012).

6 From now on referred to as the Commission.

(Council of the European Union 2003, p. 5). It repeatedly emphasized both the necessity of the CA to international crisis management and the EU's ability to implement such an approach.

It is a popular and often repeated assumption that the EU's advantage compared to other international actors, most notably the United Nations (UN) and the North Atlantic Treaty Organization (NATO), precisely lies in this wide array of instruments and manifests itself by the combination of humanitarian, development, military, police and diplomatic elements under one umbrella. The EU is anticipated to be particularly well equipped to meet the challenges of the contemporary crisis management environment using a wide range of complementary instruments, from civilian tools to military means, and mixing them in terms of a CA (Brauss 2006, p. 51).

> "Given the broad range of instruments it has at its disposal, it appears that the EU is virtually meant to act comprehensively – that it is simply fulfilling some sort of teleological drive that originates in its history and experience as an organization" (Gebhard/Norheim-Martinsen 2010).

On closer inspection, however, the picture is more complicated and the challenge of comprehensive crisis management is intensified by a complex multilevel structure. EU crisis management activities are subject to different national and institutional frameworks. Responsibilities are shared between various crisis management institutions. In CA terms, such divisions obstruct the proposition that crisis management instruments should be closely coordinated and seem contrary to the envisaged CA. Therefore, apart from providing it with an advantage, the EU's wide array of instruments and structural diversity can also become a limitation. It is certainly true that the EU and its member states posses a wide variety of instruments for the management of crises and conflicts, which make them unique and capable actors in crisis management. The tremendous challenge, however, is to employ them all in a coherent way and to coordinate across bureaucratic, organizational and functional boundaries. A remarkable amount of effort has to be spent to secure coordination and coherence at all levels.[7] In addi-

7 As also pointed out by Norheim-Martinsen (2010) and Drent and Zandee (2010). This set of problems has been well described by Gebhard (2011, p. 4) who states that "in practice the functional fragmentation of the EU's institutional structures (still) keeps it from performing unitarily on the global scene. And this is what places the issue of coherence at the centre of any assessment of the EU's foreign policy. In a way, the overall success of the EU in the world stands and falls with its political and administrative capacity to cope with the very core features of its distinctiveness as an international actor: its versatility and holistic predisposition, which in turn is based on its multi-faceted structural character, and thus, its institutional complexity."

tion to its internal institutional fragmentation, another one of these levels is the EU's interface with its member states as their material and ideal commitment is crucial for the Union's crisis management and, consequently, for the successful implementation of the CA.[8] The member states and their institutional crisis management structures and arrangements play a critical role in comprehensive EU crisis management. National crisis management coherence is a necessary precondition for international coherence and not only mirrors the need for coordination, but is a prerequisite for its success and for the accomplishment of the CA.[9] Governments who demand that the EU coordinates its crisis management activities better must do their share to create the foundation for it. When seeking to achieve national coherence, also coherence on the international level must be in the minds of policy-makers. While no European state can effectively and meaningful deal with conflict and crisis unilaterally, the individual crisis management policy of EU member states remains important and various national conditions and constraints shape the way a country engages (Houben 2005, p. xiii).[10] Thus, in order to assess European crisis management coherence, it is vital to look not only at the EU level CA efforts and coherence and coordination between EU-institutions, but also to analyze how the CA is dealt with in member states. Both horizontal and vertical coherence are essential for the further development of Europe as a comprehensive crisis management actor.

In this regard, on the national level, comparable policy coherence challenges to the ones experienced by the EU exist.

> "Almost any governance system is faced with the challenge of having to reconcile concurrent policy spheres, and of managing the very implications of structural complexity, which are, essentially, an enhanced need for coordination and the necessity of closing ranks across a growing range of functional areas" (Gebhard 2011, p. 3).

8 Policy implementation in this regard focuses on the link between the expression of intention and the achievement of appropriate action. Implementation as used in this dissertation thus refers to the set of actions taken towards the achievement of the CA mandate.

9 See also Zürn and Herrhausen (2008, p. 281), Major and Mölling (2009, p. 27) as well as Major et al. (2011, p. 137) who emphasize the importance of member state commitment to make the CA happen. Also Gebhard and Norheim-Martinsen (2010, p. 8) underline the significance of EU member-state efforts to coordinate inter-institutional security arrangements which ultimately "determines the EU's performance and success at delivering a comprehensive 'European' approach". This approach is also advocated by Teuten and Korski (2010) with regard to the UK. On operational level, Grönberg (2010, p. 29) states that "[a] comprehensive approach in a crisis area requires a similar approach within the contributing countries."

10 Houben (2005, p. 7) concludes that although the nation has been declared increasingly irrelevant as a political entity as well as a unit of analysis, international crisis management remains a largely intergovernmental affair and its analysis needs a certain focus on the individual nation state.

The complexity of crisis management objectives and the multitude of tasks imply that diverse governmental crisis management instruments and actors have to be involved and coordinated (Egnell 2009b, p. 4). Crisis management affects the responsibilities of a number of ministries, as it necessitates diminishing existing divisions and mandates of crisis management actors to combine and coordinate military and civilian policy means. Such a change essentially requires institutions that are accustomed to act largely autonomously to cooperate and coordinate their policies (Weiss/Spanger/Van Meurs 2009, p. 9). This has considerable policy implications as these requirements fundamentally contradict the national governmental "culture of departmental stove-piping" (Egnell 2009b, p. 4) established in most political systems.[11] Moreover, it implies not only advancing their ability to coherently implement crisis management policy but also managing the tensions between various actors spanning the traditionally independent spheres of diplomacy, development and defence.[12]

Thus, in Europe, the CA idea and need highlight the demand to develop coherence at different levels in parallel. As elaborations show, any assumption of a European CA in international crisis management has to deal with an underlying double challenge of providing crisis management coherence across the EU-member state interface and to also ensure coherence between different crisis management actors on EU and national level individually, as shown in figure 1.

11 As observed by Bourgon (2011, pp. 61/62) "[g]overning in the 21st century entails dealing with complex issues in an uncertain environment. The main difficulty for government is that public organizations were not devised or designed to deal with complexity and with high levels of uncertainty. Their strength is in providing stability and predictability. What is an asset in some circumstances becomes a liability in others. Relying on traditional approaches, government runs the risk of being left in a reactive position, unable to anticipate or detect emerging patterns in a fast-changing landscape and therefore unable to intervene in a timely way."

12 The three key words diplomacy, development and defense do not cover all aspects of crisis management which, for instance, could also involve the judiciary, the police and other civilian expertises which should also be an integral part of contemporary crisis management. Though it is recognized that crisis management policy coherence may also involve other actors, such as finance, interior and trade ministries in country case study capitals, the focus of crisis management coherence, and the CA as the concept and term is used in this dissertation, is primarily on the roles of three main sets of actors focusing on crisis management challenges: development ministries, foreign ministries, and defense ministries and respective institutions on EU level.

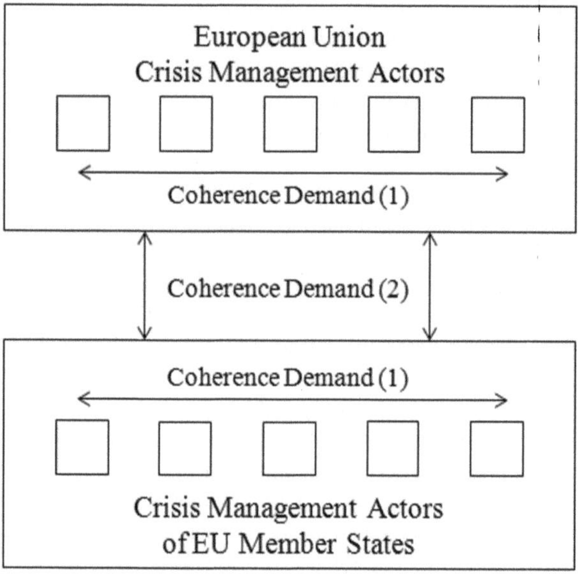

Figure 1: The Double Challenge of European Crisis Management Coherence *(Source: Own Design).*

European crisis management is an essentially complex creation of political and bureaucratic structures that are located in member states' national capitals and emerge from intergovernmental and supranational EU-structures resulting in a double problem of coherence, both in EU institutions and national governmental actors where civilian, diplomatic, development, humanitarian and military actors seek to operationalise the concept. Comprehensive crisis management in Europe has to be multi-institutional on EU level, multi-sector on national governmental level and it has to be multi-nationally managed between these two levels. It is this fragmented context that the reforms introduced by CA must be understood in.[13]

Against this background, the EU and individual national member states are in an ongoing process of specifying the CA and introduced the concept as a guiding idea to international crisis management. In recent years, numerous reform

13 This dissertation considers crisis management coherence between actors on state-level and EU-institutional level a priority while largely disregarding aspects of coordination between governmental and non-governmental actors which in theory are also part of CA multi-dimensional crisis management.

efforts in domestic government institutions of the EU member states and within the EU administrative apparatus have been observable and CA debates have resulted in some outcomes. In this regard, the latest reforms of the Lisbon Treaty[14] have provided the EU with new opportunities to improve crisis management coherence and implement the CA. Also member states' crisis management has been subject to innovations to support CA ambitions and governments undertook various efforts to effectively implement the CA and reorganize national crisis management policies.

At the same time, however, it has been broadly observed that only minor progress has been made toward actual crisis management coherence. The CA is so often referred to, both on EU and on domestic level. Its practical successful application, however, is limited. So far, it can be claimed that neither the EU nor its member states are able to apply their seemingly comprehensive crisis management toolbox. While the CA and the need for crisis management coherence are widely uncontested, implementation of the CA proves difficult and has not yet resulted in sufficient effects (Gauster 2012, p. 75). Even as the idea of the CA and of linking various crisis management tools and coordinating available instruments and actors in a common approach to address today's complex crisis is convincing in theory, its track record leaves a mixed picture and room for improvement (Van de Goor/Major 2012, p. 1). Moreover, despite an overall CA agreement, within the EU and also among the EU member states, there has been no compliance on what European comprehensive crisis management actually is or should look like (Johannsen 2011, p. 17). Political, conceptual and organizational obstacles to the CA continue to be immense.[15] The observation that there is a challenging and problematic gap between the available concept of the CA and its implementation and organizationally coordinated crisis management practice serves as the starting point of this dissertation.

1.2 Research Questions and Purpose

Above observations raise several complementary research questions. As the previous chapter demonstrated, there is a significant gap between, on the one

14 Shorthand for the Treaty of Lisbon amending the Treaty on European Union and the Treaty establishing the European Community.

15 See, for instance, observations by Nilsson et al. (2008, p. 13), Weiss, Spanger and Van Meurs (2009, p. 20), De Coning et al. (2010, p. 1), Gebhard and Norheim-Martinsen (2010), Hanssen (2010), Gebhard (2011, p. 4), Gauster (2012, p. 75) as well as Van de Goor and Major (2012, p. 1).

hand, the extent to which the benefits of crisis management coordination are upheld and on the other hand, continuing problems to actually achieve crisis management coherence. Although the CA concept is widely accepted, as it brings key crisis management actors together, it has proven to be problematic both on national and on international level. This observation raises an overall question about why the EU as well as individual states experience such difficulties in effectively implementing a CA.

The described set of problems can be approached from two inquiring angles. The first one holds that difficulties of CA implementation are caused by inadequate policy arrangements in terms of crisis management structures. From this point of view, CA challenges can be answered by organizational reforms. From the perspective of their organizational implementation, concepts such as the CA are understood as tools for addressing current problems that require comprehensive mechanisms. The second approach argues that the gaps between CA rhetoric and practice are triggered by inconsistency such as contradicting interests and values of crisis management actors. Both perspectives are viewed to be partly interdependent in this thesis and both find reflection in the analysis.

However, the analysis of comprehensive European crisis management has to proceed along another level of the debate: the fragmentation between the EU and member states are a central factor in European crisis management and cannot be ignored. As mentioned above, taking into account the realities of crisis management in Europe, it is necessary to consider that behind the EU's crisis management stand 28 member states that need to approve crisis management activities and coordination at the national level and whose material and ideal commitment is crucial for the successful implementation of the CA. The EU developed its crisis management architecture as mechanisms to harmonize, coordinate and to some extent even integrate national policies. However, implementation of relevant decisions remains with the member states and gives national policies and crisis management systems a privileged position. On the other hand, legitimacy has moved more and more to the international level (Zürn/Herrhausen 2008, pp. 280/281). Since the increasing demand for international crisis management is unlikely to lessen, it is obvious that governments contributing and the EU serving as a framework need to increase their coherence. To analyze CA implementation on national governmental level, the United Kingdom (UK), Germany and Sweden, three member states that have been very involved in the development of both EU crisis managements and the CA concept, serve as cases to probe CA implementation on the member state level.[16]

16 More details on the selection of cases can be found in chapter 1.4.3.2 of this introduction.

As shown in figure 1, the dissertation looks at crisis management coherence across two levels: firstly, across the institutions of the EU and the national governments themselves; secondly, internationally, across the EU and member states. The explicit European approach taken in this thesis is a particular interesting area for researching the targeted crisis management comprehensiveness, precisely because the institutional logic of coordinated conflict management policy fails to result in coordinated and concerted action. As this dissertation aims at providing a comprehensive analysis of this complex and new EU and member state policy field, central research questions are elaborated in a subject-driven way. European crisis management in this context is considered as a combination of policies integrated in the EU's community structures, policies only partly integrated in the sector covered by CFSP/CSDP, and additionally, national crisis management policies.

To answer the question of why the CA is so hard to apply, a number of issues and sub-questions are raised. As outlined above, the appeal of the CA is common to most governments and IOs. However, a considerable amount of uncertainty still exists with regard to the 'How?'. The dissertation assumes that the essential issue is not whether Europeans will engage in crisis management, which is considered a question that can be affirmed; rather, the critical question that it has to address is how the Europeans will organize themselves to do so. Questions relating to this aspect on European crisis management are the focus of the following analysis. Relevant aspects looked at concern the characteristics of CA policies on EU and on country case study level and its conceptual and institutional evolution and implementation. In doing so, this thesis aims at identifying the results and direction of their particular CA agenda as well as the concept's scope, challenges and limits. The overall purpose of this dissertation is to investigate how far the EU and these three member states have come in applying a CA to their crisis management systems and to highlight the problems facing the EU and national governments in the development of their comprehensive crisis management policy, in particular with regard to the inherent conceptual and organizational difficulties associated with the rise of complex crisis management. As mentioned above, the continuing working process has brought up specific documents, instruments and models for the enhancement of coordination and resulted in the creation of specialized institutions. But to what extent and with what effect has the CA been implemented?

The emergence of the CA at the EU and national governmental level and its subsequent translation into organizational practices is traced to identify specific areas that represent a challenge in achieving a CA, this thesis analyzes what steps have been taken by the EU as well as by member states to improve the coherence of their crisis management institutions and what concepts and coordination mechanisms look like concretely. Thus, the analysis starts by looking at the particular crisis management approaches and the background that has influenced the development of the CA with the aim to map the different conceptual profiles of the CA. What are the specific CA conditions that can be interpreted from the CA definitions and interpretations? Analyzing distinctive CA concepts, an inquiry into the difference and specificities of approaches to crisis management coherence is given.

Beyond a better understanding of the conceptual issues and motivations of actors to implement a CA, the analysis attempts to catalog the practical strides made in advancing crisis management policy coherence and the instruments available to implement the CA. In this regard, the dissertation takes into account that the way in which the CA is implemented and exercised is conditioned by the formal structuring of the respective organizational apparatus. It does so by asking, how the CA concept has been translated into organizational practices, what specific measures have been taken to improve crisis management coherence and how current approaches can be improved? To what extent have existing structural conditions facilitated or hampered the implementation of the CA? In this context, this thesis also addresses what type of organization and coordination mechanisms have been developed to improve crisis management coherence. What different forms can coordination in the name of CA application take, and what are determinants of these coordination modes? What constitutes a particular incentive with regard to coordination in crisis management between relevant actors to occur? The analysis seeks to indentify key lessons with a view to advance future CA attempts, taking into account recent reform efforts. Knowledge of these questions ought to allow a better understanding how coherence and coordination of crisis management mechanisms could enhance the European overall crisis management capacity. The aim of this evaluation of the CA is to assess current government and EU approaches to the CA and examine the design and impact of selected reforms and instruments developed to organizationally implement the CA to provide an overview of lessons learned and recommendations for the way forward. What is recognized as good CA policy and good implementation practice? What is essential to successfully implement a CA, and what is merely desirable? Above questions are answered by referring to arguments from organization and public administration studies which allow for a focus on the aim of coherence and mechanisms that are supposed to enable the CA.

A key institutional challenge on EU level has long been to overcome the discrepancies between the Commission in the first pillar and the second pillar's CFSP and ESDP. With the implementation of the Lisbon Treaty, the EU's crisis management has undergone an overhaul in substance and in institutional framework. The Lisbon Treaty reforms formally ended the pillar structure of the EU, which caused tense relations between EU crisis management institutions. But in spite of these clear structural changes, the overall question remains whether the new organizational model of EU crisis management is appropriate to improve coherence and to effectively implement the CA. Thus, this dissertation shifts the former CA research focus from the EU's pillar structure to the changes introduced by the Lisbon Treaty. It considers the treaty's impact factors on the EU's CA. The aim of research in this direction is to assess the extent to which the latest reform developments have enhanced the EU's ability to implement a CA, while ultimately evaluating the appropriateness of the EU's CA organization. The same setting applies for member states' crisis management structures. Have developments and institutional innovations carried out to achieve greater crisis management coherence had the expected effects? This should bring further understanding to research already available, lead to further areas of research and show the way toward a *Whole-of-Europe* approach in crisis management.

In doing so, the thesis is also directed toward discovering common denominators and generic principles of the various frameworks for the CA on European level. These can in turn be valuable in informing further conceptual developments and in making the CA more applicable. The central thesis is that the CA of the EU's crisis management must be analyzed and judged upon a basis other than the usual concentration on EU structures only by shifting the focus to the member states level. With that approach in mind, it is possible to assess how and why the CA and coherence is undermined within the European crisis management aggregate that creates it. Member states' individual foreign and security policy and crisis management apart from the EU is still enormous. The analysis of comprehensive European crisis management therefore calls for an investigation of the question on how much room individual member states actually allow for the EU in their crisis management policy and what role they aim at with EU crisis management. In order to investigate the nature and strength of EU and member states' connectivity as well as the means by which the 'EU-factor' in crisis management varies in member states, sub-questions on member state commitment toward the EU are explored by employing the concept of Europeanization. From this perspective, interest in the following questions is triggered: to what extent have member states adapted to the EU's crisis management machinery and an EU manner of the CA? A question linked to this issue is to what extent EU crisis management has influenced member state policy in terms of procedures and substance. This question is also posed the other way around. What impact did member states have in developing the EU's crisis management approach? To what extent did member state crisis management approaches and CA implementation influence respective EU structures? The analysis of these questions should ultimately reveal whether the EU offers a different logic of comprehensive crisis management to that found at the member state level (Gross 2009, pp. xii/xiii). How do national and international European agendas toward a CA support each other? In what regard are national and EU approaches compatible?

Observing these questions serves to contextualize current discussions about the issue, and to explain continuing challenges of CA implementation despite all the political and academic awareness of it. This thesis intends to clarify concepts, mechanisms and structure at both the national and the EU levels to understand the dynamics that explain why actors have not lived up to the expectations mostly outlined by themselves. In doing so, the dissertation seeks to contribute to a consistent understanding of respective terminology and to clarify different concepts and terms with regard to the CA in the European context to address the concept more systematically and help to focus the debate. In this context, the dissertation also has a theoretical aim by examining what guidance organization and public administration studies as well as the concept of Europeanization can provide for dealing with this research subject and for approaching posed research questions. It thereby aims at combining an academic interest, to critically review and further develop the discourse regarding the conceptualization of comprehensive crisis management with the empirical and practical objective of identifying the institutional arrangements that set limits to the successful implementation of the CA.

Summarizingly structuring the above elaboration, each of the cases looked at, the one on the EU as the IO analyzed and the three country case studies, are based on three related sets of questions which are discussed at different levels. The first concerns the particular crisis management approaches and security cultures. The second deals with the orientation of the conceptual and political agenda of the CA and crisis management coherence. Conceptualization and strategies to increase crisis management coherence are evaluated. In the third dimension, structures, organization arrangements and coordination mechanisms are looked at and its organizational implementation and institutionalization is examined. In the country case studies this set of questions is complemented by a fourth section analyzing the role of the EU in this context.[17]

This subject is clearly a very broad one to which it is difficult to do justice,[18] however, it serves as an attempt to thoroughly analyze European comprehensive crisis management to date and to contribute to debates by applying a policy-oriented perspective. According to King, Keohane and Verba (1994, p. 15) two criteria exist which make a research topic relevant. Firstly, “a research project should pose a question that is ‘important’ in the real world.” How CA is interpreted has an effect on how crisis management actors perform. The CA is regarded as an essential tool for achieving crisis management effectiveness and sustainability. Exploring these questions is necessary and the criterion is even

17 These analytical dimensions are elaborated in greater detail in chapter 2.4 which concludes the analytical framework.

18 Hence, some limitations are necessary and given word in chapter 1.4.4.

more fulfilled the more crisis management difficulties and incoherence in this area are revealed. The research subject thus has great real-world importance. Secondly, it should make a contribution to existing scholarly literature "by increasing our collective ability to construct verified explanations of some aspect of the world" (King/Keohane/Verba 1994, p. 15). This dissertation also fulfills this latter criterion since it assesses the development of new and highly relevant crisis management issues. There is no shortage of potential crisis creeping, or even rapidly emerging, onto the EU's and its member states' agenda. Events show that the challenge of coordinating the European crisis management instruments is acute (Juncos 2008, p. 2). At the same time, expectations from the European public and external demands for a greater involvement and effective response in world affairs have continued to grow and require expertise. Crisis management draws more and more political and public attention, as its successes or failures not solely affect states in the conflict region; it has also become a credibility test of European policy. Furthermore, the timeliness of the subject is clear given initiatives and reform underway in the EU and member states. The current context of efforts toward improving crisis management coherence, reform implementations following the Lisbon Treaty and national efforts make the research subject highly interesting. Has this been done in the spirit of moving toward a Whole-of-Europe approach? Beyond the relevance of European crisis management at large, the approach to the research subject that this thesis takes provides it with research originality on several accounts. These aspects are taken up in the next chapter.

1.3 Current Research and Literature Review

As crisis management became an expanding policy area and the scope of EU and member state activities widened and complexity increased, the poor sustainability of these activities and diverse actors' inability to undertake meaningful coherent crisis management became evident. Becoming more pressing, the need for substantial research on various levels was incited. Thus, the overall availability of material has been good regarding underlying aspects of this research subject, such as the analysis of the crisis management environment giving explanations on the necessity for crisis management coherence. The changing crisis management environment provoked a huge bulk of literature on issues such as the widening of the security agenda and strategic, organizational, and operational implications of these developments (Gebhard/Norheim-Martinsen 2010, p. 3). Also, shortcomings in contemporary crisis management have been analyzed extensively by academia, independent commissions and involved actors themselves. By the late 1990s, numerous research institutions dealt with issues of crisis management producing influential reports.[19] A number of reviews and studies in the context of peacebuilding and conflict prevention, such as the Joint Utstein Study of Peacebuilding,[20] have been used as helpful benchmarks for this dissertation. Substantial research has also been undertaken by the Organization for Economic Cooperation and Development (OECD) which focused its analysis on links between development cooperation and conflict prevention (OECD 2003). Other studies were useful in giving initial overviews on the subject and starting points for further investigation. In this regard, reports with a practical focus have been particularly relevant for this thesis. Evaluations undertaken identified the failure to coordinate various dimensions of crisis management systems as a serious cause for concern. In doing so, the complexity of the present crisis management scene and the CA as a conceptual answer has been noted in almost all articles. Most of the studies on crisis management coherence recognize that some progress has been made, but at the same time identify substantial limitations. While acknowledging the important results already achieved and calling for further development of concepts and structures, establishing a successful CA is seen as challenging if not unrealistic. On these matters, the sheer quantity of research report and think tank papers is overwhelming and challenges prioritization.[21]

Also on the broad level of national governmental coherence, the challenge of ensuring that its diverse parts operate in a coordinated way to achieve their goals is far from new and various studies in different contexts reflect respective difficulties.[22] Already in the 1970s, scholars observed that "[n]o suggestion for reform is more common than 'what we need is more coordination'" (Pressman/Wildavsky 1973, p. 133). Debates on policy coherence and coordination have produced an enormous amount of literature. An important line of argument unfolded in studies on interorganizational coordination with early works by Scharpf (1976, p. 5) which "necessarily proceeded from the assumption that

19 Like the Carnegie Commission's Final Report on Preventing Deadly Conflict (Carnegie Commission on Preventing Deadly Conflict 1997) or the International Crisis Group's (ICG) reports on the EU's Crisis Response Capability (International Crisis Group 2001, 2005). See also Lund (1996) for the United States Institute of Peace. The European Commission, for instance, sponsored the Conflict Prevention Network of academics to develop conflict prevention policy recommendations and to bridge different approaches.

20 The Joint Utstein Study analysed 336 peacebuilding projects executed by the United Kingdome, Norway, the Netherlands and Germany and discovered a lack of coherence at strategic level as one of the major impediments to sustainable peacebuilding. See Smith (2004).

21 Numerous publications issued over several years by De Coning, Friis and Jamyr, Gebhard, Norheim-Martinsen, Major and Hull have provided particularly valuable insights.

22 Among many others see also Wollmann (2003), Bogdanor (2008), Hustedt and Tiessen (2006), as well as Bach, Fleischer and Hustedt (2010).

policy making might be improved through reorganization." In hand with a renewed theoretical interest in national public organization, also "international organizations as organizations" (Bossong/Benner 2010, p. 1076) have been rediscovered.[23] However, still only little information about the effects of different organizational forms can be found when looking at studies of public organization and it can be claimed that organizational structure is that aspect of public organizations that is most manipulated but least understood (Christensen et al. 2007, p. 144). In particular, literature regarding foreign policy and crisis management coherence and coordination has been limited. There is a lack of analysis across sectors, countries and relevant IOs that debates whether and to what extent organizational crisis management patterns have been changed, how sustainable these reforms are, whether these are changing in a similar manner and for similar reasons. Toonen (2003, p. 465) notes with regard to administrative reform that studies on the focus of such reforms in terms of the idea and concept that stands behind them are much more common than the analysis of the implementation and locus of reform. This observation is highly applicable with respect to the implementation of the CA as well. Studies mostly focus on the feasibility of the CA concept, its pros and cons or the general theoretical approach and paradigm behind it. There has been considerable analysis of the theory of the CA, both in academic and political circles. This research aims very much at trying to define the CA and is largely focused on the conceptual side and resulted in a state in which the CA, though a relatively new concept which only recently emerged on the research agenda, as conceptual topic is well covered, if not overdone.

At the same time, it is this 'buzz' generated by all CA ideas which merits a closer look. One of the deficits of the current debate is that it is primarily focused on conceptual aspects of the CA. Research, criticism and support of CA conception and reforms are often raised without analyzing the degree to which the concept is pursued in practice and has become a reality. Due its implementation focus, the thesis can build on existing research only to a limited extent. As Wendling (2010, p. 11) fittingly observed "[w]hat is surprising, when looking at these [Comprehensive Approach, S.P.] concepts that are subject to debate in government ministries and within international organizations, is the scarcity of academic publications dealing directly with the topic." While some reports on the CA both on EU as well as on country case study level exist, these analyze the

23 For an outline of evolutionary pattern in organization theory see Calabrò (2011, pp. 22-24), who finds that in the last two decades *New Public Management* and network theories have been the prevailing lens of analysis. Calabrò's (2011) study provides a systematic literature review on the topic of changes in the research agenda on governance structures and mechanisms that analyses 184 journal articles published from 1979 to 2009 quantitatively.

CA in a rather sketchy, conceptually oriented manner and limit their focus to the theoretic level, lacking an in depth examination of the CA in its multidimensionality and practice. Only few actually analyze what is "happening around this fashionable concept on a practical level" (Wendling 2010, p. 11). Consequently, it is important to further probe this issue. This dissertation has to be understood as an effort to bridge the gap between the theoretical side of the CA and its practical implementation. In addition to the extensively covered conceptual side, it is concerned with what kinds of different organizational changes lead to which kind of coordination, and which kind of coordination and adjustments works best toward crisis management coherence.

While few academic books analyze the actual implementation of the CA, it is mainly Non-Governmental Organizations (NGOs) and think tanks which work on this subject. Analysis of the CA and related conceptions is coined by a kind of 'Nordic intellectual leadership' as lot of research and evaluations resulting in comprehensive reports on the issue take place in Scandinavian think tanks.[24] As the UN is seen to be the organization where most progress in achieving a CA has been made (Friis/Jarmyr 2008, p. 9), generally more literature is available about UN than about EU efforts. From a methodological perspective, however, analysis that has been conducted with regard to UN peacekeeping coherence proved helpful as research approaches could partly be transferred to the research subject at hand.[25] De Coning (2008c, pp. 57/58) observed that two schools of thought have materialized in response to the issues of crisis management coherence and coordination between the security and development dimensions. The first school of thought asserts that coherence and coordination can be increased by establishing more order and command and by centralizing coordination. The second school refers to coordination in terms of consensus. In a more minimalistic manner, centralized coordination or integration is rejected and the importance of independence in humanitarian action emphasized. As outlined before, the analysis combines both schools by looking at both concepts and structures.

24 In particular by researchers working at the Norwegian Institute of International Affairs, the Crisis Management Center Finland, the Danish Instiute for International Studies, the Swedish Defence Research Agency, and the Netherlands Institute for International Relations Clingendael. Also, the journal *Cooperation and Conflict*, a valuable source for various aspects taken up in this thesis, is a publication of the Nordic International Studies Association.

25 See, for example, De Coning (2008a) and Friis and Jarmyr (2008). As Lipson (2005) observed, issues of coordination are central aspects in the field of public administration and organisation theory. He therefore adopts organisation theory to understand and improve coordination between organisational actors involved in complex peace operations. In her work *Organizing Peacebuilding* Herrhausen (2009) analyses coordination between different UN peacebuilding entities from the perspective of organisation theory. Also Martens (2010) enriched her dissertation by looking at organization theory.

Furthermore, it can be observed that, in research on policy coherence in the crisis management field, a lot of literature specifically concentrates on civil-military relations and how to achieve a culture of understanding between civilian and military crisis management actors. In addition, a large amount of research focuses on coherence in development policy and practice. These strands of literature focusing on development, however, exclude linkages of security and development. In addition, literature examining policy coherence focuses on existing policy approaches to precarious statehood and failed states.[26] The analysis in this dissertation can partly relate to this branch of literature. In particular studies concerned with the stabilization of fragile states deal with contemporary crisis management and reflect new challenges of a different crisis management environment with unique features demanding new policy approaches (Weiss/Spanger/Van Meurs 2009, p. 13). Thus, these studies' research approaches and findings are partly relevant in the comprehensive crisis management context.[27]

Of these, some works are of particular relevance exploring the outlined research questions. The most comprehensive and authoritative reviews of the practical CA implementation are the studies by Patrick and Brown (2007a, 2007b), a seminar publication by the Crisis Management Initiative (Rintakoski/Autti 2008) and a 2007 report by the OECD's Development Assistance Committee (DAC) on Whole of Government Approaches to Fragile States. The assessment of efforts toward policy coherence in fragile states of individual donor governments by Patrick and Brown (2007a, 2007b) comparatively studies the experiences of seven Western governments with the quest for coherent approaches toward failing states and examines efforts to promote policy coherence by seven donor governments. As noted by the President of the International Peace Academy Terje Rød-Larsen in the report's preface,

> "[t]his timely book provides the first independent, comparative assessment of recent efforts by individual donor governments to integrate their defense, diplomatic, development, and other policies in engaging weak, failing, and war-torn countries in the developing world. (…) Beyond a better understanding of the conceptual issues and motivations of the donor community, this study attempts to catalog the practical strides made in advancing policy coherence, the resources and instruments available to implement joined-up approaches" (Patrick/Brown 2007a, p. vii).

This introduction underlines the value of this book for this dissertation's research subject in taking a similar approach. Similarly, Stepputat's (2009a) valuable

26 This focus may also be the result of the OECD's focus in this area. See OECD (2006, 2007).

27 See, for instance, works of Patrick and Brown (2007a, 2007b), Drent and Volten (2008) and Weiss, Spanger and Van Meurs (2009).

study looks into the forms and dynamics of attempts to develop coherent and coordinated approaches in peace operations in the UK, the Netherlands and Denmark and analyzed individual actor's CA efforts. Also, the volume edited by Weiss, Spanger and Van Meurs (2009) on diplomacy, development and defence policy coherence provided highly valuable input and can be considered one of the rare studies applying a similar focus as this thesis, as it is concerned with the peculiarities of institutions and policies of national governments as well as the EU and the UN in their efforts of a coherent policy approach to fragile states. This dissertation has also benefited from a report by Paris (2009) who emphasizes an organizational perspective. In his work, Paris (2009) emphasized the need for greater coordination and proposed that peacebuilding actors should be organized as a network. He additionally points to potential disadvantages of hierarchical organizational structures and centralized coordination in peacebuilding. Analyzing the UK, in particular an evaluation of the UK's cross-departmental work and inquiry into CA challenges by the British House of Commons Defence Committee (House of Commons 2010) provided very useful insights and a great source of material.[28] The detailed report and large document includes formal minutes as well as written and oral evidence given by a wide range of subject matter experts from within and outside government following extensive interviews. Based on some fifty interviews, Daniel Korski, one the experts also questioned by the Defence Committee, subsequently published together with Richard Teuten another in-depth study (Teuten/Korski 2010) examining how the British government and institutional structure can best organize itself to manage conflicts in fragile states. The focus on the CA was recently renewed in diverse publications on its different dimensions. The second 2012 issue of *Sicherheit + Frieden*, a German journal for peace research and security policy, examined different aspects of the CA, such as security-development perspective (Alamir 2012) and its implementation on the ground in Afghanistan (Gauster 2012). At the same time, studies on the Comprehensive Approach have begun to question the concept, as for instance De Coning and Friis (2011) did.

This may also be the result of the fact that the CA in crisis management and its implementation bring about a number of challenges among actors involved but also in the research community. First of all, dealing with the outlined subject raises numerous conceptual and terminological questions. In both official and public discourse, terms like crisis and conflict tend to be used almost interchangeably, as are prevention, management and response. Academics as well as practi-

28 The House of Commons specialist Select Committees have been increasingly engaged since their creation in 1979 to evaluate the conduct of external policy and their often highly critical reports are widely appreciated and respected (Self 2010, p. 269).

tioners define crisis management in a number of different ways.[29] While analysts have revised and formulated terms to spell out what they mean, the effect has rather been confusion than clarification. Besides the puzzling number of terms used or associated with conflict and its management, a multitude of different and partly contradictory notions for the CA concept have emerged, making it appropriate to speak in plural about Comprehensive Approaches (Hull 2011, p. 7).[30] Given the conceptual comprehensiveness and intricacy of the underlying practical problem, even the expert community has become infested with substantial terminological ambiguities about the matter (Gebhard 2008, pp. 2/3). The reviewed literature in the context of this dissertation largely revealed the elusive and contested nature of the CA. Consequently, substantial need exists for structured terminological analysis about the matter in order to secure conceptual clarity.

Similar to the research on the CA, the explosion of the EU's interest in crisis management has been followed by a rise in academic interest in the Union's crisis management activities. Publications on EU crisis management, especially on conflict prevention and military conflict management, mushroomed after 2000.[31] The issues of coherence and coordination in EU external relations policy in general were identified as problematic back in the early days of European political cooperation (Stewart 2006, p. 62). Clearly, the problems of policy coherence have long been on the agenda of the EU and the respective research community. The EU's institutional system is a special challenge for the imperative of an integrated security policy and a CA in crisis management, both formulated in theory and demanded by practical concerns. At the same time, the "alleged holistic predisposition as a security political actor has turned the EU into a popular subject for the study of comprehensive security actorness" (Gebhard/ Norheim-Martinsen 2010, p. 2) and the case of the EU has initiated extensive interest concerning its crisis management activities. Thus, the availability of material on general aspects of EU crisis management and in particular CFSP structures has been good, on the more specific issues of the CA and crisis management coordination, however, rather limited.

---

29 In their book *Contemporary Conflict Resolution* Ramsbotham, Woodhouse and Miall (2011) offer a comprehensive study of key concepts, developments and challenges as well as latest thinking and debates in the field of conflict studies. These issues are also elaborated further in chapter 2.1, which clarifies relevant crisis management terms.

30 Different CA terminology and concepts are more fully discussed in chapter 2.2.

31 Great contributions to the study of the EU within the *International Relations* (IR) field are made by textbook style volumes which reflect recent developments and diverse issues in the broader EU context and consider questions of both theoretical and practical importance. See in particular Bache, George and Bulmer (2011) who incorporated the changes brought by the Lisbon Treaty in their volume.

Furthermore, a large amount of literature about the EU's CA is rather selective. Most contributions focus on specific aspects of comprehensive actorness without contextualizing the matter within the broader picture and existing studies tend to illuminate very specific aspects within EU crisis management structures only. In-depth analysis of actual developments pertaining to the EU's organizational and institutional implementation of the CA and the role of member state approaches to enhance coordination in crisis management are lacking. Studies on EU crisis management often limited their scope to the examination of coordination and coherence between ESDP actors, not including the Commission instruments. Some focus on the evolution and prospects of the largely intergovernmental ESDP, other scholars only look at the civilian parts of crisis management.[32] Also, as truly pointed out by Bossong and Benner (2010, p. 1075), "EU studies as a whole can certainly not be accused of neglecting the 'inside' of institutions." Indeed, of the EU's civilian crisis management apparatus, the Commission "is one of the best researched international bureaucracies" (Bossong/Benner 2010, p. 1075).[33] In addition, a growing number of mission case studies offer snapshots of the EU's crisis management problems with regard to the specific planning issues. The lack of coherence within and among EU bodies during missions has been quite well documented in various reports and studies (De Coning/Friis, 2011, p. 259).[34] On the conceptual development and the related evolution of the EU's CA however, only little research is available.[35] This propensity to focus on narrow areas of EU crisis management has led to a considerable gap of broader research of the EU as a comprehensive crisis management actor. Questions about EU CA characteristics remain unexplored in the literature. While various issues are dealt with in terms of CA matters, few investigations try to grasp the EU's CA as a whole. The same applies to the national governmental level.

However, due to the research subject's underlying concept of comprehensive crisis management analyzing just one of the involved EU or member states actors would appear incomplete and flawed or even contradictory toward the

32 See inter alia (Missiroli 2001), Duke (2002), Dwan (2001), Mustonen (2008) and Drent and Zandee (2010) who analysed how the EU aimed at establishing a comprehensive civil-military approach to security.

33 The Aide à la Décision Économique (2011) study, for instance, analyses to what extent the means of the Commission facilitated the implementation of an integrated approach to conflict prevention and peacebuilding.

34 See, for instance, Grevi, Helly and Keohane (2009).

35 Johannsen's (2011) study represents an exception. Johannsen's (2011) in-depth study covers conceptual and structural CA adaptations of the EU's institutional pre-Lisbon set-up. However, as Johannsen (2011, p. 350) points out in her study, it would be interesting to analyze the impact of the Lisbon Treaty on the EU's ability to implement its comprehensive crisis management ambitions. This is the objective of this dissertation.

overall intended comprehensiveness in theory and praxis. Hence, it is assumed that the EU as an IO just like its member states is "a vast conglomerate" (Allison/Zelikow 1999, p. 143) of loosely linked sub-organizations among which primary responsibility for particular tasks is divided and which have their own organizational behavior the kind of explored by Graham Allison[36] in reference to the US government borrowing arguments from studies of organization. Bureaucratic as well as organizational theory and studies in IR are intrinsically tied to the works of Allison and his enduring finding that policy makers are influenced by embedded organizational habits. Though Allison focuses on the governmental organization and bureaucracy of a state, his analytical approach can guide studies on IOs as well as it was his work that opened a preoccupation of foreign policy analysts with exploring linkages between actors, policy processes and policy outputs (Jäger/Oppermann 2006, p. 109).

In sum, several gaps in the pertinent literature are detectable. At present, evaluations largely remain at an abstract level and provide a theoretical framework for understanding the CA. Implementation oriented studies either focus on specific aspects of crisis management coordination, such as civil-military coordination, or represent broad evaluations of crisis management mission, that only touch on CA issues. In particular, as the Lisbon Treaty reorganization took place only recently, so far only few studies exist which have switched from the now non-existent pillar structure to the new organizational reality. Also on member state level, CA reforms are constantly under way. Thus, the requirements for a CA are on the rise and pose a tremendous challenge not only to the EU but also for the research community to clarify concepts, structures and procedures. Consequently, this dissertation aims at filling a research perspective on the CA as well as EU and member state crisis management which has been lacking so far.

1.4 Research Design

The following subchapters introduce the dissertation's research design. Following the outline of the structure of this thesis, theoretical and methodological considerations are addressed. The presentation of the theoretical framework comprises initial reflections on main theories of IR and theoretical challenges resulting from the research subject. Subsequently, the methodology used is presented and the selection of cases chosen for this study is justified. Lastly, challenges and limitations of the research conducted are described.

36 Allison's (1971) book *Essence of Decision: Explaining the Cuban Missile Crisis* was released in an updated and revised second edition in 1999 (Allison/Zelikow 1999). For a detailed discussion of Allison's work see Gross (2009, pp. 10-13).

1.4.1 Structure

This dissertation is organized in a total of five chapters. Following the outline of the research field and context, this introductory chapter has provided the development and presentation of the study's research questions and objectives. The chapter has then presented the state of research in connection with a review of the most relevant literature. After the layout of structure in this chapter, the next subchapter sketches theoretical deliberations and concerns with regard to the analysis. After clarifying the methodological approach followed in this dissertation and the analytical rationale behind the case selection, the last subchapter identifies challenges and limitations of the research project.

For the then following analysis it is useful to take a four step approach, distinguishing between: firstly, the development of the analytical framework; secondly, comprehensive crisis management on EU level; thirdly, the analysis at country case study level; and fourthly, concluding remarks on the central research questions. The analytical framework of this dissertation presented in chapter 2 begins with clarifying and defining relevant terms of international crisis management. As an analytical starting point, it is important to examine the complex crisis management environment as the context the CA concept has to be understood in and to observe developments that led to the idea of the CA. The chapter thus continues by highlighting the changing nature of contemporary conflict. The section introduces crisis management, both as a concept and as a practice in contemporary perspective and points out key concepts used in current debates on the topic. Subsequently, the Comprehensive Approach is analyzed in detail in order to determine precisely what an analysis focusing on the concept actually examines. Presenting the foundations of the CA, a conceptual overview and presentation of key dimensions and rationales behind the concept and its modalities are clarified and discussed. As a concept study, the chapter also looks at the use of the CA by different actors and at related ideas before substantiating it in this dissertation's context. After the analysis of the concept underlying this thesis, the theoretical approach taken is developed. In doing so, this dissertation takes a step back from the crisis management focused debate in order to obtain a more consistent and systematic view on coherence and coordination in general. The next part presents the theoretical lens of this thesis by clarifying the use of a perspective based on organization theory and public administration studies as well as the concept of Europeanization. It separately maps out the development of each theory and highlights main uses and definitions before outlining the approach as adopted in this dissertation. Building on this theoretical outline, which guides subsequent chapters, the next section details how the study is empirically undertaken in different analytical dimensions looking at crisis management approaches, the conceptual and the organizational dimension of CA implementation as well as the European dimension and EU-Interface. Imposing the same chapter structure in the empirical analysis on EU and on country case study level ensures that the analysis is carried out in a systematic manner. The structuring in a common format also serves reasons of analytical clarity and functionality. Length and complexity of the chapter dealing with the analytical framework has made a concluding synopsis at its end necessary.

The subsequent empirical part analyzes the means applied to implement the CA by examining the practices of the EU as well as the UK, Germany and Sweden. In doing so, the EU is considered the contextual dimension of this dissertation. Consequently, in order to solve the puzzle of comprehensive European crisis management, it is essential to first portray the emergence and development of the CA on the EU level and to have a closer look at the EU's crisis management approach and structures. Historical and contractual developments to build the EU's crisis management policy are covered. Subsequently, the Lisbon Treaty's institutional adjustments and impact on the subject are analyzed. The EU's institutional set-up, crisis management bodies and instruments are examined by looking at the key structures of EU crisis management. Subsequently, CA efforts by EU member states are analyzed. The case studies are conducted according to the dimensions of analysis outlined above.

Building on the analysis and research conducted, the conclusion of the dissertation presents a summary of findings bringing together previous research steps to derive conclusions with regard to the possibilities and restrictions of CA implementation in the EU and country case studies and draws some central lessons for the future of European crisis management. It identifies areas that constitute a challenge to achieving a CA and further steps that need to be taken toward successfully implementing the concept. On this basis, the dissertation's conclusion provides a set of recommendations for future CA efforts. These are formulated against the theoretical background taken in this thesis. In addition, the conclusion discusses the limits of the CA and European crisis management coherence and indicates directions for further research. The dissertation concludes by drawing some final personal reflections on the research subject and developments toward a *Whole-of-Europe Approach.*

1.4.2 Theoretical Considerations

As the dissertation's research subject is at the intersection of numerous theoretical fields, this chapter reviews various influential theoretical strands of IR as theoretical points of departure. These are briefly considered with the objective to present the abilities of these theories *vis-à-vis* the analysis of European comprehensive crisis management and to assess what guidance they can give to the analysis at hand.

Analyzing the EU poses a major challenge to IR theory. There are numerous theoretical models aiming at explaining the nature of European integration processes. Most early theories dealing with the EU focused on the nature and problems of European integration and the transfer of national competencies onto the supranational EU level. Debates in this context also dealt with issues of EU actorness and the question whether it can be classified as an intergovernmental, neo-functional or *sui generis* organizational entity, and moreover whether European integration processes include a security dimension (Jonson 2006, p. 28).[37] While EU theorizing moved from these early approaches, academic work on the EU still is framed by the discussion on EU characterization by either intergovernmentalism or supranationalism (Bache/George/Bulmer 2011, p. 3).[38] Also the issue of how to analytically approach the role of member states in the EU has been controversial for scholars of European foreign and security policy (Gross 2009, p. 3). Yet, EU scholars undertook a 'governance turn' with theories focusing on the consequences of European integration and understanding the EU as a political system (Bache/George/Bulmer 2011, p. 21).[39] In this regard, also conflict, crisis and their management and resolution have been the subjects of great theoretical debate in IR. However, while sticking to no particular classic IR theory, the thesis inevitably makes some initial theoretic assumptions which draw on established concepts.

Several theoretical venues are possible to pursue in order to shed light on the research questions. Generally, classic theoretic debates in IR focus on realist, liberal, constructivist and institutionalist approaches.[40] Both *Realism* and *Liberalism* are state-centric theories. Corresponding neorealist-neoliberal debate in IR theory

---

37 On problems associated with conceptualizing the EU as an international actor and for an overview of different theoretical concepts and the realist-liberal divide in IR-theory see Hill (1993) and Ginsberg (1999). More detailed remarks on 'EU crisis management actorness' can be found in chapter 3.1 on the EU's crisis management approach.

38 For a detailed overview on European Integration 'grand theories' and IR theories on European Integration see Dyson and Sepos (2010, pp. 17-19) as well as Bache, George and Bulmer (2011, pp. 3-20). Also Jonson (2006) gives an overview on classical theories of IR and theoretical perspectives on European integration.

39 As advocated in Featherstone (2003), Bache, George and Bulmer (2011, pp. 21/22) as well as Dyson and Sepos (2010, pp. 3/4). For an in-depth elaboration on EU governance which includes conceptions of network governance and multilevel governance see Bache, George and Bulmer (2011, pp. 27-36).

40 For a detailed elaboration of individual theoretical approaches to contemporary security such as Realism, Liberalism and Constructivism and respective sub-theories see various contributions in Dunn Cavelty and Mauer (2010).

concentrates on how member states can enforce their interests and preferences through strategic interaction within the structure of IOs. These theories assume that states act through IOs and that these do what states want (Barnett/Finnemore 1999, p. 704; 2004, p. 2). Discussing IOs as conflict managers in the context of these two dominant perspectives, the neo-realism argument focuses on states as central actors in international crisis management as it is the individual state's decision that leads to conflict management activities from IOs.[41] For neo-liberals, managing conflicts using IOs decreases the cost of management for individual states and also creates legitimacy (Frazier/Dixon 2009, pp. 49-51). While liberalism underlines the relevance of the state in the international system, it also argues that there are other factors such as international institutions and international norms that impact considerations of states (Jonson 2006, pp. 30/31). Underlining the importance of structural features, these approaches refuse the idea that the EU can ever develop into a full actor because it lacks state actorness. States as the only crucial actors follow a rational cost-benefit analysis in which from realist perspective loss of sovereignty due to EU cooperation is compensated by advantages of such cooperation. EU membership in this regard facilitates national politics under the normative EU cover. From this perspective, also, the EU's CA in international crisis management can be considered as a highly positively coined label behind which national interests can be followed (Johannsen 2011, pp. 40/41).

In contrast, *constructivists* emphasize the role of rules and norms in international politics. Consequently, a security identity can be constructed by shared values and norms as well as common experiences. A European CA from this perspective can be regarded as a result of shared values (Johannsen 2011, p. 41). In examining the organizational policy coordination, one should agree with constructivists that IOs have an institutional 'life of their own' with distinct institutional characters and that priorities do not exist solely to serve the interests of their member states, as claimed in neorealist interpretations of IR. Constructivists assume that because IOs diffuse certain norms, they influence and even change state interests and shape international politics. However, certainly it cannot be claimed that IOs are actors entirely independent of their member states. States constrain organizations, yet they cannot control their behavior or institutional processes (Barnett/Finnemore 1999, pp. 700-707). This dissertation is embedded in an overall interest in institutions based on the assumption that 'institutions matter' and that "the organization of political life makes a difference"

41 For classic neorealist interpretations see inter alia Waltz (1979), Grieco (1988) and Mearsheimer (1995).

(Olsen 1991, p. 95).[42] The analysis thus draws loosely on tenets developed in *institutionalist* theories. The immense amount and diversity of institutionalist literature provides various foci, interpretations and analysis and a variety of different approaches to institutional issues.[43] However, the different institutional perspectives do have some common features that make an approach to political questions peculiarly institutional. According to Peters (2005, pp.18/19),

> "the most important element of an institution is that it is in some way a structural feature of the society and/or polity. That structure may be formal (a legislature, an agency in the public bureaucracy or a legal framework), or it may be informal (a network of interacting organisations, or a set of shared norms). As such, an institution transcends individuals to involve groups of individuals in some sort of patterned interactions that are predictable based upon specified relationships among the actors. (…) A second feature would be the existence of some stability over time. (…) The third feature of an institution (…) is that it must affect institutional behavior. (…) Again the constraints may be formal or they may be informal (…). [T]here should be some sense of shared values and meaning among the members of the institution."[44]

A central assumption of institutionalism is that structure matters and determines behavior. Institutions affect available policy-making options as well as the selection of choices and change (Olsen 1991, p. 95). "[A]n institutional perspective requires a careful delineation of the nature of particular institutional arrange-

42 The term institution is contested in academic discussion and used loosely in political science "to mean everything from a formal structure like a parliament to very amorphous entities like social class, with other components of the socio-political universe such as law and markets also being described as institutions" (Peters 2005, p. 29). From sociological perspective, the word institution is often used interchangeably with the word organization. While acknowledging the need for greater specificity (Peters 2005, p. 29), it is appropriate to use the term institution loosely in a rather vague sense for the form of analysis aimed at in the dissertation, thus concentrating on institutions as formal government structures. This means that the term institution is used as a synonym for political bodies and not in terms of norms, rules and routines. For an introduction to this debate see Peters (2005, pp. 74/75).

43 See Peters (2005) for a comprehensive overview of institutionalist theory. He distinguishes various different approaches to institutional theories like normative institutionalism, rational choice institutionalists, historical institutionalism, empirical institutionalists, international institutionalists and societal institutionalists (Peters 2005, pp. 19-21).

44 The last of the features in the above quote is central to the *New Institutionalism* approach advanced by March and Olsen (1989) who place an emphasis on the norms of institutions "as a means of understanding how they function and how they determine, or at least shape, individual behavior" (Peters 2005, p. 19). According to new institutionalists it is not formal rules that are important but the way informal rules such as norms and culture affect behavior. For an overview on differences in perspectives of organizational structure between Institutionalism and New Institutionalism see Powell and DiMaggio (1991, p. 13). An outline on New Institutionalism in the context of political research can be found in Olsen (1991, pp. 95-102).

ments" (Olsen 1991, p. 95). Institutionalist approaches thus allow for a focus on structure and organization but also on reflecting underlying norms. Considering the purpose of this dissertation, such approaches allow to emphasize institutional features of the crisis management system. In addition, institutionalist theory emphasizes the chances and benefits of international cooperation. It is contended that institutions do matter, have become significant in world politics and security relations and play an important role upholding or even creating international values and commitments (Haftendorn/Keohane/Wallander 1999, pp. 3-5). From an institutionalist perspective, it is of less importance whether politics occur within or among states. It rather matters that it takes place within a framework of mutually understood principles, rules and procedures within an institutional context (Jupille/Caporaso 1999, pp. 431/440). It is particularly the institutionalist school's emphasis on the "'underlife' of the organization" (Perrow 1986, p. 176) and the focus on its latent functions and unplanned aspects of complex systems that contribute to theoretically framing the research subject. An awareness and appreciation of institutionalism thus is helpful in understanding the following analysis and is vital in the CA context.

Analytical reasoning regarding the development of the CA concept refers to numerous theoretical fields and tenets that provide additional insights into comprehensive crisis management and that debate CA relevant developments from various angles and theoretical perspectives. Thus, some theoretical strands in addition to the above 'classics' are important to mention. First of all, it seems appropriate to place the analysis on comprehensive crisis management within the broader field of research on *global governance* which gestures toward new kinds of relationships and blurring of boundaries between previously distinct functions.[45] Notions of governance have become central in political science as they can be attributed on a domestic and an international level (Bache 2008, p. 21). Governance definitions emphasize that the mechanisms of governance are not solely exercised by governments but also include semi- and non-governmental actors (Hofmann/Türk 2006, p. 4). In this regard, the more defined concept of *multilevel governance* refers to the particular problem of coordinating the activities of different layered dimensions of governance and authority crossing traditional boundaries. "[T]his can involve relations between local, regional, national, supra national, and transnational bodies" (Thompson 2003, p. 159). Multilevel governance thus is seen to hold particular promise for studying developments in European crisis management where boundaries between domestic and international politics have blurred. Also *security governance* has often been argued to

45 As undertaken in studies by the OECD (1997, p. 15), Ling (2002, p. 624), Benner, Mergenthaler and Rotmann (2009, p. 229) as well as Dingwerth, Kerwer and Nölke (2009, p. 21).

provide main arguments for comprehensive security and describing different dimensions of EU security policy. Since IOs assume an expanding range of tasks in the international security context, global governance research increasingly draws on IOs. (Bogumil/Jann/Nullmeier 2006, p. 16).[46]

By interlinking multilevel governance as an analytical perspective with security governance as the relevant level, a useful framework for understanding and explaining distinct forms of political interactions between a diverse number of actors and for conceptualizing EU policy-making in crisis management can be established.[47] Webber and his co-authors link governance and security in this way, arguing that "the validity of a governance approach lies in its ability to locate some of the distinctive ways in which European security has been coordinated, managed and regulated" (Webber et al. 2004, p. 3). The concept of governance can provide a framework for the analysis of the changes in security threats and for studying the dynamics that these changes bring about in the management of conflicts. It helps to deal with the increasing number and types of actors and their ways of participation. Moreover, it forces a consideration of the limits and necessity of institutional coordination (Ehrhart 2007, p. 173). Governance concepts thus provide an interesting analytical starting point in the CA context and this dissertation is in line with its core declarations as these bring structures cutting across levels of governments into focus.

However, though numerous researchers have turned to the concept of governance to theorize their analysis and findings, conceptions of governance do not qualify to theorize analysis in this thesis because of its heavy emphasis on non-state actors. Moreover, most governance definitions imply a transformation in the public sector that minimizes the role of formal governmental actors and that expects that governments lose their capacity to govern and to control (Peters 1998a, pp. 8/9).[48] While it is recognized that the reality of public administration is more complex and includes the participation of a multitude of actors in different roles and the linkages between public and private sectors, it is focused on formalized governmental bodies in this thesis. Non-state actors do not form a subject of investigation in this dissertation. As governance approaches shift "attention away from institutions and processes toward how societies organize" (Mannin 2010, p. 32), such an approach does thus not provide the appropriate perspective. Analytically, the research questions have to be approached from a

46 For further details on governance concepts see, among many others, Rhodes (1996, 2007), Smith (2003), Schröder (2006), Jonson (2006, pp. 37-39), Bache (2008), Gebhard and Norheim-Martinsen (2010, pp. 12/13), Norheim-Martinsen (2010) as well as Borrás and Radaelli (2011).

47 See Ehrhart (2007, pp. 172/173) for a similar approach.

48 As also underlined by Martens (2010, p. 16).

perspective that still sees the state as a relevant actor. Governance perspectives focus very much on the role of organizations in the international order and do not allow an investigation of the role of particular crisis management instruments. Furthermore, the concept does not encourage the detailed and practical analysis aimed at as governance is a useful term to display general trends in government practices but does not help practitioners to grasp day-to-day problems linked to the practical reality of governing a multilevel world. As pointed out by Smith (2003, p. 620), the concept of multilevel governance "is undertheorized and as such often leads researchers to set up research designs that almost inevitably lead to conclusions that are empirically fragile and excessively general."[49] Hence, while acknowledging the broad utility of governance theories in CA contexts, the analysis undertaken chooses a different approach that concentrates on central institutional actors at different levels within the European crisis management governance system.

Furthermore, the concept of *Securitization* can provide insights with regard to the development of CA agendas. The notions of Securitization and *De-*Securitization deal with the conceptual widening and functional expansion of security due to political dynamics after the end of the Cold War and respective changes in the international policy environment. The approach disconnects security studies from the state as the main reference object of traditional national security and focuses on the military to move it to other actors and values as objects to be protected. Seeing Securitization as "a more extreme version of politicization" (Buzan/Wæver/De Wilde 1998, p. 23) the approach emphasizes that theoretically any public policy and social issues can be defined in terms of security challenges and dealt with respectively. Scholars identify environmental, economic, societal, and political sectors that security is dissected in. By lending increased importance and urgency to a concern, this may legitimize the use of measures that go beyond the usual bounds of political measures to deal with it. The Securitization concept thereby "raises analytical awareness about the political and intellectual sensitivity in simply tacking the 'security' label onto an ever-wider range of issues" (Gebhard/Norheim-Martinsen 2010, p. 11). Buzan, Wæver and De Wilde (1998, pp. 4/5) aim at finding "coherence not by confining security to the military sector but by exploring the logic of security itself to find out what differentiates security and

49 See also Borrás and Radaelli (2011, p. 468) on the value of multilevel governance. In fact, the academic debate holds a wide spectrum of conceptual variations and definitions for the term governance. As Hofmann and Türk (2006) observe in the context of EU administrative governance, governance can be defined in terms of exercise of public power of governing institutions. On the contrary, governance stands for a particular form of steering, "in which public and private actors interact in an open way in order to reach common public policy aims" (Hofmann/Türk 2006, p. 4).

the process of securitization from that which is merely political." Securitization and CA concepts are thus interlinked when it comes to the issue of thinking about security in cross-sectoral terms and patterns of mutual references among sectors, units and actors (Buzan/Wæver/De Wilde 1998, pp. 167-169). The CA can be approached in terms of Securitization, as it securitizes dimensions of a crisis that until then were dealt with by other policy fields and not as part of comprehensive crisis management as a new dimension of action.[50]

In view of the complexity of European crisis management, this dissertation holds the view that the application of one specific theoretic approach does not permit investigating the research subject in all its dimensions. As Christensen and Lægreid (2006, p. 9) observed when analyzing CA-like *Whole-of-Government* (WoG) approaches, "there is no overriding theory that captures all the key aspects". Instead of competing, it is assumed that different theoretical paradigms explain different aspects and attributes of the research subject.[51] Thus, theoretical assumptions are not regarded as exclusive, but rather as permeable frameworks with partially conforming assumptions and diverging ones elsewhere. This wide theoretical and policy-oriented approach results from the broad setting of research questions interlinking various elements and dimensions of European crisis management such as the EU and the member state level, conceptual as well as structural dimensions.

With regard to the research questions at hand, the study adopts an interdisciplinary theoretical approach that departs from traditional IR and political science theory. Though above-outlined theoretic approaches offer interesting insights to analyzing CA implementation in Europe, they cannot cope with the requirements of complexity of the research subject. Theoretically, the research conducted needs an approach which structures the analysis' diverse elements as part of a broader, more holistic process. In doing so, this thesis seeks to utilize insights drawn from organization theory and public administration studies as well as the concept of Europeanization in order to capture the range of CA activities in Eu-

50 Though there are many scholars involved in the so called Copenhagen School, the concept of Securitization was particularly coined in the influential seminal writings of Ole Wæver, Barry Buzan, and Jaap de Wilde. See, for instance, Wæver (1995) as well as Buzan, Wæver and De Wilde (1998).

51 With this multi-theoretical and pluri-conceptual approach the dissertation follows the opinion that today's political complexity can only be captured adequately by using different approaches of analysis which combine and integrate different causal factors. See also Jäger and Oppermann (2006, p. 106), Wendling (2010, p. 18) as well as Vanhoonacker, Dijkstra and Maurer (2010, p. 15) who use multiple theories in parallel to account for individual aspects of complex subjects and argue respectively to get a hold on diverse actors and levels of European foreign policy.

rope.[52] These are viewed as the most tangible way to exploit the subject and most directly concerned with the research questions at hand. Aiming at an investigation of the broad conflict management setting, this 'toolkit-approach' has the benefits of identifying and addressing the various dimensions of this complex subject.

Organization and public administration studies appear to provide the most extensive attempt to deal with the composition of CA implementation in the EU and member states. A central aim of this dissertation is to clarify how the organizational context and mechanisms influence CA implementation. Thus, an organizational perspective offers a framework within which reform efforts are interpreted as organization theory "draws attention to the role of political institutions, to the questions of how they are organized, what their causal impact is, and how they change" (Windhoff-Héritier/Czada 1991, pp. 16/17). Dynamics due to organizational characteristics are more complex and organizationally contingent than assumed by the theoretical IR models and paradigms. This gap is reflected in a general division between sub-disciplines such as public administration and organization theory (Trondal et al. 2010, p. 3). In contrast to theory-oriented analysis, applying insights from organization studies allows working from an empirical and policy oriented perspective. Thus, this theoretical approach proves particularly relevant with regard to practical questions on prerequisites for coordination, different coordination forms and modes as well as specific steps that have been taken to improve crisis management coherence. In doing so, this thesis shares the view that "[i]nstitutional and organizational capacity provides the starting point for preparing governments to be fit for the future" (Bourgon 2011, p. 92). By choosing this theoretical approach, the attempt is made to confront issues of practice and demonstrate the relevance of academic research to the day-to-day issues of dealing with CA challenges.[53]

52 Considering above theoretical elaborations it is worth mentioning that the roots of the institutionalist approach are especially evident in organization theory and the institutional school served organizations theory well in various aspects. The institutional school in organization theory largely deals with values in organizational processes (Perrow 1986, pp. 157-158). It focuses "on actual organizations with histories and functions" (Perrow 1986, p. 164). A comprehensive overview of Institutionalism and organization theory is provided by Scott (2001). For an elaboration of the institutional school's contributions to organizational theory see also Perrow (1986, pp. 164-166). Furthermore, Europeanization "is institutionalist by nature" (Bache 2008, p. 12). Scholars of Europeanization traced the concept back to the spectrum of theories that fall under the umbrella of institutionalism. The concept highlights the role played by European institutions and in this sense can be located in institutionalism. See Börzel and Risse (2000), Wong (2005) as well as Vink and Graziano (2007).

53 To some extent, this thesis thereby follows the call for a third generation of IO studies by Trondal et al. (2010) which aims at explaining international dynamics by formal organizational features and concentrates more on "the actual workings of international bureaucracies" (Trondal et al. 2010, pp. 205/206).

Europeanization fits neatly into the research subject as the concept provides a useful instrument to assess EU influences on a national level as well as national member state influence on the European level and seeks to explore adaptational pressures. In doing so, this dissertation perceives EU and national crisis management structures as part of the same system. The concept of Europeanization has been increasingly applied by scholars of EU foreign and security policy "to analyse the way in which 'Europe matters'" (Gross 2009, p. 13) and allows taking a "'European' route" (Vink/Graziano 2007, p. 4) to the study of crisis management policies. The arrangement of combining the organizational perspective with the concept of Europeanization allows unpacking CA implementation efforts at the national level but also at the EU level and to additionally explore links between and across the European crisis management system. The concept of Europeanization provides a useful theoretical lens to evaluate the change of a nation state taking the EU into account and capturing the transition of domestic policies by putting it into a dynamic perspective with the EU dimension. Yet, it goes beyond the European level concentration of classic integration theories by shifting the focus to the domestic level (Vink/Graziano 2007, p. 4).[54]

Both theoretical approaches, organization studies as well as the concept of Europeanization serve more in terms of an "attention-directing device" (Jones/Clark 2010, p. 1) to clarify the scope of research than as a strict theoretical system that covers the research object in all its complexity. The theoretical approaches are applied in order to interpret different cases and findings and to function as lenses in order to gather observations. The theoretical framing is seen in a descriptive rather than an explanatory way that allows mapping the debate on the existence of a *Whole-of-Europe* approach in crisis management.[55]

---

54 Some Europeanization scholars suggest that the research agenda of Europeanization provided European Integration Theory with a *second image reversed*, tracing Europeanization literature back to Gourevitch's (1978) seminal article on the international sources of domestic politics. The tradition of *second image* (Waltz 1959) and second image reversed (Gourevitch 1978) traces down domestic policy change to international dynamics and examples (Grabbe 2006, p. 47).

55 On the meaning and utilization of theory in this dissertation see also chapter 1.4.4. Martens (2010) applies institutionalism in this manner in her dissertation.

1.4.3 Methodology

1.4.3.1 Methodological Discussion

The outlined approach is highly empirical, essentially inductive and the research undertaken has rather an explanatory than a confirmatory character.[56] In this regard, the research design of this thesis is not based on generating testable hypothesis or isolating specific variables, it rather addresses a wider set of research questions as outlined above. Methodologically, the research for the thesis consists of two components of qualitative analysis: firstly, the descriptive elaboration of the security political background and international crisis management context, the survey of theories and concepts forming this dissertation's analytical basis and lens, as well as relevant terminological clarifications. Secondly, the EU's and selected member states' CA in international crisis management and the overall EU-factor in this policy field are analyzed.

Due to its flexibility, a case-study approach presents the suitable methodological approach for this dissertation.[57] It is mainly case study research that has the advantage of modeling multifaceted relationships and interactions (Bennett 2004, p. 19). This allows an analysis of the different ways how the CA has been implemented. Though multiple case studies are criticized for suffering from a lack of detail and depth (Eriksson 2006, p. 11), defining best practices is essential, as *learning by doing* has been a significant factor explaining the adoption of new initiatives in European crisis management in order to improve coordination and cooperation (Juncos 2006, p. 6). Moreover, as Kettl (as cited in Bourgon 2011, p. 1) advocating case study analysis states, "[p]ublic administration without a guiding theory is risky; administrative theory without connection to action is meaningless". The analysis of the CA on EU level and in the UK, Germany and Sweden can demonstrate in exemplary manner diversity and differences of the CA implementation. In doing so, the dissertation asks *why*-questions to discover reasons for certain political outcomes, by analyzing *how* the CA is implemented in the EU and the country case studies. In looking at each case, insights

56 Inductive in terms that rather than forcing findings into preconceived theoretical categories, realities are analysed and these are in turn related to categories of empirical description. See Martens (2010, pp. 19/20) on this differentiation. Generally, organisation theory and public administration studies as domains have been portrayed as very descriptive genres (Thoenig 2003, p. 127).

57 Case studies of both IOs and nation states have been widely conducted in studies relevant for this dissertation, for instance, by Patrick and Brown (2007a), Drent and Volten (2008), De Coning (2008a), De Coning et al. (2009, 2010) and Friis (2010). See also Rieker (2006b) who argues similarly in his study.

into CA motivations, assumptions, and rationales behind its implementation and respective reorientation, as well as the conceptual and organizational challenges that have to be dealt with are investigated.

Time-wise, the analysis covers the development of CA concepts since the end of the Cold War and respective developments on national and EU level since the end of the 1990s until mid-2012. Of particular relevance in the EU context are the changes resulting from the adoption of the Lisbon Treaty in 2009 and its impact on EU crisis management coherence. Thus, the analysis on EU level structurally considers the post-Lisbon crisis management architecture, while covering conceptual development since the CFSP/ESDP creation in 1999.[58] On the country case study level, the analysis follows respective developments within the same time frame. Discussing these in the UK case starts in 1997 with the Labour government coming into power. The period of the German case study starts roughly in the same year as have developments in Sweden. Considering the different dates of EU accession of the selected countries, this, however, means that the Swedish case study also touches on the country's accession to the EU. Those aspects are not considered in respect to the longer EU member states Germany and the UK.

The research is based on various sources and multiple data was gathered to perform the outlined analysis. In this regard, mainly three bodies of literature inform this dissertation: first, primary sources such as legal texts, official documents, statements and press releases have been helpful in mapping policy evolution and serve as a frame of reference for the identification of conceptual developments and characterization of European comprehensive crisis management models. Being aware of the CA concept, the EU and its member states have striven to improve the coherence of their activities. Consequently, what governments want to be able to do in crisis management and what they actually do to implement the CA can be fairly precisely re-examined on paper. Documents analyzed are treated as clear statements of will. Furthermore, conducting a content analysis of official material allows merging and systemizing the different levels of analysis. In doing so, it can be analyzed how both the CA and the EU are considered to be of importance. Another benefit of such sources is that they are not based on other sources and interpretations by others. Yet, some deliberations are not available to the public. Thus, secondly, published secondary academic sources, existing policy-oriented and research literature and relevant re-

58 Developments of EU crisis management institutions before the entry into force of the Lisbon Treaty are, however, looked at when deemed necessary in the respective context. Giving a full and detailed chronology of the political evolution of EU crisis management would lose sight of the key research objectives of this dissertation.

ports centering on the key research questions have been used and analyzed. Secondary sources have been predominantly exploited in the context of contextualization of the subject and in order to set the historical background. Due to the topicality of the research subject and the need of up-to-date information, the analysis furthermore relies on a rather large number of online sources such as the homepages of relevant institutions and think tank reports. This particularly applies to the analysis of CA developments on EU level due to the recent implementation of the Treaty of Lisbon.

Due to the contemporary character of the research subject, research questions have been additionally explored using qualitative interviews to complement abovementioned sources. A total of 48 interviews were conducted in 2011 and 2012 on EU level in Brussels and in the country case studies' capitals London, Berlin and Stockholm. Extensive consultations took place with officials of EU-institutions and of these governments, as well as discussions with knowledgeable experts and observers in academia and think tanks outside the official institutional structures. Background expert interviews on the CA as a concept also took also place in Washington DC. Interviews lasted from 30 minutes up to 3 hours. The interviews were granted on the condition of anonymity and data was agreed to solely be used anonymized. The conduction of interviews in terms of confidentiality has also been selected to allow for open expression of assessment and considerations other than official policy statements. Consequently, interviews are referred to without providing any additional information on the interviewee throughout the running text. However, a list of interviewees appears in the list of references.

In addition to their availability, the selection of respondents was based on their involvement and proximity to the key debates surrounding the research subject. The face-to-face interviews were semi-structured, following a predefined set of open-ended questions that enabled checking specific points. This guided approach was intended to ensure that the same general areas of information were collected from each interviewee. At the same time, this type of interview technique allowed interview partners to talk freely, reply in their own words and add comments beyond the set of questions in order to ensure an appropriate openness of the research process for new aspects and additional concerns. The semi-structured character of the interviews allowed posing follow-up questions and of adjusting the questionnaire as information was obtained. In the usage of the interview data, the interviewees are utilized as informants and information sources that provide inside knowledge and analysis from within their organizational structures, administrative units and regarding the research questions. Practitioners' first hand assessments were of particular value to enable better insight into events, and helped to ultimately propose policy recommendations. Furthermore, the interviews provided insights into areas of policy development that were difficult to take hold of from official documents and secondary sources. In this regard, the interviews were utilized to find and confirm information and thus also served to put the analysis of official material into perspective. Consultations also helped to further clarify the variety of existing documents. Finally, the interviews proved highly valuable due to the topicality and the explorative ambitions of the research subject and respectively lacking academic analysis in some areas.[59]

1.4.3.2 Selection of Cases

As stated by Martens (2010, p. 19), "[w]hen we are conducting empirically motivated studies, we would like to know more of, or better understand, a concrete phenomenon." The case selection here serves to illustrate how actors adapt to the CA and help to create a body of evidence demonstrating how crisis management coherence can be achieved. It is thus highly empirically motivated. The selection of the EU as the overarching framework and the UK, Germany and Sweden as the country cases to be analyzed is based on the evidence collected on the CA as a research subject laying the groundwork for this dissertation which showed that certain IOs and states are more prone to the CA in international crisis management than others. In this regard, the EU "is the international organization with the most elaborated institutional framework" (Weiss/Spanger/Van Meurs 2009, p. 20) and, next to the UN, the most relevant and paradigmatic organization when it comes to the CA. Even more so, as crisis management experiences of this comparably new actor has reached a critical mass. The overarching analysis of the EU in terms of an international European dimension is particularly interesting as it developed its overall crisis management architecture also as a mechanism to harmonize individual member state policy and its own organizational actors. Furthermore, the EU's special character in terms of comprehensive crisis management given the range of economic, diplomatic, humanitarian, development and military tools at its disposal, gives it a unique position for the realization of CA visions. Its ability to draw on a wide range of instruments theoretically allows diverse crisis management actions in terms of different sectors. Thus, the EU in this thesis is seen to be in a better place for CA implementation and respective research since it has a more comprehensive toolbox.

The state of European crisis management coherence is analyzed by also investigating CA realization at EU member state level. The decision for an analysis of the CA at the EU level and at the domestic level instead of an analysis of the

59 For an overview on qualitative methodology in organizational research and the role of expert interviews see Kühl, Strodtholz, and Taffertshofer (2009, pp. 19/20) and Liebhold and Trinczek (2009).

operational field level by drawing on EU mission case studies is based on the assumption that the CA needs to be developed within these organizational and conceptual structures first before it can be effectively applied in operational crisis management activities. Because of the partly intergovernmental nature of EU crisis management, respective policies of member states are highly relevant as they are key actors not only for the development of the institutional framework of EU crisis management, but also for its application. Consequently, research in this dissertation advances from the assumption that member states' crisis management policy and structures are crucial for the effective implementation of crisis management policies adopted by the EU. These factors justify a focus on the EU and the member state level, without the investigation of the implementation of the CA in an additional mission case study. Another rationale for analyzing the EU and the selected member states is the idea that the organizational implementation of the CA reflects less the international conflict environment in which they are embedded but much more the organizational variables of the actors themselves.

To keep research manageable, a choice of member states was required. While approaches and initiatives toward the improvement of crisis management coherence cannot be identified in many states at present (Major/Bail 2011, p. 22), some member states are attempting to improve the coherence of their crisis management policy. The United Kingdom, Germany and Sweden are among them and were chosen as country case studies on the grounds that they are active or innovating in this area, albeit at different intensities and ways, whilst also representing different ends in the spectrum considering the EU as a crisis management actor. The three states are regarded as pioneers in promoting the CA (Tannous 2008, p. 34),[60] but they are also included because they play a key role as initiators of crisis management developments in Europe, although in different ways. The selection is thus a reflection of the assumption that the countries substantially affect the direction and pace of EU crisis management. Thus, apart from their general influence in the CA setting, crisis management policies and structures of the UK, Germany and Sweden are of particular relevance in analyzing European crisis management due to their substantial involvement in EU crisis management and their different approaches toward the EU foreign and security institutional framework. A study of these countries is rewarding since they differ not just in their positions toward the EU but also with respect to their individual approaches to international crisis management. Thus, the cases have also been chosen to represent a broad range of preconditions based on institu-

60 See also assessment by Korski and Gowan (2009, p. 13) with regard to EU civilian crisis management capacities.

tional architectures, political traditions and strategic preferences. As there is no best case scenario, analysis and results relevant for improving European crisis management coherence are to be gained from the examination of different concepts, structures and processes that where put in place to implement the CA on the domestic member state level. Furthermore, the cases provide practical examples on where improvements are needed. All three member states have been accorded particular focus due to the size and complexity of problems. They represent comprehensive cases, which cover a broad range of crisis management issues and help to review them. The selected cases also allow to systematically addressing the outlined problems and questions in a structured and focused analysis.[61] However, the cases were also chosen for pragmatic reasons, such as the access to sources, the major languages used in these countries.

1.4.4 Challenges and Limitations

From the outset, this dissertation has been characterized by the challenge of analyzing an extensive and highly complex research area. A particular challenge of the thesis and its subject lie in their potentially all-encompassing scope and broad approach that had to be adopted as a consequence. Indeed, the CA by definition implies a focus on different dimensions and ideally would mean a paradigm shift whereby virtually all activities and actors relevant for a crisis or a conflict region should be considered relevant in terms of comprehensive conflict management. Thus, a too narrow focus on specific parts of crisis management policy would have contradicted the underlying concept of the Comprehensive Approach. The notion essentially obliges a perspective that looks beyond boundaries and focuses on the overall outcome instead of the efforts of individual actors. Moreover, such an analysis has to address various sectors and high-level issues. The challenge of complexity has been dealt with through the very detailed structuring of the analytical framework in the next chapter. This has also necessi-

61 At first sight, the added value of considering Sweden as a relatively small member state of the EU in the same analysis as two of the largest and most powerful member states might seem unusual in this regard. However, with regard to Swedish effort towards comprehensive crisis management it is particularly the country's size and respective foreign and security policy which poses an argument in favor of Sweden as a case study. Furthermore, the selection of the Swedish case is driven by the country's political culture and political-administrative system, which differs in numerous ways from that in other countries. Another interesting dimension is the Swedish tradition as a militarily non-aligned state. Considering Sweden as a case of Europeanization is interesting as the country is a fairly new member of the EU which was highly skeptical of the development of the Union's crisis management dimension.

tated cautious considerations concerning the selection of the actual research subject and questions in relation to the overall topic to keep the study reasonably clear and focused. Accordingly, the empirical focus delimits the scope of inquiry to some extent and the analysis has to be partly selective in order to ensure adequate presentation. Thus, a few words should be said about what this dissertation does not include.

First of all, this thesis is not about the EU's and member states' particular crisis management instruments and actions nor does it seek to discuss specific interventions and efforts in particular conflicts. It deals with overall European crisis management coherence and not with the effectiveness of specific crisis management policies or instruments. The actual content and the policy produced are not assessed and interaction of crisis management actors in the field is not investigated. Though these are to some extent picked up when relevant, the research focuses on actors, interaction and the organizational crisis management design as a whole. It is about what the EU and member states can do to enhance the delivery of their policies. Though the focal points of this dissertation are on the issues of coherence and coordination in international crisis management, it should be said that this focus does not mean that coordination is considered as the only and most significant factor of success in international crisis management. Using problems of crisis management coordination as a catch-all phrase for deeper divergences and indecision on different levels of crisis management entails several risks. It could not just raise expectations concerning the capability of organizational fixes to reduce crisis management complexity and overcoming conflicts, but defining crisis management dilemmas in this manner could also draw attention away from deeper problems or other priorities (Paris 2009, p. 60).[62] In spite of its significance, incoherence is not the only cause of failure in crisis management. Clearly, other factors like the severity level of the conflict or actions of other actors involved have some weight as well. However, coordination is the issue that lies in the domain of international competencies and responsibility, "it can be influenced by the internationals by way of institutional design and thus is practically more relevant" (Zürn/Herrhausen 2008, p. 280).

Also, the fact that restructuring efforts after the EU's Lisbon Treaty reforms were almost constantly underway at the time of writing this thesis needs to be considered. Relevant reorganization was at a transitional stage with details not yet defined. EU crisis management is in a period of transition and the state of

62 See also Teuten and Korski (2010, p. xiv) for a relativization of the problem of crisis management coordination. This issue is also picked up in the conclusion of this dissertation.

play at the time of writing is not the state of play a year later.[63] Thus, any evaluation is preliminary due to the short existence of new structures. Also interviewees frequently pointed out that their statements would be influenced by the nature of the changing situation which could result in speculative assessments. Consequently, a thesis on a topic like this might appear as overtaken by events of constantly changing crisis management architecture in Europe. Yet, it is still promising to analyze the way taken with the Lisbon Treaty changes, the direction in which EU crisis management has developed so far and the degree to which the new provisions embody certain types of coordination. As the first months of new institutions are important from an organizational perspective, an investigation is possible and necessary, in particular as EU crisis management has been in a constant process of change since its invention. Also at a country case study level, crisis management in the UK, Germany and Sweden is a 'moving target' in the CA context as states are trying to adapt to challenges and to adjust their particular systems to the changing crisis management conditions. Of course, speculative assumptions are avoided; however, these are to some extent inevitable in the face of the situation of transformation at the time of writing. EU institutions have evolved significantly and the degree of cooperation varies immensely depending upon the policy area or respective institution studied. The same applies to national crisis management structures. The complexity of procedures and the number of actors involved, make the European continent the institutionally most complex region in the world (Stewart 2006, p. 91). An added challenge existed concerning the availability and accessibility of sources. In this regard, the complexity of the EU system is undoubtedly a setback for researchers, contributing to the vast quantity of conflicting accounts that anyone researching the EU comes across, starting with the accessibility of EU documentation.[64] As an ICG report concludes, "the EU's conflict prevention and management resources remain unknown to most and mysterious to the few who try to penetrate them" (International Crisis Group 2005, p. 52).

Another point to consider is the financial and economic crisis that struck the global economy and started spreading across Europe in 2008. The Euro-crisis has forced itself to the top of the agenda, overshadowing virtually all other policy areas and is omnipresent in academic and public debate. Moreover, it questioned the effectiveness of the Eurozone and continues to do so at the time of writing as member states struggle to find tools to respond. Nonetheless, implica-

63 Post-Lisbon Structures became operational in 2010/2011 and are scheduled to be evaluated for the first time 2013/2014 (Major/Bail 2011, p. 18).

64 Johannsen (2011, p. 31) was confronted with similar issues of limited availability of information in the context of ESDP.

tions of the persistent financial crisis and the continuing debt crisis in Europe are largely left out in this dissertation. While it is acknowledged that financial burdens as a consequence of the crisis heavily impact and constrain the ability of all actors to respond to international conflict, this limitation is necessary due to reasons of practical feasibility.

The fact that this thesis has a rather large and detailed theoretical framework does not mean that it is a theoretical study. Though organization theory, public administration studies and Europeanization are set out in a detailed way below, this dissertation does not seek to make a contribution to IR theory or strictly apply existing theories to the cases. Rather, the theoretical framework is used to focus, structure and guide research and to also limit the empirical analysis undertaken. As articulated by Martens (2010, p. 19) "[t]he empirical phenomenon is at the centre of the analysis, and theory is primarily applied in order to shed light on and explain this specific phenomenon." In line with this statement, the central purpose and larger part of the thesis is the empirical analysis of European crisis management coherence. It is the practice and not the theory with which this dissertation is concerned.

As shown later in this dissertation, CA origins and meanings vary and "there is a risk of wanting to compare the use of the concept in its widest sense, without paying attention to the specific context in which it is implemented" (Wendling 2010, p. 25). In order to avoid such a concept stretching, the analysis concentrates on taking into account the particular nature of the EU concepts as well as the national concepts. In these terms, the thesis presents an inclusive review of the meanings given to the CA by the analyzed crisis management actors rather than being a comparative study. Though each national case study and the analysis of the EU dimension are structured in a similar fashion and invite analytical comparisons, it should be noted that the examination is executed through the prism of each particular context. The different scale and context makes it difficult to consistently compare the cases in a meaningful way. Also, CA differences make sense in different settings. However, a comparing summary of findings in the last chapter of this dissertation seems nonetheless possible and the final section of the thesis provides a comparative look at the similarities and differences of the approaches. In this regard, the comparison is not between the cases as such but rather between the instruments, debates and issues surrounding European CA debates.

2 Analytical Framework

The following chapters contextualize the subject of this dissertation, clarify important terminology and synthesize relevant theoretic and conceptual aspects. In doing so, more order is brought into the terminological and conceptual confusion while at the same time reflecting on established knowledge and the overall research background.

Before taking a closer look at the CA, it is essential to examine relevant terms and the setting associated with the concept's application in the context of crisis and conflict. There is little consensus in academic literature, official and public discussion on how relevant terms like conflict and conflict management, or other key variables in the dissertation's context like coordination and coherence, should be defined. Thus, a close examination of the terminological debate is necessary. Providing the initial background, firstly, terminological issues related to crisis management are addressed and clarified. The CA concept and its analysis need to be seen in the context of increasingly complex and interdependent international crisis management scenarios. Thus, patterns and the nature of conflict as well as changes in international crisis management after the Cold War are analyzed. To provide an insight into broader debates, the subchapter is concerned with the changes that took place in the international security environment after the end of the Cold War and its consequences for international crisis management. It also discusses the necessity and intricacies of taking a broad view on security and considers links between security and development as well as other relevant dimensions of the contemporary international security environment. Remarks about the specific character of today's conflicts and respectively changing parameter of contemporary crisis management are given. The chapter continues with a detailed examination of the CA as the concept on which this thesis' research subject is built on. Before analyzing implementation challenges that arise when trying to put the CA into practice, it needs to be explored what the concept actually means. Consequently, key dimensions and motivations as well as different conceptual approaches are mapped out that allow for a further specification of an analytical approach.

Conceptual assumptions are then extended to include perspectives of organization theory and Europeanization as the theoretical approach put forward to analyze research questions and to analytically complement the thematic background given by the CA concept. Before these two analytical lenses are elaborated in detail, a clarification of the key terms coherence and coordination is carried out. Finally, the rather abstract tenets of organization theory, public administration studies and the concept of Europeanization are translated into more specific operational parameters to allow theoretical assumptions to be connected and transferred in the specific context of research questions and empirical observations. In doing so, elaborations of the preceding chapters are clustered around the four recurring dimensions: the crisis management approach, the conceptual as well as the organizational dimension of CA implementation, and the European dimension and EU-interface. Lastly, for reasons of better overview, the analytical framework that is applied in order to shed light on the above developed research question is summed up.

2.1 Crisis, Conflict and Preoccupation with Conflict

2.1.1 Terminological Clarifications

Despite the widespread usage of the expression *crisis management* by governments, IOs and NGOs, there is by far no agreement on the notion's precise meaning. Crisis management is often used very loosely and interchangeably with other phrases such as peacebuilding, peacemaking and peacekeeping as well as crisis response, conflict resolution and conflict prevention. Depending on the disciplinary background, scholars offer different definitions.[65] The same applies for the two terms *conflict* and *crisis*.

To begin with, most definitions of conflict do not imply that conflict in terms of an opposition of incompatible interests and goals is undesirable *per se* as it is a universal feature of human society and necessary phenomenon of change. Rather, it is the resolution of conflict through violence which is destructive and impedes lasting solutions (Ramsbotham/Woodhouse/Miall 2011, pp. 7/8). While the term conflict was long used synonymously with the term war, this changed with the end of the Cold War and the increasing use of the expression for various disputes. To underline the nature of conflict, supplementary terms were added such as the specification of *violent, armed* or *deadly* conflict when the use of force was involved, or *acute* and *chronic* to describe its nature (Johannsen 2011, p. 51). Former UN Secretary-General Boutros Boutros-Ghali

65 Koops (2009, p. 3) points out that the prominent role of the concept results precisely from its broad applicability that appeals to many actors in the peace and security field. See also the International Crisis Group (2001, pp. 2-6), Johannsen (2011, pp. 50-53) as well as Ramsbotham, Woodhouse and Miall (2011, pp. 30-32) for additional remarks on this matter.

differentiated in his *Agenda for Peace* (United Nations 1992, para. 20/21) between disputes as non-violent conflicts, whereas conflicts imply violence. According to his definition, disputes escalate into conflicts. Depending on the particular criteria applied, the definition of the interdisciplinary expression crisis varies as well. In the context of IR, crises "are first and foremost regarded as destabilizing characteristics of the international system on global and regional level" (Johannsen 2011, p. 50). The term thus describes a state of tension, which demonstrates a dynamic that indicates the increasing potential of escalation and the beginning or the extension of an armed conflict and war (Johannsen 2011, p. 50). Houben (2005, p. 12) describes the contemporary concept of crisis "as a situation *jenseits von Krieg und Frieden*: where there is a crisis, there is no war and there is no peace." When developing stages of peace or conflict, Lund (1996, p. 39) defines crisis as a "tense confrontation between armed forces that are mobilized and ready to fight and may engage in threats and occasional low-level skirmishes but have not exerted any significant amount of force." This situation of very high tension, but not yet war,[66] implies that there is only limited time and a significant amount of uncertainty (Wedin 2005, p. 11).

Comparing both terms, conflict is often referred to in a wider manner to describe situations where priority values, interests and goals are threatened on the national or international level. The word crisis in this regard is used in more acute situations (Blockmans/Wessel 2009, p. 10). As such, conflict and crisis are dynamic in their escalation and complex in the interplay of different conflict parties, levels at which the dispute takes place as well as the matters fought over which additionally vary over time (Ramsbotham/Woodhouse/Miall 2011, p. 8). As both terms are used inconsistently both in relevant official documents as well as academic research and literature, they are applied in the same interchangeable manner in the following analysis to cover the widest set of conflictual circumstances and respective responses.[67]

The terminological confusion when referring to the *management* of crisis and conflict is even more problematic. Terms for the various kinds of intervention into conflicts can be arranged roughly along two parallel lines: the first line contains notions which are generally employed in discussions closely associated with the UN such as peacekeeping and peacebuilding, the second line is general-

66 International law does legally not require states to formally declare war or their intention to use force. Thus, it is possible for states to participate in an armed conflict without recognizing that they are formally at war. Also, the fact that the state of war can only be declared by sovereign states to sovereign states poses a legal problem concerning non-state actors such terrorist groups (Houben 2005, p. 12).

67 This lack of consistency becomes apparent in particular when looking at the use of the two terms on EU and member state level, there is a lack of consistency (Johannsen 2011, p. 51).

ly preferred in the academic literature and includes terms such as conflict prevention, crisis response, conflict management and conflict resolution (Lund 1996, p. 40).[68] Narrowly defined approaches view crisis management as one of many other approaches to realizing and maintaining security. Crisis management in this regard deals with situations, when preventive measures have already proved insufficient (Houben 2005, p. xii). In contrast to conflict prevention, crisis management is perceived more short-term and contingency-oriented.[69] Crisis management instruments may entail a more direct use of military and negatively perceived means such as sanctions, embargoes or the freezing of relations. It contains crisis assessment, crisis response and post-conflict peacebuilding (Missiroli 2001, p. 8). In this understanding crisis management is conceptually very different from any preventive action and involves immediate efforts to manage tensions that have reached a high level of confrontation and violence (Lund 1996, pp. 42/43). In other words, narrowly defined crisis management can only exist if a conflict has reached a violent stage. More broadly defined, crisis management in the international arena is taken to refer to any attempt by third parties to resolve crisis and prevent further escalation of conflict. It is seen as a multilayer process which ranges from long-term and structural policy to project stability to short-term operational civilian or military crisis management policy. In this approach, crisis management activities cover a broad range of methods which can be separated into categories with varying degrees of third-party commitment attached to managing conflict: verbal expression, diplomatic approaches, judicial processes, administrative assistance and military involvement (Frazier/Dixon 2009, pp. 52/53).[70]

However, despite differing definitions and use, these concepts were developed on the basis of a broad overall consensus between states and IOs regarding

68 See various chapters in Ramsbotham, Woodhouse and Miall (2011) for a differentiation of these concepts. Generally, crisis management can be grasped on various levels of analysis: at the personal, group or community level or else at national or international level. Crisis management takes on various forms. It can be unilateral, where for instance one party withdraws from conflict, bilateral and multilateral, where outside parties intervene (Ramsbotham/Woodhouse/Miall 2011, p. 7). For an overview of the various ideas in the field of crisis management such as direct, structural and cultural violence, negative and positive peace as well as conflict and conflict escalation models see Ramsbotham, Woodhouse and Miall (2011, pp. 10-16).

69 According to Teuten and Korski (2010, p. xi) "[c]onflict prevention involves mitigating the underlying causes of conflict and potential triggers for escalation through a range of instruments that may include fostering political dialogue, social and economic development, developing conflict resolution mechanisms, and building accountable and effective security institutions."

70 Concepts of third-party interventions into conflict are elaborated in detail in Ramsbotham, Woodhouse and Miall (2011, pp. 21-24).

methods and purpose of international crisis management (Tardy 2011, p. 87).[71] Also, the CA context of this thesis dissolves the separation of above-mentioned concepts of how to deal with crisis and conflict (Wendling 2010, p. 16). Although, it is not intended to suggest that conflicts follow linear patterns, different phases of conflict require different approaches which cannot be kept separate and may be required at the same time. Consequently, this dissertation takes a broad definition of crisis management which contains above-mentioned concepts and involves both long-term conflict prevention and short-term stabilization.[72] In line with the definition developed by Major et al. (2011, p. 91) for their Crisis Management Toolbox, crisis management is viewed as

> "the commitments made by civilian personnel, police and military, within a bi- or multilateral framework, to build peace and stability in crisis regions, by using various instruments. These instruments include measures for crisis prevention, for the resolution of acute and lasting armed conflicts, and for the consolidation of peace."

Thus, crisis management is used as an overarching expression for all types of activities and engagement carried out to respond to crisis and conflict. It is understood "as the coordinated actions taken to defuse conflicts or crisis, prevent their escalation into an armed conflict and contain hostilities if they should result" (Rintakoski/Autti 2008, p. 10). This covers activities aiming at the prevention of conflict as well as responses to a crisis that already developed into armed conflict to prevent further escalation and contribute to a conclusion. Moreover, this definition entails activities that range from humanitarian work and development cooperation to military engagement and covers terms and concepts of different crisis management actors as set out above combining the multidimensional elements most crisis management efforts actually entail. It is "used as a generic term to cover the gamut of positive conflict handling" (Ramsbotham/Woodhouse/Miall 2011, p. 31).

While it could be argued that enlarging the concept of international crisis management in such terms in practice risks diluting it to such a degree or stretching it to include all foreign policy, the adoption of this definition should be understood

71 However, Tardy (2011, p. 87) also observed that pushing the limitations of the traditional principles of impartiality, consent of the host state and use of force as the ultimate mean increasingly undermines this broad legitimacy of crisis management operations. The increasing presence of emerging powers contributing to crisis management enterprises may also lead to tensions over norms and aims. Tardy (2011, p. 87) thus concludes that this consensus is increasingly undermined by questions about what respective operations can and should achieve in accordance with what principles.

72 Teuten and Korski (2010, p. xi) define stabilization as "the process of establishing peace and security in severely conflict affected countries where the level of insecurity makes many normal development activities impossible."

as "a cycle of responses, corresponding to the different chronological stages through which crisis and conflicts tend to move" (International Crisis Group 2001, p. 3). This may encompass a wide range of activities, but at the same time captures the overall complexity of the international environment while mirroring the broad and comprehensive features of existing structures. In doing so, crisis management is contextualized on the international level within the larger external foreign and security policy domain.[73] The internal dimension of crisis management such as different civilian protection efforts, emergency planning or national security mechanisms to fight organized crime are largely left out.[74]

2.1.2 Characteristics of Contemporary Crisis Management in Today's International Security Environment

The international security environment as the framework for international crisis management changed drastically since the end of the Cold War. While on the one hand, the last two decades have been a period "of unprecedented security for the citizens and states of the West, at least in traditional military terms" (Edmunds 2012, p. 266), initial optimistic expectations that this might be the *End of*

73 Douma, Van de Goor and Van Klaas (1998, p. 80) argue similarly when taking on a broad definition for the concept of conflict prevention arguing that such a broadly defined approach "reflects more accurately the current practices of international actors". Also Banim (2009, pp. 305/306) understands the term EU crisis management as "any policy instrument the EU deploys during a crisis." According to Ramsbotham, Woodhouse and Miall (2011, p. 9) it does not matter which of the various labels is used "as long as the field itself is coherent enough to contain the substance of what is being advocated in each case." Though termed differently, all concepts effectively engage in the same enterprise. This approach and argumentation has also been adopted by the ICG (International Crisis Group 2001, pp. 2/3) and their decision of the broad usage of the term crisis response can partly be drawn on for the purpose of this dissertation. Drawing onto the ICG's Crisis Response Cycle, this includes different crisis management phases of building, maintaining, restoring and enforcing peace (International Crisis Group 2001, p. 4). It nonetheless needs to be noted that in the context of this dissertation, the conflict cycle terminology, which is convenient for a division of labor on operational level by pointing out in which phase of a crisis different actors engage and instruments are utilized, is misleading as it is contrary to the coherent response in concrete crisis situations. A crisis management cycle illustrates crisis phases as ideal types which in practice merge into each other. At this point of the dissertation, the model is nonetheless a clarifying analytical tool. Details on the crisis management cycle and its different phases can also be found in Major et al. (2011, pp. 92-94).

74 While such a distinction in internal and external crisis management dimension is useful for analytical reasons, it has become less clear in reality. Conflict prevention efforts in third countries, for instance, are increasingly used to reduce the spread of international crime (Rieker 2004, p. 370). For more details on the dissolution of the internal-external policy divides see the next chapter.

History,[75] a peace dividend and the start of a peaceful era quickly were disillusioned. In terms of frequency, trends show a rise at the beginning of the post-Cold War era making the scale of crisis management witnessed since 1990 a rather new political phenomenon (Houben 2005, p. 12). Attention moved from the Cold War bi-polar superpower rivalry to the outbreaks of various new conflicts and wars all over the world which resulted in significant crisis management engagements, a multitude of different peace support and crisis response operations as well as stabilization and reconstruction efforts.[76] As the amount of different names and labels for crisis management operations indicates, these could not be clearly defined anymore and confronted the international community with some major difficulties (Felberbauer/Feichtinger/Schmidl 2011, p. 5).[77] As described by Crocker, Hampson and Aall (2011, p. 41) "[t]he dominant powers in a bipolar international system sought to 'manage' their conflicts in order to avoid a loss of face or strategic setbacks and to prevent their conflicts from escalating 'out of control'". Yet, one experience from recent crisis management engagements is that strength in terms of military means alone are less conducive than civilian means when it comes to creating stability. Also, traditional, state-centric understandings of conflict have been questioned. Security and stability hreats cross national state borders, affect more than one state at time and dissolve the separation of domestic internal and external security provision (Schröder 2006, p. 1). As argued by Grönvall and Lundberg (2005, p. 40)

75 Fukuyama (1992) attracted particular attention with his book *The End of History and the Last Man* in which he argued that the Cold War's end also means the end of history as great ideological conflict ended with a victory of Western liberal democracy leaving the rest of the world to be ordered in a liberal democratic way. Experiences of the last two decades, however, have largely shown the limitations of Fukuyama's highly debated assumption. Yet, it reflects the widely shared point of view on profound changes worldwide.

76 Almost 80 percent of UN authorized operations were conducted between 1988 and 2007 (Johannsen 2011, p. 18). Global crisis managemement continuously grew and in 2011, more than 263000 peacekeepers were deployed for crisis management through the UN, NATO and the EU (Annual Review of Global Peace Operations 2012, p. 2). On the contrary, data also shows an overall global reduction of armed conflict. See, for instance, respective chapters and tables on *Trends in Armed Conflict* and *Patterns of Major Armed Conflicts* in various yearbooks of the Stockholm International Peace Research Institute (SIPRI) since the end of the Cold War or the Carnegie Commission's 1997 report on the Prevention of Deadly Conflict (Carnegie Commission on Preventing Deadly Conflict 1997). See also SIPRI Yearbook chapters on *Multilateral Peace Operations* for greater details on developments in peace operations regarding the number of personnel deployed and conducting organizations. However, the classification of conflicts to be included in data collections and data on peace and conflict itself is highly disputed. On the difficulties of measuring peace and dealing with conflict data see Ramsbotham (2010, pp. 33-36) and in greater detail Ramsbotham, Woodhouse and Miall (2011, pp.63-93).

77 "[O]ne power's 'peace enforcement' is another power's 'peace making'" (Felberbauer/Feichtinger/Schmidl 2011, p. 5). Terminological and definitional problems outlined in the previous chapter reflect this complexity.

> "the abundance of vulnerabilities that come with globalization, hyper-complex interconnected technological systems, and an ever faster pace of innovation and evolution of science and technology have consequences that spill over jurisdictional, sectoral and geographical borders." [78]

A fundamental change that took place as a consequence of globalization and which further contributed to complexity is the dissolution of the policy divide into internal and external security. Domestic political processes and attitudes are extended into the international environment and domestic politics have become internationalized. At the same time, international politics have become increasingly domesticated (Houben 2005, pp. 14/15). It is thus unclear into which political sphere they fall and clear boundaries between institutional responsibilities get blurred (Schröder 2006, p. 6). Traditional differentiations have become obsolete as challenges contemporary crisis management is confronted with are multi-dimensional and interdependent mirroring both the causes and symptoms of crisis. Conflict today rarely stands for disputes between relatively similar parties such as states but between unequal and diverse actors. Classic interstate conflicts as an example of a symmetric conflict between actors of parity are the exception nowadays.[79] Instead, the international community is faced with numerous complex, asymmetric crises such as regional conflicts, civil wars and insurgencies. The roots of such asymmetric conflicts are rarely found in one particular issue but in the various characteristics and interrelationships defining them (Ramsbotham/Woodhouse/Miall 2011, p. 24). Also, classical crisis management measures

78 For a separate analysis of these security threats, contemporary security topics as well as regional security challenges from China to the Middle East see contributions by different authors in Dunn Cavelty and Mauer (2010). Some scholars argue that these security challenges rather present a range of risks as no single threat dominates in this complex and interdependent international environment. According to this differentiation, threats stand for clear dangers that are identifiable and thus specific and known. The logic of risks, on the contrary, is that of unclear potentially dangerous uncertainties instead of imminent danger (Edmunds 2012, pp. 266/267). Other scholars, however, argue that the discussion on new threats emerging is just 'new wine in old bottles', and that, for instance, the debate concerning state failure "amounts to little more than new terminology for an older conflict-prevention approach focused on tackling the root causes of insecurity through addressing socioeconomic needs and promoting protection of human rights" (Banim 2009, p. 303).

79 Though statistics on conflict and peace often vary depending on definitions and datasets used and they must be carefully read, as mentioned above, a majority of respective research confirms a decline of interstate wars (Ramsbotham/Woodhouse/Miall 2011, p. 70). According to Gill (2011, pp. 6/7), for instance, only two out of 29 major armed conflicts from 2001 until 2010 were carried out between states while all other conflicts were fought within states involving a state actor and at least one armed non-state actor.

and ways to resolve conflict do not apply, suggesting more nuanced concepts and forms of crisis management.[80]

These developments resulted in wider notions of security which gained an extended meaning including economic, financial and development issues as well as social, cultural, humanitarian and ecological features besides the traditional core aspects of security and defence (Schröder 2006, p. 1). Justifications arguing in favour of wide security approaches mainly point out ethical reasons with an intervention ending the suffering of people. But also arguments in a narrower security political sense of threats due to terrorism and the spread of weapons of mass destruction are put forward (Weiss/Spanger/Van Meurs 2009, pp. 11-13).[81] In this context, atrocities by states encouraged the development of ideas toward a *Responsibility to Protect* aiming at preventing most serious violations of human rights. The Responsibility to Protect norm prompted efforts to establish an international consensus claiming the right to intervene when states are not able or not willing to protect their own citizens as part of their sovereignty. International engagement in crisis situations was no longer about intervening on ceasefire but moreover on incorporating human aspects and a normative shift toward a focus on the individual (Wendling 2010, p. 14).[82] The expansion of the concept of security shifted the focus of security and the entity to be secured from the state to the individual human being. Against this background, the *Human Security* concept emphasizes the individual's freedom from fear and freedom from want. The widening of the security paradigm with integration of human aspects is closely linked to a more comprehensive crisis management approach.[83] Complementary

80 See Ramsbotham, Woodhouse and Miall (2011, pp. 25-30) for models and figures illustrating crisis management feasible for asymmetric conflicts.

81 Widened notions of security questioned the legal foundations of the international system of states by including non-traditional security categories. Thus, in the process of the security concept's broadening became a battleground in and of itself (Smith 2005, p. 57).

82 The Responsibility to Protect norm is linked to the debate on humanitarian interventions and was formally acknowledged by the UN member states at the 2005 UN world summit (Major et al. 2011, p. 101). Adopted in 2011, the UN Security Council Resolution 1973 on Libya demonstrates an example of implementation of the Responsibility to Protect norm, although officially not explicitly referring to it. However, the expansion of NATO's initial military objective of enforcing a no-fly zone, enforcing an arms embargo, and protecting civilians to also include the provision of air support for rebels and targeting Muammar Al-Qaddafi resulted in disputes over the resolution and set countries which continue to place greater value on sovereignty against the norm. A future consensus on Responsibility to Protect engagements will thus be difficult to forge and the principle continues to be heavily discussed concerning its practical interpretations and legal consequences. On the Responsibility to Protect debate and practices see Kerr (2010, pp. 117-120) who distinguishes between the responsibility to prevent, the responsibility to react and the responsibility to rebuild.

83 The Human Security concept is commonly associated with the UNDP's 1994 Human Development Report (United Nations Development Programme 1994). The report defines Human

to this approach, the importance of *Local Ownership* which embraces the recognition that crisis management engagements have to be adequately accountable to the local population they are affecting and that local actors should have a say as a prerequisite for crisis management sustainability has been increasingly acknowledged (Major et al. 2011, pp. 97/98).

These developments led to considerable changes in the perception of both security and development when actors from both began to increasingly interact. Development actors became aware that efforts undertaken to assist development are ineffective or even counterproductive without basic security. In turn, security actors realized that short-term or *ad hoc* military reactions are insufficient in complex and multidimensional crisis situations (Tannous 2008, p. 33). Thus international crisis management began to increasingly being analyzed under the recognition that development and security are closely linked. Summarized with the paradigm 'No development without security, no security without development',[84] this so-called *Security-Development Nexus* has been high on the international agenda and a central focus of the international crisis management debate since the early 2000s. The nexus denotes an increasing consensus on the causal relationship between development and security and promotes more integrated and comprehensive approaches to conflict. An emphasis on the link between security and development, in particular in the CA context, has become commonplace (Alamir 2012, p. 69).[85] From the perspective of this thesis' research subject, it thus needs to be pointed out that the security-development nexus involves practical challenges for development actors and the military in the framework of common efforts.[86] The wide-ranging debates associated with the question of the

Security as the summation of seven distinct dimensions of security: economic, food, health, environmental, personal, community and political (United Nations Development Programme 1994, pp. 24/25). The concept is still debated due its vagueness and 'impracticability' as well as a fear of securitization of international politics (Major et al. 2011, p. 97). For an overview on the Human Security norm, its evolution and debates see Knight (2008, pp. 15-22). In her analysis, Kerr (2010, pp. 115-118) distinguishes between an external Human Security debate on the one hand, which takes place between supporters of the concept and advocates of the traditional state-centric approach, and an internal debate between different human security schools.

84 This particular sentence has been frequently repeated both in the academic as well as in the political sphere (Alamir 2012, p. 69).

85 Klingebiel (2006, p. 1) points out that debates on the relationship between security and development are not a new issue. Debates, however, changed from an abstract level to directly discussing their practical and conceptual convergence.

86 For details on these practical challenges see Alamir (2012, pp. 71-73). Alamir (2012) analyses the security-development nexus comparatively from the perspectives of both the military and the development communities and argues that, though the two sides increasingly adapt to the nexus-reality, particular challenges remain regarding strategic frameworks and decision-making processes. Doing so, Alamir (2012, p. 71) points out that the actual impact of the security-development interconnection is not generalizable but largely depends on the practical viewpoint.

relationship between development and security in contemporary conflict are also concerned with the interactions of military and civilian actors. The security-development nexus illustrates the impact insecurity has on the individual, socio-structural, the governance and the international dimension of development.[87] Shortfalls in development impact military stabilization efforts by negatively affecting technical, logistical and operational aspects of the operational capability, communication and cooperation as well as the risk and threat environment. Moreover, from military perspective underdevelopment impacts on all these dimensions as they ultimately shape opportunities for military exit (Alamir 2012, pp. 69/70). The security-development nexus' underlying questions are clearly reflected in this dissertation's research question as the interdependency of security and development ultimately leads to the policy puzzle of how to coordinate them. Importantly from the CA perspective is that crisis management efforts are not considered sequential steps anymore. Rather than chronologically, the relationship between security and development initiatives has come to be viewed as functionally complementary (Hull 2011, p. 3).

Increasing numbers of crisis management operations involving a wide range of actors signify greater need for cooperation and coordination between the civilian and military aspects, and respective instruments. Consequently, the dividing lines between the two dimensions have become increasingly unclear and the former sequence of military intervention followed by civilian engagement no longer applies (Drent/Zandee 2010, p. 2). Crisis management experiences showed that diverse instruments are important from the beginning and resulted in a growing importance of non-military crisis management instruments. This development also points to the above elaborated widening of security as "an 'encompassing' feature, including civilian, police and military aspects" (Felberbauer/Feichtinger/Schmidl 2011, p. 5). The importance of civil-military relations for the effectiveness of crisis management is broadly acknowledged in this regard. Moreover, "that increasing effectiveness of armed forces in contemporary peace support operations is not primarily a military endeavour" (Egnell 2009b, p. 138).[88]

As demonstrated above, conceptions of security and defence changed from territorial defence to crisis management where the territory or sovereignty of a

87 Impact of insecurity on the individual development dimension for example means impeding people's freedom of movement, creating an atmosphere of fear and impacting on people's well-being. The socio-structural dimension is affected as insecurity leads to flows of refugees and internally displaced people. The governance dimension is concerned with insecurity that undermines regime stability by weakening and delegitimizing government institutions. Internationally, insecurity influences inter-state relations, undermines a state's international credibility and pushes the proliferation of arms and weapons (Alamir 2012, pp. 69/70).

88 Civil-military relations in the CA context are discussed more fully in chapter 2.2.3 on CA understandings.

state itself it hardly ever in direct danger. Hence, participation in crisis management is not necessarily required and most time presents an engagement of choice. Perceived national responsibility thus plays an increasing role in today's international crisis management (Houben 2005, p. 14).[89] In this context, crisis management engagements changed "from traditional mandate-based missions, to missions involving the possibility of proactive use of force" (Rieker 2006b, p. 107). Changes and the complexity of international crisis management are "a direct consequence of the nature of the problem" (Houben 2005, p. 18), as the characteristics of a conflict impact on states' abilities to deal with them. Several crucial characteristics can be identified that deal with the multidimensionality of crisis management. First of all, the crisis itself mostly consists of various different facets due to a combination of interrelated economic, social, ecological and security aspects which have *direct* consequences such as victims or economic losses as well as *indirect* effects such as flows of refugees or the instability of a region. Even more so, they are characterized by geographic diffusion. Crises easily spill over into other states and are hard to contain thus contributing to regional security interdependency (Houben 2005, p. 19). On the basis of the understanding that external assistance comes with side effects, the *Do No Harm*-approach as a guiding crisis management principle been developed. The principle demands that external crisis management should be conflict sensitive and minimize negative effects (Major et al. 2011, p. 96).

Furthermore, international crisis management since the end of the Cold War is characterized by an increasing spectrum of tasks, expanding timelines and a growing number of actors. Contemporary crisis management tasks range from state-building, promotion of rule-of-law and humanitarian aid to establishing secure living as well as stable economic and social structures. As its range and scope has enhanced, the array of actors has grown. While states remain the dominant crisis management actors, non- and quasi-state actors become increasingly important. While these actors on the one hand can "contribute more to peaceable outcomes (…), [o]n the other hand, other non-state and quasi-state actors have a fundamentally debilitating effect on peace and security" (Gill 2011, p. 6). IOs and nation states alike have to provide for a large range of potential security challenges instead of restraining their effort to traditional conventional security political concerns of state-centric and military nature. This means that

89 Houben (2005, p. 14) makes the distinction between *wars of necessity* and *wars of choice* "based on the presence and absence, respectively, of a compelling security imperative to act or participate". He also points to the increasing demand this means for governments to justify their engagement to parliament and the domestic public.

> "[t]oday, the position is far more complex given a broader range of governmental and non-governmental policy actors engaged in complex interaction within a more fragmented and 'compartmentalised' policy environment, in which a diverse range of pressures are brought to bear on the formulation, implementation and presentation of external policy" (Self 2010, p. 257).

The recognition of intra-state conflict being a threat to international peace and security led to greater international attempts to intervene and a greater interest in crisis management methods. Increasing involvement of state and non-state actors adds a further layer of complexity to crisis management efforts. The number of actors involved has increased as much as broader expertise and specific tools are needed which can only be provided by different international actors (Mölling 2008, p. 1). A major reason for this development is that the rather clear international order of the 20th century has been substituted by a global increase of institutions and actors of which states are just one factor (Edmunds 2012, p. 266). Furthermore, actors have acknowledged that stable conflict solution demands long-term commitment that covers the whole conflict cycle. Short-term humanitarian engagements have to be followed by long-term development efforts or structural measures such as *Security Sector Reform* (SSR) (Major/Schöndorf 2011, p. 1).

Additionally, political objectives standing behind efforts have become more ambitious and the intricacy of contemporary crisis management introduces many conceptual and organizational changes. Different actors engage in crisis management, undertake a wide range of activities and operate on several different levels and in different formations. Connectivity and complexity are the key words for the contemporary international environment. Actors are not faced with a single omnipresent conflict or threat, but a complex international security environment. Complexity in this regard is not just caused by the nature of crisis and the international security environment itself but largely by the nature and number of actors. The international sphere has seen a growing number of different organizations, from NGOs, regional organizations and private military companies, playing more and more important roles. At the same time, national state governments and international agencies established specialized units and departments for certain crisis management scenarios (Paris 2009, p. 56). But while on the one hand each actor contributes valuable resources and expertise as well as a distinctive approach to crisis management, on the other hand the risk of duplication of efforts and sometimes even competition raised. Dealing with the results of this expansion can be referred to as "complexity management" (Mölling 2008, p. 1), a term that has also been used to describe the challenge officials and departments in IOs are faced with at the working level on a daily basis (Trondal et al. 2010, p. 206).[90] As Gordon (2006, p. 340) states,

> "the political objectives underpinning interventions have become more ambitious, resulting in a greater degree of complexity in the institutional responses of states and international organisations. The convergence of the human security and the traditional, narrowly defined state security agendas, the gradual importation of 'political economy' approaches to conflict analysis, state and IOs' pursuit of multidimensional missions and policy 'coherence' between the various aspects of these interventions have each been significant factors in this evolution."

In this context of overall globalization, contemporary crisis and conflict have also been grasped with the notion of *new wars*. Describing the changing nature of war in the post-Cold War era, the characterization as new describes how civil wars and low intensity conflicts imply transnational connections and a complexity that make the maintenance of traditional divisions in internal and external or regional and global difficult. In contrast to traditional wars, new wars are largely about identity politics instead of ideological or geopolitical goals. Also the ways through which these new wars are fought changed. Drawing on experience of both guerilla warfare and counter-insurgency, the goal is not to capture territory by military means but through political control of the population. Consequently, most violence in these new wars is directed at the civilian population with a dramatic increase of human rights violations, refugees and displaced persons. Also the way these new wars are funded changed. New war economies are mostly decentralized and heavily rely on external resources which can only be sustained through continued violence so that the logic of war is incorporated into the economic functioning (Kaldor 2007, pp. 7-10).[91]

In the last two decades, these changes of international crisis management revealed some powerlessness and incapability of the international community to deal with emerging conflicts in an effective way. The complexity of above challenges fosters the need to bring different competencies and institutions both on the national as well as international level together and to reexamine traditional

90 To see what this entails, one should recall the set-up of the current UN system. As the Report of the High-level Panel on System-wide Coherence points out, "the UN system now encompasses 17 specialized agencies and related organizations, 14 funds and programmes, 17 departments and offices of the UN Secretariat, 5 regional commissions, 5 research and training institutes and a plethora of regional and country level structures" (United Nations 2006b, p. 9).

91 The first edition of Kaldor's book was published in 1999 under the title *New and Old Wars: Organized Violence in a Global Era*. Another key scholar of the New Wars discourse is Münkler (2005). The concept has been challenged by critics on empirical grounds by pointing out that a lot of 'new war'-characteristics are actually not novel and have been present in various conflicts over the past.

distinctions of responsibility. Already by the early 2000s, awareness grew that the complexity of intertwined problems such as human security and socio-economic underdevelopment demands policy coherence and that great coordination problems within the international crisis management machinery exist (Paris 2009, p. 56). Linear models of international crisis management with strict functional divisions of, for instance, civil and military reaction have been abolished (Johannsen 2011, p. 18). As a result, the crisis management setting for state governments has experienced two transformations: apart from the structural changes due to the broadening of tasks as explained above, they have also undergone internationalization and decentralization processes. National crisis management is marked by an increase of issues, actors, channels and procedures. "The traditional foreign policy apparatus centered upon the foreign ministry is increasingly unable to control effectively the ever-broadening and intensifying external involvement of the state" (Ekengren/Sundelius 2004, p. 112).

The dynamics, developments and experiences elaborated above spurred the conceptual debate on the appropriate mix of crisis management instruments and raised significant questions on the management of complex and multidimensional crises in terms of a process of crisis management as a whole. They put forward concerns and questions about how to secure coherence and coordination in crisis management on the national and international level and made concepts to approaching crisis management comprehensively a high demand.

2.2 The Comprehensive Approach

The transforming international crisis management environment of the last two decades and revisions of international crisis management have made the realignment of traditional concepts possible and necessary. The changing nature of international conflict in terms of actors, time frames, norms, instruments and tasks has required the reconsideration of crisis management responses and conceptual approaches of the international community to handle these challenges (Johannsen 2011, p. 18). The CA is one of the answers developed in this regard and the outcome of the most topical developments in the international arena as outlined above.

2.2.1 Conceptual Overview and Key Dimensions

Contemporary crisis management is a complex undertaking. A distinctive feature of this complexity is that the large number of actors it entails are independent in the broad range of tasks they carry out and in their legal status as autonomous agents with own organizational objectives, resource access and decision-making capabilities. At the same time, none of them can achieve the ultimate goal of sustainable crisis management on its own, as each actor contributes only a part of the whole (De Coning/Friis 2008, pp. 6/7). The ability of individual crisis management actors to accomplish their objectives depends on other organizations involved and thus, coordination is necessary for comprehensive crisis management to succeed (Lipson 2005, p. 2). "The success of each individual activity is linked to the success of the total collective and cumulative effect of the overall undertaking" (De Coning/Friis 2008, p. 7). Crisis management today involves a significant overlap of work by different actors which makes a joint approach essential. In consideration of complex crises, enhanced coordination of civilian and military, international and local actors and instruments is required. Operational experiences have shown that neglecting, overestimating or not coordinating different crisis management instruments damages overall crisis management efforts (Gauster 2012, p. 75). A one-dimensional or single-faceted crisis management approach proves counterproductive in dealing with complex crisis as it only addresses some conflict dimensions rather than dealing with them in a comprehensive manner (Wendling 2010, pp. 8/9).

With the changing nature of conflicts and the aim of resolving these deficiencies and to improve crisis management effectiveness, governments and IOs have developed a range of concepts (De Coning/Friis 2011, p. 246). In addition, most IOs, NGOs and national governmental ministries have developed specialized capacities to deal with multifaceted dimensions of these complex conflict systems. Still, the lack of coherence and coordination among these crisis management capabilities and among diverse actors is one of the most cited factors explaining the poor success rate of international crisis management (De Coning 2009, p. 117).[92] The CA in international crisis management is supposed to render such and the need for a more coherent approach to crisis management policies has been highlighted in numerous documents and international meetings. The

92 Various studies analyze the failure of crisis management and testify that this result can partly be attributed to a lack of coherence and poor coordination between the actors engaged. See, for instance, Collier et al. (2003), the Utstein Peacebuilding Study (Smith 2004), which analyses various peacebuilding projects, Paris and Sisk (2007, pp. 6/7), De Coning and Friis (2008, p. 2), Herrhausen (2009, p. 14) and Paris (2009). Research on the success rate and sustainable peace and stability of crisis management operations indicates a relapse of 23% up to 50% of all peace processes into conflict or the freezing of conflict. Collier et al. (2003, p. 83) estimate a 44% risk of returning to conflict within five years after a country reached the end of civil war. Also De Coning (2008c, p. 52) refers to the 50% figure. However, Collier, Hoeffler and Söderbom (2006, p. 9) revise the risk of returning conflict to a figure of 23% as the short-term risk and the estimate of 40% for the ten-year risk. Also, Suhrke and Samset (2007) argue for a figure closer to 20% when reviewing previous estimates of risk of countries sliding back into war. On the difficulties of measuring peace and dealing with conflict data see Ramsbotham, Woodhouse and Miall (2011, pp. 63/93).

quest for a CA has become integral part of contemporary crisis management operations (De Coning/Friis 2011, p. 245). The implicit rationale and theoretical assumption of the CA is that crisis management has a

> "more meaningful impact, when different actors engaged have a common strategy, based on a common understanding of the problem, a common theory of change, and an agreed synchronized plan for implementing and evaluating such a strategy" (De Coning/Friis 2011, p. 251).

Proceeding from above elaborated crisis management complexity, both, on the international political as well on academic level a wide consensus exists "that peace and stability operations are managed more effectively when the interdependency and interconnectedness of the political, security, governance and development dimensions of these operations are recognized" (De Coning et al. 2010, p. 1).

Thinking about a more integrated approach in the international community began during the 1990s with regard to development cooperation (Aide à la Décision Économique 2011, p. 5).[93] The recognition of a gap between humanitarian emergency relief and long term development efforts encouraged debates on coordination and required relief organizations to include longer term perspectives into their concepts and actions. Furthermore, humanitarian and development organizations more and more operated in areas of armed conflict with the consequence of intensified interaction with armed forces (Stepputat 2009b, p. 15).[94] It was often also major crises that served as a catalyzing force of the CA (Wendling 2010, p. 19). In this regard, the need for better crisis management coordination became apparent in consequence of the failures and setbacks in the UN peacebuilding framework and traditional peacekeeping operations during the early 1990s (Franke 2012, p. III). In particular, operational experiences such as the

93 As pointed out by Wendling (2010, p. 11) historical foundations of the concept and related terms are debated and range from seeing their origin within the UN context and UN documents of the late 1990s to others who present NATO as the defining organisation. Origins of the CA on different levels are discussed in depth by the Advisory Council on International Affairs (2009, pp. 14/15) and Wendling (2010, pp. 13-26). See also studies by Gebhard and Norheim-Martinsen (2010) and De Coning and Friis (2011). For a detailed analysis of concepts developed by different actors see chapter 2.2.3 which analyses different CA understandings. It has also been mentioned that the new label describes an issue which has always existed and that the challenge of improving policy coherence and coordination is far from new. "[T]he idea must be rejected that there was once a simpler age, when policies could be enacted that would require a single agency to carry them out, acting alone" (6 2008, p. 44). Yet, this dissertation argues that due to the major changes in international crisis management the CA concept can be examined as an innovative approach.

94 Also see Macrae and Leader (2000) on issues of crisis management coherence and humanitarian action.

crisis management engagements in Kosovo, Bosnia, Iraq and Afghanistan largely contributed to the emergence of the CA and informed the concept's development (Major et al. 2011, p. 111). Failures of isolated humanitarian, development and diplomatic efforts led to a push of concepts describing coherence and the coordination of these different efforts as essential (Nilsson et al. 2008, p. 28). Mentioned in UN documents by the late 1990s, the CA shortly afterwards appeared in documents by other IOs. In a "pendulum movement" (Wendling 2010, p. 20) the CA has been increasingly used on the ground in specific crisis and at the same time formalized in doctrinal attempts in documents adopted IOs or state governments. In a broad international understanding, the concept has been widely used by states and IOs since the early 2000s (Wendling 2010, p. 20).

However, the CA takes on different meanings for different actors and is far from being a single clear cut concept. There is no commonly agreed CA model and a single definition cannot be found. Its terminological roots are debated which results in terminological confusion, complicates international comparison and makes the term "notorious for being a catch-all phrase" (Drent 2011, p. 4). While an international or scholarly consensus on the meaning of the CA has not been developed yet, all approaches are based on two basic points of departure. Firstly, there are many actors involved in crisis management efforts which are, to different degrees, interdependent in their individual efforts. Secondly, these dependencies have to be managed through harmonisation (Hull 2011, p. 4).

Thus, all notions encompass the harmonization of various crisis management actors when crisis occurs and the view that effective crisis management requires enhancing complementarity and coherence at various levels in an approach that coordinates military, political and diplomatic as well as development and humanitarian efforts. "Despite the lack of a common definition, there is a general understanding that the CA concept entails managing the interdependencies that exist between the various issue-areas that need to be addressed in international peace support efforts" (Hull 2011, p. 4). In doing so, the CA aims at encouraging the external and internal coordination of crisis management instruments and the coherence of common objectives between different actors in an all-encompassing response to a crisis (Major/Mölling 2009, p. 21). Also De Coning and Friis (2011, p. 271) define the CA in the context of international peace and stability operations in the broadest possible way as a process directed toward facilitating system-wide coherence by combining security, governance, development and political dimensions. Common to the variety of CA concepts is their long-term and holistic approach to cross-cutting international crisis management policy issues. Another commonality is the importance given to output-oriented goals, and the re-organization of existing organizational structures. These modifications of crisis management intend to bridge the institutional and policy gaps of the security-development nexus, at integrating the diverse civilian and military instruments, and at preparing more integrated strategies to react appropriately to a crisis by improving links between short-term and long-term crisis management measures (Tannous 2008, pp. 33/34). The matter thus essentially deals with restoring security, governance and development through an inter-institutional, inter-ministerial or inter-organizational approach (Wendling 2010, p. 1). According to these elaborations, from CA perspective, coordination and coherence of different dimensions of activities are crucial preconditions for crisis management to be effective and the concept aims at providing the necessary conceptual and organizational basis to this end (Major et al. 2011, p. 111).[95]

In consideration of these elaborations and to clarify it despite its conceptual softness, the CA can be broken down into different key dimensions which deal with the actual meaning of the concept and which refer to different categories in which coherence in the CA context is aimed at, as shown in figure 2 below.

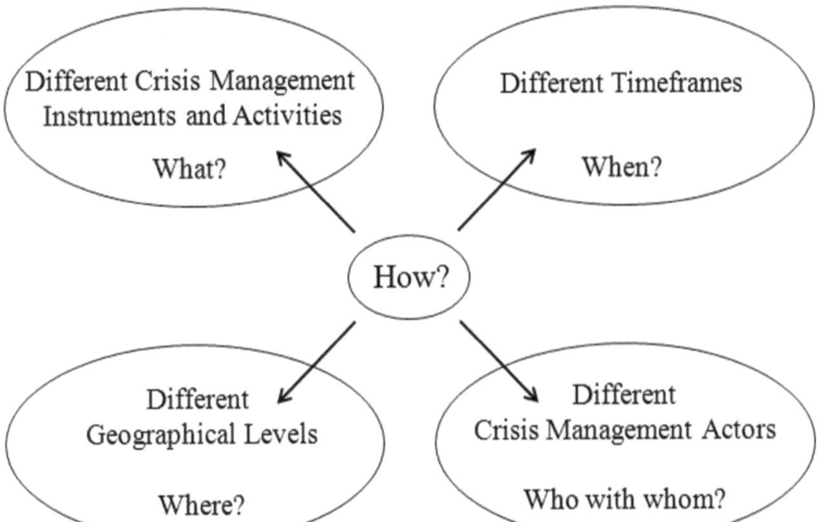

Figure 2: Dimensions of CA Coherence *(Source: Based on Aide à la Décision Économique 2011, p. 12; Modified by S.P.).*

95 The meaning of coherence and coordination in the dissertation's context is examined in detail chapter 2.3.1.

The first CA dimension applies to the coordination of different types of crisis management activities such as development, political or security means. A second dimension deals with the different timeframes of comprehensive crisis management and asks how short-term and long-term crisis management instruments can be linked. A third dimension which can be observed refers to different geographical dimensions of conflict and its potential international, regional and local levels. The fourth dimension of elements to be coordinated refers to different crisis management actors either within a system or an organization or with other actors and to the extent and with what effect their activities are coordinated. Finally, the question how different means are coordinated is relevant for all dimensions and also between them (Aide à la Décision Économique 2011, p. 12).[96] These dimensions of CA coherence exist at four levels of crisis management: firstly, at the field level, between the various international actors concerned with crisis management tasks and other actors of all kinds within a crisis; secondly, within governmental crisis management systems of nation states; thirdly, within the system of IOs; and fourthly, at the NGO level between all crisis management organizations.

Thus, CA implementation can be interpreted with a focus on managing interdependencies that exist between the various issue-areas addressed in crisis management efforts. This "system-aware approach" (Hull 2011, p. 4) acknowledges that the various actors in international crisis management which are to varying degrees interdependent need to be managed through various degrees of harmonisation. From the systems perspective,

> "[h]ow harmonisation is de facto applied is (…) less relevant than an awareness of the system and understanding one's own role within the system. CA is, as such, not a 'method' or 'tool' for interaction. It is also less about the instruments and resources used than the attitudes and approaches an actor adopts towards the broader peace effort. At is most basic, CA is *thinking of ourselves as part of a system* and understanding the gains that can be made both at an individual and system-wide level by acting comprehensively" (Hull 2011, p. 5).

The second perspective on CA implementation focuses on organizational changes for coherent policy and action. It does so by interpreting it in terms of structures and processes that can be organizationally employed to advance cooperation between the actors engaged. From this view, the CA is closely linked to the interaction between various actors and organizations aiming at generating coherent policy (Hull 2011, p. 5). However, a CA does not imply that all actors and institutions should always engage and all activities have to be jointly implement-

96 While helpful on conceptual level, these dimensions overlap in practice. The objective of this differentiation rather than providing strict categories thus is to offer an understanding of different levels in the overall CA complexity.

ed. CA thinking rather focuses on the agreement of strategic objectives, shared perspectives, tools, knowledge and experiences (Department for International Development 2010, p. 6). Thus, by ensuring coordination of different crisis management dimensions and tools, the CA also aims at supporting the definition of common objectives (Major et al. 2011, p. 111). Coherence in this regard "is supposedly achieved through harmonization of the strategic processes, planning and objectives across all instruments and agencies" (Nilsson et al. 2008, p. 28). Importantly, the various instruments

> "are not just added to each other but must be conducted as a part of an overall plan in which they support and reinforce one another. Therefore, each actor must understand the big picture, share the understanding of the conflict, agree to a common planning procedure, and take the requirements of the other actors into account when conducting its own activities" (Stepputat 2009a, p. 13).

Within a CA framework, crisis management actors are supposed to set common objectives of their activities at an early stage. On this basis, the appropriate deployment of measures and resources in the different stages of the conflict cycle should be decided. Their implementation is then agreed upon between the different actors. This includes the coordination between civilians and the military as much as the coordination between different civilian engagements such as humanitarian and diplomatic actors (Major/Schöndorf 2011, p. 2). Whichever crisis management instruments are chosen, "the full range of these elements should entail both so-called 'hard power' and 'soft power' which in combination are sometimes called 'smart power'" (Schnaubelt 2009, p. 7). The interpretation of the CA in terms of a 'smart power' approach and above conceptual clarifications show that

> "[t]he merits of the CA are so obvious that one might be excused for modifying the term slightly and making reference to a 'common sense approach' in the light of the fact that within any organisation there is a need to work in a cooperative and collaborative manner if processes and tasks are to be completed efficiently and effectively" (Neal 2011, pp. 5/6).

As stated by Weiss, Spanger and Van Meurs (2009, p. 43) it is "hard to argue against policy coherence as a strategic principle". However, the concept remains particularly difficult to realize in practice.[97]

---

97 Various reports which analyze IO and national efforts come to this conclusion and highlight several key issues and requirements necessary to implement the CA in its ideal form. See, for example, Drent and Zandee (2010) with regard to the EU, De Coning (2008a) with regard to the UN and Jakobsen (2008) analyzing NATO efforts. See also studies by Friis and Jarmyr (2008), the UK House of Commons Defence Committee's study (House of Commons 2010a) as well as De Coning and Friis (2011) on a more general level and without a particular focus.

2.2.2 Motivations and Challenges

Motivations behind the CA are multisided and there are several reasons that explain why actors ascribe importance to the concept. They can be largely generated from the above elaborated complex crisis management environment and are mainly driven by changes in the external environment of crisis management actors as well as consequential necessities to act evolving from within themselves (Haugevik 2007, p. 8). In looking at the logic that underpins the CA, the concept can be broken down into several justifying aspects which host debates on coherence and incoherence in international responses to conflict situations. Most of these aspects deal with functional issues related to the performance of crisis management. Besides, some are also concerned with political considerations which favor CA promotion.[98] As shown above, the concept is designated to the multitude of developments in the area of crisis management and there are several drivers behind this increasing demand for coordination. Complex crises cannot be dealt with by one actor solely. In addition, the scope and frequency of crisis management engagements is of such that actors need to engage in complementary manners (Nilsson et al. 2008, p. 29). In this regard, the need to implement a CA originates from two primary rationales: the mutual dependency of security and development endeavors and the understanding that managing these interdependencies is crucial to achieve sustainable and meaningful impact (Hull 2011, p. 4). Besides this security argument viewing conflict as a security risk even when far away, the major overall motivation to implement the CA is the search for greater effectiveness and efficiency which are assumed to be the outcomes of increased policy coordination and coherence (Weiss/Spanger/Van Meurs 2003, p. 43).[99]

The underlying rationale of the effectiveness motivation is that activities of crisis management actors have more impact and are more sustainable when relevant institutions involved are better coordinated, follow a common strategy of handling issues, share a common understanding and perspective on the problem

98 Haugevik (2007, p. 8) distinguishes motives of materialist and ideational nature for inter-organizational security cooperation. The materialist category is based on the idea that actors cooperate because of material gains they expect to obtain. Reasons in the ideational category result in an increased cooperation based on values and principles considered the right thing to do in a given context.

99 In the development policy context, the OECD's Glossary of Key Terms (OECD 2002, pp. 20/21) defines effectiveness as "the extent to which the development intervention's objectives were achieved, or are expected to be achieved, taking into account their relative importance." Efficiency refers to "a measure of how economically resources/inputs (funds, expertise, time, etc.) are converted to results." These definitions are applicable to the use of both terms in the international crisis management context.

and furthermore agree with respect to the implementation of such a strategy (De Coning et al. 2010, p. 1). Both scholars and policy-makers alike base their call for the CA on the assumption fragmented and inconsistent crisis management policies result a higher risk of working at cross purposes due to different or even contradicting priorities and objectives. The lack of coherence poses a risk that measures or decisions in a given sphere will counteract or nullify attempts and progress. In the large sense, this rationale is also concerned with the tensions and trade-offs between crisis management objectives which need to be dealt with when aiming at reaching consensus on crisis management priorities and strategies (Department for International Development 2010, p. 6). At the same time, more than the avoidance of overlap and duplication, increased coordination is expected to create positive synergies. A lack of cooperation in this regard translates into lost opportunities for synergy. Also the sensitive political context international crisis management takes place in demands a single approach based on shared analysis, perception and ultimately crisis management objectives. Because of this sensitive environment, increased efficiency and positive crisis management results also avoid a relapse into conflict (Aide à la Décision Économique 2011, p. 58). In addition, apparent ineffectiveness may also lead to the frustration of the crisis management personnel (Rintakoski/Autti 2008, pp. 11-12).

The CA as an efficiency measure is driven by the aim to avoid the inefficient use of resources and the duplication of effort which wastes resources. The efficiency argument can be applied in particular in times of declining resources and budgetary restrictions but also on a broader level as resources, both financial and human, are limited considering the enormous level of need in conflict (Aide à la Décision Économique 2011, p. 58). With regard to resources and means, the competition for funding and resource asymmetries pose a major obstacle for coordination efforts (Nilsson 2008, p. 44). In this context, the motive of resource dependency gains importance. As actors are increasingly uncapable of meeting the demands on their own, they recognize the benefit from gaining access to each other's resources. CA motivations are thus also based on the comparative advantage an increased coordination and cooperation provides (Haugevik 2007, p. 11). As crisis management actors overlap in geographic and functional competences, the CA also offers opportunities to share and shift burdens. In this context, "organisational survival" (Haugevik 2007, p. 9) can also be a rationale for implementing a CA. Cooperation and coordination from this perspective serve as a way of maintaining relevance and basis for existence in the international crisis management arena (Haugevik 2007, p. 9). Ultimately, CA efforts can result in organizational learning. Organizational learning as a motive in the CA sense needs to be understood in the sense that one actor functions as a role model and thus offers opportunities for enhancement and growth. "In practice, this means that one organization observes another, and chooses to imitate or model that organization's structures, procedures or values. This, in turn, may contribute to establishing a basis for further cooperation" (Haugevik 2007, p. 15).

More effective crisis management is also justified on the grounds of clarifying objectives and of bringing diverse actors together around compatible goals and programmes due to an existing need for consultation. A related aspect in this regard is that the CA can be a way of neutralizing competition between crisis management actors and their overlapping and hence potentially competing functions and competences (Haugevik 2007, p. 10). The lack of coherence between crisis management actors from this perspective mainly results from inter-institutional rivalry, working at cross-purposes and competition for funding. The CA serves to reduce such turf mentalities. From the functional perspective, importance in civil-military interaction is particularly emphasized as a reason behind the CA. Argumentation in this context derives from the security-development nexus and the recognition of civil-military dependence which argues that sustainable development and reconstruction activities can only be successful if a minimum of security has been achieved. On the operational level, the civil-military dimension of the CA is also driven by the logic of force protection which represents a rather military-centric reason to justify CA activities (Ankersen 2008, p. 5).[100]

While commitment to the CA largely stems from the above elaborated crisis management requirements in an increasingly complex international environment, the concept's implementation is also stimulated by political interests and priorities. "[I]mplementation of the CA and its structures has to be grounded in the political realities surrounding any major operation" (Neal/Wells 2011, p. 2). Domestic issues and political imperatives do not only drive or constrain timing, scope and organization of CA implementation "but they also have the tendency to change rapidly as part of either the domestic issues of the day or the response from the international community" (Neal/Wells 2011, p. 2). The debate on the CA can thus not be limited to discussions on efficiency and effectiveness but ought to also focus on matters of legitimacy. In particular, as international crisis management increasingly receives public and media attention, the need to demonstrate the legitimacy of action raises. The higher a certain case is on the political priority list, the more important comprehensive working becomes (OECD 2006, p. 20). As such, questions of political effectiveness also gain relevance with respect to specific policy goals which contain the use of particular policy instruments in a way that increases the chance to reach and thus to justify the defined policy objectives (Héritier 2003, p. 113). In this regard, it needs to be pointed to the role coherence plays for taxpayers and other actors which are

100 See previous chapter on the security-development nexus. Civil-military relations in the overall CA debate are explained in greater detail in the next chapter.

interested in the application of a coherent strategy. The realization that crisis management efforts often do not have the results intended leads to fresh urgency of CA issues from a political perspective (De Coning/Friis 2008, p. 3).

In addition, the CA is anticipated to increase the legitimacy on various other levels of crisis management by adjusting the interaction and interdependency of tasks and actors (Friis/Jarmyr 2008, p.3). While above-mentioned need to demonstrate crisis management effectiveness and thus the legitimacy of governmental action are closely related, for some actors it is merely the coordination itself that helps to enhance the legitimacy of crisis management action. The more actors involved, the more the political and moral legitimacy increases. In particular, multinational cooperation in cases potentially involving the use of force is essential, not just in order to acquire adequate resources. A wide participation of different crisis management actors not just adds to the burden sharing but also largely increases legitimacy. As observed by Bailes (2008, p. 142), policy correctness "does not lie so much in the content of a policy but rather in the way it is formed and managed, including the openness of its owners to necessary adjustment and re-thinking." In this context, some observers contend that organizations and bodies involved in crisis management want to look good and represent their crisis management activities in a way in line with what public and politicians want. In that case, the CA, especially its civil-military dimension, is seen as an opportunity to bolster or at least maintain political and financial support (Ankersen 2008, pp. 4/5). At the same time, "[a] certain degree of solidarity may also be built, making it somewhat easier to sustain temporary setback" (De Coning/Friis 2008, p. 3). Shared values can be a driving force and important factor for cooperation between different crisis management actors. These legitimizing effects on the actions of a crisis managing actor make the former an appropriate partner which contributes to the normative credibility of both their crisis management activities (Haugevik 2007, pp. 12/13). "Values can also strengthen the grounds for cooperation by 'marshalling moral convictions and emotional preferences to its support'" (Haugevik 2007, p. 14). In this context, an actor promotes the CA because it believes in the ideals and values it embodies.

Contrariwise to these motivating incentives to implement the CA, several challenges have been recognized which hamper crisis management coherence and coordination and pull actors at every level away from the CA. Some of the reasons why coordination has not worked effectively have already been identified, especially with regard to the coordination in UN peacebuilding.[101] The most prominently offered explanation points to the larger number of organizational actors involved in crisis management. The more actors involved the more com-

101 See, for example, De Coning (2008a) and Herrhausen (2009).

plexity increases and coordination becomes more difficult. Furthermore, the widening security agendas also resulted in the creation of new procedural and structural links. Another reason that has been identified is a lack of a common vision and an agreement on goals. As De Coning (2004, p. 42) argues "[o]ne of the reasons why it has been so difficult to achieve coherence is because the different agencies lack a shared understanding of the need for coordination and a common vision of how it can be achieved." Furthermore, friction occurs around resources and financing mechanisms (Herrhausen 2009, p. 32). Obstacles also occur due to the conflicting mandates various actors and organizations may have as a possible formal and legal impediment to collaboration. Different working styles and organizational cultures add to this complexity. Most prominently cited in this regard, though mainly playing out in the field, are the differences between civilian and military actors (Herrhausen 2009, pp. 31/32). Consequently, a major challenge arises from different organizational and professional cultures and mindsets which lead to prejudices. In particular the humanitarian and military actors have the tendency to view each other with suspicion but also between civilian actors like diplomats and development officials the variety of values, principles, objectives and approaches can lead to turf wars. Difference in the organizational culture and goals is additionally manifested in a different use of relevant terminology which results in misinterpretations (Nilsson et al. 2008, p. 43). Different organizational structures and composition obstructs coordination and causes confusion. In this context, bureaucratic rigidity and hurdles due to the organizational system of one actor may not enable the needed flexibility to adjust to the requirements of other actors (Nilsson et al. 2008, p. 44). Additional potential obstacles are the different time frames actors operate in. Major differences in this regard again exist in particular between military and humanitarian actors concerned with immediate relief.

While these differences are legitimate from various perspectives, they nonetheless result in a protection of organizational prerogatives. Institutions have a propensity to guard their core missions and their organizational culture and structure from being changed in the name of coherence. Each actor seeks to preserve its autonomy and independence which makes its organizational routines and procedures difficult to synchronize and coordinate in particular when organizational goals differ. The organizational consequences of this travel from the national to the international level and are additionally reflected in crisis management engagements on the ground. Both national and international crisis management actors have to deal with political and organizational incentives that support rather narrow thinking over advancing coherence. As these elaborations demonstrate, the CA aims at improving both the management and the impact of crisis management. In order to address these crisis management challenges and shortcomings which give reason to the CA, various crisis management actors, independently from each other, have started to explore models and mechanisms and aim at improving the overall coherence of their crisis management systems (De Coning 2009, p. 117).

2.2.3 Understandings

As the above analysis shows, ideas of the CA and crisis management policy coherence have been driven by diverse factors and evolved at a time of essential redefinition of the rules and context of international affairs. Against the background of the complex security environment, the aim of providing crisis management in a comprehensive way has "become what could be called the gold standard in international security affairs" (Gebhard/Norheim-Martinsen 2010, p. 1). It is regarded as the key to successful crisis management operations and various models have been developed. Since the 2000s, CA ideas have been increasingly used in official policy documents and strategies. "The concept seems to travel from one state to another, from one international organization to the next" (Wendling 2010, p. 20). Crisis management actors, which range from IOs and states to NGOs, have increasingly tried to implement Comprehensive Approaches and are in the process of developing reforms in order to improve and better coordinate their response to conflict situations. But while the fundamental objectives and challenge are the same, a multitude of concepts and terminological variations appeared at the national and international level to describe the need for greater coherence. Yet, looking at the concept's diverse and sometimes even contradictory origins, it is not surprising that the conceptual evolvement and the implementation of the CA prove inconsistent. Relevant actors involved in international crisis management, ascribe a range of different terms to the concept which reflect their individual conceptual and structural origins (Advisory Council on International Affairs 2009, p. 13). Characteristics, notions and implications additionally vary as these conceptions have been shaped by the specific historical background and organizational context in which they have arisen (Hull 2011, p. 4).

On a conceptual level, the field of *civil-military interaction* represents a major reference concept of the CA in international crisis management. Also referred to in terms of civil-military coordination and civil-military cooperation, the notions aim at defining the relations between the military and civilian actors.[102] Civil-military interaction is a requirement of comprehensive crisis management as the concept suggests coordination and common efforts anywhere military and

102 The terms civil and civilian as used in this dissertation take their connotation from the fact that they are not military and thus entail a range of institutions and actors, including governmental and non-governmental as well as political, humanitarian and development actors.

civilian actors are jointly engaged. At the operational-tactical level, civil-military cooperation broadly deals with the interaction of civilian and military actors in the crisis management field and is usually abbreviated to CIMIC. The CIMIC doctrine describes occasions in which elements of armed forces engage together with civilian entities such as other government institutions, local authorities or NGOs. Originally, CIMIC addressed the interaction of national armed forces and national civil institutions as well as civil society in the context of territorial defence. In the new crisis management environment, however, CIMIC deals with the improvement of the different civil-military interfaces in international military interventions such as the relations of the military with governmental or non-governmental civilian actors as well as local actors involved in a conflict region. As CIMIC is a military doctrine, the major aim of the concept is the support of military success and the purpose of force protection. The development of relations between the military actors and their civilian counterparts serves this objective. Key functions of CIMIC are civil-military liaison, support to the civil environment by the temporarily taking over of civilian tasks by the military and support to the military force by the utilization of civilian resources (Johannsen 2011, pp. 199/200). From CA perspective, in particular regarding the issue of different working cultures, lessons can be drawn from CIMIC actor coordination experiences (Wendling 2010, p. 15).

Depending on their mandate, different types of NGOs are confronted in varying degrees by CA matters. In particular in the field of humanitarian aid, NGOs view the development and respective involvement in CA efforts rather critical as the participation in any political or even military strategy conflicts with the principle of humanitarian neutrality of their commitment (Wendling 2010, p. 83).[103] In particular CIMIC aspects of the CA terminologically emphasize military control and military mission purposes. Thus, for many actors in the civilian sector CIMIC's preoccupation with military goals is highly problematic. A lot of NGOs and civil society actors argue that the CIMIC label entails a militarization and military domination of crisis management. Particularly humanitarian aid organizations fear that cooperation with the military endangers the principles of independence, neutrality and impartiality and the blurring of boundaries between the mili-

103 The memorandum by *Care International* to the UK House of Commons Defence Committee's CA study underlines the growing insecurity for NGOs in the CA context and notes that a too integrated approach "threatens the space for NGOs or other agencies to deliver independent, neutral and impartial humanitarian assistance" (House of Commons (2010a, Ev 137). In an earlier study, Macrae and Leader (2000) analyze the role policy coherence plays for humanitarian actors. For analysis of issues regarding CIMIC in the humanitarian aid context in particular with view on the principles of neutrality, impartiality and independence as adopted by humanitarian organizations see Johannsen (2011, pp. 200/201).

tary and neutral civilian actors in conflict, thereby increasing the risk for civilians to become the target of attacks themselves. It should be noted in this regard that CIMIC is mostly limited to its operational framework dealing with the theatre of operations in contrast to the all-embracing CA concept. In terms of the CA, CIMIC covers comprehensive crisis management only incompletely as it does not include problems of civil-civil interaction. While civil-military interaction is a crucial element of the CA, coordination between the various civilian components is as important. The CA denotes a wider perception and should not be equalized with the CIMIC doctrine (Wendling 2010, p. 14). Ultimately, from CA perspective, CIMIC can only be applied partly in a meaningful way because the concept is only one component embedded in the complex, multidimensional CA context. Comprehensive crisis management entails a much greater range of activities and forms of cooperation than those labeled as CIMIC.[104]

The UN is considered as having developed the most advanced CA to its crisis management engagements.[105] Confronted with the fundamental changes of crisis management since the end of the 1980s, the UN's multidimensional missions gradually included political, administrative and societal dimension in addition to the traditional peacekeeping tasks (Major/Schöndorf 2011, p. 6). With then UN Secretary-General Boutros Boutros-Ghali's *Agenda for Peace* (United Nations 1992) the UN's multidimensional missions were conceptually enshrined by affirming that

> "[p]eacemaking and peace-keeping operations to be truly successful, must come to include comprehensive efforts to identify and support structures which will tend to consolidate peace and advance a sense of confidence and well-being among people" (United Nations 1992, Section VI/55).

The document addresses preventive diplomacy, peacemaking, peace-keeping and post-conflict peacebuilding together. Demanding a comprehensive and widened conception of security, it calls for respective design and planning of UN missions. By now, "[i]t has developed a sophisticated multidimensional and Integrated Approach model that has been refined over the last two and a half decades in more than 20 peacekeeping operations" (De Coning 2009, p. 119). Both, on the strategic and operational level, the UN further advanced and developed its CA model under the notions of *Complex Peacekeeping* and *Complex Peace Op-*

104 For detailed elaborations on civil-military coordination and CIMIC-doctrines see Gordon (2006), Ankersen (2008), Egnell (2009b) and Pradetto (2011).

105 View, among other observers Mölling (2008, p. 2), Drent and Zandee (2010, p. 18), De Coning and Friis (2011, p. 246) as well as Major and Schöndorf (2011, p. 6) who come to this conclusion when comparing international efforts of CA implementation.

erations.[106] Conceptually, the organization's CA is mainly directed by the challenges resulting from the security-development nexus.

By the late 1990s, the UN had commissioned various high-level working groups and panels to reflect on various aspects of the diverse, complex challenges posed to the organization in its efforts and to consider strategic and operational coordination models. Under the slogan 'Delivering as One', diverse recommendations toward UN-system-wide coherence have been elaborated by these high-level panels (De Coning 2009, pp. 119/120). In 2001, the UN Security Council recognized that "peacemaking, peacekeeping and peace-building were often closely interrelated and stressed that the interrelationship required a comprehensive approach" (United Nations 2001). The statement continues and explains that

> "experiences of the United Nations and regional organizations and other actors in peace-building pointed to the need for formulating a strategy based on the interdependence between sustainable peace, security and development in all its dimensions. To be successful, such a strategy should meet basic criteria, such as: relevance, coherence and consistency of programs and actions; the consent and cooperation of the authorities of the State concerned; continuity in and conclusion of the process; cooperation and coordination among organizations and other actors involved; and cost-effectiveness of the overall peace-building operation" (United Nations 2001).

UN official wording then started to use the expressions *Integrated Missions* and *Integrated Approach* when referring to CA efforts.[107] While the UN at first referred to its CA efforts under the notion of Integrated Missions, respective coherence initiatives were later refined with the more encompassing Integrated Approach. The Integrated Mission concept was established as the guiding principle for the structure and implementation of complex post-conflict UN operations at the UN World Summit in 2005 (United Nations 2006a). Its major objective has been to ensure both strategic and operational partnerships between different UN agencies, funds and programmes in a given UN operation. The structure of specific Integrated Missions differed depending on the country. Implying that there is no template-like single structure and form which all UN missions can follow, the phrase that 'form must follow function' has been coined in the UN's

106 Both terms were prominently employed in the 2000 Report of the Panel on UN Peace Operations (United Nations 2000), better known as the 'Brahimi Report', after the panel's chair Lakhdar Brahimi. On the report as an impetus for greater integration between different UN entities see Herrhausen (2009, pp. 50-52).

107 To give an example of the terminological confusion, the Advisory Council on International Affairs (2009, p. 15) points out that the term 'Integrated' in the CA context can be misleading as integration in terms of merging various instruments into a whole is mostly not what is intended with respective efforts.

Integrated Mission context. Structural arrangements of each mission have to be based on the specific context (Eide et al. 2005, pp. 17-19). Integration in this regard grounds on the independence of individual actors as an overarching principle. As a part of its CA efforts, the UN also initiated the Integrated Mission Planning Process that seeks to both internally integrate UN departments, funds and programs and also to allow the UN to improve UN reach-out to international partners in 2006 (Major/Schöndorf 2011, p. 6). The Integrated Mission Planning Process is concerned with the proper sequencing of plans as well as the coherence in recognizing needs, objectives and results and puts respective mechanisms in place to generate and sustain such (Drent/Zandee 2010, p. 19). This process was underlined with the decision to set up the Peacebuilding Commission which reflects the UN's crisis management expansion to post-conflict situations and integration efforts at the intergovernmental level. The Peacebuilding Commission aims at developing integrated strategies and implementation plans in conflict-prone states, mobilize resources and to coordinate long-term engagement in specific countries as a potential instrument to increase peacebuilding coherence (Major/Schöndorf 2011, p. 6).

Introducing the expression of the Integrated Approach in 2008, UN Secretary-General Ban Ki Moon has applied an approach to operational processes and arrangements that not necessarily requires structural integration but refers to a strategic partnership between the UN mission and the UN country team and is mainly reflected in country level arrangements (United Nations 2008). In doing so, the UN applies its Integrated Approach to "a specific type of operational process and design where the planning and coordination processes of the different elements of the UN family is integrated into a single country-level UN system, when it undertakes complex peacekeeping operations" (De Coning 2009, p. 120). The UN's Integrated Approach calls for a shared vision of strategic objectives, aligned or integrated planning, agreed results, time lines and division of responsibilities as well as agreed monitoring and evaluation mechanisms (Drent/Zandee 2010, p. 20). This approach covers the whole crisis management cycle from strategy and planning to the evaluation of activities. The concept intends to "create a balance between the need for a security response and the necessity to consider all perspectives of crisis management" (Major/Mölling 2009, p. 24) and contains political, security, development, human rights, rule of law and humanitarian dimensions. With its reform initiative *A New Horizon for UN Peacekeeping* (United Nations 2009) the UN aims to assess policy and strategy dilemmas facing UN Peacekeeping and to refresh the dialogue with stakeholders on possible solutions to UN crisis management to meet contemporary requirements. For the UN, the CA refers predominantly to the coordination of the various UN agencies and bodies, though it also includes the objective of a more coherent thinking encompassing other international actors. Its major difficulties with regard to its implementation result from the organization's size and the parallel structure of its subordinated and specialized organizations (Mölling 2008, p. 2).[108]

NATO faces similar demands to develop a CA to its interventions and its efforts are mainly coined by the alliance's operational experiences (Drent/Zandee 2010, p. 18). The CA within NATO primarily emerged as a reaction to its engagement and problems during the state-building process in Afghanistan.[109] NATO acknowledges the need for coherence in its activities and announced its intention to develop the CA in a wide range of political statements and documents. The alliance increasingly became aware that its objectives cannot be achieved without cooperating with civilian actors since the 2005 Copenhagen meeting on the *Concerted Planning and Action Initiative* (Stepputat 2009b, p. 22). To this effect, NATO declared the CA as a guiding principle in its 2006 *Riga Summit Declaration* (NATO 2006) in which the alliance declared to extend the military planning process to include civil aspects and to strengthen cooperation with external civil actors.[110] Acknowledging that a military operation cannot achieve sustainable peace and stability on a concrete level, NATO introduced Provincial Reconstruction Teams (PRTs) within its International Security Assistance Force in Afghanistan, where civilian actors are embedded in military units (De Coning/Friis 2011, p. 249).[111]

However, as NATO is essentially an intergovernmental military alliance, the organization is unable to implement the CA concept and to obtain system-wide effects in its own structures. NATO does not have civilian instruments, means and capacities that a fully-fledged CA requires at its disposal. Accomplishing the concept for the alliance thus relies on partnerships with other international actors and, more generally, coordination with civilian actors outside NATO. Consequently, NATO's "identity as a largely military crisis reaction

108 The UN's CA has been deeply analyzed in various studies. See, for instance, Wendling (2010, pp. 49-56). An early study on the CA in the UN context and the Integrated Missions concept has been undertaken by Eide et al. (2005). For a detailed analysis of the UN's Integrated Approach particularly with regard to the Integrated Mission Planning Process see Rintakoski and Autti (2008, pp. 56-74) as well as Herrhausen (2009, pp. 52-56). Also De Coning (2008a) provides a detailed analysis of distinctive features of the UN's CA. For summary of the core features of the Integrated Approach see analysis by De Coning (2009, pp. 120/121).

109 For an in depth analyses of NATO's CA efforts see Jakobsen (2008). NATO's engagement in Afghanistan from a CA perspective has been studied by Wendling (2010, pp. 43/44). NATO's CA and Afghanistan as a case that illuminates the problems built into NATO's CA is also dealt with by Rynning (2011).

110 "Experience in Afghanistan and Kosovo demonstrates that today's challenges require a comprehensive approach by the international community involving a wide spectrum of civil and military instruments" (NATO 2006, paragraph 22).

111 The role of PRTs in the CA debate is clarified in detail later this chapter.

structure" (Gordon 2006, p. 348) causes most obstacles for the alliance concerning the establishment of a more civilianized crisis management and a CA itself. The alliance's CA concept centers on its relationship with other international and civilian actors involved in a given conflict environment.[112] This has been reflected in NATO's CIMIC doctrine which largely coins the alliance's approach to policy coherence. NATO defines CIMIC as "the coordination and cooperation, in support of the mission, between the NATO Commander and civil actors, including national population and local authorities, as well as international, national and non-governmental organizations and agencies" (NATO 2002). In this context, NATO also developed the CA as part of an operational strategy based on its *Effects Based Approach to Operations* (EBAO) which means to also consider civilian aspects in the planning and conduct of NATO operations with the ultimate goal of a unity of effort of all instruments which have to be incorporated into a comprehensive purpose (Wendling 2010, p. 22).[113]

In line with this approach to crisis management policy coherence, the *Bucharest Summit Declaration* (NATO 2008) declares that NATO can only contribute to a CA through a broad partnership with the wider international community. In addition, its 2008 *Comprehensive Approach Action Plan* has aimed at better realizing a CA in crisis management from the strategic to tactical levels by guiding the direction for a CA of NATO member states based on joint planning and training, strategic communication and network building (Gauster 2012, p. 75). NATO reaffirmed its CA efforts at the Strasbourg Summit in 2009 as well as other official texts and Rasmussen, in his first statement as NATO Secretary General, declared that "we need a comprehensive approach, a reinforced interaction between our military efforts and our endeavours with regard to civil reconstruction" (NATO 2009). Also in its *New Strategic Concept* (NATO 2010) the alliance identifies the importance of a CA, active coordination with international partners in order to maximize coherence and effectiveness of crisis management efforts and underlines lessons learned from past operations. According to the New Strategic Concept, NATO aims at developing its own civil planning capabilities and to further identify civilian capabilities.

> "To be effective across the crisis management spectrum, (...) [NATO will, S.P.] form an appropriate but modest civilian crisis management capability to interface more effectively with civilian partners, building on the lessons learned from NATO-led operations. This capability may also be used to plan, employ and coordinate civilian activities until conditions allow for the transfer of those responsibilities and tasks to other actors" (NATO 2010, para. 25).

112 For a detailed elaboration of this aspect see Wendling (2010, p. 39).

113 The EBAO in the CA context is explained in greater detail later in this chapter.

Nonetheless, the creation of moderate civilian capabilities as provided by the New Strategic Concept is not supposed to lead to the institutionalization of the civilian capacity required for comprehensive crisis management.[114]

This again shows that the development of civilian capabilities as part of the CA is largely directed toward its international partners. Rather than competing with the UN and the EU within the civilian range, the organization strives for the CA in coordination with other IOs, through closer links and liaison.[115] The CA from NATO perspective can only be effectively implemented when other actors contribute. NATO's role thus remains that of a military alliance providing defence support to CA efforts and not that of a comprehensive crisis management organization itself. Despite its efforts, particularly civilian actors continue to view NATO as a purely military alliance without granting any awareness of the significance of the civilian crisis management dimension to the alliance (Major/Schöndorf 2011, p. 5). From their perspective, NATO as an international crisis management actor cannot provide the incentive and legitimacy for comprehensive crisis management operations. This indicates that NATO's CA efforts and thinking how to best integrate the CA largely implies a quest for legitimacy.[116]

Though NATO has taken steps to improve its cooperation with the UN, the EU and other crisis management actors, the alliance continues to be confronted with practical challenges in the field. As a consequence of these operational experiences, the development of NATO's CA is mainly evolving along the line of developing an operational concept.[117] The conceptual development of the CA

114 For a detailed analysis of the CA in NATO's core documents, particularly the new Strategic Concept, and strategic thinking in general see Kamp (2011).

115 Suggestions in this context range from the establishment of a civil Security Committee to the institutionalization of arrangements with the EU to use EU civilian experts such as a 'Berlin+ in reverse' (Wendling 2010, p. 45). According to Hull (2011, p. 6), advancement of the CA within NATO has experienced difficulties due to the requirement of member state consensus.

116 However, Wendling (2010, pp. 41/42) observed that there is no consensus on the CA within NATO. Major disagreement in this regard occurs on the issues of civilian capabilities and consequences for NATO's mandate and crisis management role as well as on financial issues which would go hand in hand with the alliances expansion of tasks. Also Jakobsen (2008, pp. 3/4) points out that "[w]ithin the Alliance the adoption of CA is being hampered by disagreements over NATO's role in world politics, disagreements over the extent to which NATO should use its military capabilities to fill the gap if civilian actors are incapable of carrying out their CA tasks, an instinctive military reluctance to engage in civilian gap-filling and finally disagreement over what CA means and how it should be implemented in Afghanistan."

117 Concrete organisational steps underlining NATO's aim to integrate a civilian dimension in its internal planning processes and to facilitate cooperation include the appointment of a Civil-Military Interface Advisor in the NATO Headquarter whose task it is to establish and maintain contact to civil actors and to develop interaction procedures and the creation of a Comprehensive Approach Team. The main instrument in charge for the NATO CA is its Allied Command Transformation ensuring CA requirements and standards. In March 2012, NATO agreed to re-

as a part of NATO largely depends on the alliance's future role in crisis management being either that of a military alliance or a manager of all kind of crisis. As a military alliance it continues to define its role as providing defence support to CA efforts. From this view, a 'NATO Comprehensive Approach' cannot exist, but NATO contributions to CA efforts by the international community can (Neal/Wells 2011, p. 2). The alliance is thus confronted with the dual challenge of finding a joint understanding of its own role in a CA while being faced with the need to develop cooperative relationships with other international actors who provide civilian capacities (Stepputat 2009b, p. 22).

Also other international actors started to develop and use specific CA models. The OECD has emphasized the importance of greater coherence between development and security policies. In doing so, the OECD's Development Assistance Committee established a Fragile States Group with the task to build up principles for good engagement in situations of state fragility and to promote improvement of national WoG approaches (Stepputat 2009a, p. 12).[118] OECD activities also focus on aid-related issues, such as the improvement of aid effectiveness through coherence, aid impact and donor coordination. In this context, the OECD is mainly concerned with national government actors to ensure the coherence of funds allocated to development efforts against state failure. Thus, in terms of the CA, the OECD positioned itself largely in the context of state fragility and better policy for development (Wendling 2010, p. 59).[119]

As already briefly mentioned in the context of the OECD above, the CA not only appeared within IOs but also within governments. Many governments started to develop initiatives to improve coherence across ministries and government departments involved in international crisis management. On the national level, different crisis management policies are usually generated separately by ministries and government departments with diverse organizational cultures, goals, languages, methods and approaches as well as a divergence of time horizons and

view the list of tasks to update its 2008 Comprehensive Approach Action Plan. The tasks are planned to be implemented by a civilian-military task force with the objective to improve planning, implementation and lessons learned of operations (NATO 2012).

118 See also activities such as the OECD's Network on Conflict, Peace and Development Cooperation (CPDC) and its Fragile States Group (FSG).

119 Among the IOs active in the CA debate, though to a lesser degree, is also the Organization for Security and Cooperation in Europe (OSCE) which promotes a CA that relies on a very wide concept of security which includes the respect for basic liberties, socio-economic and environmental factors. The OSCE also particularly underlines the necessity of conflict prevention by promoting early warning systems and also is a major driver of putting the topic on its member states' agenda (Wendling 2010, p. 57). For a detailed analysis of the OSCE as a crisis management actor see Rintakoski and Autti (2008, pp. 75-83). Also African organizations with crisis management relevance such as the African Union are involved in CA debates (Rintakoski/Autti 2008, pp. 27/28).

mission frameworks (Weiss/Spanger/Van Meurs 2009, p. 19). In this context, the CA "requires institutions that once acted largely autonomously – defence, diplomacy and development – to exchange information, share resources and cooperate in strategy development and implementation" (Weiss/Spanger/Van Meurs 2009, p. 9). Such efforts to combine political, security and development crisis management dimensions and actors from one country are referred to under the Whole-of-Government notion. According to the OECD, a WoG approach is

> "one where a government actively uses formal and/or informal networks across the different agencies within that government to coordinate the design and implementation of the range of interventions that the government's agencies will be making in order to increase the effectiveness of those interventions in achieving the desired objectives" (OECD 2006, p. 14).

WoG approaches thus seek to harmonize and bring coherence to all government activity with crisis management relevance and relate to the inter-institutional coordination of government activities. With the aim of ensuring a nationally consistent approach, the *Whole-of-Nation* approach presents an extension of the WoG which additionally entails non-state actors (Gauster 2012, p. 79).

WoG approaches also derive from the understanding that incoherencies in domestic actors' positions and policies hamper not only a coherent national policy but also represent a major issue for the acknowledgement of internationally coherent crisis responses (Mölling 2008, p. 2). In some cases, the WoG approach is stretched to not only include the office of the head of government, the foreign ministry and the ministries of defence and development, but also at the least the ministries responsible for justice, police and finance to be actively engaged (Hull 2011, p. 8).[120] Acknowledging the links between diplomacy, defence and development (3D), Canada's 3D concept stands for one of the most advanced WoG approaches (De Coning/Friis 2011, pp. 249/250). The 3D concept has served as the foundation and model for approaches taken by other actors and developed into a catchphrase in the CA context as it "so concisely captures the main security development axis and the need for inter-connectedness among these different dimensions of Government" (De Coning 2009, p. 118).[121]

---

120 For an overview on different national WoG approaches see Rintakoski and Autti (2008).

121 Coherence of the three policies diplomacy, defence and development is regularly expressed in the notion of the 3D concept. For a terminological elaboration of the 3D approach see Advisory Council on International Affairs (2009, pp. 15/16). In addition to the 3D abbreviation a number of other acronyms have been employed to refer to the different dimensions of the CA. Examples include acronyms such as DIME (diplomatic, informational, military and economic elements), DIMEFIL (diplomatic, informational, military, economic, financial, intelligence and law enforcement elements) or MIDLIFE (military, intelligence, diplomatic, law enforcement, information, finance and economic elements) (Wendling 2010, p. 10).

In the context of public administration and organization studies, the efforts of individual governments to achieve policy coherence can also be attributed to so-called *Joined-Up Government* strategies.[122] Analyzing public sector reforms in the UK, Ling (2002, p. 616) views Joined-Up Government "as a group of responses to the perception that services had become fragmented and that this fragmentation was preventing the achievement of important goals of public policy." Bogdanor (2008, pp. 1/2) more broadly defines Joined-Up Government as "a strategy which seeks to bring together not only government departments and agencies, but also a range of private and voluntary bodies across organizational boundaries towards a common goal". As respective reform initiatives are based on the position that a policy objective can neither be delivered by strictly separated organizations nor by the creation of one super-agency they aim at connecting activities of formally separate organizations toward particular policy objectives. Joining government up seeks to increase coordination between actors and activities without eliminating boundaries themselves. Consequently, "'joined-up government' is an umbrella term describing various ways of aligning formally distinct organizations in pursuit of the objectives of the government of the day" (Ling 2002, p. 616).[123] In Joined-Up Government context, public administration studies describe lacks of coordination and not being joined-up as *silos* which result from the division of labor.[124]

As both the CA and the Joined-Up Government concept are closely related, Joined-Up Government explorations provide useful insights in the CA in international crisis management context. Both slogans present overlapping terms for previously existing problems of how to coordinate government policies as a response to the problem of increased fragmentation

122 The terms Whole-of-Government and Joined-Up Government are just two of a whole group of terms which prescribe the challenge of improving coordination across government policy sectors. Other terms such as *networked government* or *connected government* are also used to describe the variety of responses to the problem of fragmentation of the public sector and the wish to increase coordination (Christensen/Lægreid 2006, p. 9).

123 As Ling (2002, p. 617) points out, "[c]learly, in the sense that they seek to coordinate the work of different agencies, all governments attempt to 'join-up'. Just as the functional separation of state agencies is a necessary part of managing complexity and is a characteristic feature of the modern state (…), so too is the development of strategies to deal with the problems this creates."

124 *Turf wars* are probably the most extensively mentioned silo mechanism stopping respective institutions from working together. Turf means that organisational units aim at maintaining or extending their own interests and area of responsibilities. *Budget protection* forms another silo mentality that can lead to inter-organisational conflicts. The silo concept also observes that different actors perceive the same issue from completely different standpoints. 'Where you stand depends on where you sit' is a frequently quoted summary of this. *Ignorance* relates to a lack of awareness of an overlap of interest and activity of actors in the same issues which not automatically has to be a sign of incompetence but also result from policy complexity (Page 2008, pp. 141-148).

> "in order to eliminate situations in which different policies undermine each other, to make better use of scarce resources, to create synergies by bringing together different stakeholders in a particular policy area, and to offer citizens seamless rather than fragmented access to services" (Christensen/Lægreid 2006, p. 9).

Respective reform initiatives can be a group, a locality, or a policy sector and can include all levels of government and also involve groups outside government (Christensen/Lægreid 2006, p. 9). Although referring to a wide set of reforms covering the government and public sector as a whole, the slogan reflects the CA context as it identifies that the issues governments are confronted with today do not suit strict ministerial boundaries anymore. Particularly international conflict issues today seem resistant to be tackled by the traditional departmental approach. Behind both concepts lie two common problems: firstly, the problem of coordination in terms of encouraging actors to follow the same objective not to undermine each other's efforts. Secondly, the problem of organization and aligning structures and cultures to deal with cross-organizational tasks (Mulgan 2008, pp. 175/176).

Apart from the above approaches some compare and link the notion of the CA to the military ideas of the EBAO and *Network-Centric Warfare* (NCW). While NCW deals with questions how new information technology allows increased integration and to conquer frictions between crisis management actors, the EBAO centers around the question how military and non-military means can be utilized to achieve effects. In this context, the EBAO aims at executing a systems approach which implies that the target to be affected is approached from different perspectives which involve a particular focus on political, military, economic, social, information, and infrastructure aspects (Borchert 2006, p. 7). Though it is a military concept, the EBAO can be considered closely linked to the CA as both "are based on systemic methods of analysis attempting to integrate the complexity of crisis management" (Wendling 2010, p. 22). The EBAO has been developed to address the changing security situation and, while it is not as inclusive as the CA and much more detailed and programmatic, both approaches share the same underlying rationales and objectives. In particular with regard to NATO, the EBAO has helped to launch the CA on the doctrinal level (Wendling 2010, p. 22).[125] In a similar context, the notion of NCW led the force transformation agenda by concentrating on the principle of network centricity in the application of military instruments. Network centricity in this regard underlines the

125 As pointed out by the UK Ministry of Defence (2006, p. 8), the EBAO can be "applied as the military aspect of the CA, in conjunction with the instruments of power and employing a wide range of coordinated activity, could produce more coherent and desirable effects and thus deliver the desired outcome with greater assurance."

> "deliberate act of linking goals, capabilities, processes, structures, and capacities of security-relevant state and non-state actors in order to coordinate, harmonize, and integrate their action. Network centricity thus refers to the close interaction between different levels of planning, decision-making, and implementation and various actors working together to achieve different tasks by using a wide spectrum of instruments of power" (Borchert 2006, pp. 12/13).

Also mirroring CA objectives, SSR resonates with the CA concepts as it underlines the necessary alignment and coordination of efforts as well as the interconnectedness of security sector instruments and institutions, both civilian and military, to promote stability.

At the field level, the CA is commonly associated with PRTs which can serve as a CA example in practice and an attempt to apply the WoG approach to nationally deployed entities. The PRT concept has unfolded since 2004 in operations in Afghanistan and Iraq as an instrument of structured cooperation between the diplomatic, political, military and the development dimension of the engagement. The development and adaptation of PRTs in the operational field context can be viewed as a "laboratory of civil-military relations" (Stepputat 2009a, p. 57) as PRTs involve a military component responsible for security-related issues as well as development, political, police and judicial advisors. As there is no predetermined standard structure, different models exist, resulting in different challenges and degrees of coherence. PRTs largely differ in their composition, objectives and measures according to local deployment circumstances but also depending on the preferences of the participating nations (Major/Schöndorf 2011, p. 5).[126] The main differences between PRTs are related to the relationship and mix of military and civilian elements with regard to leadership and engagement in crisis management tasks (Stepputat 2009a, p. 58). As the PRT concept provides the combined deployment of relevant government institutions it is "a Whole-of-Government experiment" (De Coning 2009, p. 118) and expected to advance WoG coordination and to lead to better crisis management results in the mission due to the collocation of civil-military elements (Hull 2011, p. 8). From CA perspective, PRTs can be considered an accomplishment as they take into account diverse crisis management aspects such as the diverse operational environments in the field, the domestic political requirements of contributing countries as well as the combination of military and civilian aspects.[127]

126 The Norwegian and the two German PRTs, for instance, combined defence and development in a 'two-headed' leadership, while the UK tries out a PRT with a high-level diplomat in the directing role (Stepputat 2009a, p. 58).

127 PRTs have also been criticized on many levels, among other things for being too small and too few (Major/Schöndorf 2011, p. 5). Eronen (2008) gives a detailed overview on the history of PRTs, their mission and tasks as well as different PRT models and challenges.

Though it largely differs from the organizational and conceptual approaches by individual actors described above, the Multinational Experiment 5 (MNE 5) is closely connected to CA efforts in the context of multinational operations. MNE 5 was an international experiment which ran from 2006 until 2009 with the participation of Canada, France, Germany, Norway, the UK and the US under the umbrella of NATO. The experiment's objective was to develop better methods and processes to utilize the CA in the planning, conduct and evaluation of complex multinational and multifunctional peace operations through a series of joint workshops and exercises (Jakobsen 2008, p. 19). This experimental approach was intended as a comprehensive international program to further and link conceptual development and cooperation and to define common challenges with regard to the CA. In doing so, MNE 5 used the CA in the broad sense

> "to describe the wide scope of actions undertaken in a coordinated and collaborative manner with the affected nation(s). Coordination and collaboration include coalition civilian government agencies and their defence and security forces, international and intergovernmental organisations, non-governmental organisations and the private sector to achieve greater harmonisation in the analysis, planning, management, and evaluation of actions required to prevent, ameliorate, mitigate and/or resolve the conditions precipitating the crisis" (Rintakoski/Autti 2008, p. 16).[128]

Above elaboration shed light on the diversity of CA interpretations. To be truly successful, however, the CA has to be implemented by the widest possible group of actors in the international crisis management community. The optimum approach and ultimate goal would thus be a *Whole-of-System* approach which brings together relevant ministries and agencies within governments as well international multilateral organizations. The Whole-of-System approach thus presents the highest level of coordination in terms of a conceptual hierarchy within the CA-spectrum (Gauster 2012, p. 77). This all-embracing meaning that under the paradigm of the CA ultimately all actors should strive for coherence and coordination can also be termed *Whole-of-Effort* approach (Stepputat 2009a, p. 13). At this level, the CA not just seeks coherence between governmental actors or within one organization but considers and deals with structures, relationships and interdependencies that "exist among and between the plethora of international and local actors and organizations engaged in a given context. This type of international, all-encompassing CA has rarely – if ever – been fully oper-

128 The multinational experiment was criticized for being military-led while relevant civilian actors were underrepresented, as well as for its insufficient engagement with the UN (Rintakoski/Autti 2008, p. 16).

ationalised" (Hull 2011, p. 9). Unlike WoG approaches or efforts undertaken by IOs a concretization of the CA at this level is most inherently complex.

2.2.4 Application in this Dissertation

The different approaches to the CA and the objective of crisis management coherence illustrate a large variation of terms and ideas as well as the struggles of actors which try to impose the appropriate policy. The conceptions that materialize from different actors and fields demonstrate different variants in their forms of coordination. However, as mentioned above, they are all based on the assumption that a more coherent approach and the provision of crisis management activities and instruments in a more coordinated manner increase crisis management effectiveness. While acknowledging the articulation of this assumption in many different expressions, the term 'the Comprehensive Approach' is adopted for the analysis in this thesis to underline that the concept as used here does not refer to any particular crisis management phase, type or actor. It is viewed as an instrumental idea targeted at improving crisis management coherence and the coordination of different national and international as well as civilian and military crisis management actors and instruments for the reasons illustrated above. In order to avoid conceptual misunderstandings, the notion is used as an umbrella term to cover efforts to increase crisis management coherence that have emerged on the EU level and on the national level in the country case studies. As analyzed above, the CA in crisis management is required on different levels. It can be understood in terms of activities conducted in the field, the integration of different time-frames and geographical dimensions but also in terms of coordination and coherence between the different actors involved on the national governmental and the international level. As highlighted in grey in figure 3 below, it is this last dimension this thesis is concerned with.

The focus is on inter-organizational relationships and respective attempts *how* to solve coordination challenges and to achieve crisis management coherence between different actors and institutions. Within the multi-actor environment of the CA, various categories of relationships can be distinguished. In this regard, intra-institutional relationships address the relationship between individual departments of a larger institution; inter-institutional is concerned with the cooperation between and coordination of different actors (Nilsson et al. 2008, p. 29). In these terms, comprehensive crisis management is understood as the end-state of an actor's crisis management policy achieved through the coordinated application of the multiplicity of its crisis management components and instruments.

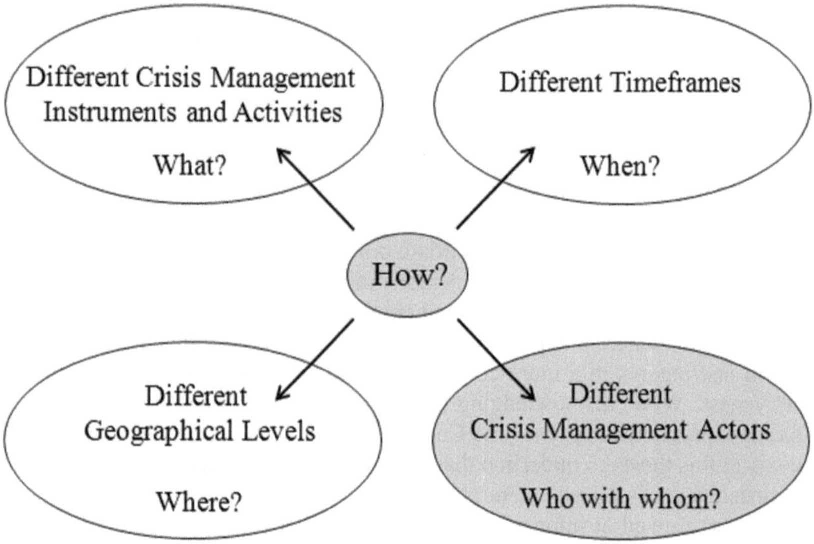

Figure 3: Dimensions of CA Coherence Analyzed in this Dissertation

A CA ensuring coordination and coherence between actors goes beyond the objective of increasing crisis management effectiveness. It rather is an objective *per se* as the provision of a coherent response to conflict ultimately is key to sustainable peace and stability. A successful implementation of the CA would facilitate the definition of common objectives by crisis management actors as well as the coordination of diverse actors' activities. This understanding covers different CA models used by different actors and is understood as a question of enhancing coherence in two dimensions: first, conceptual approaches and strategic efforts toward crisis management coherence; second, organizational efforts to improve crisis management coherence and inter-institutional coordination between the various actors. Both dimensions apply to the national and international level and cover the doctrinal and strategic implementation in relevant documents as well as organizational provisions and mechanisms to improve crisis management coherence and coordination between relevant bodies.[129]

129 The conceptual and the organizational dimensions as a framework for analysis are explained in detail in chapter 2.4.

2.3 Theoretical Approach

Though "the CA may not require explicit, highly granular definition before it can be used" (Hallett/Thorngren 2011, p. 36), the following chapters nonetheless set some conceptual boundaries. They present the theoretical approach put forward in this dissertation to analytically complement and specify the thematic background given by the CA concept. The following theoretical elaborations are based on the assumption underlying the CA, namely that certain modifications and adjustments in concept and strategy, structures and procedures lead to better organizational coordination. Better coordination leads to greater coherence and coherence in turn is necessary for the successful implementation of the CA.[130]

Indeed, the CA is directly linked to the catchphrases of policy coherence and coordination. Furthermore, the two central CA ideas also form main concepts of public administration and organization studies and are central aspects in studies of Europeanization. As analyzed by Gebhard (2011, p. 4) "[t]he proliferation of concerns over 'coherence' and 'consistency' in the political debate has not exactly helped to clarify the actual meaning of the concept and its political implications." As "the subject is a nebulous one" (Hood 2008, p. 20), the following chapter firstly examines the meanings of these two central terms and related expressions. Different dimensions of coherence and coordination are identified that allow for substantiating the analytical framework and to position them in the context of this dissertation's subject of comprehensive European crisis management.

To further approach the research subject, the analysis is based onto two different theoretical approaches which are analyzed in the then following chapters. Firstly, insights from organization and public administration studies as the main analytical approach allow for a focus on the aim for coherence and coordination as well as structures and mechanism that are supposed to enable the CA. The second analytical approach is based on the concept of Europeanization and deals with the impact of EU crisis management on domestic member state crisis management policies and *vice versa*. Applying these theoretical lenses allows an analysis of comprehensive crisis management that includes both the EU level and member state structures. The benefit of adding the Europeanization concept to the organizational perspective is that it links the two levels of European crisis management and allows for an analysis of both vertical and horizontal patterns

130 However, the difficulty of multiple causality needs to be considered in this regard, namely that an assumed effect of a specific variable can actually be caused by another hidden variable. Issues of causality are further explored in the respective theoretical context of organisation studies and Europeanization. The difficulty of proving this causality is also described by Herrhausen (2009, pp. 63-69) in the context of coordination and UN peacebuilding performance.

of interaction in the complex European crisis management system. Both theoretical lenses mutually complement each other.

2.3.1 Coherence and Coordination – Semantics and Conceptualizations

Both terms, coherence and coordination, are used interchangeably in treaty language, political practice and corresponding literature.[131] The attempt of precisely defining them is thus ascribed a minor importance in the remainder of this dissertation. However, analyzing and dividing the broad terms of coherence and coordination in separate conceptual components "helps to untangle the complex set of issues" (Gebhard 2011, p. 9).

2.3.1.1 Coherence

One idea, practitioners and scholars of comprehensive crisis management can settle on is that *coherence* is crucial for effectively implementing the CA and ultimately increases the effectiveness of crisis management activities. Coherence can be regarded as the underlying principle of the CA concept. It is referred to as the answer to deal with the implications of the contemporary crisis management complexity and to handle related institutional and organizational implications (De Coning/Friis 2011, p. 251).[132]

Coherence has been a great issue and lengthly been dealt with in IR literature and debates. In particular with regard to the EU, the concept of coherence has a long history and has conceptually evolved as one of the most discussed implications of the Union's complexity.[133] Coherence is one of the most frequently evoked phrases seen to improve the EU's political and bureaucratic performance. Though the problem can be found in most areas of EU activity, with special regard to foreign policy making, coherence can be understood as "the challenge and ambition to

131 In particular in EU legislation and translations, the two terms consistency and coherence are used synonymously, as Blockmans (2011, pp. 6/7) observes in the Lisbon Treaty context. See also Gebhard (2011, pp. 7/8) on this issue.

132 See, for example, De Coning (2004, 2008b, 2008c) who analyses coherence in the context of various international peacebuilding efforts or Giorgis (2010) with regard to G8 (Group of Eight: France, Germany, Italy, Japan, UK, USA, Canada, Russia) and EU peace and security support in Africa.

133 See, for instance, articles in Missiroli (2001) and Christiansen (2001). This chapter does not discuss the term and concept of coherence extensively in all its theoretical dimensions. It rather intends outline aspects of relevance for the CA and European crisis management to complement the analytical framework. A detailed overview of scholarly ideas on coherence and related debates can be found in Bertea (2005).

coordinate the multiple parts of the EU's international relations in order to increase both strategic convergence and procedural efficiency across several strands of EU external action" (Gebhard 2011, p. 2). The same applies on national level where it can be perceived as the need to bind together domestic institutions and procedures both with regard to crisis management coherence and internationally. However, though being widely used, coherence "is among the most frequently misinterpreted and misused concepts" (Gebhard 2011, p. 31). The term is essentially vague by definition since there are several other expressions such as harmonization, complementarity or comprehensiveness that have similar connotations.[134] While no consensus on specific features of coherence exists, there is some agreement on conceptual differences and respective implications.

Policy coherence as such was a historically largely issue-based subject linked to topics such as trade, corruption and migration. Up to the early 1970s coherence was mainly employed along with the term of *cohesion* and used to refer to the benefits of common activities by states on certain political matters (Gebhard 2011, p. 5). Rather than being an organizational matter, coherence was a question of political solidarity. Resulting from the growing difficulties in the institutional implementation of the coherence principle, the need for coherence was subsequently formulated in a more technical manner and in this sense became "the art of managing the interface" (Nuttall 2005, p. 95). *Congruence*, as another frequently used term in the coherence context, is essentially linked to analyzing the 'fit' between different organizational and institutional structures.[135] Terminological confusion also appears with regard to the term *consistency*. Yet, while consistency is understood as the absence of contradiction, coherence entails positive connections. "The former is more about compatibility and making good sense, the latter more about synergy and adding value" (Missiroli 2001, p. 4). Thus, whereas consistency is considered least advanced, merely involving the absence of contradictions and disagreement, coherence constitutes enhanced systemic synergy that even "involves 'positive connections' between several factors" (Gebhard 2011, p. 8). Pointing to the considerable different meanings of the terms and degrees of coherence, Gebhard (2011, p. 8) moreover sees consistency as related to the nature of a condition or outcome. In turn, coherence

134 Whether conceptual differences between these terms matter and conceptual differentiation is needed for analysis is contended in literature. For instance, Nuttall (2005, p. 93) states that "attempts to distinguish them risk ending up in linguistic pedantry". In this dissertation, however, the conceptual discussion, differentiation and clarification of coherence as the key concept behind the CA is useful for analytical purposes. See also Missiroli (2001, p. 4) on this issue.

135 For instance, when considering the fit of the institutional architecture of the EU and that of the member states. The term *congruence* is further conceptualized in the context of Europeanization in chapter 2.3.3.

deals with “the quality of a process, in which ideally the single entities involved join together in a synergetic procedural whole” (Gebhard 2011, pp. 8/9). While coherence is a matter of degree, consistency is a static concept. Compared to the notion of consistency “[t]he multi-layered concept of coherence is wider and relates to the construction of a united whole” (Blockmans 2011, p. 7).

Thus, a way of differentiating these notions is the question of the degree in terms of ascribing different levels of interaction toward coherence to the individual terms. Coherence is considered a more advanced dimension of structural harmonization that demands cohesion, comprehensiveness, continuity, and completeness as “secondary conditions or requirements” (Gebhard 2011, p. 8). In this sense, coherence is more than merely the absence of contradictions between policies, instruments, institutions. The notion entails that the overall impact of different actors and policies needs to be more than the sum of the individual impact and that actors not only agree but complement each other (Odén 2009, p. 14).

Reflecting this understanding and its prominent place in many official documents, the notion of coherence is highly positively coined. In addition to the negative connotation of incoherence as the logical alternative, the positive expectation behind this quest for coherence translates in the assumption that its enhancement results in improved efficiency, effectiveness and ultimately success of the actor. However, certain aspects should be considered with regard to this normative vision of coherence which by definition awards the term with a affirmative value as something necessarily to be aimed at. Coherence can be conceived in various ways. It can be approached with regard to its strategic or policy related dimension that deals with contradictory objectives and incompatible agendas. Alternatively, it can be understood in terms of its functional dimension that refers to structural and organizational implications of coherence (Gebhard 2011, p. 9). While the first dimension deals with coherence in terms of systemic outputs and the question whether the substance of different policies form part of a coherent whole, the second dimension focuses on institutional processes by which policies are formulated (Christiansen 2001, p. 747).[136] The focus in the latter is on the degree to which institutions operate in a coordinated way. It is in this functional dimension in which the value of coherence can lack substance. Seeking coherence in this regard might run the risk of taking out functional depth and begs the question whether coherence is always something to be desired (Gebhard 2011, pp. 13-15).

In this regard, it is important to be clear about the policy around which policy coherence is sought. In the context of comprehensive crisis management, coherence can be understood as “the effort to direct the wide range of activities undertaken in the political, development, human rights, humanitarian, rule of law

136 These dimensions are further operationalised in chapter 2.4 outlining the dimension of analysis.

and security dimensions of a comprehensive approach system towards common strategic objectives" (De Coning 2008a, p. 8). Coherence thus refers to the obligation of ensuring synergy between different crisis management instruments and fields and, furthermore, forms a matter of quality of interaction between different organizational entities.

To guide thinking, special levels of crisis management coherence have been categorized by several authors which are useful to keep in mind and which further differentiate the double challenge of European crisis management coherence as shown in figure 1. According to De Coning and Friis (2011, pp. 253/254), crisis management coherence combines four dimensions. *Intra-agency coherence* refers to the "consistency among the policies and actions of an individual agency" (De Coning/Friis 2011, p. 253). It is concerned with the internal consistency of specific crisis management programmes and policies such as the internal coherence of a ministry for foreign affairs or a certain Directorate General (DG) of the EU Commission. *Whole-of-Government Coherence* is the consistency between different government entities of a country, for instance between the ministry for foreign affairs, the ministry of defence, the ministry responsible for international development assistance. On the next level, *inter-agency coherence* deals with consistency among the activities and policies of the various international actors in a given context, such as coherence between the EU and the UN, or the UN and NGOs. *International-local coherence* applies to consistency between the activities and policies of various internal and external actors in country contexts, for instance, between the host nation and international actors. Similarly, Odén (2009, pp. 3) refers in his analysis to five coherence areas in the development field where it has been argued that without improvements the UN's Millennium Development Goals are unlikely to be achieved. These areas entail *internal* coherence, *intra-government* coherence, *inter-governmental* coherence, *multilateral* and *donor-recipient* coherence.

With regard to the EU, Nuttall (2005, pp. 97/98) specifies a different set of categories of coherence that distinguishes between *horizontal* coherence as entailing the compatibility of the whole range of different EU policies in general and *institutional* coherence as a sub-category of the horizontal level that deals with the two different bureaucratic frameworks of EU foreign and security policy namely the supranational Community level and the intergovernmental level.[137] *Vertical*

137 A prominent example of horizontal inconsistency in this sense is the coexistence of the EU's common agricultural policy, which subsidizes exports and protects the European market with trade barriers, and the Union's development policy, which aims at the elimination of poverty and encourages developing countries to develop their agricultural sectors with a view to exports. For further information on inconsistencies between EU agricultural policies and development policy objectives see the project *Policy Coherence – Exploring Links between EU Agricultural Policy an World Poverty* of the Trinity College Dublin (Trinity College Dublin 2011).

coherence relates to coherence between EU and national policies. Also Gebhard (2011) identifies four levels in which the concept of coherence gains relevance in the EU context: just as in Nuttall's categorization, *vertical* coherence is understood as the harmonization of member state policies regarding EU level policies and positions. This implies compliance with EU political commitment thus affecting member state solidarity. It also includes functional compatibility of policies. Coherence on the *horizontal* level in Gebhard's (2011) categorization deals more narrowly with the coordination between the intergovernmental and the supranational areas of EU crisis management and accordingly with coherence between the European Commission and the European Council as the main institutions governing them, including corresponding bureaucratic machineries.[138] Horizontal coherence is mostly concerned with institutional issues. Coherence within the two foreign policy domains, namely within the intra-institutional mechanisms of the CSDP and ESDP and within the external policy dimension of the Commission is termed *internal* coherence. Internal coherence of EU foreign policies is concerned with challenges resulting from the functional and procedural dimension and to coordinate inconsistent bureaucratic machinery. *Inter-organizational* (or *external*) coherence is associated to EU relations to other actors in the international system, such as the UN, NATO, the OSCE or key partners like the US, and the coordination of respective positions toward them. Determined by functional internal coordination processes, inter-organizational coherence also deals with the credibility of EU external actions (Gebhard 2011, pp. 9-12).[139]

Accordingly, coherence can be aimed at across various dimensions, at different levels and among different actors. However, while being managed through different decision-making structures, the types of coherence are strongly interrelated and decisions that concern coherence in one dimension also affect coherence in the other dimensions. Notwithstanding such a categorization, coherence does not occur on each level separately. Consequently, the degree to which activities of a crisis management actor can be considered coherent to a greater or lesser extent is a factor of all levels of coherence. Thus, coherence cannot be placed in just one of

138 Before the entry into force of the Lisbon Treaty, horizontal coherence could also be referred to as inter-pillar coherence, describing the coordination challenge between the three pillars of EU policy.

139 Both, Nuttall (2005, pp. 96/97) and Gebhard (2011, pp. 15-18) categorize coherence in a further restricted manner which additionally identifies the *banal*, the *benign* and the *malign* sub-level of coherence. The banal understanding refers to the rare case that policy-making takes place in one single pillar and inconsistencies are therefore inexistent. The benign level deals with active interaction Community instruments and the CFSP; and the malign level is about internal power struggles between institutions. The understanding of coherence in this limited sense refers to discussions about the interface between the intergovernmental CFSP and the EC only and is thus not of interest in the context of this thesis.

the dimensions as outlined above and this thesis is based on the assumption that different levels of coherence in European crisis management are linked and mutually influence each other.[140] Likewise, an extended categorization in various analytical coherence dimensions runs the risk of complicating the picture. A categorization in the context of this thesis needs to be relevant on various levels of European crisis management while also considering the specifics of crisis management environment itself. The most relevant dimensions in accordance to the above categorization in this dissertation are the horizontal and vertical level which can usefully sharpen the meaning of coherence. While the conceptualization of vertical coherence deals with coherence between EU and member states' concepts and structures, horizontal coherence refers to the crisis management policy worked out by the EU and the member states for their institutions individually.[141] For the ease of distinction, differentiations of crisis management coherence are only undertaken when relevant in the respective analytical context.

It is also important to note that in practice due to the dynamic nature of complex systems and the political environment, coherence can never be fully achieved and is an unattainable state. Nonetheless, the use of the term is appropriate as a guiding principle, as what can be achieved in practice is the compensation of incoherence through various modes of interaction (Gebhard 2011, p. 32). Moreover, "even though coherence can never be fully achieved, it is possible to distinguish between systems where there is little coherence and ones where is more coherence" (De Coning 2008c, p. 55). It is coordination which in this sense is the essential instrument used for achieving coherence and consequently the best indicator to assess coherence in terms of degree not end-state (Johannsen 2011, p. 35).[142]

2.3.1.2 Coordination

"If 'coherence' is the aim, then 'coordination' is the activity through which coherence is pursued" (De Coning 2008b, 92). Managing the implications of structural complexity by aiming at greater policy coherence essentially stands for an

140 De Coning's and Friis' (2011) coherence model in particular views the coherence of crisis management efforts as a whole strictly as a matter of the specific degree of coherence in each dimension.

141 Considering the following theoretical considerations, insights from organisational studies are mainly applied to the horizontal level while the concept of Europeanization mainly helps analyzing coherence on the vertical level to reflect the importance of the political will of member states.

142 De Coning and Friis (2011, pp. 255-257) developed a scale breaking coherence down on different levels representing different degrees of coherence running from competition to unity.

enhanced need for coordination and "the necessity of closing ranks across a growing range of functional areas" (Gebhard 2011, p. 3). Just as coherence, the term coordination has been used in different ways, given the number of actors that can be coordinated and the multitude of different strategies that aim at achieving it.[143] Coordination always raises the subtle question about which actors are coordinated and which do the coordination.[144] At the same time, coordination is a factor which is heavily influenced by the actors involved. Differences also occur on the issues which form it should take and how it can be achieved best (Herrhausen 2009, pp. 15/67). Given the practical confusion surrounding the notion, some clarification is necessary of what is meant by the coordination that is so often called for and of its specific meaning in the CA context.

Generally, both in official documents and academic literature, other terms such as cooperation, collaboration or alignment are used in the same manner and being ascribed characteristics similar to coordination leaving it with a rather vague meaning (Herrhausen 2009, p. 66).[145] However, coordination always deals with the interaction between different organizational parts. In abstract terms, it entails the creation of a common order to adjust a number of separate elements that are initially different, but also linked with respect to their purposes, policy goals and tasks. Coordination thus implies that distinct components are interrelated by superior values and share a purpose. From this perspective, coordination first and foremost involves the creation of a common order for various separate elements. At this point, the function of coordination is to secure that outputs of a range of entities are not in elemental conflict (Ekengren/Sundelius 2004, pp. 112/113).

Scholarly efforts show that coordination largely depends on the context. Some studies dealt with coordination in a framework similar to this dissertation's subject.

143 On this issue see, for instance, Mulford and Rogers (1982, p. 9).

144 The question who should be affected by coordination and how it should be organized reflects very well in the saying that 'everybody wants to coordinate but nobody wants to be coordinated' that succinctly frames the paradox and that frequently has been stated in the interviews conducted for this dissertation.

145 Some authors subordinate cooperation under coordination, while others understand the two as analytically distinct (Hall 1991, p. 237). According to the Oxford Advanced Learner's Dictionary of Current English (Wehmeier 2000, pp. 276/277) cooperation means "working together for a common purpose" whilst coordination deals with "making parts of sth, groups of people, etc. work together efficiently and in an organised way." In case of a common purpose, but conflicting goals of actors, coordination can also become another term for coercion. If differences can be solved by compromises on the goals, coordination can also be another term for consent (Pressman/Wildavsky 1973, p. 134). Collaboration denotes activities which take place between actors of different mandates acting strongly independent but who on the contrary, share interests and objectives (Nilsson et al. 2008, p. 30). While these elaborations serve to demonstrate the terminological difficulties, an in depth analysis of these different terms goes beyond the scope of this dissertation.

While coordination as a basic principle is generally upheld and carries strong positive connotations, the actual organizational implications of the CA in international crisis management in these studies is rather vaguely referred to as the 'challenge of coordination' or the so often identified 'lack of coordination' (Herrhausen 2009, p. 67). Drawing from theory as developed in business management, Seybolt (2000) reviews the organization of relief efforts in crisis situations. He defines coordination as "a search for efficiency in an organic whole, where diverse parts operate within a shared set of environmental, structural and procedural constraints and opportunities" (Seybolt 2000, pp. 2/3). Analyzing the UN's peacebuilding organization, Herrhausen (2009, p. 72) defines coordination more broadly as "the act of bringing different elements in a system into alignment." According to Hall (1991, p. 237) "[c]oordination involves a process of concerted decision making or action in which two or more organizations participate with some sort of deliberate adjustment to one another." These definitions recognize the importance of structural interdependence. Following a more process oriented conceptualization in the context of a CA system, according to De Coning (2008a, p. 8) coordination essentially means "to incorporate developing strategies, determining objectives, planning, sharing information, agreeing on the division of roles, responsibilities and tasks, and mobilizing resources." Coordination is thus not just viewed as the outcome of an inter-institutional relationship, but is understood as the effort to optimize coherence across policies in order to overcome sectoral fragmentation (Wollmann 2003, p. 594). A key factor of viewing coordination as a process is that it is deliberate and, moreover, involves a collective goal (Hall 1991, p. 237),[146] which avoids that increased coordination poses a threat to a program or institution as every actor retains decision-making control. This is particularly important with regard to the organizational cultures in the context of comprehensive crisis management in which coordination has to be non-threatening. To be non-threatening for CA actors, coordination has to be voluntary.[147]

With regard to central government coordination in the public-sector's interorganizational context, Peters (1998b, p. 5) refers to the need of coordination as an instrument to ensure that the various organizational bodies responsible for delivering public policy work together and do not generate either redundancy or gaps in services. In a similar vein, Hustedt and Tiessen (2006, p. 5) refer to coordination "as the attempt to produce coherent government policies and achieve a state that

146 According to Hall (1991, p. 237) this is the main aspect differentiating coordination from cooperation. Mulford and Rogers (1982, p. 13) underline that it is the outcomes of coordination and cooperation that are different. While actors cooperate to achieve their individual objectives, with coordination, joint decisions and actions lead to a joint outcome.

147 For more and different definitions see, for example, Ranade and Hudson (2003, p. 33).

minimizes conflicts that arise between different government organizations over programs, proposals or legislative bills." The emerging coherent government policy shows in political programs without contradictions both within individual policy areas as well as in the whole of governmental programs. Coordination is thus understood as "a process of inter-organisational conflict-resolving, which becomes manifest in certain coordination techniques" (Hustedt/Tiessen 2006, p. 5).[148]

The "coordination imperative" (Dimitrakopoulos/Passas 2003, p. 442) also is a highly relevant concept in the EU context. The assumption that coordination automatically results in a more efficient and effective EU is used like a mantra among scholars and practitioners and it seems to offer "soothing simplicity in the face of disturbing complexity" (Paris 2009, p. 53). In this regard, coordination also includes both a horizontal and a vertical dimension. The horizontal dimension deals with the coordination of organizational actor's activities at the same administrative level with the aim of promoting a coherent set of policy preferences for the agenda-setting in EU policy or of EU actors and institutions. The vertical dimension is viewed as the need of the member states' government units to be aware of EU policy implementation (Dimitrakopoulos/Passas 2003, p. 442). Pointing to the importance of coordination both within member states and on the European level, Kassim (2001, p. 7) notes that "coordination refers to the action of, and interaction between, the individual departments, agencies, politicians, and officials, involved in a particular policy area or with respect to an activity or task."

Similar to coherence, different dimensions of coordination can be distinguished. Intra-institutional or intra-ministerial coordination refers to the coordination of relevant activities within an institution or ministry of crisis management actor. Inter-institutional and inter-ministerial coordination deals with the coordination across different institutions and ministries.[149] Furthermore, "coordination can be strategic, directed toward the common pursuit of an overarching objective, or selective: issue-oriented, reactive, or concerned only with certain

148 As underlined by Gebhard and Norheim-Martinsen (2010) a fundamental difference exists between coordination at political level and coordination at administrative level. While administrative coordination is mainly concerned with keeping up procedural flows or functionally enabling an organizational process, political coordination is strictly speaking not an organizational process in itself. While it entails administrative activities, the subject of coordination is not an organizational one. According to Gebhard and Norheim-Martinsen (2010) is thus "important to differentiate whether 'coordination' means 'organizational enabling' or 'searching for convergence'." See also Peters (1998b, pp. 15/16) who points out important differences between policy coordination and coordination of administration.

149 In the context of humanitarian aid, Weiss, Spanger and Van Meurs (2009, p. 17) also distinguish inter-donor coordination between aid agencies and donor approaches as well as donor-partner coordination which demands alignment between donors and the needs and priorities of partner countries.

stages of the policy cycle" (Kassim 2001, p. 8). This differentiation highlights that coordination can on the one hand be directive and designed to achieve a particular policy goal, or on the other hand can be procedural to ensure that the issues are dealt with the right machinery (Kassim 2001, pp. 8/9).

Some additional forms coordination takes on can be differentiated. *Positive* coordination means the sustained effort which aims at assuring that all parts actively engage with each other. In contrast, *negative* coordination merely involves strategies to mutually avoid each other to reduce conflict as shown in figure 4 below.

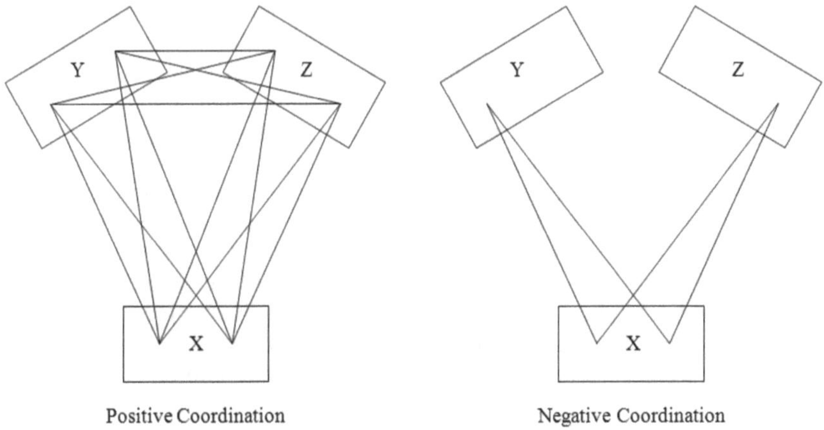

Figure 4: Positive and Negative Coordination *(Source: Adapted from Scharpf 1993, p. 144).*

The differentiation and concept of positive and negative coordination has been popularized by Scharpf (1993) who developed the terms in the context of an expanded scope of multilevel coordination. According to Scharpf (1993, pp. 143/144), the goal of positive coordination is the maximization of effectiveness and efficiency of a policy by searching and using joint strategy options of the actors involved. On a procedural level, positive coordination involves multilateral interactions on multiple levels between relevant actors simultaneously. Negative coordi nation, on the contrary attempts to ensure that new policy initiatives in a sector do not interfere with policies and interests of other actors and mainly deals with potential objections. Limited in aspirations, negative coordination is procedurally dominated and initiated by one actor specialized in the policy area at hand and typically involves bilateral interaction between affected actors. Yet, figure 4 also makes clear that negative coordination significantly decreases the overall complexity of coordination. On the downside, however, the possibility for policy innovations is reduced when the organizational status-quo policy set is given and policy options are only explored by the initiating actor (Scharpf 1993, p.144). In addition, coordination and respective organizational mechanisms can be characterized more or less *formal*. It may be based on formal conventions or *informal* practices. Formal mechanisms can be outlined in organizational charts as well as routines and rules that specify procedures, methods and responsibilities of organizational entities. In contrast, traditions and individual organizational cultures have informal characteristics and are established through experiences (Christensen et al. 2007, p. 15). Formal coordination involves the clear assignment of authority, responsibility and tasks, as well as the specification of procedures. It consists of any type of planned and formally established activity and is linked to structural positions. Communication occurs via official channels which are described by formal documents. Informal coordination, on the contrary, evolves through social mechanisms and *ad hoc* responses to interdependence. It often occurs spontaneously outside formal structures on a personal level and communication is not linked to structures and formal positions (Lipson 2005, p. 14). Informal coordination increases communication and thus offers even more opportunities to cross departmental boundaries because it is less bound to the formally established links. However, both forms of coordination are partly complementary and depend on the other to function effectively. "Informal coordinating activities can defeat the intended purposes of formal measures, and dysfunctional formal arrangements can inhibit potentially salutary informal responses in addition to failing to achieve their own goals" (Lipson 2005, p. 14). These explanations on formal and informal coordination also point to the characteristic of coordination being for instance based on legal requirements and thus *mandatory* or happening on a non-binding footing.

Hall (1991, pp. 228-231) identifies four foundations for inter-organizational coordination. Firstly, these can take place on *ad hoc* basis when a specific need or issue arises among organizations without or with only little previous patterning in their relationships. Secondly, on *exchange* basis, inter-organizational coordination occurs as a voluntary activity between organizations leading to actual and anticipated consequences for the realization of their objectives. The exchange basis can be understood as a form of bargaining "in which each organization seeks to maximize its advantage in acquiring resources from another organization" (Hall 1991, p. 229). However, this becomes less of a factor, when interactions between organizations become, thirdly, *formalized*. Formalization of organizational interaction is the "degree to which the interdependency among organizations is given official sanction by the parties involved" (Hall 1991, p. 229). Formally agreed, inter-organizational relationships are based on specific patterns that seek to simplify interactions. A very different source of inter-organizational interaction is when these are, fourthly, *mandated* to interact. As such, "[m]andateness refers to the extent to which relationships are governed by law or regulations" (Hall 1991, p. 230) which are imposed from outside to bring actors together because of a legal political mandate. As Hall (1991, p. 230) underlines, "mandated relationships may lead to conflict, since organizations may be forced to interact even with domain dissensus, interpersonal animosities among members and so on."

Considering above elaborations and coordination forms, coordination in this dissertation is seen in concrete organizational terms focusing on the organizational implementation of the CA. Mulford and Rogers (1982, p.12) describe "interorganizational coordination as the process whereby two or more organizations create and/or use existing rules that have been established to deal collectively with their shared task environment." This characterization proves particularly useful in the CA context because it emphasizes the significance of the shared task environment, it views coordination as a new rule which can be established and furthermore underlines the role of commonality and the achievement of joint objectives (Mulford/Rogers 1982, p.12). Coordination is understood as an instrument to enhance coherence and organizational efficiency of crisis management actors and to effectively implement the CA. For the purposes of this analysis, coordination is broadly viewed "as the extent to which organizations attempt to ensure that their activities take into account those of other organizations" (Hall et al. 1977, p. 459). It takes place as different actors and organizations try to attain their goals or carry out their tasks and at the same time attempt to deal with their respective environment. In addition to this rather loose characterization, a perspective by Warren et al. (1974, as cited in Mulford/Rogers 1982, p. 16) is considered which concentrates on structures and processes and describes coordination as "concerted decision making or action wherein the decisions or actions of two or more organizations are made simultaneously in part or in whole with some deliberate degree of adjustment to each other." This allows a focus on formal structures, relations between organizational entities and to discuss conceptual and organizational modalities from an analytical perspective focusing on the coordination-related organizational settings.

Keohane (1988, p. 380) suggests that the impact of cooperation could be evaluated "by measuring the difference between the actual outcome and the situation that would have obtained in the absence of coordination." The effect of coordination or a certain organizational form in the context of this study, however, cannot be measured this way. The outcome of a certain policy in the absence of certain organizational means cannot be predicted as too many other factors and actors on different levels influence the outcome of crisis management as well. The cause-and-effect-chain of the effective implementation of a Whole-of-Europe Approach cannot consistently be proven from efforts to enhance coherence on the member state level as well as the EU level of crisis management. To this effect, the policy coordination scale developed by Peters (1998b, pp. 6/7) facilitates to measure and compare coordination. It lists coordination opportunities available to actors starting on the lowest level that increase along the scale as shown in figure 5.

At one end of the scale, actors engage autonomously, at the other end based on consensus. Between the two ends options for positive coordination like a search for agreement and for negative coordination like avoiding divergences exist. However, a strict analysis along the Policy Coordination Scale is too demanding in the context of the following analysis. However, the scale presents a helpful mechanism to systematize CA implementation efforts and coordination mechanisms. While it becomes possible to declare more precisely how the CA and crisis management coherence are implemented, this dissertation rather seeks to analyze the organization of coordination and than examining its degree. To this end, organization theory and public administration studies are drawn on to understand different types and methods of coordination suited to the circumstances of international crisis management.[150]

Development of a Common Overall Strategy
Establishing Policy Priorities
Setting Organisational Parameter
Settlement/Arbitration of Policy Differences
Search for Agreement Among Actors
Avoiding Divergences
Consultation with Other Actors (Feedback)
Communication with Other Actors (Information Exchange)
Independent Decision-Making by Different Actors

Figure 5: Policy Coordination Scale *(Source: Based on Peters 1998b, p. 7; Modified by S.P.).*

150 Coordination is not necessarily a tool for direct organizational improvement nor does every organisational reform that aims at improving coordination enhance coherence. Assumptions that view coordination as a good on its own right have been questioned (Hall 1991, p. 273) and an emphasis on improving coordination in crisis management appears "strangely anodyne and technocratic" (Paris 2009, p. 58). Peters (1998b, pp. 24/25) also underlines this aspect. Indeed, it is important to note, that the coordination problem cannot serve as a catch-all which would result in overlooking deeper shortcomings by redefining them in merely procedural-technocratic terms. See argumentation by Paris (2009). However, rather than perceiving coordination as an unalterable good itself, this dissertation seeks to analyze the balance that has to be struck in coordination.

2.3.2 *Organization Theory and Public Administration*

As elaborated in the introduction of this thesis, various theoretical approaches are of limited applicability in the comprehensive crisis management context and do not fully qualify for addressing this dissertation's research questions.[151] So far, while studies on the CA as a concept as well as on its practical implementation identify necessities of coordination, they rarely analyze the organizational realities. In this regard, findings from organization theory and studies of public administration provide insights on the implication of the CA in concrete organizational terms and arrangements.

2.3.2.1 Overview

Organization theory and public administration studies are integrating disciplines that are only partially rooted in political science, drawing on strands of sociology, psychology, economics and management theory (Christensen et al. 2007, p. 2). The wide and complex body of literature of both strands makes it difficult to isolate organization theory in public administration (Christensen 2003, p. 109). As Hall and Quinn (1983, pp. 7/8) describe the interconnections of organization theory and public administration, "[o]rganizations are the means by which public policy is implemented. (…) Public policy is formulated in organizational settings. (…) Organizations are the object of public policy." This heterogeneous field comprises a number of different theories expounded both separately and in combination, and public administration scholars draw on organization theory for insights and guidance (Hult 2003, p. 150).[152] Considering the increased complexity of political-administrative organization, organization theories for the public sector became increasingly complex as well and reflecting on organization theory takes place across disciplines without a clearly and strictly defined object of study (Christensen et al. 2007, p. 14). The approaches are both termed more by the subject-matter of analysis than by a particular discipline or a single dominant analytic paradigm

151 See chapter 1.4.2 which considers diverse theoretical approaches with CA relevance.

152 Originally, organization theory did not provide a basis for studies of public administration but over the past decades an organization theory geared to studies of public administration has been developed, which acknowledges the close connection between public administration practice and organisation theory. In this context, Olsen (1991, p. 88) states that "[o]rganization theory can no longer be seen as a subfield of public administration which, in turn, is seen as a subfield of political science." See, for instance, Christensen (2003) and Christensen et al. (2007). For an early but in-depth review on the question of mutually subordinating both theories to each other see the whole volume on organization theory and public policy edited by Hall and Quinn (1983).

(Bogdanor 1987, p. 398). It is thus useful to focus on possible contributions of specific considerations of organization theory and public administration studies for this dissertation rather than elaborating their various forms.[153]

Organization theory and public administration studies are both rooted on a similar basis of theorists such as early public administration theorist Luther Gulick who was one of the leading figures developing general administrational principles.[154] Also, both disciplines traditionally stand upon a foundation build by theories of Max Weber (1946, 1966) and his analysis of authority and bureaucracy. Major aspects of Weber's characterization of organization are that due to mechanisms of inclusion and exclusion organizations have boundaries and are designed to serve a certain purpose with their activities. Weber also introduces the idea of order that suggests that organizations contain a division of labour, structured interaction and hierarchy of authority (Hall 1991, p. 29). While Weber stresses the system of organization, Chester Barnard, another classical writer in the field, focuses on the role of the members of the system. Herbert A. Simon, who built his understanding around Barnard's approach, developed a theory about administrative decision-making behaviour. This behavioralist understanding of organization has been viewed rather critical by Philip Selznick who gives a priority to organizational processes and sees organizations primarily shaped by their environment (Knudsen 1995, pp. 148/149).

153 Widely rooted in different disciplines both bear internal debate on their links to other disciplines as well as their overall extent and scope (Bourgon 2011, pp. 8/9). Organisation theory traditionally has been developed and associated with studies in the context of private organisations, especially business firms. It has been dominated by a focus that seeks to analyze how organisational units should be assigned to tasks in order to increase output, effectiveness and efficiency in the business context (Christensen et al. 2007, p. 2). In particular the theory of Transaction Cost Economics aims at explaining organizational structures as a consequence of economization of costs by creating contracts governing exchange (Lipson 2005, p. 14). The contributions of economics' Transaction Cost Economics and its value for organizational analysis are examined in Perrow (1986, pp. 236-247) and Herrhausen (2009, pp. 76/77). Besides this economics based theory strand, sociology based theories of organization examine linkages between formal rules and informal group behavior, motivational effects of alternative organizational design or how sociological factors shape organization. Numerous studies based on organization theory and public administration studies also analyze socialization processes such as shift of loyalty or senses of community of the personnel within an organization (Trondal et al. 2010). Regarding political science and organization theory, Olsen (1991, pp. 87/88) states that despite significant parallels in agendas, both "have been in a state of mutual disregard for years" (Olsen 1991, p. 88). Also Rhodes (2007, p. 1258) points to "the convergence between political science and organizational studies." Observing a renewed attention paid to the relationship of the two fields, Olsen (1991, p. 88) at the same time points to the difficulty of discussing this interdisciplinary relationship due to the diversity of meanings ascribed to both.

154 Gulick identified seven major elements and functions of administrative managing namely planning, organizing, staffing, directing, coordinating, reporting, and budgeting (Gulick 1937, p. 13).

The different utilization of its insights show that the field is fragmented. Generally, public administration deals with administrative structures and concrete political bodies as well as organigrams, constitutions and procedures that conceptually bind them.[155] The public sector is build up by a set of specific organizational entities such as ministries and agencies as well as public servants (Thoenig 2003, p. 127). Their management is different from other types of organization because public organizations are located in "a complex policy and political environment, under formal control of politicians, and is subject to a high degree of scrutiny and accountability" (Calabrò 2011, p. 10). Public administrations shape concepts and values thus forming the role of the government and defining the functioning of public institutions and organizations (Bourgon 2011, p. 7). A major objective of administrative studies is the analysis of multiple structures of administrative governance in order to systemize and compare their common solutions and problems. In this regard, public administration studies deal with issues how to sustain the machinery of government and ensure that policies are put into practice (Chandler 2000, p. 2).[156]

Organization theory has brought forward various theoretical approaches and models, and has undergone different permutations. As Shafritz and Ott (1992, p. 4) note "there is no such thing as the theory of organizations". Organization theory, literally, "is the result of academic study of organisations and their constituents" (Herrhausen 2009, p. 71) and refers to "all artificial or consciously designed units

155 Public administration is both a domain of practice and a field of study. According to the Blackwell Encyclopaedia of Political Institutions (Bogdanor 1987, p. 504), public administration abbreviated in the lower case, p.a. stands for "institutional arrangements for the provision of public services", while in capital letters P.A. describes the study of those institutional arrangements. To avoid confusion, this dissertation sets both above mentioned abbreviations aside and clearly refers to public administration. Regarding the distinction between public administration and the term public management some scholars view both terms synonymously but also argue both ways seeing public management as the more general concept as well as public management as a segment of the broader field of public administration, which is the prevailing viewpoint (Lynn 2003, pp. 14-16). The term 'public' in this context is not treated in the sense that the use of public power is involved but basically refers to government administration that is concerned with international crisis management (Bogdanor 1987, p. 504). For more penetrative comments on the characteristics of public organizations and differences to private organizations see Christensen et al. (2007, pp. 4-10, 179). A vastly investigated aspect of public administration studies is the link between these institutional arrangements and society. However, this dimension is not relevant here and will largely be left out. In doing so, this dissertation focuses on formal governmental status and structures that are also referred to as the 'core executive' (Hult 2003, p. 150). Rhodes (1995, p. 12) defines the core executive as "all those organisations and procedures which coordinate central government policies, and act as final arbiters of conflict between different parts of the government machine". The core executive thus refers to the key institutions at the centre of government as well as less formalized structures, departments and coordination mechanisms such as interdepartmental committees in this thesis.

156 See also Chandler (2000) for details on public administration as an academic subject.

of human organization" (Bogdanor 1987, p. 398). March and Simon (1993, p. 35) define organizations as "systems of coordinated action among individuals and groups whose preferences, information, interest, or knowledge differ". Organization theory in this context can explain "the delicate conversion of conflict into cooperation, the mobilization of resources, and the coordination of effort that facilitates the joint survival of an organization and its members" (March/Simon 1993, p. 35). This is achieved mainly by control and channelling of information, the shaping of identities, goals and loyalties, the creation of common beliefs and behaviours and the creation of incentives that facilitate respective efforts (March/Simon 1993, p. 35). Organization theory intends to identify issues generic to all formal organization, such as the structure of communication or the division or integration of tasks and responsibilities (Bogdanor 1987, p. 398).

But organizations also operate as active participants in change processes and are change agents themselves. As Hall (1991, p. 19) points out "in order to achieve change, there must be organization." Thus, dynamics of interactions with the environment influence both the organization itself and its environment (Hall 1991, pp. 17-19). However, apart from organizational contributions to change processes, organizations can also actively resist change, especially when it is introduced from outside.[157] Organizations change on a long time basis and changes therefore may not be in the same phase as the changes in political thinking or the government in power. This organizational conservatism contrariwise proofs to be a source of stability as a major organizational outcome. However, organizational modes of the core executive are not neutral technical questions but are also highly political. Organization influences the selection of actors, problems and solutions. It furthermore defines which groups, values and interests are taken into consideration. Formal organization offers different kinds of possibilities and constraints (Christensen et al. 2007, p. 9). As Thompson (2003, p. 154) assesses, policy-making systems "involve a complex and sometimes unstable combination of competition and coordination, negotiation and the exercise of power, independence and interdependency, and consultation and the exercise of pressure politics."

Just as organization theory examines organization in terms of the individual, the community, the society, economics and the international scene, organization theory in public administration focuses on respective dimensions. One differentiation can be drawn by the level on which analysis focusses on. While the micro-level framework focuses on *individual* decision makers and the extent to which organizations impact in individual actors preferences and interests, the *organiza-*

157 Using government offices' civil service system as an example of organisations as resisters of change, Hall (1991, pp. 22-24) identifies a strong tendency for an organisation to work and function as it has done in the past as it contains rules and procedure that developed along the way as well as members that are trained to follow this system.

tional framework is given by the formal structure in which these actors operate. Another focus can be the *task environment*. The *cultural institutional* perspective deals with the gradual development of public organizations into institutions by introducing values to the formal framework (Christensen 2003, pp. 110/111). *Instrumental* perspectives in organization theory see organizations as instruments. Instrumentality from this point of view is demonstrated in the structural design of an organization which affects organizational behavior (Christensen et al. 2007, p. 21).

2.3.2.2 Analytical Key Factors

An important insight drawn from organization theory is that structural organization form and the type of coordination are essentially interlinked. "[T]here is an automatism between organization form and organization mechanisms" (Herrhausen 2009, p. 190). Thus, a key assumption of organization theory and public administration studies is that organizational form influences policy content (Christensen et al. 2007, p. 9). "Coordination cannot be discussed without referring to the organization form in which coordination is supposed to take place" (Herrhausen 2009, p. 72). The respective organization form determines how and in which form coordination occurs. On the contrary, choosing a particular organization form limits the set of available coordination mechanisms (Thompson 2003, p. 22). The design of structural organization is anticipated to channel information exchange, coordination processes and conflict resolution. Organizational boundaries created through its structure affect which problems and solutions at which level which policy makers become aware of (Egeberg 2003, p. 120). While formal organizational structures do not automatically show how members in an organization behave, it constraints how tasks are carried out and puts limitations on the choice of action (Christensen et al. 2007, p. 9). Formal organizational structure thus provides categories that structure complex information, offers procedures for reducing overlaps and costs, regulates, sets up physical boundaries and temporal rhythms (Trondal 2009, p. 240).[158]

---

158 Formal organizational structure or merely organizational structure is a "structure that consists of positions and rules for who shall or can do what and which defines how organisational tasks should be executed" (Christensen et al. 2007, p. 21). A focus on the formal structure allows for an impersonal view as coordination mechanisms exist independently of the individual. Formal organizational structure in the context of this dissertation means institutions and ministries that share characteristics as formal organizations such as special tasks, relatively stable patterns of behaviors and resources connected to their activities (Christensen et al. 2007, p. 9). Though this dissertation focuses on formal organizational mechanisms, informal norms and coordination forms are taken into account in the relevant context when they become important for CA ac-

Organizational structure itself has various dimensions such as its size which indicates the number of roles the organization has to fill and points to its capability to initiate, develop and implement policies and decisions. A size of an organization implies that regular attention and support in the policy process can be assured to various concerns embedded in organizational units (Egeberg 2003, p. 120). Drawing on the widely accepted differentiation, essentially three ideal type organization forms and modes of coordination are distinguished, namely hierarchies, markets and networks (Wollmann 2003, p. 595). These "three pure routes to governance" (Calabrò 2011, p. 10) describe certain types of interaction and structures and also provide a useful point of departure for the analysis of CA implementation.[159]

According to Wildavsky (1987, p.6), "[h]ierarchy is institutionalized authority" that involves some form of explicit rule-driven design and direction (Thompson 2003, p. 22).[160] The hierarchical model presents the traditional way of core executive organization with clearly defined boundaries, setting vertically delegated authority of organization, and units which are grouped according to responsibility, expertise or function (Bourgon 2011, p. 48). *Hierarchies* are a form of organization suitable for recurring processes that can be standardized. Such standardization through determined rules makes hierarchies very effective forms of organizations. Conversely, for the same reasons critics argue that hierarchies are rather static and inflexible, leading to slow adjustments to external changes, being inefficiency-prone and fail to draw on their own professional expertise (Peters/Pierre 2003, p. 5). At the same time, hierarchies are viewed to be also particularly useful in coping with situations of complexity, novelty and uncertainty as they are able to transfer issues into familiar levels of action by falling back on habits and routines (Paris 2009, p. 59). Organization in a hierarchy results in moving processes of coordination, conflict resolution and information exchange upward in the organizational structure. The low-level splitting of issues in hierarchies results in situating coordi-

tivities. Methodologically, informal organization forms are difficult to grasp as they have a lot to do with socialization (Christensen et al. 2007, p. 38).

159 Thompson (2003, p. 14) points out that this association represents a rather crude cut. Nonetheless, he assesses that "it is a robust enough one to enable a demarcation in the different logics of the three (…) modes of coordination and as concrete expressions of organizational difference." Traditionally, only hierarchies and markets were differentiated in organization theory. However, "it has now become standard practice to refer to markets, hierarchies and networks" (Herrhausen 2009, p. 72). The amount of research on network forms of organisation has increased considerably since the 1990s. In this regard, a debate developed whether networks have to be viewed as a combination of elements of hierarchy and market, or whether networks should be better understood as unique form in their own right. See, for instance, Powell (1990) and Rhodes (1996).

160 In organisation theory as well as in public administration studies, hierarchy is sometimes used interchangeably with bureaucracy. This is historically founded, as hierarchy became the preferential organisational form when public bureaucracies were organized (Peters/Pierre 2003, p. 5).

nation responsibilities at higher echelons (Egeberg 2003, pp. 117/120). Thus, to ensure coordination in a hierarchy, leader's control and rational calculation are key factors. Furthermore, the characteristic of hierarchical relations depending on command from superior to subordinate contribute to affirm the politically accountable government that lower levels of government carry out policies in a coordinated manner (Wollmann 2003, p. 595).

Characteristic of hierarchies is that they distinguish according to functional tasks and capabilities, called *horizontal differentiation*, and according to command and control, called *vertical differentiation*. At the same time, *functional differentiation* ensures that a specialization can occur by grouping individuals with the same functional expertise in the same functional unit. Different functional units are connected by a common superior authority and thus ordered in a sequence of superordination and subordination of power which leads to a great degree of vertical coordination. In addition to its characterization by superior and subordinate positions, hierarchies entail various vertical levels. The superior level is responsible to instruct, delegate and command by using downward channels of communication. Hierarchies are characterized by a high level of specialization, division of labour as well as routines, rules and procedures in written documents (Christensen et al. 2007, p. 24). If there is a need for coordination, the impetus has a top-down direction in which central administrative and political figures take the lead in creating the needed coordination among organizations. Consequently, commitment and support by organizations involved is less significant than authority relationships and the hierarchical approach to coordination presents "a highly centralizing way of solving the problem" (Peters 1998b, p. 17). Coordination can be considered in terms of administrative and regulated measures, which "require 'political' decisions in the first instance, which are then followed up by directive action to implement those decisions" (Thompson 2003, p. 23). Hierarchical models are thus marked by clear lines of accountability and supervision. Coordination in hierarchies is realized through characteristic features such as vertical integration, clear-cut areas of authority and leadership as well as an emphasis on rule, routines and procedures (Calabrò 2011, pp. 8-11).

At the other end, the *market* as the second ideal-type organization form consists of an indeterminable number of organizations without any organizing or directing centre resulting in largely decentralized decision making (Thompson 2003, p. 24). Each organization pursues its own activities. They are not subordinated to each other while each is entitled to engage. Relations are largely impersonal and based on the chance of making an exchange that is perceived to be advantageous (Peters 1998b, p. 19). Linkages and coordination occur irregularly and only for the time acutely needed. Coordination is primarily the outcome of the rules of the market and the competitive process between organizational entities. In markets, coordination is based on *competition* and *exchange* between actors, aiming at creating performance incentives. Thus, organizations are motivated by considerations of survival and growth and act out of their own self-interest in policy-making and implementation (Herrhausen 2009, p. 74). Coordination in market-like organization forms takes place in the self-regulating nature of the market setting. In an ideal-type abstraction, the market model is thus fundamentally based on the actor's willingness to compete, trade and exchange resources (Wollmann 2003, p. 596).

Advocates point out that the adoption of market principles would make the government more efficient and reduces costs (Calabrò 2011, p. 11).[161] Contrariwise, difficulties that occur in the flexible market organization form mainly result from a reduction of accountability and poor responsiveness to echelons of government. Furthermore, it has been pointed out that efficiency should not be the main value of public sector organization but effectiveness and accountability (Peters/Pierre 2003, pp. 5/6). Besides, from a strict organizational perspective, by introducing more actors into the respective policy arena such a break up into a market form of organization makes coordination and the steering of such a system in a coherent direction rather more difficult than less. Accomplishing a holistic approach becomes even more complicated as competition promotes self-interest instead of common interest (Calabrò 2011, p. 11).

161 Over the past decades, reforms in public administration focused on managerial characteristics of government and pressures on the public sector to mirror the way the private market sector conducts business grew. Efforts to make government more efficient concentrated very much on the role of market forms of organisation as examples for good management and an alternative to more traditional hierarchical forms (Peters/Pierre 2003, pp. 5/6). Government reforms since the 1980 focused on improving organizational capacities and on making the government productive (Bourgon 2011, p. 94). In this context, the topic of *New Public Management* (NPM) became increasingly important in theory and practice. The increasing significance of NPM scholarship followed the widespread perception that "government needs to give more attention to achieving effective management, often through the adoption of management procedures or arrangements resembling those of business firms" (Rainey 2003, p. 11). NPM themes involve advancing the productivity of government activities by professional management, by using economic market like strategies, the decentralization of responsibilities and a focus on output and outcomes by introducing explicit standards and measures of performance (Dimitrakopoulos/Passas 2003, p. 445). The NPM key message is that these market forms of coordination should replace hierarchical coordination and that organizational models used by private organizations should be adopted by public organizations. NPM assumptions have been followed by a reform wave in Western states but also in developing and transitional countries (Verhoest/Bouckaert/Peters 2007, p. 226). See also Wollmann (2003, p. 596), Christensen (2004, p. 24) and Christensen et al. (2007, p. 128). Some scholars see WoG initiatives as a reaction to the downsides of NPM reforms. See, for instance, Christensen and Lægreid (2006). Joined-Up Government in contrast to NPM indeed focuses on the holistic character of complex matters that includes a social and cultural dimension in addition to insights from economics (Bogdanor 2008, p. 1).

The *network* as the third ideal organization form known by organization theory and public administration studies can be used in various contexts.[162] As an analytical dimension, networks are classified as an organizational entity with loosely linked elements. Single elements are autonomous and not fixed together (Hofmann/Türk 2006, p. 3). A network "exists when organisations, though independent, repeatedly transact with one another and thus form permanent, albeit loose ties" (Herrhausen 2009, p. 74). They are built up and maintained voluntarily and rest upon shared values, trust, and solidarity. Thus, "[n]etworks are heavily dependent upon consensus, and assume that all the actors involved in the network are in basic agreement about the issues as well as the appropriate policies for the area in which they are involved" (Peters 1998a, p. 17). Each organization administers its particular service or product to the output of the entire network without a particular hierarchy (Wollmann 2003, p. 595). Thus, a network keeps together without organizational authority because the organizations' individual goals match the overall common goal of the network.[163]

Coordination in networks is very much based on social control mechanisms such as reputation or organizational culture that can over time become structurally tightened. These informal coordination mechanisms a network relies on lie in rich interaction on the personal level such as 'ringing up', talks at lunch, or 'dropping in'. Consequently, organizations in networks have more than just the functional information about other network partners; they also know about working styles, resources and potential cultural constraints (Herrhausen 2009, pp. 74/75). Factors that facilitate coordination in a network are fixed principles of interaction, common tacitly understood rules, ideas and values that can conceptually be described as an organizational culture.[164] Networks occur spontaneously and rely on personal contact in flat organizational structures with formal equality between actors. Coordination of activities is realized through the identi-

162 For instance, metaphorically used, the notion merely presents any non-hierarchical structure.

163 Network governance perspectives have gained high explanatory prominence, especially concerning EU governance. However, these network focused approaches often implicitly assume that networks and the respective forms of coordination represent the most characteristic feature of the emerging organisational structure of the EU and generally present the solution to complex situations. Approaches solely focusing on network governance merely concentrate on positive effects of horizontal coordination and partly leave aspects of efficiency unexplained (Schröder 2006, pp. 4/5).

164 Organisational culture can be embodied in certain coordination routines and a shared understanding of basic principles, assumptions and expectations. Furthermore, occurrences are interpreted in similar ways and translated in similar rules of behavior. Organisational cultures result out of long-term repeated interactions and are sustained by an institutional infrastructure (Herrhausen 2009, pp. 85/89). Varying organizational and working cultures manifest themselves in different attitudes towards coordination and planning procedures (Wollmann 2003, p. 595).

fication of common purposes and interest to work for a collective result. Loyalty and a problem-solving attitude are important attributes. Networks favor positive coordination while at the same time negative coordination is likely to work in network structures as much in hierarchical structures (Scharpf 1993, pp. 156/157). As outlined, one main feature of the network is that its individual organizations share an overall objective. Actors participate in a network to achieve this common goal, though they may have distinctive approaches when trying to achieve this goal or to solve collective problems (Schröder 2006, p. 4). Network-type coordination mechanisms focus on incentives for other actors and in doing so rely on soft-power coordination mechanisms such negotiation and persuasion (Herrhausen 2009, pp. 93/94). Coordination in networks is largely based on mutual interdependence. However, to interlink with other actors, individual organizations firstly need to be aware of the functional differentiation and the resulting interdependency with other organizations. To achieve this there must be some relation between the organizations as well as familiarity and awareness about other organizations. Networks then have to maintain these interorganizational links, for instance, by manifesting channels across which information and resources are exchanged. Communication in general plays an essential role in networks.[165] Hierarchies, markets and networks as the three ideal-type organization forms are shown in figure 6 below.

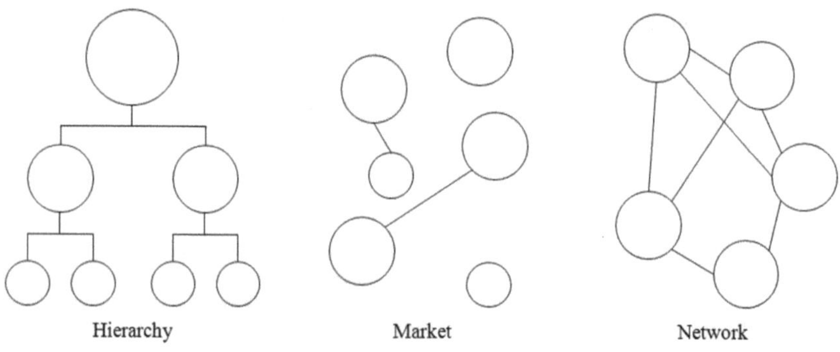

Figure 6: Ideal-type Organization Forms *(Source: Adapted from Herrhausen 2008, p. 76).*

165 However, as Thompson (2003, p. 232) points out, paradoxically at the same time networks bridge boundaries it also creates new ones. Thus, it should be kept in mind that "[a] network is a way of reducing the effects of certain boundaries by creating other ones" (Thompson 2003, p. 232). As actors orient themselves in the same direction, interdependence between them increases. At the same time, other alignments are weakened and loosened.

These three modes of organization can be conceptualized by linking hierarchy with the category of command and authority, market with competition and price, and network with cooperation, consensus and mutuality (Thompson 2003, pp. 13/14).[166] Thinking about organization theory in the political domain usually takes place via the view on a hierarchical set of layered governmental institutions and various forms of government bodies such as a set of central government bodies, administrative departments, the government itself, the Cabinet, and also the parliament. Organized hierarchically, these central government institutions possess complex relations of subordination and superordination between them. Against this traditional conception, networks stand for parallel, somewhat different mechanisms that are produced through negotiation, committees and agencies, experts and specialized policy analysts (Thompson 2003, pp. 150/151).

Regarding the different organization forms of hierarchies, networks and markets, scholars of organizational studies extensively examined questions that deal with the choice of organization form or the connection of the individual organization form with its environment. For instance with regard to networks, Powell (1990, p. 322) argues that "the open-ended quality of networks is most useful when resources are variable and the environment uncertain." In cases problems or conflicts arise, organizations in a well-functioning network access joint problem solving measures. In networks, "the achievement of a goal is more important than the completion of the transaction according to previously specified parameters" (Herrhausen 2009, p. 78). Furthermore, networks better qualify to achieve coordination between conflicting interests by relying on trust and collaboration than hierarchies or markets. In cases problems or conflicts arise, organizations in a well-functioning network access joint conflict resolution measures that allow for problem solving and positive-sum-solutions (Wollmann 2003, p. 595). On the contrary, actors in market type organization forms would end interorganizational coordination. In contrast to hierarchies, where splitting divisions result in moving coordination upwards, network participants are expected to push processes of coordination downward. Coordination is based on flexible horizontal mechanisms, and allows for articulation of interests, compromises and negotiation between actors (Schröder 2006, p. 4). As Powell (1990, p. 303) notes "[n]etworks are 'lighter on their feet' then hierarchies". Compared to hierarchies, networks are more flexible as interactions of organizations are not subordinated to fixed rules or timeframes but on a voluntary basis in the pursuit of their shared goals (Paris 2009, p. 61). Compared to the anarchy of the market and the centralized and rigid hierarchies, network forms of organization are seen

166 Verhoest, Bouckaert and Peters (2007, p. 332) provide a good overview on different coordination forms in the three types of organization.

better suited to meet the challenges of complexity (Calabrò 2011, p. 8). Network approaches are particularly needed "to support government-wide priorities, to bridge policy domains or to compensate for the disaggregation that the fragmented approach among public agencies may cause" (Bourgon 2011, p. 49).

However, in reality there are plenty of mixed organizations that can exist in parallel in complex sets of relationships in different setting or overlaid on each other (Ranade/Hudson 2003, p. 36). As Perrow (1986, p. 255) notes, the strict contrasting of organization forms may be delusive. For instance, though being neither organized like a market nor being formally sanctioned by hierarchical structures, networks may partly involve these mechanisms (Thompson 2003, pp. 30/31).[167] While this seems to overdraw above elaborated distinctions, different organization types nonetheless remain a valuable tool. First of all, "theories simplify" (Perrow 1986, p. 258) and secondly is not possible to discuss the imperative of coherence and coordination in the CA context without structuring its organizational modes. Consequently, modeling organization along these three forms is a simplification and certain elements and coordination characteristics can coexist with each other and overlap (Bourgon 2011, p. 48). Organizational form and structure can be reorganized through mergers, the division of organizational entities or by moving them horizontally or vertically. Reorganization also occurs by changes in procedural rules. Furthermore, collegial organization forms such as committees or boards set up in addition or instead to hierarchical leadership can be used to cut across organizations and improve coordination (Christensen et al. 2007, p. 26). Especially in processes of reorganization, the importance of informal norms begins to show. Such reform processes that involve and reorganize organizational entities threaten dominant informal norms and often resistance and opposition to reform is formed (Christensen et al. 2007, p. 44).

According to organization theory, the way coordination evolves depends on several factors in addition to the form of the organizational system such as the nature and consistency of the respective organization or the organizational context and task environment. In this context, Egeberg (2003, pp. 117-119) has developed criteria for selecting explanatory factors being relevant for behavior and action of organizations. He refers to institutionalization, organizational demography, the organizational structure and the organizational locus as the four key variables of an organizational perspective. Institutionalization means that organizations add characteristics and include informal norms and practices that go beyond the technical requirements of the task at hand. To become an institution the organization develops a distinctive, separate identity with its own values and

167 "There are strong elements of markets within hierarchies. On the other hand, markets have strong elements of hierarchies within them" (Perrow 1986, p. 255).

ways of acting. Organizational demography deals with the composition of an organization in terms of basic factors such as education, ethnicity, age and sex. Studies on organizational demography closely relate to socialization of individuals. The organizational locus refers to the physical dimension of organizations. Features of an organization's location and physical space affects contact patterns and coordination behavior. Overlapping organizational structures that are locally separated are mostly also disjoined time wise and evoke different role perceptions and identities.[168]

Organizing from the perspective of formal organizational structure and design also occurs through *specialization* (Christensen et al. 2007, p. 24). As Mulford and Rogers (1982, p. 11) note "[i]t is interesting (...) that, as the environment becomes more complex, organizations become more specialized; and this increased specialization leads to a greater need for increased (...) coordination". Specialization takes place in a horizontal form as the divide of organizations at the same level, or in a vertical form as a split of responsibility on hierarchical levels. *Horizontal* and *vertical specialization* demonstrates how different areas are decoupled or disconnected from each other. Organizational reforms often have to cope with the balance between specialization and coordination. If actors concentrate more on coordination, fragmentation is reduced. On the contrary, with a focus on specialization, fragmentation may be increased.[169] Horizontal specialization is expected to need less control, to make it easier to reach goals and to ensure accountability. On the downside this can lead to higher levels of fragmentation and require more coordination, attention and resources from the top of the system (Verhoest/Bouckaert/Peters 2007, p. 226). Government organizations can also be vertically specialized into a hierarchy of separate institutions. Vertical specialization tends to weaken and reduce political steering and control

168 Classic organization theory generally distinguishes between institutions and organizations. In organizational research, "all institutions are organizations, but not all organizations are institutions" (Egeberg 2003, pp. 118/119). In an institution, norms grow in a gradual process, while an organization is characterized by formal, mechanically developed norms and instruments (Christensen et al. 2007, pp. 37/38). From this perspective, a formal organization that develops additional informal norm thereby acquires institutional features and becomes an institutionalized organization. However, in the following analysis the term institution is used as a general term describing the complex of ministries, departments, agencies, organisational entities, formal rule structures which forms the organizational means of public services (Bogdanor 1987, p. 504).

169 To avoid misunderstanding, it should be noted here that the perceived problem to which greater coherence and coordination are proposed as solutions is not specialization as such. Rather, than the trade-off between specialization and function, it is fragmentation or inadequate structures that are meant. As 6 (2008, p. 52) observed, "[d]espite the rhetoric sometimes used of 'overcoming barriers' or 'breaking down the boundaries' (...), it is always about attempts to put boundaries in different places to create border crossings suitable for particular vehicles."

capability and can result in a loss of political direction as the distance between the different levels increases. This can be compensated organizationally either by duplication or by integrating and transforming a unit into a ministry department (Egeberg 2003, pp. 122/123). An example of vertical specialization is the so called agencification where a subordinated organizational unit is separated from a ministerial department.[170] Other vertical specialization efforts include the creation of centralized mega-departments that are supposed to increase efficiency (Radin 2003, p. 610). The decentralization of tasks to semi-independent bodies that are situated outside the core of the politico-administrative system means that they become less hierarchical and more fragmented (Martens 2010, p. 3).

2.3.2.3 The Comprehensive Crisis Management Context

Above outlined theoretical insights allow explaining formal organizational dimensions of crisis management coherence and structures that aim at enabling the CA. By emphasizing the importance of formal structuring, this approach offers an understanding of the organizational setting of CA implementation and how structures can be formed and mobilized for action on behalf of comprehensive crisis management. The approach helps to conceptualize the implications for crisis management organization and coordination stemming from the CA concept (Schröder 2006, p. 5).

When approaching questions of organizing comprehensive crisis management, the application of this theoretical approach needs to consider factors which are directly relevant to the organization of crisis management coherence. A potential limitation which should be kept in mind considers that this theoretical approach

170 Agencies differ across national cultures, legal and political systems in their size, function, rationale for creation, funding, legal form and relationship to the respective ministry (Martens 2010, p. 3). A detailed debate and clarification on the term can be found in Christensen et al. (2007, p. 131) as well as Bach, Fleischer and Hustedt (2010, pp. 10-14). While integrated bureaucratic models combine autonomy and task specialization in a narrow range of policy issues, agency structures result in a twofold organizational division. "[T]here is separation both on a vertical dimension between the politico-administrative leadership and the agencies and on a horizontal dimension between different agencies responsible for different tasks" (Martens 2010, p. 5). With regard to coordination, single purpose agencies can improve the coordination of different actors in the respective policy field, but may also neglect policies and objectives outside the organizations main purpose. As every agency is likely to be largely interested in accomplishing its own performance, it may tend to ignore or even counteract the objectives of other agencies, and hence to hinder coordination (Wollmann 2003, p. 600). Agencification is said to raise the importance of professional considerations and to diminish political interference in administrative policy-making (Martens 2010, p. 5).

does not offer completely transferable analytical concepts and models. A restriction in this sense, for instance, is that though improved coordination in crisis management is said to avoid the waste of resources, incentives in form of benefits compared to business organizations are not immediate. Furthermore, it needs to be considered that it might be the search for organizational legitimacy rather than criteria of effectiveness and efficiency which push organizational change and restructuring in this policy field. Political factors also in the form of symbolic reform efforts "strongly condition institutional reform and organizational learning" (Bossong/Benner 2010, p. 1081). The most significant implication of the CA and the expanded security concept from organizational perspective is the impact of comprehensive actorness on interorganizational processes between the actors within IOs or states involved in international crisis management. Actors have to redefine their position and situate themselves within the new conditions of the crisis management set-up (Gebhard/Norheim-Martinsen 2010, p. 14). Complexity plays a major role in organization theory and public administration studies which acknowledge the need of a holistic understanding and multifaceted approach as well as the contribution of multiple actors. Complexity displays a dynamic due to intertwined interactions across systems that "can only be resolved by looking at the system as a whole" (Bourgon 2011, p. 21) and with an understanding of the whole of environment (Bourgon 2011, pp. 21/22).

To encourage successful policy implementation from public administration and organizational perspective stands for inducing coordination. It is important from the organizational perspective taken in this dissertation, to organize systematic attention around organizational reforms. Motivation for coherent action in government ministries and departments in this regard are authority, at least one organization that feels obligated, common interest and exchange. Hierarchy as a means for the CA signifies the creation of an identifiable coordinator with the authority to modify inputs and activities of actors (Hood 2008, p. 28). While hierarchically formed organizations can rely on authority to secure coordination, this incentive hardly ever occurs in the interorganizational context between government ministries. Therefore, coordination across organizational borders must to a greater extent rely on the other incentives. This means to find or motivate common interests between organizational units and to develop and maintain exchange between them (O'Toole 2003, pp. 237-238).

CA implementation in an institutional setting that requires the coordination of various organizations poses a particular challenge due to different routines, specialized languages and differing and probably conflicting views. O'Toole (2003, p. 236) identifies several factors that contribute to facilitating policy implementation to structurally complex situations that seem especially relevant with regard to comprehensive crisis management. To maintain or increase common interest, firstly, it is important that the reform policy background and relevant facts are clear to all involved and that everyone shares the commitment, so called *signaling*. Another factor is the *framing* of interorganizational efforts that highlights key points of common interest and focuses the respective organization's perceptions. *Iteration* is the third factor that can help to encourage interorganizational coordination by predictable and repeated interaction to increase understanding (O'Toole 2003, p. 240). *Exchange* as the other "(…) kind of interorganizational 'glue' that can congeal cooperation (…)" (O'Toole 2003, p. 241) can refer to the use of pooled funds to strengthen coherent action as the most evident kind of exchange to consider. But exchange can also take the form of input that can range from political support, human resources to information (O'Toole 2003, p. 241).

Organization theorists view organizations as open systems that respond to changes in their environment, develop new mechanism and can change their goals accordingly. Viewing organization's operational mode reflects external pressures and demands (Christensen et al. 2007, p. 9). However, as Patrick (2009, p. 62) pointed out, there is an awareness among scholars of organization theory that "institutions of government are notoriously resistant to change". He identifies three key ingredients institutional innovation thus largely relies on. Firstly, a major policy failure that questions the usual working mechanisms and procedures facilitate reforms. The emergence of a new and alternative set of beliefs provides the second trigger. Thirdly, political support of a well placed champion is needed to achieve organizational change of government organizations (Patrick 2009, p. 62). Also Olsen (1991, p. 94) observed that changes in organizational identity and procedures mainly result from a performance crisis in terms of a "gap between the level of aspiration and the level of performance" (Olsen 1991, p. 94).[171] These triggers are also helpful to look at in the analysis of CA implementation.

Organization theory and assumptions drawn from public administration studies provide the basis from which to analyze coherence of European crisis management and which offer insights into the means of coordination in this regard. Looking at the idea of comprehensive crisis management through the three main organization forms provides a useful way of mapping the efforts applied to achieve crisis management coherence. They additionally allow examining to which extent crisis management coordination structures match each other and which organization form and type of coordination proves to be suitable to effectively implement the CA in the European crisis management system and reveal potential advantages of specific models of CA implementation.

171 Olsen (1991, p. 94) also observed that organizations often pay more attention to their failing to meet targets than to successful activities.

However, this dissertation assumes that European comprehensive crisis management cannot be analyzed by looking at either the EU level or the member state level of crisis management. Activities must be coordinated at least at two levels: the EU level and the domestic level. EU and national organizational crisis management structures are considered as part of the same system and the organizational approach is applied to examine organizational CA features at both levels and to unpack links between different actors in the European crisis management system. To be better able to analyze vertical coherence, the approach is complemented by the concept of Europeanization.

2.3.3 Europeanization

An organizational perspective that looks solely at the capacity of actors to act according to CA principles does not capture all factors relevant in European crisis management. The question on the international sphere can be better grasped with reference to the role of IOs on policy change at the national level. A country that is internationally active has to consider which IOs are suitable to use for receiving input and for directing outputs it is ready to make. Thus, while arguments drawn from organization studies and public administration serve as the organizing theoretical framework to identify and analyze ways the CA is implemented, they are analytically complemented by assumptions from the concept of Europeanization.[172] The increasing role of the EU in international crisis management and its respectively growing relevance in the crisis management policy of European states is the broader context for the following theoretical considerations that serve as an additional theoretical framework and analytical lens to take the EU dimension into account and to underline the EU perspective of this dissertation.

2.3.3.1 Overview

European states increasingly have to include an EU dimension in their crisis management policy and have sought to advance their performance through conceptual, institutional and procedural changes. However, notwithstanding the

172 As a matter of principle, Europeanization is understood as EU-ization. While keeping in mind that there are other institutional tools and international organisations than the EU at the disposal of European states in matters of international crisis management, the analysis relates to the impact of evolving EU crisis management.

rising importance of EU crisis management institutions, the intergovernmental character of EU foreign and security policy and the distinctive nature of international crisis management mean that also the role of nation states continues to be a determining factor in the implementation and formulation of EU crisis management policies (Gross 2009, p. xii). As Kassim (2001, p. 3) argues from a public management perspective, the capacity of the EU and "its ability to implement common policies, crucially depends on national administrations." Hence, member states matter in European crisis management and at the same time it is impossible for them to neglect the EU crisis management context in their own policy.[173] Furthermore, rising EU engagement leads to increasing demands for coordination in member states' core executive. Though particularly in the context of foreign, security and defence policy there has been no formal transfer of sovereignty to the EU, it exerts an omnipresent influence due to the fact that states form a group which impacts on their behavior. "EU membership is normative, even without formal transfer of competences" (Houben 2005, p. 18).

The concept of Europeanization broadly relates to the impact of the EU on its member states and policy changes caused by this impact. The essential proposition of the Europeanization concept is that membership in the EU increasingly influences member state policy (Wong 2005, p. 152). The concept is used to examine the degree to which national policies and structures adjust to the development of EU policies and to assess the changes that occur as an answer to policies institutionalized at the EU level (Miskimmon 2007, p. 2). Analytically, the concept stands for processes in which European structures, mechanisms, rules and norms interact and thereby transform their domestic correspondents. Thus, Europeanization is characterized "by interrelationships between the various layers constituting the European multi-level structure" (Noutcheva et al. 2004, pp. 20/21). It stands for dynamics of political, societal and economic transformation in contemporary Europe and permits assessing the choices for or against a larger role of EU crisis management. The concept of Europeanization thus allows for conclusions on the place of the EU and European influences in national crisis management policies and CA implementation.

Depending on the research purpose different typologies of Europeanization exist and Europeanization as an analytical concept has been employed in a variety of ways, in particular, to examine the consequences of integration in communitarian EU policy areas.[174] Yet, theorizing Europeanization and modeling the concept's dynamics remains to be a major challenge (Vink/Graziano 2007, p. 7)

173 This is also underlined by Houben (2005, p. 7).

174 For an overview of developments in the Europeanization field, different approaches and relevant literature see Olsen (2002) as well as Bache, George and Bulmer (2011, pp. 59-61).

as it "has become the buzz word for everything related with the impact of the European integration process on its member states (and even on non-members)" (Torreblanca 2001).[175] Still, the concept has been widely applied in recent years which can be attributed to its utility in two areas: firstly, Europeanization considers parallel and interrelated processes at both the EU level and the national level. Secondly, the concept highlights the interface of institutions and identities (Wong 2005, p. 149). It is these two factors that make it particularly relevant in the context of this dissertation.

Starting with Ladrech's (1994) often cited article which investigates the Europeanization of politics and institutions in France, the concept has been used to embody transformations in the context of integration within the EU, its enlargement and the accession of new member states. Ladrech (1994, p. 69) defines Europeanization as an "incremental process reorienting the direction and shape of politics to the degree that EC political and economic dynamics become part of the organizational logic of national politics and policy making." It is understood as a process in which member states adjust their domestic policies as a reaction to the constraints and requirements of European institutions. A multi-state study by Cowles et al. (2001) focuses on the EU effects on two particular categories of domestic structure. Firstly, the impact of Europeanization on policy structures defined as the content of policies is investigated; the second part concentrates on system-wide domestic structures related to changes in the nation state, its society and economy (Risse/Cowles/Caporaso 2001, p. 5). The authors understand Europeanization "as the emergences and the development at the European level of distinct structures of governance" (Risse/Cowles/Caporaso 2001, p. 3). Mapping out an institutional approach, Olsen (2002) makes a distinction between five possible meanings of Europeanization. In doing so, Europeanization can firstly stand for changes in the external territorial boundaries of Europe, such as EU-enlargement; secondly for the development of institutions of governance at the EU level; it may in a third meaning also refer to pervasion of national and sub-national systems of governance; fourth, to the export of forms of distinctively European political organization and governance beyond the territory of the EU; and finally to a political project aiming at a unified and politically stronger Europe. To substantiate this survey in the framework of this dissertation, the defini-

175 This caused scholars to point out that the term Europeanization might be in danger of being overstretched and to warn that the term can easily become a catch-all explanation for any process of domestic change. Challenging debates about the concept's purpose and utility make it heavily contested as to its value for the analysis of European politics and for understanding the construction and operation of the EU. This view is particularly shared by Radaelli (2003), Wong (2005), Bache and Jordan (2006), Lodge (2006), Lehmkuhl (2007), Vink and Graziano (2007) and Bache (2008). This issue is elaborated in more detail below.

tion of Europeanization presented by Radaelli (2003, p. 30) is highly appropriate to complement the organizational focus on a general level. He refers to Europeanization as

> "processes of (a) construction (b) diffusion and (c) institutionalization of formal and informal rules, procedures, policy paradigms, styles, 'ways of doing things' and shared beliefs and norms which are first consolidated in the making of EU public policy and politics and then incorporated in the logic of domestic discourse, identities, political structures and public policies."

This description very well reflects the different analytical dimensions of this thesis.[176]

As above definitions show, a lot of studies of Europeanization stress EU policy processes and limit their analysis to Europeanization effects in EU member states.[177] However, some studies emphasize the export of particular features of EU governance beyond its borders and examine impact and mode of European integration dynamics. Grabbe (2006) for instance brings forward arguments for a strong Europeanization effect through conditionality on accession countries in Central and Eastern Europe. Also, Juncos (2011) examines Europeanization in relation to the case of police reform in Bosnia and Herzegovina. Other scholars focus on Europeanization as a theoretical framework for addressing conflict resolution. Concerning this matter, Coppieters et al. (2004) explore the relevance of Europeanization for the settlement of secessionist conflicts at the periphery of the EU. In this case, the concept of Europeanization is perceived as a process and change that is triggered by European institutions by attaching and conditioning the outcome of a conflict to the extent of integration into the European structures on the legal and administrative level as well as in economic, social and security dimensions. The EU takes on the role as an actor in the conflict or a framework for the conflict setting (Noutcheva et al. 2004, pp. 20-23). In the context of the studies by Coppieters et al. (2004) and Grabbe (2006) the concept of Europeanization is dealt with in terms of a foreign policy instrument of the Union.

Definitions on Europeanization given by Tonra (1999, 2001b), Bache and Jordan (2006) as well as by Wong (2007) are most directly relevant to also clarify its meaning in the given crisis management context. Tonra (2001b, p. 229) defines Europeanization in the field of foreign policy as

176 See the next chapter on the analytical dimensions utilized in this dissertation.

177 See for example works on Europeanization by Ladrech (1994), Tonra (1999, 2001b), Wong (2005) or Gross (2009).

> "a transformation in the way in which national foreign policies are constructed, in which professional roles are defined and pursued and in the consequent internalization of norms and expectations arising from a complex system of collective European policy making."

This definition underlines three aspects: the planning of foreign policy, the role of officials and the absorption of values and expectations. Wong (2007) considers Europeanization under CFSP more broadly as a process of mutual foreign policy convergence being a "dependent variable contingent on the ideas and directives emanating from actors (…) in Brussels, as well as policy ideas and actions from member states capitals" (Wong 2007, p. 323). Drawing a multidimensional model of Europeanization, he proposes that Europeanization evolves in three separate processes through which EU institutions influence national foreign policy: a top-down process of policy convergence explaining domestic adaptation and processes in response to EU demands; a bottom-up and sideways process referring to projection of national models on the EU level and to other member states; and Europeanization as identity reconstruction involving a process of foreign policy identity and interest convergence (Wong 2005, pp. 141/142; 2007, p. 325).

The works discussed above show that Europeanization has been approached in a variety of ways adding to the vagueness of the concept. Yet, as Featherstone (2003, p. 19) argues it "is precisely the breadth of application and the demanding explanatory framework needed that attests to the value and importance of the term." Indeed, also in this dissertation's context the contemporary reality requires an investigation that "acknowledges the dynamism, imbroglio, and limits to determinism in present-day Europe" (Featherstone 2003, p. 19). Europeanization as an analytical framework allows stressing key changes in the European crisis management framework in the broadest sense thus covering the different dimensions of complexity of this policy field and the different CA levels.

2.3.3.2 Analytical Key Factors

There are multiple examples of different combinations of these various perspectives in Europeanization research which resulted in a variety of debates.[178] The downside of the broad applicability of the concept is mirrored in the numerous efforts of scholars to put different approaches on Europeanization in order. These attempts are particularly helpful to set limits to the meaning of Europeanization in

178 See Wong and Hill (2012b, pp. 4-11) for an overview on different debated questions in the context of foreign policy Europeanization.

the context of this thesis. In this regard, Dyson and Goetz (2003, pp. 13-20) distinguish between two strands of Europeanization research. Though definitions of Europeanization are often narrowed to an article's or book's specific topic, in its initial form Europeanization has been considered as a vertical *top-down* process of downloading EU institutional structures and norms to the domestic level describing the first strand of Europeanization research.[179] Domestic change in this approach derives from the EU; travels down to the member state level and is then incorporated in the logic of domestic discourse, political structures and policies (Bache/Jordan 2006, p. 21). These studies emphasize the reactive nature of adaptation. Adaptation is understood as a shift toward greater consistency by the actors to maintain a balance between needs and demands. Poor adaptation on the contrary implies a retreat to national solutions or the focus on crisis management activities outside the EU framework (Miskimmon 2007, p. 5). While early studies concentrated on the above downward direction from the EU to the domestic member state level, later on the *bottom-up* process of national projection with member states also having a major impact on European policy formation was increasingly highlighted. The unidirectional perspective on Europeanization was challenged by the view that sees Europeanization as a circular two-way interaction between the EU and its member states, compromising both top-down and bottom-up dynamics, which is the second variation of Europeanization research.[180]

In a similar attempt to systematize Europeanization, Bache and Jordan (2006, pp. 20-23) differentiate four supplementary ways in which Europeanization can be organized in more detail. Besides highlighting the top-down impact, the bottom-up process or two-way interaction between the EU and its member states as explained above, Europeanization is, firstly, categorized as the accumulation and creation of policy competences at the EU level. In this approach, Europeanization is used almost synonymous with European integration. The second category views Europeanization as the growing importance of the EU as a reference point for national and sub-national actors. Considering the EU as an arena for interstate communication, Europeanization thirdly deals with the horizontal transfer of concepts and policies between member states. Finally, Europeanization can also be accounted as a discursive frame of reference for national actors. Grouping Europeanization's different meanings, Wong (2005, pp. 136-140) also identifies five categories. Besides the widely received conceptions of the concept in terms of top-down national adaptation and bottom-up national projection he

179 Works by Ladrech (1994) and Kassim (2001) fall in the category of first generation research on Europeanization.

180 For a detailed exploration of the top-down and bottom-up terminology in research on Europeanization see Lodge (2006) as well as Bache, George and Bulmer (2011, pp. 61-64). See also figure 7 below.

underlines *identity reconstruction* as a third meaning, which deals with the redefinition and convergence of identities and interests in the European context. According to Wong a fourth meaning of Europeanization is the political, economic and social *modernization* in the context of EU enlargement. *Policy isomorphism* as the fifth meaning highlights the degree of convergence.[181]

Analyzing the link between member states and the EU as an interactive two-way relationship also focuses on the European impact on interests, values and ideas. Europeanization in this context also stands for the adaptation to potentially new normative frameworks (Wong 2005, pp. 140-142).[182] Such conceptions follow the observation that Europeanization in the political and administrative world is put into practice and carried out as both top-down and bottom-up (Lodge 2006, p. 67). An incentive for EU member states to upload "is that if they can get their existing policy preferences accepted as the preference of the EU, they will have less trouble adapting to the EU policy when it comes into force" (Bache/George/Bulmer 2011, p. 61). They do so, by in particular controlling initial phases of policy development and aiming at convincing other actors of their preferences. Thus, member states can initially set off the policies they later adapt to in a process of uploading and downloading. Key issues of the concept are mechanisms of change due to Europeanization and the causal processes through which EU level processes lead to domestic change. One line of analysis points out that adaptation problems particularly emerge, when EU requirements *misfit* with member state domestic practices. In particular, Cowles et al. (2001) emphasize the importance of the degree of fit between EU level changes and existing member state policies. Such approaches stress that the degree and nature of adaptational pressures due to Europeanization depends on the *goodness of fit* between the domestic level and the European level. Compatible structures constituting a good level of fit mean weak pressure to adapt while a poor level of fit implies strong adaptational pressure (Risse/Cowles/Caporaso 2001, pp. 6-9). In general, the conception of misfit applies more to the top-down accounts of Europeanization (Bache/Jordan 2006, p. 18). Rather than characterizing Europeanization to matters of misfit only, Lehmkuhl (2007, p. 340) ascribes the concept an additional "sensitivity to institutional and ideational factors". This reflects that while some Europeanization analysts concentrate on the formal and observable

181 Knill (2006, p. 5) relates isomorphism in Europeanization to its original definition "as a process of homogenization that 'forces one unit in a population to resemble other units that face the same set of environmental conditions'". The notion refers to mechanisms through which organization become more similar. While isomorphism thus describes the tendency to become alike, diffusion, on the contrary, deals with the spread of ideas (Eriksson 2006, p. 56).

182 This approach to Europeanization is also advocated by Bache and Jordan (2006, p. 22), Lehmkuhl (2007, p. 353), Miskimmon (2007, p. 6) and Gross (2009, p. 16).

consequences of EU membership, such as changes in political-administrative structures, others emphasize the less formal content of national policy such as internalized European norms and policy paradigms (Bache/Jordan 2006, p. 25).

Depending on the impact of Europeanization pressures, different degrees of domestic change can be measured. Börzel and Risse (2003, pp. 69/70) categorize three degrees of domestic change: *Absorption* is the lowest degree of domestic change. Though including European requirements into their structures, existing national processes, policies and institutions are reframed only superficially and not substantially modified. *Accommodation* goes beyond the level of political rhetoric and occurs when member states transform existing policies and institutions while leaving their essential features intact. The overall adjustment is modest. *Transformation* is the highest degree of domestic change. It takes place when member states fundamentally alter or replace their existing policies, processes and institutions. This differentiation and a wide subsequent debate whether the presence of a misfit is a sufficient condition for Europeanization has triggered the greater discussion about the role of different intervening variables of Europeanization. Bulmer and Radaelli (2004) relate Europeanization to different modes of governance and hypothesize that in areas with a clearly defined EU-model flowing downwards from the EU level, greater pressures for domestic change are generated. This *governance by hierarchy* is marked by a high degree of power delegated to supranational institutions. On the opposite, in less hierarchical areas of EU policy where national governments are the key actors, such as EU foreign and security policy, Europeanization tends to take on a less coercive form with less pressure for domestic change which they call *facilitated coordination*. Besides distinguishing between governance by hierarchy and facilitated coordination as settings in which Europeanization occurs, Bulmer and Radaelli (2004) also identify *governance by negotiation* that relates to the process of agreeing to the rules at the EU level.[183]

Another serious analytical issue is the nature of the dependent and independent variable as Europeanization "blurs the boundaries between cause and effect" (Wong 2007, p. 330). Bache and Jordan (2006, pp. 29/30) identify four ways to disentangle the EU's effects from other causes of domestic change. The first suggestion is to conduct a careful historical analysis of connecting change and EU-membership. Secondly, using a bottom-up approach that starts on the domestic level to trace changes back might help disentangling causes. On the opposite, the third approach asks, what would have happened without the EU. A fourth way to handle the issue is to compare EU member states with non-

183 A problematic aspect in this approach is, however, that not all EU policy areas can be categorized in such a definite and clear-cut way (Bache/Jordan 2006, pp. 24/25).

members. It is thus challenging to assess the impact of Europeanization on each individual level as the concept "makes for neither a neat independent nor intervening variable when explaining domestic change" (Noutcheva et al. 2004, p. 23). EU policies and institutions can be considered "both as modifying policy preferences at the national level while at the same time originating at the national level, and therefore can potentially serve as a dependent or an independent variable" (Gross 2009, p. 18). In practice, these Europeanization processes are interlinked in ongoing and mutually constitutive processes that are characterized by a certain degree of flexibility that makes it problematic to prescribe concrete templates or specific forms of domestic adaptation (Miskimmon 2007, p. 9).[184] In particular in the crisis management context, other variables at the global, national and sub-national level interact in complex manners, which make it difficult to claim pressures from Europeanization as the dominant causal variable.

2.3.3.3 The Comprehensive Crisis Management Context

Above overview of the Europeanization concept and research shows that further limitation in the particular context of this study is necessary as Europeanization depends on the sector concerned (Eriksson 2006, p. 53). Thus, aspects of Europeanization need to be applied to the specific subject of international crisis management. While at first mostly been used for studying various EU supranational policies and their impact on member states, the concept of Europeanization in recent years has also been applied to the EU's foreign and security policy. However, compared to the broad interdisciplinary use of the Europeanization concept in 'classic' European policy domains,[185] only few scholars tried to establish its applicability in this policy area that frames international crisis management (De Flers/Müller 2012, pp. 20/21).[186]

The major difficulty occurring in this regard is that foreign and security policy remains a field of activity and responsibility of sovereign national governments and thus primarily knows states as actors. Apart from the debate on the actual existence of its foreign and security policy,[187] the EU does not have a centralized

184 See also Gross (2009, p. 18).

185 See, for instance, the empirical chapters in Cowles, Caporaso and Risse (2001) which focus on transport policy, telecommunications, and environmental policies.

186 See studies by Tonra (2001b), Wong (2005, 2007), Rieker (2006a, 2006b), Miskimmon (2007), Gross (2009), Jokela (2011), De Flers and Müller (2012) as well as contributions on different countries in Wong and Hill (2012a) which are among the few. See Wong and Hill (2012b, pp. 2-4) for an overview of the concept of Europeanization in foreign policy studies.

187 See chapter 3.1 which takes this debate up.

state-like decision-making executive in this policy area. The intergovernmental nature of the EU's CFSP does not create adaptation pressures of the compulsory and legally binding type like supranational policy areas. Hence, the adaptation to EU crisis management structures is looser than in other EU policy areas due to the lack of a concrete institutional model. European crisis management institutions only have a limited ability to change domestic structures as they lack robust institutional structures (Daehnhardt 2012, p. 40). "Member states themselves (…) shape the decisions they are later affected by" (De Flers/Müller 2012, p. 21). When it comes to foreign and security policy it is evident that the European dimension and EU level crisis management is comparably far from replacing the national one.

But still, the EU crisis management dimension does exist and is incorporated in the given national state composition depending on the degree of resonance the EU meets. Moreover, it gives member states a comparatively high room for manoeuvre. Domestic and EU crisis management policies, institutions and structures are becoming increasingly mutually constitutive. Furthermore, EU foreign and security policy has become part of the European integration process due to the very fact of institutionalizing CFSP and CSDP structures (Gross 2009, pp. 14/15). Consequently, Europeanization can be considered a relevant concept in the area of international crisis management and can serve as a dimension and category of international crisis management analysis in Europe (Gross 2009, p.19). Though acknowledging challenges above, it is assumed that the EU has a decisive influence on member state crisis management. The EU does have an impact in this field and is active as an actor itself and through its member states (Wong 2005, p. 144). The Europeanization approach actually allows for a middle path that recognizes that member states adapt to EU crisis management structures and principles, while also accepting that at the same time it is them actively shaping these structures and norms. Furthermore, considering the broad view on crisis management taken on in this dissertation, the EU's humanitarian assistance and development policy are part of the supranational European Commission policies. Above conceptualization facilitates the capturing of Europeanization outside the community area and is particularly motivated by the character of the empirical parts of the dissertation.

Yet, the conceptualization of Europeanization in international crisis management does pose additional challenges. As mentioned above, an issue that is strongly contested when employing the concept is how to dissolve the influence of the EU from other triggers of domestic change like the effect of globalization or important national developments in the respective policy area. The impact of Europeanization is just one of the many effects in domestic policies and processes of EU member states (Lehmkuhl 2007, p. 342). This matter is of particular thematic importance as the EU is just one of several international crisis management actors that its member states are actively involved in. Consequently, the analysis carries the danger of being biased toward EU level explanations, where in fact other factors caused shifts in the political systems of the member states. It is critical to bear in mind other important international actors that states interact with in the context of international crisis management, in particular, other multilateral organizations such as the UN, NATO and the OSCE, and other factors in the international security environment. Other crisis management actors either complement or compete with the processes of Europeanization and the receptiveness to particular processes varies from each member state. Taking this problem into account is important with a view to analyzing crisis management Europeanization and to conducting the aimed at analysis on a realistic level. What makes Europeanization different in this regard, according to Ladrech (1994, p. 71), is its geographic delimitation and the distinct nature of the framework member states provide to mediate adjustments. However, in the area of EU crisis management including the largely intergovernmental CFSP and CSDP, Europeanization processes may involve the EU without being directly caused by it. The EU, in this regard provides an arena, a framework and a reference point for actors (Aktipis/Oliver 2012, pp. 90/91).

This thesis tries to strike a middle path which "accepts that member states adapt to CFSP decision-making structures and norms, while at the same time recognizing that these same member states are themselves actively involved in creating these structures and norms" (Wong 2007, p. 323). While assuming the continuing existence of national foreign and crisis management policies, the essential questions center on the extent to which EU crisis management has led to change in procedures and structures of national crisis management policies. The question in this regard is not how the EU matters but *to what extent* it matters in member state international crisis management and when trying to implement the CA.

For the purpose of the analysis, the scope of application of the concept of Europeanization has to be restricted in several ways. Most Europeanization scholars look at specific mechanisms of Europeanization like political incentives or normative pressures (Grabbe 2006). The following analysis is primarily concerned with questions of conceptual and structural adaptation of member states to EU crisis management and with the extent to which crisis management policy choices presented by the EU are compatible with those of the member states. As the concept of Europeanization is adopted to complement insights from organization studies, its conceptual application stresses the reorientation in the organizational logic of national crisis management policy making. It focuses on the reorientation of national policies by way of organizational change though EU membership and international crisis management options. Changes in the organizational logic in this regard apply to the adaptive processes of relevant institutions such as ministries to EU crisis management and responds to changes that occur in crisis management principles and institutional design. Europeanization is understood as a shift of attention of national crisis management institutions and their increasing participation in the EU decision-making cycle. It is analyzed whether the EU is a relevant and important factor leading to adaptation and change in domestic institutional arrangements. Thus, the dissertation considers the impact of Europeanization upon the organizational bodies involved in comprehensive international crisis management.

Furthermore, this thesis adopts the view that the concept of misfit that emphasizes the goodness of fit is useful only in areas where the EU actively promotes clear obligations and has strong capabilities to act. In policy areas such as foreign and security policy, in which the EU's competence is weaker, more flexible approaches to Europeanization seem appropriate (Bache/Jordan 2006, p. 23). Instead of strictly deploying the fit or misfit approach of Europeanization to the analysis, notions of congruence and convergence as the broader usage of this approach seem to be better suited for considering respective institutional features of European crisis management. While a high degree of congruence does by no means imply the unproblematic accommodation of specific EU crisis management commitments, major policy misalignments are unlikely and convergence increases. For the purpose of this dissertation, the research task considering Europeanization of crisis management is instead concentrated on detecting the linkages between the national and European level.[188] It thereby focuses on the policies toward a European comprehensive crisis management approach rather than on the policies within European crisis management as an external policy.[189]

188 Congruence in this context is understood as "the broader systemic alignment of the supranational level of governance, on the one hand, and patterns in the member states, on the other" (Bulmer/Jeffery 2010, p. 115). Policy convergence is defined "as any increase in the similarity between one or more characteristics of a certain policy (e.g. policy objectives, policy instruments, policy settings) across a given set of political jurisdictions (supranational institutions, states, regions local authorities) over a given period of time" (Knill 2006, p. 5). For a deeper examination of policy convergence see various contributions on cross-national policy convergence in the book of the same name edited by Knill (2006), in particular the introduction which clarifies policy convergence and associated concepts and different types of policy convergence as well as mechanisms and causes which explain the similarity of policy changes (Knill 2006, pp. 1-11).

189 To analyze the congruence and convergence of institutional and administrative characteristics of the EU and national governments and to describe the increasing congruence of administrations and administrative practices between member states and the EU level towards a common European model, the expression *European Administrative Space* has been used, for instance, by Olsen (2003), Hofmann (2008) and Trondal (2009). The development of a European Administrative Space is marked the mutual fusion of administrative systems across the different government levels. One feature of this transition is the raising integration of domestic member state ministries and international administrative organization on the other (Trondal et al. 2010, p. 9). Hofmann and Türk (2006, p. 3) refer to the term of the *EU administrative network* in this context to describe the development of institutional structures and *ad hoc* mechanisms demanded by coordination. The notion of the EU administrative network is used in a metaphoric way to express the administrative system is not a hierarchy in the strict sense. While these expressions seem inappropriate to describe organisational structures in the sensitive area of for-

While some definitions imply a top-down approach seeing Europeanization as the domestic effects of engagement with the EU, this is an overly narrow usage of the concept with regard to the research subject. Recognizing the argument that "the EU itself may best be understood as both a feature and cause of Europeanization" (Bache 2008, p. 9), the concept in this dissertation is conceptualized as a two-way interaction as outlined above and shown in figure 7 below. As demonstrated above, it appears evident that Europeanization is inherently both a top-down and a bottom-up process in terms of the ways the concept has been studied by scholars and in terms it has been practiced in political reality.

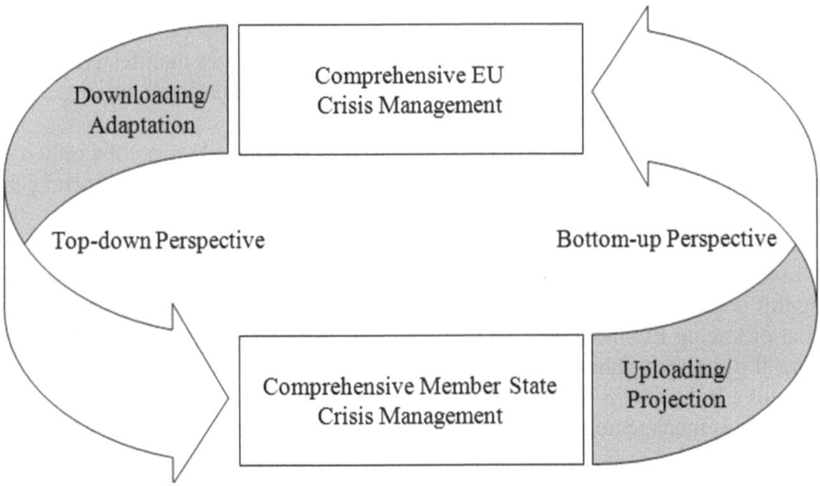

Figure 7: Europeanization Processes *(Source: Based on Miskimmon 2007, p. 8; Modified by S.P.).*

Building on broad conceptualizations presented above and on definitions adopted by scholars of Europeanization of EU foreign and security policy,[190] top-down national adaptation is considered as a reactive process that initiates change of policies, institutions and preferences from the EU level to the domestic member state level. This includes institutional processes as well as informal structures. As such, Europeanization describes "the reorientation or reshaping of politics in the

eign and security policy, they nonetheless illustrate the integrated approach and a high degree of close coordination between respective levels of member states' administrations. Especially, as organisation forms, coordination and coherence are the substance behind the theoretical notions of the previous theoretical arguments drawn from organisation studies.

190 Such as Torreblanca (2001), Wong (2005) and Gross (2009).

domestic arena in ways that reflect policies, practices and preferences advanced through the EU system of governance" (Bache/Jordan 2006, p. 30). Europeanization as bottom-up national projection is understood as a process where member states export domestic policy models and structures to the EU level. In practice, these uploading and downloading dimensions are interlinked in an "ongoing and mutually constitutive process" (Gross 2009, p. 18). The outcome of the downloading dimension is adaptation and change on the national level in terms of crisis management objectives and substance of crisis management responses. In the context of this thesis, thus the second generation of Europeanization gives the right direction, particularly as the EU from this perspective can be understood as an arena rather than an actor (Wong 2005, p. 142). With this definition, Europeanization serves as a concept encompassing both the changes manifested as policy congruence in terms of top-down dynamics as well as national policies influencing EU policy as bottom-up projection.[191]

This conceptualization also takes into account that the EU is not the only reason and explanation for international crisis management changes. The second generation perspective on Europeanization also appears appropriate to analyze the complexity of the dynamics and recognizes the important role of the member states in European crisis management structures. At the same time, the rationale of adopting this understanding is to put limits to what is already a complex task of research: "that of tracing EU-member state relations" (Bache 2008, p. 9). However, considering the research subject it is interesting and necessary to assess mutual transformative influence in order to analyze limits of European congruence in international crisis management and thus the limits toward the aim of a Whole-of-Europe Approach.

2.4 Analytical Dimensions and Parameter

After above terminological clarifications and after having explained the conceptual and theoretical approaches providing the background to examine comprehensive European crisis management, this chapter translates these core theoretical concepts into more observable forms. Given the complexity of levels and linkages in the empirical analysis aimed at, it is essential to concentrate on selected issues that enable a focus on particular dimensions where indicators of crisis management coherence and explanatory factors can be found. As stated above, it is essential to examine challenges of comprehensive European crisis

191 Horizontal, cross-loading interaction between member states is not taken into account in this perspective.

management both at the member state domestic level and at the EU level involving vertical and horizontal levels of European crisis management. However, adopting the CA presents a multilevel challenge to the actors involved with various requirements necessary for its realization (Jakobsen 2008, pp. 9/10). For analytical reasons and for the purpose of structure it is thus useful to distinguish these levels of analysis into further determinant dimensions.

Amplifying these dimensions in detail, the following chapters develop above CA deliberations, abstract tenets of organization theory and public administration studies and the concept of Europeanization into more specific operational parameters. This allows theoretical assumptions to be connected, transferred and aggregated in the specific context of this dissertation's research questions and aimed at empirical observations. Thus, each dimension maps and helps locating specific issues of European comprehensive crisis management provisions. In doing so, the following chapters also underline purpose and nature of the evaluation as well as major themes that it investigates. The dimensions represent different interpretive lenses through which efforts to implement the CA are viewed thus helping to avoid single-factor explanations. It allows for a possible cross-case comparative exercise about the factors determining distinct CA dynamics and effects. The following research dimensions furthermore serve as a way to guide the analysis of the subsequent chapters and structure research findings. Methodologically, these dimensions present a way of standardization, focusing and placing the analysis on a systematic footing.

The aim of this dissertation is to map the different approaches taken and to analyze various instruments developed to achieve greater crisis management coherence in Europe. Importance has to be paid to analysing responses and tools, strengths and weaknesses associated with every actor and to understanding the particular organizational nature and culture as well as strategic concepts of the diverse settings (Nilsson et al. 2008, p. 48). Thus, the theoretical elaborations of the previous chapters are clustered around the following four dimensions that cover vital elements that need to be taken into account when analyzing comprehensive European crisis management: the individual actor's overall crisis management approach, the conceptual dimension as well as the organizational dimension of CA implementation, as well as the European dimension and EU-interface.[192]

192 The categorization in these dimensions reflects and combines approaches taken in relevant literature to deal with the research subject's complexity. On a general level, Borrás and Radaelli (2011, pp. 470-472) distinguish between an ideational and an organisational component of government architectures. Also Hustedt and Tiessen (2006, p. 10) note that investigating changing coordination requires an identification of the basic feature of the institutional context, stages of the coordination process as well as the use of coordination techniques. Concerning WoG interpretations and measures, Christensen and Lægreid (2006, p. 6) distinguish between a

2.4.1 Crisis Management Approach

The surrounding broader foreign and security policy field entails a particular policy environment which makes it necessary to not only consider the policy toward crisis management coherence itself but also thinking behind it. The specific policy context needs to be considered. The basic assumption behind this dimension is that the crisis management approach of a country or organization does affect its efforts toward crisis management coherence.[193] Traditions of policy-making cannot be ignored and specific attributes influence changes by altering the capacity to act (Miles 2000, p. 189). This assumption is mirrored in the works of several relevant scholars. As Hull (2011, p. 4) observes regarding different CA conceptions, "[t]he exact nature of each of these concepts has, however, been primarily shaped by the organisational context in which they have arisen." She further observed in the CA context that "[t]he opportunities for formal cooperation have nevertheless been dependent on the organisational context within which CA is implemented" (Hull 2011, p. 5). Coordination in the complex crisis management environment runs into the problem of having to cope with the issue of different security as well as organizational cultures.[194] The basic logic here is that political-administrative systems develop a distinctive culture that adds to their formal structure when building up. As Lenschow, Liefferink and Veenman (2006, p. 42) argue, "the general cultural framework will be the dominant factor structuring domestic adaptation to foreign input at the level of ideas and principles." Also Hustedt and Tiessen (2006, p. 6) note that

---

structural, cultural and myth-based perspective. In a similar manner, Drent and Zandee (2010, p. 7) state that "[t]he EU's comprehensive approach to security should be discussed at different levels: the strategic level, the institutional level, and the planning level." Nuttall (2005) divides the EU's approaches to deal with the problem of inconsistency along the three dimensions of legal instruments and practices, structures as well as obligations. While the first two dimensions of Nuttall's categorization are reflected in the conceptual and organizational dimension, his third dimension on obligations does not form an extra one. Also Gebhard (2011, pp. 18/19) differentiates three kind of measures which can be employed for enhancing coherence in the EU context, namely legal remedies, institutional reforms, and political initiatives, and which overlap with the conceptual and organizational dimension as developed in this dissertation. This approach partly reflects issues taken up in the OECD's study on WoG approaches to fragile states (OECD 2006) and the House of Commons' (2010a) CA report.

193 Kickert (2008, p. 2) argues analogous with regard to the question how the political-administrative context of a country influences public management reforms.

194 This does not only show for the case of civil-military interface or long-term and short-term crisis management efforts, but applies for all bodies within the European crisis management system.

> "before slicing the process of coordination in different stages, we need to clarify the institutional context in which coordination takes place. (…) [T]he specific historical, constitutional and institutional setting of a country sets up the conditions under which central governments work, thereby framing the broad manner in which coordination proceeds."

A broader country's culture is important due to the policy-specific discourse in terms of the ideas and narratives which allow an understanding how certain policies are interpreted and developed (Lenschow/Liefferink/Veenman 2006, p. 38). This dimension of analysis thus looks at security and crisis management cultures as

> "the lens through which national authorities refract the structural position of the state in the international system; it explains the subjective understanding of objective threats to national security, the instruments relied upon to meet those threats, and the preference for unilateral or multilateral action " (Sperling 2010, p. 11).

Four criteria can be defined which are loosely touched upon in the analysis undertaken in this dimension which are the view on the external crisis management environment, national security identity, instrumental preferences, and interaction preferences (Sperling 2010, pp. 11/12).

Transformations and adjustments in foreign policy are thus analyzed through a portrayal of the respective system. The impact of domestic factors such as history, informal traditions, norms, and objectives shaping a country's international crisis management policy is vital. Security identity, key actor specific and contextual factors and features play a role.[195] This dimension of analysis thus focuses on normative elements, belief systems and the political landscape which channel crisis management policy and which are important prerequisites for political action. It lays out dominant crisis management discourses and the context within which the particular crisis management policy is articulated and developed.[196] Furthermore, relevant factors of political culture are identified.[197] An

195 The national security identity of an actor is understood here as "the dominant national discourse on security" (Rieker 2004, p. 371).

196 Discourse in this dissertation is understood as "'the overall understanding of security' presented by the political leadership and especially through the official statements" (Rieker 2006b, p. 195).

197 The impact of political culture on coherence efforts have also been noted by Windhoff-Héritier and Czada (1991, p. 14), Forster (2000, p. 48), Pradetto (2006, p. 15), Lenschow, Liefferink and Veenman (2006, pp. 38/41), Mannin (2010, p. 311) and (Wendling 2010, p. 25). Political culture can be viewed "as being made up of a multitude of attitudes, values, and feelings that give the political process order and meaning; political culture is thus the subjective psychological dimension of politics" (Pradetto 2006, p. 21). It can also be understood "as the broad context in which individual and collective identities are linked producing shared meanings that influence the framing of political action" (Aggestam 2000, p. 66). Political culture thus refers to a particular set of ideas that is drawn on to mobilize and legitimize a particular policy direction.

underlying question in this dimension is whether the crisis management, security and foreign policy traditions are compatible with the CA and respective reforms.[198] Christensen (2004, p. 37) envisages that if they are incompatible, cultural resistance may occur. On the contrary, when compatible an implementation of reforms is more easily achievable. If only partly compatible, "it stands to reason that only the compatible elements will be implemented" (Christensen 2004, p. 37). Observing crisis management agendas, the empirical chapters on the *crisis management approach* deal with the complex set of background factors surrounding and involved in comprehensive crisis management. Just as the CA as a concept cannot be understood without a previous examination of the international security environment and contemporary crisis management, the actual implementation of the CA on the EU and national level needs to be seen in the same broader context from the particular case study perspective.

2.4.2 Conceptual Dimension

While the analysis of the crisis management approach as outlined in the previous chapter considers overarching policies relating to crisis management, security and conflict, the *conceptual dimension* of analysis examines relevant strategies and official written documents as instruments of the CA.[199] In the conceptual dimension, coherence can be operationalised by exploring the anchorage of the CA within documents and strategies on the EU and the member state level. Explicit policy statements on coherence transform external policy pressures into a declaration of what the actors intend to do and provide a guiding focus. It moreover points to the

---

Applied to the context of international crisis management, foreign policy culture comprehends "the 'totality of the historically grown, (foreign) politically important, and over time relatively stable attitudes and modes of behaviour of society' vis-à-vis foreign policy" (Pradetto 2006, p. 21), legitimizing and determining the direction and orientation of respective crisis management actors.This dimension refers to the task environment as elaborated in the context of coordination in organization theory and public administration studies. It also reflects Europeanization studies such as those of Ladrech (1994) and Cowles, Caporaso and Risse (2001) on the importance of national political culture shaping the effects of Europeanization.

198 As Paris (2009) points out, not just an organisational but also a normative coherence dilemma can exist. This observation is mirrored with this dimension of analysis.

199 The meaning of the term *strategy* in IR has departed from its traditional, military based definition. While the notion originally referred to the application of military forces to achieve a certain goal, the term strategy now more broadly stands for rather politically than legally binding adopted public documents of a state or an organization (Bailes 2009, pp. 2/3). Accordingly, when used in this dissertation the term strategy is understood in the wider sense of a "concept of a 'statement of intent'" (Bailes 2009, p. 3). For a detailed and timeless overview on the basic principles of strategies see May (1973).

strategic importance of CA in policy statements (Johannsen 2011, p. 36). The conceptual implementation of the CA is viewed to create the principals for interaction between crisis management actors. Furthermore, taking into account the particular nature of EU and national CA conceptualizations in each case avoids the risk of concept stretching (Wendling 2010, p. 25). In doing so, the conceptual dimension aims at avoiding defining potential cultural and conceptual problems of CA implementation in entirely organizational terms and as pure coordination problems. Therefore, both, on the EU level as well as the domestic member state level, crisis management conceptualizations and, in particular, direct CA notions and conceptual variations are highly relevant. In this regard, the conceptual dimension allows "a panoramic view of the various meanings given to the concept" (Wendling 2010, p. 25) by the analyzed actors and, moreover, ways in which they use the CA to position themselves (Wendling 2010, p. 25).[200]

The set of questions in the conceptual dimension deals with the rise of the CA and crisis management coherence on the political agenda. As a political consensus on comprehensive crisis management is widely recognized, this dimension analyzes guiding lines of the CA's conceptual developments and views CA implementation as matter of conceptual and rhetorical reflection of the aim for greater crisis management coherence. In doing so, the conceptual dimension focuses on the particular ideas, strategic intention, objectives and vision as well as conceptual rooting behind coherence and coordination on document level.[201] It analyzes the tone and nature of the individual approach to comprehensive crisis management, strategic components, and doctrine as well as CA principles.

However, strategies and concepts can also imply a CA without explicitly using the notion (Mölling 2008, p. 2). To deal with this risk of terminological variations leading to obscurity when intending to link or compare particular approaches, the conceptual dimension analyses some additional components, such as a common and comprehensive understanding across government of what the CA involves. Shared understanding between parties is essential to optimize the effectiveness of their various capabilities. This understanding entails an acknowledgement of the multidimensional and complex quality of issues, the

200 See Christensen and Lægreid (2006, p. 13) for a similar argumentation underlining the importance of what they call a cultural perspective to WoG efforts. Also Lenschow, Liefferink and Veenman (2006, p. 40) note that ideas and principles that are found in documents "establish basic notions on the nature of the problem and legitimate means to handle it."

201 As Christensen et al. (2007, p. 79) explain, the term objective can be understood as a definition or concept of something that should be achieved or realized in the future. Objectives offer guidelines for the actors' activities and as organizations are collectively oriented towards achieving them, coordination and specialization is required and the structure of formal organizations is influenced.

expansion of actors both in terms of objects and subjects as well as the broadened scope of responses (Drent/Zandee 2010, p. 8). In terms of the changes in international crisis management and approaches to security this also means that traditional security concerns become de-securitized, while new issues get securitized and integrated into crisis management structures.[202] "The wider the range of challenges that a given state – or international institution – includes and recognizes as 'security' ones, the more comprehensive and multi-dimensional its security policy will be" (Bailes 2008, p. 141). Also the hierarchical organization of ideas and norms as well as their linkage with actors and instruments play a role (Borrás/Radaelli 2011, p. 470).

The CA can also be expressed in terms of the responsibilities official government documents expect its crisis management actors to perform. The prerequisite for acting coherently on the conceptual level is that involved actors have to adjust their understanding on the CA and recognize the interdependency between crisis management actors.

> "At its most common denominator, CA is a mindset. It includes recognition of oneself as part of a system and an understanding that effectiveness, efficiency and sustainability can be achieved if the interdependencies that exist within this system are responsibly managed" (Hull 2011, p. 14).

Individual actors must view their contribution to crisis management in connection with other crisis management actors and the sustainability of crisis management efforts and should not measure their success by maximizing their role and image.[203] As noted by Friis and Jarmyr (2008, p. 17) the recognition of interdependency entails that each actor "contributes to achieving a specific effect that makes sense only if one takes into account that others are contributing towards other effects, and that the total combined effect is necessary to achieve momentum."

Another key component of the conceptual dimension concerns the development of relevant strategies. Essentially, such a document today "demands cooperation and obedience from just about every ministry in a government (…). The new breed of documents therefore has a strong coordinating or community-building function that goes across traditional military-civil (…) divides" (Bailes

202 "Novelty [of the respective security or crisis management issue, S.P.] does not mean necessarily that they are created from scratch. They often bring about a novel understanding or re-interpretation of existing issues, sometimes with different interpretations associated with different actors, in the context of a new political thrust of international and supranational politics" (Borrás/Radaelli 2011, p. 470).

203 As formulated by the OECD (2006, p. 23), "the different actors need to understand the importance of their involvement." See also Norheim-Martinsen (2010, p. 13).

2008, p. 143). Cross-government strategies "produced by the cumulative and collective planning efforts of all the units, offices and programmes in the system" (De Coning 2009, p. 124) thus present in important CA element. Regarding the role of cross-government strategies, agreement appears to exist that, while their value should not be overstated, joint strategic planning "can make a difference determining a feasible set of coherent objectives and enhancing the likelihood of success by promoting clarity of responsibility and unity of effort in their implementation" (Teuten/Korski 2010, p. 25). By taking into account policy coordination in policy formulation, the development of joint strategies can thus be viewed a coordination tool itself (Johannsen 2011, p. 36). This involves a comprehensiveness of targeting the strategy to key political actors.[204]

However, it also needs to be considered that when it comes to values, principles and mandates, concepts partly have to be intrinsically incoherent as actors exist in specific contexts such as military, diplomatic, development, or humanitarian ones, with different value systems and principles specific to that background (De Coning/Friis 2011, p. 263). This leads to the prioritization of different aspects and results in different approaches. Such differences cannot be resolved by coordination but rather need to be negotiated in the specific crisis management context. Hence, also the degree of animosity and reluctance between the different crisis management bodies presents an important indicator of the extent to which meaningful coherence can be achieved. Crisis management actors are likely to act incoherent when value systems and organizational principles that guide them are averse to each other.[205]

2.4.3 Organizational Dimension

While the conceptual dimension analyzes the particular strategic background, the *organizational dimension* examines the way in which CA implementation is organized in existing structures. This dimension addresses factors of the institutional set-up that aim at enhancing policy coherence and inter-institutional coherence, concerns the level of institutionalization of CA issues and how the CA is translated into the current institutional arrangement, and looks at new procedures to enhance coherence. It analyzes whether the EU and country cases in their quest for crisis management coherence have been able to transfer strategic

204 See also the arguments for integrated and joint country specific strategies put forward by Teuten and Korski (2010, pp. 17-19).

205 Christensen and Lægreid (2006, p. 13) speak about a "cultural compatibility test" proposed reforms have to go through.

concepts into practice and studies the institutionalization of the need to have a shared understanding and agreement on strategy. Ultimately, the intent of this two phase approach is to investigate what is declared on paper in the conceptual dimension, and what the European Union and the selected member states actually do to increase crisis management coherence. Having gained ground on the conceptual level, attempts to implement the CA

> "have been directed at making the concept more concrete and tangible. In most cases this has been done by interpreting it as different structures and processes that can be implemented to facilitate cooperation and cooperation between the actors engaged in a system" (Hull 2011, p. 5).

However, "how these structures or processes are outlined depends on the nature of each system and what possibility there is to direct and coordinate the system" (Hull 2011, p. 14). From organizational perspective, the structural setting can be crucial for the implementation of a certain policy. A country's institutional arrangement can lead to different CA policy formation and initiation.[206]

This dimension therefore emphasizes the importance of structures and both formal and informal organizational arrangements as opportunities and impediments for CA implementation. "Institutional reforms can be thought of as one of the most direct ways of enhancing coherence" (Gebhard 2011, p. 23). It thus focuses on the particular crisis management structures, institutions, instruments and inter-organizational coordination mechanisms that have been put in place by the EU and in the member states to promote coordination and coherence in international crisis management and to resolve inconsistencies.[207] Organizational reforms undertaken as well as types of instrumental adjustments and adaptations which have been introduced to implement and institutionalize the CA are analyzed. While a lack of coordination is an indicator of failure in CA implementation, "integrating policy or

206 Implementation of the CA "requires institutions to carry the burden of transforming general policy intent into an array of rules, routines and social processes that can convert policy intention into action" (O'Toole 2003, p. 234). Structures in this regard are the framework in which such processes take place and which set limits regarding participation and perception (Christensen et al. 2007, p. 15).

207 As Radin (2003, p. 611) points out, "[f]ormal coordination mechanisms are the embodiment of the structural dimension. They serve as instruments for structural integration of units differentiated by function or geography. Properly applied formal mechanisms of interagency coordination strengthen relationships." As further elaborated by Hustedt and Tiessen (2006, p. 9) "[s]tructural coordination techniques refer to organisational structures that are established in central government in order to generate coherent government policies. The relevant actors meet under the headings of these organisational structures. Conflicts between different government organisations are to be solved here, contradictions are to be eliminated and decisions (...) be made."

procedures is the highest measure of success in implementing CA" (Hull 2011, p. 7). The dimension deals with the question, how individual organizational bodies "discharge their responsibilities and how they come together to manage areas where these responsibilities coincide" (Teuten/Korski 2010, p. 28)

An aspect that has to be taken into account in this regard considers material crisis management means like financial resources and personnel. The capacity of an organization to act depends on what resources it has at its disposal, but also on how they are coordinated (Christensen et al. 2007, p. 9).[208] Therefore, the organizational dimension also considers questions regarding the handling of financial resources and personnel in the context of comprehensive crisis management and analyzes financial and human resources as part of the administrative capacity to deal with CA implications. Budgetary constraints, for instance, can establish an incentive to enforce coordination. Highly specialized institutional grants are the most targeted type of funding but also the most restrictive (Radin 2003, p. 612).[209] Other coherence indicators on this level are processes in place that support interaction among the actors such as secondments or rotation of staff. Such organizational arrangements allow actors to exchange information, discussion of priorities and to generally manage their interdependencies and to increase inter-institutional awareness (Trondal 2009, p. 241).

The structural matters analyzed in this regard also deal with the formal roles and organization of relationships. This involves patterns of authority and leadership and the design of organization in terms of the above-outlined forms of organization, namely hierarchy, network or market. This involves looking at the main actors on the EU level and the national core executive involved in crisis management. The organization of relevant crisis management actors are also a significant pointer to the priorities of actors with regard to comprehensive crisis management. The organizational reading of this dimension allows an examination of the CA organization, how its implementation is constructed, looks at the consequences of the CA for the organization of the core institutions, for their organization function and tasks. Institutional arrangements utilized by member

208 The capability of crisis management is constrained by the size of financial resources available and the level of resources dedicated to pursue agreed policy goals also shows the extent of political will (Tonra 2001b, p. 45). However, as investigating the availability and amount of financial resources does not serve as a significant factor in the context of the research questions posed, this dissertation does not analyze financial aspects by investigating the level of financial resources available but looks at their coordinated utilization.

209 As pointed out by Herrhausen (2009, p. 16) concerning UN peacebuilding coordination, the pooling of resources can be thought of as a mechanism that necessitates strategic coordination and also ensures it for the future. See also report by the House of Commons (2010, p. 11).

states are analyzed with a view to discover whether there are indications of coherence around a common Whole-of-Europe model in crisis management.[210]

2.4.4 European Dimension and EU-Interface

The empirical analysis of this dimension is built on two analytical steps. Firstly, the EU's CA efforts are analyzed in an extra chapter along above analytical dimensions. On this ground, the EU's impact on national crisis management and CA attempts of three EU member states and *vice versa* are looked at. In this dimension, the concept of Europeanization is applied as the contextual factor of crisis management in terms of an "attention-directing device" (Dyson/Goetz 2003, p. 14) which gives analytical primacy to the impact of EU. The intention of analyzing the European dimension of crisis management is to discuss and to relate key crisis management ideas to the international level. It is analyzed, how national crisis management policies are orienting or being oriented in the direction of EU crisis management.[211]

This dimension of analysis deals with three different processes of change. The first refers to the changes in national crisis management policies and systems caused by the participation in EU crisis management from the perspective of top-down Europeanization. The second development analyzed deals with the bottom-up process in terms of the projection of national principles, ideas and preferences onto the level of EU crisis management. The third results from the above two dimensions and involves the redefinition and harmonization of identity in the context of EU and member state crisis management (Wong/Hill 2012b, p. 4).[212] The adaptation and downloading dimension of Europeanization leads to a transformation and harmonization of a member state to the needs of EU mem-

210 A lot of scholarly literature makes extensively use of organisational charts when explaining crisis management structures. While these figures in general help to understand how an organization is formally evolving, they cannot serve to show what happens behind the scenes in an institution as well as who is in favor of whom. Additionally, they cannot map influence and informal lines of communication. In this dissertation, organizational charts describing the organisational implementation of the CA in existing crisis management structures have thus not been used. This decision was supported by the constantly changing organizational structures in the EU context where new versions of organisational chart, for instance on the EEAS, were published every couple of months.

211 Consequently, this dimension is missing in the chapter on EU crisis management which serves as an overall analytical framing to allow an examination of this dimension in the country case studies.

212 View chapter 2.3.3 above for a detailed elaboration of these key analytical aspects of Europeanization.

bership. National crisis management shows an increasing salience of the EU political agenda, adherence to common objectives, the prioritization of common objectives over national targets and the internalization of EU norms and policy. In the uploading and national projection dimension crisis management policy of a member state affects and contributes to the development of a common European policy. Attempts to influence both the EU as well as the policies of other member states form an indicator in this regard. A further indicator is the use of the EU as a cover and the externalization of national foreign policy positions on the EU (Lee-Ohlsson 2009, p. 126) Policy projection in this regard deals with the projection of national crisis management preferences, policy models and ideas onto the EU level. Policy adaptation, on the contrary deals with the changes of national crisis management institutions, processes, ideas and policies due to practices, rules and objectives adopted or prescribed by the EU. However, while it is possible to identify these distinctive processes and analyze respective developments in theory, in the following empirical analysis cannot be developed to such a clear extent and observations partly overlap.[213]

Nonetheless, specific mechanisms can be identified which contribute to the uploading or downloading dimension of the Europeanization dynamics and there are a number of indicators which guide the analysis. These range from ideational support and discursive input into policy developments, institutional export of modes and practices, consistent commitment to European solutions in crisis management, relative scarcity of overtly national posturing and the capacity to utilize formal structural powers within EU crisis management policy processes, for example through agenda-setting opportunities (Miskimmon 2007, pp. 10-12). Indicators of coherence drawn from the Europeanization concept also include the degree of anchorage and saliency of the EU agenda in member states crisis management concepts, rhetorical commitment to the EU and the emergence of shared definitions of national and European crisis management preferences but also on the CA itself. This can evolve together with a contracting scope of national domaines réservés, the relaxation of traditional national policy position and priorities to accommodate progress of common policies and the use of the EU to push policies and favor EU instruments over other available possibilities. Adaptation in this sense means the shift of the domestic crisis management policies, institutions and preferences to EU crisis management policy. In contrast, maladjustment is understood as a move away from EU crisis management to national solutions. This also

213 As mentioned in the chapter 2.3.3 on Europeanization, the intergovernmental nature of EU foreign and security policy does not create adaptation pressures of the compulsory and legally binding type like in supranational policy areas. Consequently, EU influences in this policy area are rather unobvious and hard to trace directly.

points to the general importance or prominence of EU crisis management instruments and structures in national crisis management policy (Gross 2009, p. 21).

Reflecting above developed conceptual and organizational dimension by looking at both strategies and structures, certain elements of a special EU CA perspective are analyzed. On the structural level, Europeanization might entail the change of specific crisis management policies and reform of respective institutions. The increasing demands for coordination in member states' core executive due to increased EU involvement in organizational terms becomes obvious in the creation of coordinative bodies, EU sections and advisory groups (Jacobsson/Lægreid/Pedersen 2004, p. 60). It might lead to the implementation of policies that would not have been agreed on without the emphasis on the European perspective. On the conceptual level, Europeanization might result in the creation of "policy space for new initiatives on the political agenda" (Grabbe 2006, p. 51). In this regard, establishing vertical coherence between the EU and member states merely is a structural or institutional matter. Yet, the analysis in the European Dimension also builds on the assumption that in terms of European crisis management coherence it is less relevant whether a certain action is actively supported by member states as long as there is a common line and direction agreed (Gebhard 2011, p. 23). Hence, a common approach mainly is the product of a strong political commitment on the part of the member states.

2.5 Analytical Framework Synopsis

Reflecting the complexity and multidimensionality of the research subject, the elaboration of the analytical framework of this dissertation was undertaken in a fairly detailed and comprehensive way. It thus seems necessary to briefly sum up some of the above key elaborations and findings.

After a discussion of the terms conflict and crisis, crisis management has been defined in a broad manner and is referred to as an overarching expression for all types of activities and engagement carried out to respond to crisis and conflict. Crisis management instruments include measures for crisis prevention, for the resolution of acute and lasting armed conflicts, and for the consolidation of peace. It includes a wide range of activity to capture the context of the increasingly complex and interdependent international conflict scenarios which are coined by the changing nature of international crisis in terms of actors, time frames, instruments and tasks. This has required the reconsideration of crisis management responses and conceptual approaches of the international community to handle the new crisis management challenges, resulting in the conceptual development of the CA.

Despite a lack of consensus on terminology, the term Comprehensive Approach is utilized to refer to efforts to increase crisis management coherence that have emerged on the EU level as well as on the level of the different country case studies. The instrumental idea covers the coordination of different national and international, civilian and military crisis management actors and instruments and is based on the assumption that the provision of crisis management activities and instruments in a more coordinated manner increases crisis management effectiveness. As coherence and coordination are seen as the underlying principles of the CA, this thesis concentrates on the dimension of inter-organizational relationships between the various crisis management actors. In order to place coherence and coordination in the realities of the European crisis management system, the two terms have been framed in this context. For the ease of distinction, the meaning of coherence and coordination has been conceptualized in terms of vertical and horizontal levels. Further differentiations of crisis management coherence are only undertaken when relevant in the respective analytical context. In this regard, inter-institutional is concerned with the cooperation between and coordination of different actors. Furthermore, coordination mechanisms have been characterized in terms of negative and positive coordination as well as formal and informal forms of coordination, among others.

Organization theory and public administration studies complement the analytical framework by placing the analysis firmly on understandings of actual organization forms, namely hierarchies, markets and networks, and other organization mechanisms which allow observing the translation of the CA in actual organizational structures and helps to interpret forms of coordination. Additionally they allow examining to which extent crisis management coordination structures match each other and which organization form and type of coordination proves to be suitable to effectively implement the CA in the European crisis management systems and to reveal potential advantages of specific models of CA implementation. The impact of the EU is captured through the notion of Europeanization which relates to both the uploading and the downloading dimension and serves as a concept encompassing changes manifested as policy congruence in terms of top-down dynamics as well as national policies influencing EU policy as bottom-up projection. The in-depth analysis of the CA concept, public administration and organization studies as well as the concept of Europeanization in previous chapters allowed deriving analytical parameter and determinants to analyze coherence in European crisis management on different levels of implementation. These have been categorized along the four analytical dimensions: the particular crisis management approach, the conceptual dimension, the organizational dimension as well as the European Dimension and EU-interface. The analytical framework of this dissertation and its dimensions of analysis, as summed up and illustrated broadly in figure 8, reflect the complexity of the CA in the context of European crisis management and present the point of departure for the empirical analysis in the next chapters.[214]

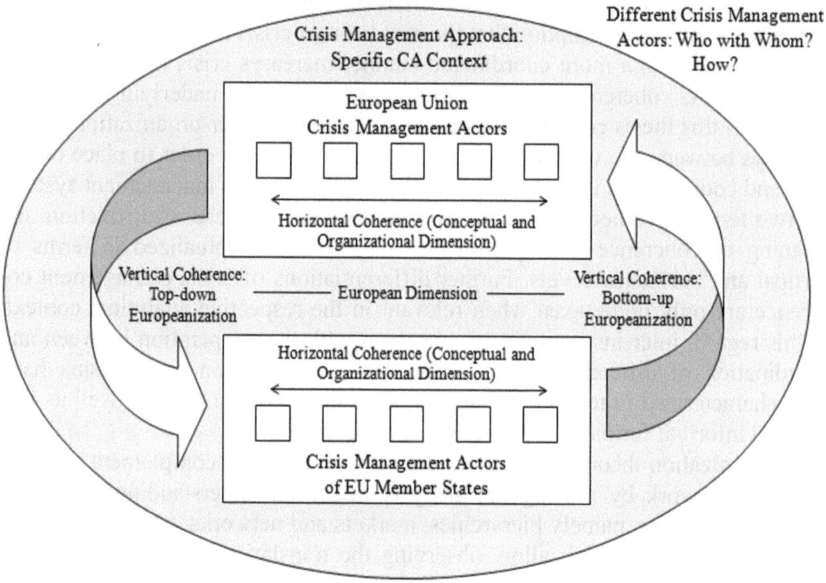

Figure 8: Analytical Framework and Dimensions *(Source: Own Design).*

214 Figure 8 displays a composition of relevant elements drawn from the previously presented figures.

3 Comprehensive EU Crisis Management

To be able to analyze the European Dimension of CA efforts on the member state level, it is necessary to first examine the EU's evolution and attempts toward comprehensive crisis management. The following analysis is thus undertaken with the aim of being able to assess congruence of EU and national efforts and to be able to analyze how the EU impacts on developments on the national level and *vice versa*. The development of the CA and its implementation in EU crisis management need to be seen in the broader context of developments in the field. The chapter thus begins by examining how the EU can be understood as an actor in the international crisis management arena. It outlines fundamental principles of EU crisis management commitments. Against the backdrop of these considerations, the Union's CA rooting and development on the conceptual level is looked at. The subsequent chapter analyzes recent restructuring efforts at the organizational level. In doing so, it is limited to the institutional set-up after the entry into force of the Treaty of Lisbon.[215] The chapter concludes with a synopsis on EU comprehensive crisis management.

---

215 While this dissertation focuses on EU crisis management structures after the Lisbon Treaty, it has to be mentioned that before its entry into force the institutional three-pillar structure of the EU, consisting of the supranational European Community pillar including the European Commission with its various civilian crisis management instruments, and the two intergovernmental pillars, covering the CFSP/ESDP and Justice and Home Affairs, marked the most obvious structural fragmentation of EU policies. The EU's pillar structure was deeply anchored in EC history and cemented by the 1992 Maastricht Treaty. For details see various chapters in Bache, George and Bulmer (2011) which explain the development of the pillar structure under EU treaties. The important consequence from this division was the institutional separation of short-term crisis management being located inside the European Council's CFSP/ESDP and long-term development activities being placed within the range of the Commission to ensure that development assistance does not run the risk of instrumentalization for security purposes. As Johannsen (2011, p. 341) observed "the original set-up and development of EU Crisis Management directly opposed the idea of comprehensive Crisis Management." Various innovations on organisational level have been introduced in the face of this institutional fragmentation to aim at greater coherence already before the entry into force of the Lisbon Treaty. These structural changes largely mirror the EU's pre-Lisbon pillar structure. While acknowledging that the new crisis management set-up has to be ascribed to its historical roots, the EU's three-pillar structure has been effectively overcome with the Lisbon Treaty. Consequently, pre-Lisbon structures are only referred to when relevant in the given context. These structures and the downsides of the EU's pillarization have been extensively analyzed elsewhere. See, for in-

3.1 The EU's Crisis Management Approach

The European Union is a fairly young crisis management actor and its crisis management capabilities and instruments have developed comparatively fast. While other existing IOs such as the UN or NATO have tried to adapt to the contemporary international environment within their existing structures, the EU is the only actor without a clear security legacy from the Cold War period and not burdened with a preexisting security political framework (Rieker 2006a, p. 512). The evolution of the EU's crisis management and in larger sense foreign and security policy has mainly taken place in response to external circumstances and own misconduct. Accordingly, after the EU's failure to act during the Balkan conflicts, its member states operationalized their crisis management commitment and developed a set of specific civilian and military crisis management capacities with the creation of the CFSP/ESDP in 1999. In doing so, the EU has acknowledged that conflict poses risks for its own objectives. Therefore, it is deeply committed to conflict prevention, and, in consequence, the use of civilian crisis management tools. Addressing the root causes of conflicts is regarded as essential from EU perspective. This particularly accounts for the field of development cooperation, in which the Union has been a major actor via the Commission for much longer. In an inclusive understanding of crisis management which acknowledges the security-development nexus, the linkages of different policy areas such as democracy, human rights, rule of law and good governance as well as long and short term crisis management measures, the EU aims at addressing all stages of conflict (Johannsen 2011, p. 70).

The ability to use civilian and military crisis management instruments can be claimed to be the "specific characteristic of the EU's approach to conflicts" (Drent/Zandee 2010, p. 1). Theoretically, the Union has both, a wide variety of civilian crisis management means such as political, diplomatic, economic, and police instruments as well as military means at its disposal. Despite the challenge

stance, Gross (2008) for an overview of individual actors and institutions relevant for EU crisis management before the Lisbon Treaty and Drent and Zandee (2010) who demand a breaking of EU pillars already in the title of their study. Also Johannsen (2011) analyses the EU's CA developments until the Lisbon Treaty in detail. Norheim-Martinsen (2010) focuses on institutional changes to particularly improve EU civil-military relations. Other important works on academic side are studies by Siedschlag and Eder (2006), Ehrhart (2007) as well as Grevi, Helly and Keohane (2009). Surveys of the EU's institutional architecture with relevance to crisis management are also provided by the International Crisis Group (2001, 2005) and Klein (2010, pp. 67-80). Even though it is precisely its minor role in EU crisis management which is heavily debated, also the European Parliaments is largely left out of the following analysis. Details on the European Parliament's role in EU crisis management after the Lisbon Treaty can be found in a report by the European Peacebuilding Liaison Office (2012a, p. 9).

of applying these capabilities in a coherent manner, the EU is often viewed as the international actor with the best ability and greatest potential to implement the CA due to this wide range of instruments covering a broad spectrum of possible crisis management tasks.[216] "It appears that the EU is virtually meant to act comprehensively – that it is simply fulfilling some sort of teleological drive that originates in its history and experience as an organization" (Gebhard/Norheim-Martinsen 2010, p. 2). Moreover, the EU in its own right perceives and presents itself as the crisis management actor capable of providing the full crisis management panoply.

Despite being at early stages, it has been tried to capture EU crisis management with several analytical labels which aim at categorizing the EU's role and actorness. Both, the assumption that the EU is an actor with a distinctive security approach as well as the existence of a European security order and culture are highly controversial. Defenders of traditional security conceptions neglect the EU as a security actor due to the lack of effective military capabilities.[217] On the contrary, understanding security in a broad sense, the EU's various security instruments "makes it one of the most important security actors of the post-Cold War context" (Rieker 2004, p. 370). As security is understood in a broad sense in this dissertation, it is particularly the EU's potential due to its various tools that makes it a major actor in the complex post-Cold-War security environment. It is this broad security perception the EU also emphasizes itself.[218] In this context, the *Civilian Power* concept argues that the EU's international influence is executed by the use of mostly 'soft power' instruments such as economic, financial and politi-

216 On the EU as a crisis management organization of comparative strength see, for instance, Drent and Zandee (2010, pp. 6/10), Hanssen (2010, p. 8), Norheim-Martinsen (2010, pp. 1/2), House of Commons (2010, Ev. 89/88), Major and Bail (2011, p. 15) as well as Major et al. (2011, p. 104). Also see respective EU documents such as the European Security Strategy (Council of the European Union 2003a) and its revision (Council of the European Union 2008).

217 Academic literature frequently refers to this 'capability-expectations gap' introduced by Hill (1993) to portray discrepancies between available EU capabilities on the one hand and the Union's ambitions as a global security actor on the other.

218 Sharing this view with regard to the existence of a European security culture Kirchner (2010a, p. 119) states that sharing of similar security interests of the member states gives rise "to the emergence of an embryonic European security culture which coexists in parallel with national security cultures". As pointed out by Norheim-Martinsen (2010, p. 2) "the EU is clearly not a state, it does embody certain non-intergovernmental characteristics that set it beyond a 'normal' inter-state organisation or alliance – the expansion of the role of the administrative level, also in the ESDP/CSDP domain, being one of them." Bengtsson (2010, pp. 4/5) links the debate on EU actorness to the discussion on EU autonomy from its member states and the recognition by other actors. On the EU's controversial role as a security and crisis management actor and the character of the Union's security identity see Bretherton and Vogler (1999, pp. 15-45), Rieker (2006b, pp. 3-6/23-50; 2007, pp. 8-10), Bengtsson (2010, pp. 4-6), Kirchner (2010a, pp. 104-106), Major and Bail (2011, pp. 18/19) as well as Wong and Hill (2012b, pp. 2-4). On the discussion of the possibility of a European strategic culture see Kirchner (2010a) and Sperling (2010).

cal crisis management tools (Johannsen 2011, p. 42).[219] Before the EU complemented its crisis management arsenal with military tools in 1999, the Union was a purely civilian institution. Moreover, the Civilian Power concept which emphasizes its civilian crisis management tools and its powers of influence and persuasion can be viewed as a function of the EU's own internal model and founding values (De Vasconcelos 2010, p. 17). The EU's gradually emerging military capabilities and missions resulted in debates about a departure of the Union from its identity as a Civilian Power. However, as the EU still strictly circumscribes the use of force as the last option in most exceptional circumstances and with broad international legitimacy only, it can be considered to remain a civilian power, despite its military capabilities (De Vasconcelos 2010, p. 3).

Separate from civilian and military power debates, the EU has been understood as an ethical foreign policy actor and *Normative Power*.[220] The Union emphasizes the spread of norms in its crisis management engagements and underlines the importance of compromise and conflict negotiation. In contrast to the Civilian Power concept, this understanding does not underline the actions of the EU, but focuses on the status of the EU as an actor of principled and value-based nature disposed to act in the respective, normative way. Thus, the nature of the EU's crisis management capabilities is characterized by activities that aim at generally improving conditions by the application of norms that are universally valid rather than being guided by any particular interest equivalent to the policies of state actors.[221] Again, these principles which are applied to its external action are rooted in the Union's internal relations and in the member states which are governed by respective norms (De Vasconcelos 2010, p. 15). Also, the EU itself

219 The notion of soft power has been coined by Nye (2004, pp. ix/x) to describe the ability to fulfill objectives by using attraction and incentives based on the strength of values, culture, and policies perceived by as legitimate and credible rather than coercion. First Civilian Power Europe debates, however, arose in the 1970s already. See Johannsen (2011, pp. 42/43) for early discussions. Also see chapter 4.2.1 on the German crisis management approach for an explanation of the Civilian Power idea in the context of Germany.

220 The term originates in Manners' (2002) idea of *Normative Power Europe*. Manners (2002, p. 239) argues that civilian power and military power can be found in the capability of an actor to use various civilian or military instruments. Normative Power, however, applies to "its ability to shape conceptions" (Manners 2002, p. 239). The perspective of the EU as a Normative Power has faced criticism regarding the definition of the concept itself which particularly questioned what is so European about the discussed norms. For details on the concept see Manners (2006). On criticism of the concept see Forsberg (2011, pp. 1186-1190).

221 Manners (2002, pp. 242-244) identifies five core norms which are peace, liberty, democracy, human rights and rule of law, together with four additional minor norms which are social solidarity, anti-discrimination, sustainable development and good governance.

emphasizes this perception.[222] Both, the Civilian and the Normative Power concept are of particular relevance in the CA context as they deal with concerns of the relation of normative and civilian strength and military force, "bringing together these two apparently bipolar features" (Johannsen 2011, p. 42).

By seeking a foreign policy that is distinctly principles-based and normative, the EU heavily emphasizes the primacy of democracy. Also, the promotion and protection of human rights is a fundamental objective and guiding principle of EU crisis management activities (Arloth/Seidensticker 2007, p. 5).[223] Other fundamental objectives include the rule of law and principles of international law (Johannsen 2011, p. 70). Furthermore, the EU underlines the importance of multilateral solutions to common problems for which the label *Effective Multilateralism* has been coined. This commitment to multilateralism is founded on the belief that different states and regions of the world share common interests. It also suggests a degree of pragmatism and compromise with partners. In doing so, the EU is an active and outspoken supporter of the UN system. The UN Security Council's primary responsibility for the maintenance of international peace and security, the acknowledgement of the UN Charter as the basis for any legitimate response to threats to international peace and security, as well as requests from the UN, regional organizations or host nation as a precondition for the launch of missions are guiding principles of EU crisis management (Johannsen 2011, p. 70). Even more, for the EU, multilateralism is not only the preferred way to engage internationally but it also largely reflects its strengths. Consequently, mulitlateralism again forms a part of the EU's own identity and model of integration. The Union's internal development between its member states represents an advanced form and system of multilateral, cooperative and collective foreign policy making (De Vasconcelos 2010, p. 16).

The specific history of EU security policy and the advancing of EU crisis management resulted in a complex policy area that spans EU institutions and decision-making procedures. Over the last few years, CSDP crisis management missions have proliferated and grown in size and complexity. Since the launch of the first ESDP mission in 2003, the EU has conducted almost 30 missions.[224]

222 As stated by then European Commission President Romano Prodi, as Europeans, "our distinguishing feature is our sense of responsibility" (Prodi 2003, p. 2). Also, the EU's then High Representative for CFSP Javier Solana underlines that this responsibility reaches beyond the EU's and that "it would be morally untenable, sometimes unthinkable, to sit idle, without reacting to so much human misery and distress. Therefore, we are compelled to act" (Solana 2000).

223 On the role of human rights in EU crisis management operations see Arloth and Seidensticker (2007).

224 An overview of ongoing and completed missions and operations of the EU can be found on the European Union's External Action Website. Also, the *International Security Information Service* Europe provides a monthly up-dated chart of EU mission (ISIS Europe 2012).

The majority of these missions have been civilian helping to build rule of law, support peace agreements or monitor borders. Being fairly small, they were mostly "manageable and with a good chance of delivering the expected results" (Johannsen 2011, p. 50).[225] As civilian crisis management lies at the heart of EU activities, it is mainly the civilian area which gained increasing importance in the EU's crisis management doctrine and practice. It is also this sector in which the EU offers an advantage compared with other crisis management actors due to the wide array of civilan instruments and mechanisms under one umbrella, in particular considering the Commission's important role in development cooperation.[226] However, since 2009 an increasing passivity and at the same time decreasing willingness to take on crisis management responsibility and burdens have become ever more evident. National interests and resentments have turned out to be increasingly characteristic of EU crisis management and the notion of EU crisis management itself continues to lack a generally accepted definition. Even more so, as a boundary line between EU crisis management and EU foreign policy activities is hard to delineate. As a result, "[t]he finalité in the EU's terminological inflation of 'crisis management' might well boil down to the external dimension of providing security, in all its cross-pillar glory" (Blockmans/Wessel 2009, p. 11). According to respondents, this uncertainty where crisis management starts or ends is one of the main problems of the EU's CA whose underlying terminology and concepts are thus largely undefined.[227]

---

225 The EU structural crisis management fragmentation is also reflected in the circumscribed nature of its CSDP missions in the field to date which follow the same pattern of civil-military separation that characterizes the EU's organizational structure. The EU has not yet set up an integrated mission with its different crisis management capabilities operating together in one mission even when deployed in the same area. EU civilian, police and military crisis management capacities have been deployed in parallel missions with separate mandates and budgets. Bosnia-Herzegovina and the Democratic Republic of Congo are examples where a single theatre contained different kinds of missions (De Coning 2009, p. 118).

226 See, for instance, the different contributions in Nowak (2006) dealing with EU civilian crisis management in detail.

227 While recognizing the confusion of terminology, the Council General Secretariat declared that "it is a matter of political appreciation depending on the circumstances, if a state of 'crisis' exists. Attempts to provide a precise definition of the term 'crises' therefore do not seem productive" (Council of the European Union 2002, p. 4). This resulted in large discretion regarding the identification of a crisis for the EU. At the same time a precise definition of what constitutes a crisis for the Union would put large pressure on the EU and its member states to react accordingly. This latter consequence also explains why EU member states refrained from agreeing on a common definition (Johannsen 2011, p. 51).

3.2 The Comprehensive Approach in EU Crisis Management

3.2.1 Conceptual Dimension

Historically, initial notions for a CA in EU crisis management go back to the emergence of the European Community's external relations and intergovernmental European Political Cooperation in the 1970s (Ehrhart/Petretto 2012, p. 9). The objective of general external policy coherence was also mentioned in the Single European Act in 1987, which introduced a coherence-consistency requirement that demanded that EC external policies and the policies agreed in intergovernmental European Political Cooperation ought to be consistent (Gebhard 2011, p. 5). Thus, policy coherence of EU external action has been a concern for decades.[228] The struggle for greater coherence continued and the need grew in parallel with institutional and legal adaptations in the field of EU external relations and emerging crisis management.

In 1999, crisis management became a new area of EU competence with the decisions of the summits in Cologne and Helsinki.[229] With the conclusion of the 1999 European Council meeting in Helsinki (Council of the European Union 1999) the heads of government gave effect to the Amsterdam Treaty and the Cologne European Council declaration by deciding to establish an EU military capacity to undertake the full range of the Petersberg tasks.[230] In doing so, the EU should "assume its responsibilities for the full range of conflict prevention and crisis management tasks" (Council of the European Union 1999, ANNEX 1 to ANNEX IV). While not part of the initial ESDP framework agreed on in 1999, during the 2000 European Council in Feira, the EU added the four civilian capacity areas of rule of law, police, civilian administration and civil protection to the military side of ESDP. Although civilian crisis management in terms of

---

228 The need for institutional and political coherence of EU matters has not only been a concern with regard to foreign policy but also in other policy fields. As Gebhard (2011, p. 3) states "coherence – or rather, the lack of it – constitutes one of the most frequently bemoaned aspects of the EU's political and bureaucratic performance."

229 For a highly detailed analysis of very early steps and essential conceptual and institutional developments with regard to EU crisis management as well as the EU's CA, beginning with the set-up of the European Communities until the 1997 Amsterdam Treaty see Johannsen (2011, pp. 71-152).

230 The Petersberg Tasks are the crisis management tasks that the Western European Union (WEU) announced it would be willing to conduct and that were included in the 1992 Petersberg Declaration as part of the reorientation of the WEU after the end of the Cold War. The tasks which included the undertaking of humanitarian and rescue tasks, peacekeeping, and crisis management, including peacemaking describe the areas covered by EU contributions to crisis management and became an integral part of the EU's ESDP (EU Glossary n.d., Petersberg Tasks).

development cooperation at this point was not a new activity of the EU, connecting it with the ESDP process meant that the topic moved up on the agenda and shaped the perception of the ESDP process as a whole (Jonson 2006, p. 204).

To fulfill personnel and institutional requirements necessary to perform crisis management missions, the EU initiated individual military and civilian headline goals to set clear targets and strengthen capabilities (Gross 2008, p. 16).[231] With respect to the creation of own EU military capabilities, the Helsinki European Council adopted a *Headline Goal 2003* which defined the military capabilities needed to rapidly deploy troops capable of carrying out the full range of the Petersberg tasks (Council of the European Union 1999, ANNEX 1 to ANNEX IV). In addition, mechanisms were set in place for handling civilian aspects of crisis management and new structures were planned to be established in order to allow the necessary guidance and strategic direction for respective engagements. The Council conclusion drew up an Action Plan to point to the way ahead and indicate the steps the EU has to undertake in the area of non-military crisis management instruments. One of the major aims spelled out in the Action Plan was "ensuring inter-pillar coherence" (Council of the European Union 1999, ANNEX 2 to ANNEX IV). Furthermore, the Action Plan demands the development of a CA for the coordination of the EU's civilian crisis management instruments which were planned to be of inter-pillar nature interlinking the Commission and Council Secretariat structures (Council of the European Union 1999, ANNEX 2 to ANNEX IV).

To reflect the evolution of the strategic environment and the EU's role as a crisis management actor affirmed in its security strategy,[232] the European Council approved the updated *Headline Goal 2010* in 2004. The new Headline Goal set the year 2010 as the target and aimed at advancing the EU's military capabilities from a qualitative perspective (Johannsen 2011, p. 183). The document demands member states to "be able by 2010 to respond with rapid and decisive action applying a fully coherent approach to the whole spectrum of crisis management operations" (Council of the European Union 2004b, p. 1). Demanding this, the Headline Goal 2010 placed a particular focus on the interoperability of different crisis management instruments.[233] Focusing on other qualitative requirements such as deployability and sustainability of military forces, the document is far more precise than its predecessor. It "gave the impression of greater

231 For EU member states, these commitments in the framework of Headline Goals are only a political self-commitment which cannot be sanctioned apart from rhetorical and moral appeals (Major/Bail 2011, p. 20).

232 The *European Security Strategy* is dealt with in detail later this chapter.

233 In doing so, it defined interoperability broadly "as the ability of our armed forces to work together and to interact with other civilian tools" (Council of the European Union 2004b, p. 2).

trust in distinct European resources" (Norheim-Martinsen 2009, p. 12), also by emphasizing that civilian crisis management instruments can be deployed together with military components (Council of the European Union 2004b, p. 4).

Also in 2004, the European Council followed up first steps formulated in the 1999 Action Plan and adopted the *Civilian Headline Goal 2008* in an individual conceptual process in order to prevent civilian crisis management efforts from lagging behind initiatives taking place on the military side (Drent/Zandee 2010, p. 9). The document entailed numerous ambitions for civilian ESDP crisis management and clarified the amount of civilian experts the EU aims at having at its disposal in civilian capacity areas by 2008. The Civilian Headline Goal 2008 clearly recognizes the need for a CA by stating that

> "[d]eveloping the civilian dimension is part of the EU's overall approach in using civilian and military means to respond coherently to the whole spectrum of crisis management tasks (…). A coherent use of Community and civilian ESDP instruments is of key importance for a qualitative improvement of the EU's capacity to act" (Council of the European Union 2004a, p. 2).

Furthermore, the document reemphasized the Union's focus on preventive crisis management activities and called for the improvement of capabilities to perform these. Mirroring the objective of the military Headline Goal 2010, the aim was set to deploy ESDP civilian crisis management missions autonomously or jointly and in close cooperation with military operations (Council of the European Union 2004a, p. 3). After the conclusion of the Civilian Headline Goal 2008, the *Civilian Headline Goal 2010* was launched for greater harmonization between civilian and military crisis management goals. Following the same methodology, the Civilian Headline Goal 2010 further increased the focus on civil-military cooperation (Johannsen 2011, p. 179). The formulation of the Civilian Headline Goals took place by mirroring the military headline goal process. This also shows in the parallel ending of both Civilian and Military Headline Goals in 2010. The earlier adoption of the Action Plan for non-military crisis management alongside the planning of the military component of ESDP shows EU attempts to mirror the military Headline Goals in documents particularly dealing with civilian crisis management. At the same time, the conceptual Headline Goal process illustrates that ESDP civilian crisis management capabilities presented an afterthought in response to developments in the military field (Drent/Zandee 2010, p. 9). However, apart from the Headline Goal process, at these early stages the need to combine the array of different conflict management instruments and the CA were mainly referred to and terminologically established in public statements and speeches rather than in substantive documents (Johannsen 2011, p. 58).

On the basis of UN and NATO developments,[234] the EU began to include CA thinking to crisis management concepts. In this regard, an essential element of the EU's CA is the Union's efforts to improve civilian and military crisis management with the UN. Before establishing a conceptual base for its own CA efforts, the EU acknowledged the UN's principal responsibility in the area of conflict prevention and crisis management. In 2001, the *Draft Council Conclusions on EU-UN Cooperation in Conflict Prevention and Crisis Management* (Council of the European Union 2001) was adopted in which the EU clarifies cooperation areas, modalities and procedures for cooperation between the EU and the UN. In this context, the EU's approach to comprehensive security is closely linked to the human security concept as promoted by the UN (Drent/Zandee 2010, pp. 10/11). In the same vein, the EU determined the need to improve donor coordination and cooperation as a crucial element of its CA. The focus of conceptual efforts in this regard, was also very much on complementarity between the EC's and member state policies on development cooperation. Cooperation between EU crisis management institutions and its member states' external activities also need to be viewed in this context. In order to accomplish coherence and efficiency, it is vital to the EU's comprehensive crisis management approach to establish coherences with partners being either IOs or its member states (Johannsen 2011, pp. 221/222).

With the adoption of the *European Security Strategy* (ESS) (Council of the European Union 2003a) in 2003, the EU for the first time takes on a comprehensive perspective on security, clarifies its policy and calls for a mixture of instruments to fight the new challenges.[235] The document highlights the EU's comprehensive crisis management instruments in several passages and also underlines that the civilian and military crisis management capabilities of both the EU as well as its member states need to be improved and further combined. While stressing that EU military capabilities for crisis management have to be better integrated with other instruments, the document at the same time makes clear that military force is viewed as a specialized tool of limited use. In doing so, it recognizes the security development nexus and the multidimensional nature of

234 See chapter 2.2.3 which describes CA developments in the UN and NATO.

235 The ESS' denotation as a strategy resulted in heavy discussions on the characterization of the document as such. For many, the word strategy was misused and misleading here as the ESS lacks the identification of concrete goals and priorities as well as a reference to the means for pursuing political end. Also, a report by the European Parliament's Subcommittee on Security and Defence has called for a European White Book on Defence as the ESS' often vague wording "fails to describe a real strategy" (European Parliament 2009, p. 7). On the strategy debate also see Bailes (2008, pp. 142/143). A detailed analysis of the document can be found in Bailes (2008, pp. 144-149).

the new security challenges by reflecting on the EU's comparative advantages in this crisis management environment (Council of the European Union 2003a, pp. 7/11/13). Although not directly mentioning it, "the document breathes an adherence to a conception of security that goes beyond the politico-military dimension of security" (Drent/Zandee 2010, p. 10). In doing so, the ESS entails a large willingness to integrate military crisis responses in its essentially civilian framework (Gordon 2006, p. 350). The strategy contains explicit references to coherence both regarding EU instruments as well as activities of the member states. Referring to the various crisis management instruments created by the EU, the ESS clearly states that "[t]he challenge now is to bring together the different instruments and capabilities" (Council of the European Union 2003a, p. 13). With this statement the ESS sketches its own complex of problems and, at the same time, the core of its approach to comprehensive crisis management. As a result, the ESS draws a CA which emphasizes the various EU contributions to crisis management and offers a wider context for ESDP. The ESS thus broadens the scope of EU challenges and places ESDP civilian and military instruments within the wider set of EU crisis management instruments. The ESS' Implementation Report five years later confirms the strategy's analysis (Council of the European Union 2008, pp. 1/2). However, among the issues recognized by the report, it notes that the EU still must improve its coherence,

> "through better institutional co-ordination and more strategic decision-making (…). Each situation requires coherent use of our instruments, including political, diplomatic, development, humanitarian, crisis response, economic and trade co-operation, and civilian and military crisis management. (…) Our ability to combine civilian and military expertise from the conception of a mission, through the planning phase and into implementation must be reinforced" (Council of the European Union 2008, p. 9).

The report points to the provisions of the Lisbon Treaty to present the framework to achieve this (Council of the European Union 2008, p. 9).

To guide the operational implementation of the EU's crisis management, several procedural and conceptual guidelines were developed such as the *Concept for EU Monitoring Missions* or the *Report on Planning and Mission Support Capability for Civilian Crisis Management*. However, the development of these concepts rarely included coordination among the different EU institutions. This resulted in concepts tailored to specific areas and pillars but largely opposing comprehensive EU crisis management, though rhetorically underlining the necessity of civil-military coordination as well as coordination with the Commission. In addition, procedurally, the European Union Military Staff (EUMS) was larger and more experienced than the civilian side in drafting documents. This is reflected in the conceptual work on the issue and early concepts were largely coined by the military side (Johannsen 2011, pp. 184-186). Another concept of the reference framework for the EU's CA is the *CIMIC doctrine* which was adopted by the European Union Military Committee (EUMC) in 2002. In this document, the EU refers to CIMIC as the

> "co-ordination and co-operation, in support of mission, between military components of EU led Crisis Management Operations and civil actors (external to the EU), including national population and local authorities, as well as international, national and non-governmental organizations and agencies" (Council of the European Union on Civil-Military Co-operation, as cited in Johannsen 2011, p. 200).

EU CIMIC explicitly stresses that its CA toward crisis management builds upon a uniquely wide array of both civilian and military instruments. The EUMS is responsible for planning and implementing CIMIC. This points to its limits with regard to the coordination of the civilian and military sides of EU crisis management operations. As the EU's CIMIC doctrine is coined by the military perspective from which it was derived, civil-military coordination is subordinated to military objectives (Hynek 2011, pp. 89/90). Strategically, the EU's CIMIC doctrine in essence is "a tactical level doctrine which provided an externally oriented mechanism for interacting with the civil environment" (Gordon 2006, p. 351).[236]

The EU's CIMIC doctrine is conceptually integrated in the Union's broader and politically further reaching *Civil-Military Coordination* (CMCO) (Council of the European Union 2003b) on a subordinated level. The CMCO concept was developed in 2003 to internally coordinate CSDP civilian and military crisis management instruments.[237] Due to its image as an essentially civilian actor, the EU aimed at further distinguishing between civil-military cooperation on the operational level with the above-mentioned CIMIC doctrine and internal civil-military coordination activities on the political and strategic level with the development of the CMCO doctrine as a concept specific to EU issues. The concept outlines an EU understanding of the CA in several paragraphs which state that

> "[t]his comprehensive approach to crisis management leads to the need for ensuring within the EU an effective co-ordination of the whole range of such instruments. This approach will have to take into account the fact that these instruments may be

236 Gordon (2006, p. 351) criticizes the focus of EU CIMIC on the operational-tactical level while largely lacking a strategic component.

237 The document dates back to successive Council meetings starting 1999 which stressed the importance of wide-ranging civil-military relations in EU crisis management to deploy all civilian and military EU instruments in a coherent and coordinated way (Johannsen 2011, p. 201). Scholars, such as Mustonen (2008, p. 21), repeatedly point to the importance of clearly distinguishing the EU's CIMIC and CMCO doctrines.

> subject to different institutional and thus decision-making processes. Civil-Military Co-ordination (CMCO) in the context of CFSP/ESDP addresses the need for effective co-ordination of the actions of all relevant EU actors involved in the planning and subsequent implementation of EU's response to the crisis" (Council of the European Union 2003b, p.2).

Underlining the availability of civilian and military crisis management instruments, CMCO considers the effective coordination of these instruments the key to successful EU crisis management and seeks to guide the EU's CA as an effort to link civil and military instruments. CMCO is supposed to ensure the CA from policy planning to implementation on the strategic policy level of the EU's pre-Lisbon structures (Major/Schöndorf 2011, p. 4).

CMCO thus represents a crucial part of the EU's CA "as the tool or method to achieve the desired 'effective coordination'" (Johannsen 2011, p. 61). Rather than focusing on structures and procedures, the document emphasizes the need for a culture of coordination based on cooperation and shared objectives, of which working closely together is an essential element (Council of the European Union 2003b, p. 2). By improving EU crisis management interaction, effectiveness should be enhanced, making CMCO "an essential element in ensuring overall coherence in the EU's response to a crisis" (Council of the European Union 2003b, p. 2). To this end, agreeing on clear objectives is regarded the "necessary basis for establishing a coherent response to a specific crisis situation, and for helping ensure unity of purpose and convergence of instruments in its crisis management activities" (Council of the European Union 2003b, p. 3). In this context, the document clearly states that the concept not just entails civil-military coordination but also comprises civil-civil coordination. Both are required at all levels of EU crisis management entailing inter-pillar and intra-pillar as well as external coordination as a prerequisite for coordination with external partners. A particular role in this regard is ascribed to CMCO working arrangements between the Council General Secretariat and the Commission during the early phases of crisis (Council of the European Union 2003b, p. 4). The CMCO document thus explicitly links community and intergovernmental crisis management tools.

Until the formulation of the Lisbon Treaty, the CMCO concept provided the most significant conceptual coordination framework in the CA context and represents "[t]he most comprehensive reference to the EU approach" (Johannsen 2011, p. 60). By emphasizing a cross-pillar crisis management approach, CMCO was the "label for any debate on coherence in EU crisis management" (Gebhard 2011, p. 27) and effectively is "EU jargon for translating its comprehensive approach to security from the strategic planning to the operational phases" (Drent 2011, p. 8). At the same time, however, the document's scope is very broad and lacks concrete action and a practical application. Though it becomes clear that

CMCO aims at creating organizational routine, the document remains rather vague on what the demanded culture of coordination actually implies organizationally. The emphasis on the culture of coordination comes at the expense of structures and procedures (Drent/Zandee 2010, p. 13).[238] As background interviews highlight, the CMCO document did not have much hands-on implementation outside its conceptual phase.

The CMCO concept was complemented by the *Crisis Management Procedures* (Council of the European Union 2003c) which describe how different levels of crisis management planning have to consider different civil or military crisis management dimensions and the overall approach of the EU to the management of a particular crisis.[239] With the objective of bringing together all relevant EU crisis management actors, the procedures outline different crisis management phases clarifying how routine policy making, decision-making and the respective implementation of decisions has to be conducted on the institutional level.[240] In doing so, they present potential procedural solutions for the particular phases of the conflict cycle, combining both Commission and Council instruments. "This was remarkable in so far as functional necessities were taken as the reference point instead of putting the onus for the comprehensive task of crisis management onto the fragmented institutions themselves" (Gebhard 2011, pp. 26/27). Encompassing all steps from the detection of a crisis, decision-making on EU-missions and their conduction to their termination, "[t]hese Crisis Management Procedures are regarded the main provisions for ensuring the implementation of the EU's Comprehensive Crisis Management Approach" (Johannsen 2011, p. 188). Though being encouraged by the military, they were also adapted by the civilian side toward facilitating comprehensive EU crisis management and thus reflect both the civilian as well as the military dimension of EU crisis management.[241]

238 See Hynek (2011, p. 90) and Gebhard (2011) for similar criticism. As Gebhard (2011, p. 28) states, "CMCO was neither expected nor probably intended to remove the inter-pillar conflicts in this specific context (…), it becomes clear that coherence was in fact intended to be achieved despite the actual gap between the pillars."

239 The *Suggestions for Procedures for Coherent, Comprehensive EU Crisis Management* are based on a reference framework paper by the then High Representative for CFSP Javier Solana and were agreed on in 2003 (Johannsen 2011, p. 187). On the Crisis Management Procedures see also Khol (2006, pp. 129).

240 For a detailed analysis of individual phases of the Crisis Management Procedures see Johannsen (2011, pp. 189-199).

241 On practical operational level, the EU's Crisis Management Procedures have hardly been followed and did thus not have much impact on the actual coordination of EU actors, as Johannsen's (2011) study revealed. As the documents lost relevance due to the Lisbon Treaty's institutional changes, the Crisis Management Procedures have been put under review at the time of writing which is expected to result in a new document in 2013. For an overview and evaluation of the review process see Mattelaer (2012).

The 2005 *Concept for EU Comprehensive Planning* (Council of the European Union 2005) forms another essential part of the CMCO doctrine and central concept. The document reemphasizes that "[c]omplex crises demand a comprehensive response including a mixture of instruments and actors" (Council of the European Union 2005, p. 4). Underlining comprehensive planning as an essential element of CMCO, it

> "is a systematic approach designed to address the need for effective intra-pillar and inter-pillar co-ordination of activity by all relevant EU actors in crisis management planning. It contributes to the development and delivery of a co-ordinated and coherent response to a crisis on the basis of an all-inclusive analysis of the situation, in particular where more than one EU instrument is engaged. It includes identification and consideration of interdependencies, priorities and sequence of activities and harnesses resources in an effective and efficient manner, through a coherent framework that permits review of progress to be made" (Council of the European Union 2005, p. 5).

Relevant for all phases of EU engagement, comprehensive planning is assumed to contribute to a coherent EU response. The coordination between the Commission and the Council is particularly stressed in this context (Council of the European Union 2005, pp. 4/5). The document stretches across EU pillars, institutions and member states to address the whole crisis management planning cycle and defines a framework for improving coordination of civilian and military actors. However, the fact that it only contains loose recommendations represents the document's main drawback (Drent 2011, pp. 8/9). In 2006, the 2003 CMCO doctrine was revised and amended against the background of the operational ESDP operations and the growing pressure for a CA in the EU. The 2006 CMCO revision proposes various possible solutions for improved management of EU crisis management operations particularly with regard to greater coherence at the planning stage, and an enhanced coordinating role for EU Special Representatives (SRs) (Council of the European Union 2006, p. 2).[242]

While above elaborated CA relevant concepts were mainly developed within the CFSP/ESDP framework of the then second pillar, conceptual adaptations also

242 "An eventual end-state would be a situation where EU crisis management operations are considered as a unified effort consisting of civilian and military missions, which exchange information and analyses on developed distributive networks" (Council of the European Union 2006, p. 3). For a detailed scholarly analysis and conceptual and terminological clarification of the EU's CIMIC and CMCO concepts and their developments within the EU, also see Khol (2006), Gebhard (2008) and Gross (2008). For an comprehensive listing on conceptual documents which aim at translating the CMCO concept into practice along specific areas of ESDP operations such as SSR, Planning and Mission Support as well as Disarmament, Demobilization and Reintegration see Johannsen (2011, pp. 203/204).

took place within the Commission's community pillar. These developments are reflected in the Commission's increasing external relations activities, and in a growing focus on international crisis management and on the security-development nexus. "[T]he Commission had slowly opened its remit to development cooperation with a political notion (...) [and, S.P.] started to develop a distinct Crisis Management or rather Conflict Prevention character" (Johannsen 2011, p. 216). Importantly, the Commission's 2001 Communication on Conflict Prevention entailed an approach to mainstream conflict prevention into the Commission's development policy thus broadening its development approach to the topic of crisis management (Aide à la Décision Économique 2011, p. 5). Underlining its cross-pillar direction, it emphasized that

> "EU crisis management, to be effective, will need to marry the full range of political and legal instruments at the disposal of the Union. This will include the forms of political co-ordination and financial assistance established under Common Foreign and Security Policy (CFSP) – declarations, démarches, joint actions and sanctions – and the legal and financial instruments developed by the European Community and its Member States in support of long term strategies of the Union" (European Commission Conflict Prevention and Crisis Management Unit 2003, p. 1).

In 2005, the Commission developed its concept on development policy coherence by issuing a *Communication on Policy Coherence for Development* (European Commission 2005), which seeks to take into account development cooperation goals in non-development policies. In doing so, the Commission aimed at applying policy coherence to 11 policy areas other than aid, considering the minimization of contradictions and building of synergies between these policies particular important.[243] Within the security area, the document focuses on the security-development nexus and states that security and development are "complementary agendas, with the common aim of creating a secure environment and of breaking the vicious circle of poverty, war, environmental degradation and failing economic, social and political structures" (European Commission 2005, p. 5). The document also refers to the ESS emphasizing that the "EU will enhance consistent and effective implementation of EU/EC external action in the area of security and development through the promotion of an integrated policy and instrument mix" (European Commission 2005, p. 10). In doing so, the Commission aimed at avoiding the creation of a hierarchy of crisis management instruments and to balance short-term responses and long-term strategies. To make progress toward these objectives,

243 The other areas were Trade, Environment, Security, Agriculture, Fisheries, Migration, Research and Innovation, Information Society, Transport as well as Energy (European Commission 2005, p. 2).

since 2005, the EU's Policy Coherence for Development (PCD) has undergone several implementation and progress reports which aim at encouraging a continual progress. In particular as of 2009, the EU's approach toward a new PCD called for the establishment of a new 'Whole-of-the-Union' approach which demands a more strategic, systematic and partnership oriented direction of the EU's PCD which better connects other policies and financial flows to development objectives (European Commission 2009, p. 7).

In addition, a work program has been developed to outline practical steps which should be taken by the Commission, the Council and the member states and the agenda has been developed further with an *EU Code of Conduct on Division of Labour in Development Policy* issued in 2007. The Code of Conduct defines complementarity between the EU and its member states as "the optimal division of labour between various actors in order to achieve optimum use of human and financial resources" (European Commission 2007, p. 5).[244] The guiding principles to achieve such a complementarity largely reflect EU CA objectives demanding that EU donors should center their activities on focal sectors on the basis of their respective comparative advantage and should also ensure an adequate EU presence in respective strategic sectors. Furthermore, EU donors obligate themselves to improve in other dimensions of complementarity primarily in the international arena (European Commission 2007, pp. 9-11). With this, the European Commission also subjected itself to a coherence strategy which aimed at increasing coherence between the various general directorates and agencies with their different instruments (Weiss/Spanger/Van Meurs 2009, p. 42). Nonetheless, a gap continued to exist between the Commission's policy commitment toward the CA and its actual implementation and the Commission has maintained a dominant developmental perspective instead of genuine crisis management and a clarification of the role it aims to play in the overall crisis management context (Aide à la Décision Économique 2011, pp. 94/95). Also respondents indicated that Commission key documents could be less exclusively development oriented and include a broader security perspective. Nonetheless, a slight shift of focus can be observed which broadened Commission activities toward more comprehensive development cooperation planning. In this context, the dimension of security is recognized as a precondition for poverty reduction and development (Johannsen 2011, p. 218). On the downside, however, the Commission also has continuously underlined the rivalry between EC crisis management and civilian ESDP instruments seeing itself in a position to conduct

244 The document contributes to the EU's efforts to implement a CA on external donor level, pointing out five dimensions of complementarity: in-country complementarity, cross-country complementarity, cross-sector complementarity, vertical complementarity as well as cross-modalities and instruments complementarity (European Commission 2007, pp. 5/6).

most EU civilian crisis management activities by its own and taking on a view seeing all civilian crisis management areas as a Community competency.[245]

In addition to the key documents mentioned above, the conceptual presence of the CA and crisis management coherence was ultimately underlined in 2009 with the adoption of the Lisbon Treaty (Office Journal of the European Union 2010) which is supposed to increase EU policy coherence, effectiveness and institutional flexibility as mentioned in the preamble already.[246] Reforms envisaged by the Lisbon Treaty are designed to improve the coordination of EU crisis management with the terms coherence and effectiveness being the two guiding phrases for reform (Drent 2011, p. 12). In doing so, the document heavily focuses on inter-institutional coordination. In accordance with this coordination-oriented approach, the document changes the EU's crisis management framework by introducing various institutional modifications to increase EU foreign and security policy coherence and by restructuring the institutional architecture of EU crisis management. It integrates the EC framework into the EU and makes the Union method the predominant decision-making character, with the exception of CFSP which remains of intergovernmental character with the authority of member states acting via the European Council. The treaty's implications relevant for EU international crisis management can be broken down into the areas of general provisions on the EU's external action, changes introduced to the pillar structure, the new post of the President of the European Council, the reinforced position of the High Representative for Foreign Affairs and Security Policy, the creation of the European External Action Service, and the provisions on coordination of member states' foreign policies (Juncos 2008). The Lisbon Trea-

245 The Commission, for instance, stated in a contribution to the EU's CMCO conceptualization that "[t]he majority of civilian crisis management operations can in fact be handled within the framework of Community instruments even if at first sight they might seem to come under the CFSP. (…) [A]ll these are covered by Community instruments" (Council of the European Union 2002, p. 11). This observation was also underlined during background interviews.

246 The EU is ruled by treaty. The Lisbon Treaty was signed by the heads of EU member states in 2007 and entered into force in 2009. It has maintained most innovations in the area of external action originally outlined in the 2004 Treaty establishing a Constitution for Europe. However, other than the Constitutional Treaty the Lisbon Treaty does not replace the former treaties, but amends them (Blockmans/Wessel 2009, p. 8). Since its implementation the EU has been governed by the Treaty on the European Union (TEU) and the Treaty on the Functioning of the European Union (TFEU). While the TEU sets out broad EU principles and objectives, the general role of its institutions, principles for membership applications and details on the operation of CFSP, the TFEU deals with the details on how the Union is to function. Both, the TEU as well as the TFEU are referred to under the heading of the Lisbon Treaty in this dissertation. For an elaboration of the rejection of the Constitutional Treaty in France and the Netherlands and the developments accompanying further developments until the ratification of the Treaty of Lisbon see Bache, George and Bulmer (2011, pp. 211-220). For an evaluation of individual treaty provisions also see Bache, George and Bulmer (2011, pp. 225-228).

ty reforms thus aim at achieving three dimensions of coherence: among EU institutions, among the member states and between EU-institutions and EU member states.[247] In addition, the Lisbon Treaty also gives the EU a single legal personality which will ultimately make it a more visible actor on international stage (European Peacebuilding Liaison Office 2012a, p. 2).

Conceptually and terminologically, the Lisbon Treaty merges the various dimensions of the EU's foreign policy under the new headline of 'External Action' and thus integrates the objectives of EU external policies, from security over development to trade and environment. Terminologically bringing together the EU's external policies under this rubric, the treaty moves away from the traditional rhetoric division between areas managed by the Commission and those of member states control in the Council toward a more unitary foreign policy and crisis management system (Juncos 2008). The documents explicitly states that

> "[t]he Union shall ensure consistency between the different areas of its external action and between these and its other policies. The Council and the Commission, assisted by the High Representative of the Union for Foreign Affairs and Security Policy, shall ensure that consistency and shall cooperate to that effect" (Office Journal of the European Union 2010, p. 29).

It thus holds a procedural obligation for the relevant actors to cooperate with each other (Blockmans/Wessel 2009, p. 28). Furthermore, the Lisbon Treaty expanded the range of the Petersberg Tasks to include

> "joint disarmament operations, humanitarian and rescue tasks, military advice and assistance tasks, conflict prevention and peace-keeping tasks, tasks of combat forces in crisis management, including peace-making and post-conflict stabilisation" (Office Journal of the European Union 2010, p. 39).

In line with this, the document *Towards a Comprehensive Approach – The Operating Implications* by the EUMS gave renewed momentum to the debate on conceptual and practical considerations of the Union's CA by demanding a policy on how the EU plans to implement its CA. The document particularly criticizes that neither an EU definition for a CA nor a conceptual framework on how the EU seeks to accomplish the CA exist (Johannsen 2011, pp. 21/22). However, it strives a characterization of the EU's CA by declaring that

> "there is broad agreement that it implies the pursuit of a methodology aimed at commonly understood principles and collaboration processes that enhance the likelihood of

247 The Lisbon Treaty's institutional reforms and structural changes with regard to EU crisis management are analysed in detail in the next chapter.

> favorable and enduring outcomes in the political, diplomatic, security, economic, development, rule of law, and human rights dimension of international engagements in pursuit of a common goal within and beyond the EU" (EUMS, as cited in Johannsen 2011, pp. 57/58).

Still, this attempt rather paraphrases existing EU declarations on the CA and the concept remains vaguely defined at EU level.

3.2.2 Organizational Dimension

Moving from the EU's conceptual framework for CA implementation at document and treaty level to organizational efforts, it is essential to note that the coordination of EU crisis management was not on the EU's immediate agenda. As the EU's crisis management operationalised with the creation of its CFSP/ESDP, it "was first and foremost concerned with the development of the EU's operational capabilities and the setting up of the necessary institutional framework" (Johannsen 2011, p. 153). Due to this initial focus on developing operational military and civilian capabilities themselves, the Union's civilian and military structure separated into different worlds shortly after their creation (Norheim-Martinsen 2010, p. 7). However, as these structural fragmentations represented a fundamental problem in achieving crisis management coherence, this focus soon shifted to acknowledging the necessity of crisis management coherence. As a result, reorganizing efforts were implemented to address structural shortcomings and the division was aimed to be loosened and bridged in structural terms with the creation of diverse institutions before the entry into force of the Lisbon Treaty.[248]

248 It was the background of first operational ESDP experiences that led the EU to create several institutional provisions which aimed at increasing crisis management coherence such as the Civil-Military Cell and Crisis Response Coordinating Teams. Both can be highlighted as innovations which aimed at bridging gaps in the crisis management planning process before the Lisbon Treaty. See Gordon (2006, pp. 352-355) on the Crisis Response Coordinating Team and the Civil-Military Cell, and Brauss (2006, pp. 52/53), Khol (2006, pp. 130-132), Drent and Zandee (2010, pp. 32-34) as well as Johannsen (2011, pp. 163/164) for an elaboration of the Civil-Military Cell. For a detailed description of individual crisis management bodies and mechanism in the process of launching EU civilian and military missions see Hanssen (2010). EU SRs which have been deployed in various crisis regions to contribute to EU crisis management efforts can also be viewed as an organizational attempt to ensure the coordination of different EU crisis management activities. However, as EU SRs serve as a coordination hub and mechanism in the operational field level, they are not analyzed in this dissertation. For an examination of the role of EU SRs as part of the EU crisis management instruments after the entry into force of the Lisbon Treaty see Tolksdorf (2012) and Peter (2012).

Though not directly reflecting the organizational dimension of CA implementation but nonetheless important on this level, is the institutional rivalry between the Council structures and the Commission which has steadily increased with the expanding crisis management activities of the Council's CFSP/ESDP. This institutional rivalry and protection of competencies became most visible in 2005 when the Commission took the Council to the European Court of Justice in the so-called *ECOWAS case* which "is a milestone in the case law of the Court of Justice on EU external competence, on the functioning of the EU system of external relations and, more generally, on the present and future constitutional architecture of the Union" (Hillion/Wessel 2009, p. 552). In the case, Commission challenged the legality of a Council Decision on EU support to fight Small Arms and Light Weapons to the Economic Community of West African States (ECOWAS). The Commission argued that the Council had encroached upon Community competencies in the field of development cooperation. In 2008, the Court ruled in favor of the Commission and "[i]n doing so, (…) bolstered the Community method in the development of the EU external action, and circumscribed the operation of the CFSP form of cooperation therein" (Hillion/Wessel 2009, p. 552).[249] In spite of their interdependency, interaction between the two institutions is characterized by differing presumptions and disputes on responsibilities and competences. While the CFSP/ESDP was build up for quick responses to acute crises with a security political background, the Commission is traditionally profoundly based in the development of political long-term commitment in terms of humanitarian aid and development cooperation. Uneven crisis management experiences due to the different institutional developments, cultures, timelines and working habits of the two institutions further stimulated the institutional rivalry of the two EU bodies (Johannsen 2011, pp. 341/342), as already indicated in the conceptual dimension.[250]

The Lisbon Treaty introduced several structural innovations to adjust the EU institutions to contemporary challenges of comprehensive crisis management and to improve the highly criticised lack of coherence in EU crisis management. Importantly, the Lisbon Treaty abandons the EU's three pillar structure.[251] However,

249 See Hillion and Wessel (2009) for greater details on the background of the case and the judgment of the Court.

250 Studies showed that politics of bureaucracy play a major role regarding the rivalry between Commission and Council bodies with respective officials largely impacting inter-institutional relations by promoting and protecting their own institution's interests. See, for instance, Juncos and Pomorska (2010) as well as Vanhoonacker, Dijkstra and Maurer (2010).

251 In this regard, the most radical change took place in the pillar of Justice and Home Affairs related to cooperation in the areas of criminal justice, terrorism, organized crime and immigration which has been integrated into community structures, replacing unanimity with qualified majority voting in this policy area. Though a topic of heavy debate in particular between member states, it is beyond the scope of this dissertation, to explain the EU's different decision-

despite the formal dissolution of the three pillar structure, the underlying principles of unanimity and intergovernmentalism of the CFSP remain as the Lisbon Treaty specifically states that CFSP "is subject to specific rules and procedures" (Office Journal of the European Union 2010, p. 30). Thus, CFSP/CSDP remain separate from the EU's other external relations policies such as trade, development cooperation with third countries, humanitarian aid or relations with IOs. In consequence, two structurally separate crisis management systems are maintained with CFSP remaining a specific policy field on its own with a separate status effectively resulting in a two-pillar structure of EU crisis management (Blockmans 2011, p. 7). Due to the remaining fragmentation of EU crisis management, it is arguably the institutional modifications and innovations of the Lisbon Treaty which should spur the drive for more crisis management coherence.

A major structural provision of the Lisbon Treaty is the establishment of a permanent *President of the European Council* for two and a half years implying the ending of the rotating six-month presidency of the European Council which is expected to improve coherence, and provide better continuity and consistency.[252] The President is appointed by agreement among the member states. As laid out in the treaty,

> "[t]he President of the European Council shall, at his level and in that capacity, ensure the external representation of the Union on issues concerning its common foreign and security policy, without prejudice to the powers of the High Representative of the Union for Foreign Affairs and Security Policy" (Office Journal of the European Union 2010, p. 23).

Being obliged to also work closely with the Commission, the treaty places the post directly at the intergovernmental and supranational interface to manage

---

making procedures and Council voting arrangements such as the qualified majority voting as well as decisions taken by unanimity. Even more as the Lisbon Treaty has not fundamentally changed the decision-making procedures relating to EU external policy and crisis management. Matters which were EU reserves of are still mainly driven by the Commission, while CFSP issues generally continue to be decided by unanimity in the Council. A general explanation of the EU's decision-making procedures including an overview of treaty articles that were switched to qualified majority voting is provided on the EU's summary of legislation (European Union n.d.) and the EU Glossary (EU Glossary n.d., Qualified Majority). On decision making procedures for CFSP see the European Peacebuilding Liaison Office's report (2012a, p. 4).

252 Under the Lisbon Treaty, the European Council is made up by the Heads of state and governments, the President of the European Council and the President of the European Commission. The European Council should not be confused with the Council of the EU which is the institution composed by member state representatives at ministerial level representing member states and a legislative body. The rotating presidency continues to exist in other policy areas other than CFSP (European Peacebuilding Liaison Office 2012a, pp. 2/3). At the time of writing, the position was held by former Belgian Prime Minister Herman Van Rompuy.

activities and to promote consensus. The President of the Council thus aims at ensuring vertical coherence among the member states' heads of state. However, in the new arrangements, member state foreign ministers do not take part in the European Council anymore and neither the Foreign Affairs Council nor the High Representative are a formal part of the preparatory process of the European Council (Helwig/Ivan/Kostanyan 2013, p. 2).

The main innovation of the Lisbon Treaty from CA perspective is the creation of the position of the *High Representative of the Union for Foreign Affairs and Security Policy/Vice-President of the European Commission* (HR/VP). The HR/VP merges and performs crisis management duties which were formerly separated between three official EU posts: the High Representative for CFSP, the respective foreign minister of the rotating presidency member state, and the Commissioner for External Relations (Hynek 2011, p. 81). A large amount of EU crisis management responsibilities are thus under the coordination of a single person. In doing so, the HR/VP is equipped with several competences and responsibilities which shall enable the post to increase EU crisis management coherence. Indeed, according to the treaty it is the HR/VP itself which "shall ensure the unity, consistency and effectiveness of action by the Union" (Office Journal of the European Union 2010, p. 31) and who should implement the EU's CFSP. Unlike other Council configurations which are chaired by the six-monthly member state presidencies, the HR/VP chairs the Foreign Affairs Council made up by the foreign affairs ministers and has the right to initiative to "contribute through her proposals towards the preparation of the common foreign and security policy" (Office Journal of the European Union 2010, p. 32). Furthermore, the HR/VP represents the EU in CFSP matters, undertakes political dialogue with third parties and articulates EU positions in IOs and conferences (Office Journal of the European Union 2010, p. 32). In crisis situations, the HR/VP shall maintain rapid decision-making and ensure the coordination of the civilian and military aspects (Office Journal of the European Union 2010, p. 39). As this enumeration of the post's responsibilities shows, the HR/VP's tasks range from the coordination of internal CFSP decision-making processes, the safe-guarding of coherence with the Commission and of member states, the support of their role as CSDP driving forces and the creation of an EU crisis management profile.[253]

253 The choice of Catherine Aston as first HR/VP was highly debated due to the rather weak EU profile, lack of support from member states, lack of ambition and the reserved acting of the former Commissioner for Trade. Discussions on her perceived lack of leadership and failure to distinguish herself as a driving force continued throughout the time of writing. In particular, as there has been no shortage of profile-making chances due to the events of the Arab Spring and the possibility to define an EU response. See also Major and Bail (2011, pp. 28/29) who argue that HR/VP's commitment is crucial to make the provisions of the Lisbon Treaty effective.

As the coordinator of the overall EU foreign and security policy, the HR/VP is envisaged to overcome the dichotomy between the Commission and the Council and to fulfill the role of a leadership figure. As mentioned above, the post unifies crisis management domains that used to be fragmented between the supranational organized Commission and the intergovernmental Council. This bridging function is reflected in the 'double-hatted' role of the High Representative also being the Vice President of the European Commission and responsible of its external action. The Commission can only put joint CFSP initiatives forward together with the HR/VP who seeks to ensure consistency between the positions of two. The Commission's function regarding CFSP tasks with most of its previous responsibilities, including the right to initiative, has thus been reduced.

The positioning of HR/VP at the interface between supranational and intergovernmental EU crisis management is of high importance from the coherence perspective. However, this provision largely relies on the assumption that the HR/VP and the Commission share the same view on a matter. Within the Commission, the HR/VP coordinates communitarian aspects of the Union's external action, such as external dimensions of policies like agriculture or environment. However, the HR/VP heads the Commission in these external dimensions only and not in trade or development issues. The Commission still has civilian crisis management instruments such as humanitarian aid, development cooperation and democracy support at their disposal. It is the President of the Commission who remains responsible for the Commission to operate in a consistent and efficient way as a collegiate body (Juncos 2008). As indicated during background interviews, the post of the HR/VP, though being established to link the Commission and the Council, is effectively trapped between two worlds and it is difficult to walk the fine line between them. Additionally, it has been critically commented that there is no explicit mechanism dealing with the allocation of responsibilities, roles and consistency between the two positions of the President of the Council and the HR/VP. This ambiguity might result in conflicting or overlapping roles of the two new posts in EU external policy.

On CFSP side, the HR/VP is responsible for the whole decision-making process from working group level up to the formation of the Foreign Affairs Council. In doing so, the HR/VP has the right to come forward with proposals to the Foreign Affairs Council which signifies a considerable advancement of the previous post of the High Representative for CFSP and allows the HR/VP to identify strategic objectives of the EU. In this function, the HR/VP replaces the rotating six month member state CFSP presidency with the objective to reduce potential inconsistencies between national presidency agendas and the continuing EU crisis management priorities.[254] Though the reform did not replace responsibilities of the member state foreign ministers or diminish the national crisis management responsibility, policy is planned to be conducted in a more coherent fashion. The HR/VP also leads the Political and Security Committee (PSC)[255] and the Committee for Civilian Aspects of Crisis Management (CIVCOM),[256] sets the agenda and chairs respective meetings (Major/Bail 2011, pp. 27/28).[257]

According to a background interview, the construction of the HR/VP post means the personification of the CA and crisis management coherence with the objective to maximize synergies between previously divided arrangements. It combines in one person the different characters, logics and lines of interests of EU crisis management institutions. The double-hat embodies inter-institutional coherence. In addition to ensuring effectiveness, it is also supposed to enlarge visibility and credibility of EU external action. As observed by a respondent, the role of the former post of the High Representative of CFSP was to give a face to EU external action. The post did not serve as an instrument to ensure coherence which is the new role to be taken up by the HR/VP. From coherence perspective and with regard to CA implementation, the HR/VP fulfills several functions. With regard to horizontal coherence, the post allows for an articulation of formerly separated crisis management policies between the Commission and the Council, but also intra-institutionally aims at merging approaches to crisis management. Vertically, the post also aims at coordinating member state and EU action in meetings of the Foreign Affairs Council.

In its coherence-related mandate, the HR/VP is supported by its own apparatus, the *European External Action Service* (EEAS) which was officially launched in 2010 (Council of the European Union 2010). The EEAS directly reports to the HR/VP and also represents the establishment of a joint diplomatic

254 As mentioned above, the position of the HR/VP has widely been described as double-hatted in public as well as in respective literature. However, as it assumes the responsibilities of the former High Representative for CFSP, the Vice-President of the Commission as well as the ones of the six-month rotating presidency for CFSP matters, including chairing the Foreign Affairs Council, it is actually at least triple-hatted.

255 The PSC is a permanent EU body for member state representatives at the ambassadorial level who discuss international security issues and developments and prepare the meeting of the Foreign Affairs Council. It constitutes a key decision-making forum offering strategic oversight and direction (European Peacebuilding Liaison Office 2012a, p. 4).

256 CIVCOM is a committee of experts advising the PSC on matters of civilian crisis management by drafting policy recommendations, accompanying capacity building, developing policy strategies, supporting EU-internal coordination as well as coordination between the EU and member states (Siedschlag/Eder 2006, pp. 11/12).

257 Furthermore, the HR/VP appoints representatives to chair other CFSP bodies and Council working groups in the area of CFSP/ESDP (Juncos 2008). The HR/VP also heads the European Defence Agency (European Peacebuilding Liaison Office 2012a, p. 5).

service for the EU. In doing so, the institution operates and cooperates closely with the diplomatic services of the EU member states, where one third of the personnel of the EEAS come from. Another third of the personnel of the EEAS is drawn from the Commission and the last third comes from the relevant departments of the Council Secretariat (Office Journal of the European Union 2010, p. 32). According to background talks, the body is expected to fill a gap in the EU's institutional crisis management machinery created by the absence of a body to coordinate EU external action. The EEAS has thus been explicitly created to improve coherence and efficiency of EU external action and to externally represent the EU. This included an entire restructuring of the Council's Secretariat and CSDP structures as well as the Commission's DGs which have been merged under the guidance of the HR/VP. Like the HR/VP, the EEAS mirrors the idea of an institutional merger between the Commission, the Council and member states. Its major task is to provide conceptual policy input as well as training capabilities and, in doing so, to lay the basis for EU administrative crisis management decision-making (Major/Bail 2011, p. 29). The EEAS also adds to the consistency of the EU's external actions by maintaining the geographical desks in the service as a source of advice and briefing on particular countries for the President of the European Council and the Commission President as well as other Commissioners.[258] Thus, the EEAS is integrated into the inter-service consultation machinery of the Commission which allows it to contribute to discussions on any policy with external relations implications. Both the Commission and the Council Secretariat have thus strengthened their services in charge of coordination of external relations issues (European External Action Service 2011). However, this also means that the EEAS as an inter-institutional body is expected to serve several masters.[259]

From crisis management coherence perspective, the EEAS encompasses military and civilian crisis management instruments such as police, judiciary and civil administration of the EU's CSDP. Within the EEAS, issues are dealt with along both geographical and thematic lines. Geographical EEAS desks are involved on a strategic decision-making level in the programming chain. On a thematic level, the EEAS takes on several former Council and Commission tasks (Hynek 2011, p. 84). A number of specialized support structures have been established to conduct the operational planning and implementation of CSDP crisis

258 In total, 937 briefing requests were dealt with by the EEAS from January 1 to September 30 2011. 243 of these were handled for the HR/VP, 67 for President of the European Council, 125 for the Commission President and 235 for the Commissioner for Enlargement. According to the 2011 EEAS review, the achievement of this coordination mechanism becomes most obvious in summit preparation and other high-level meetings (European External Action Service 2011).

259 For an assessment of the EEAS mandate, also see Blockmans (2011, pp. 9-13).

management. To this effect, a *Civilian Planning Conduct Capability* (CPCC) office to run civilian missions as well as a Joint Situation Centre for intelligence analysis and threat assessment form the civilian counterparts to the EUMS and the EUMC which provide military expertise and advice to the HR/VP.[260] The inclusion of the EUMC and EUMS inside EEAS structures augurs well for enhanced civil-military cooperation (Hynek 2011p. 93). Being attached to the HR/VP, the service has largely absorbed the organization and the chain of command from the Council Secretariat.

The opportunity provided by the EEAS as a cross-institutional crisis management body lies in the role of coordinating between institutions, member states and officials in the development of crisis management policies. From the coherence perspective, the importance of the EEAS can largely be ascribed to this unique set-up and its "specificity as common bureaucratic machinery" (Gebhard 2011, p. 31). Just as the HR/VP, so does the EEAS increase vertical coherence between the EU crisis management institutions and processes and the member state level. In addition, the EEAS plays a role as an organizational merger also on lower working levels suggesting high synergies.[261] According to interviewees, its composition and mixing of officials is an important tool for the EU's CA. The whole idea behind the creation of the EEAS has been the coordination of different institutional procedures, cultures and individuals as well as the elimination of a huge overlap of competencies. Yet, as indicated in background interviews, establishing greater coherence and coordination are both internally as well as externally challenging for the EEAS, specifically because of the different institutional cultures, management styles, fields of expertise and divided loyalties.[262] Respondents doubt that these

260 The CPCC's creation occurred in response to the lack of planning and command structures for civilian missions and by mirroring provisions on the military side to provide a civilian counterpart to the EUMS. It is responsible for conducting civilian crisis management missions and operational question concerning mission support and planning as well as monitoring and evaluation of civilian missions. The CPCC was declared operational in 2008 (Drent/Zandee 2010, p. 35).

261 As Gebhard (2011, p. 31) observed, earlier reforms to improve EU crisis management coherence largely dealt with the highest level of pillar interaction only while the respective bureaucratic communities of both institutions remained heavily disconnected.

262 As mentioned above, the EEAS was legally established in 2010 and it had to be set up at the beginning of the first HR/VP's tenure. A strategic orientation of the EEAS was thus defined rather late and the working-capacity of the EEAS and work of the HR/VP during the start-up phase was limited due to posts not being staffed by then (Major/Bail 2011, p. 28). Also see Burke (2012b, p. 2) for an analysis of difficulties during the EEAS' set-up. Interviewees perceived the period of the EEAS' set-up effectively as a time of halt coined by uncertainties about competencies and authority, placements within the new structures and practical challenges such as a separation of personnel among the different EU buildings and the non-existence of personalized e-mail-addresses. In addition, decisions on personnel and recruiting were repeatedly postponed. These circumstances seem to have effectively restricted enthusiasm with regard to the EEAS as a new institution and con-

divided loyalties within the EEAS can generate new coherence and estimate that coordination requires a large amount of high-level political impetus. In particular inside EEAS structures, disputes arrived over competencies and insufficient coordination, which impacted on its capability to act.

The EEAS entails all crisis management structures which were previously in the Council Secretariat. To further adjust EU structures to the challenge of comprehensive crisis management, the Council's DGs Defence and Civil Crisis Management were merged to a new unit. The *Crisis Management and Planning Directorate* (CMPD), which launched work in 2010 as a department of the EEAS, is supposed to ensure better coordination of civil and military planning, to integrate political-strategic direction, conduct and capability development as well as lessons learned within the CSDP framework but also with commission activities (Major/Schöndorf 2011, p. 5). Just as the EEAS itself, also the CMPD is regarded "one of the most visible material outcomes of the ambition to develop a comprehensive approach within the European institutions" (Wendling 2010, p. 28). Its creation is another institutional adaptation to achieve crisis management coherence, a major EU CA effort and decisive move toward civil-military integration in order to unify civilian and military planning at strategic level. However, the EUMS remains responsible for strategic planning of military operations. In this function, the CMPD serves as a coordinator responsible for forging links with other bodies, in particular with the EUMC, CIVCOM, the Commission and the member states. The role of the CMPD also is to ensure cost effectiveness of allocated resources and the avoidance of duplication with NATO or the UN (Hynek 2011, pp. 91/92). Mixing military and civilian staff, the CMPD is led by a civil Deputy Director General with a military counterpart (Drent/Zandee 2010 p. 36). While the CMPD does not unite the separated civilian and military worlds, it strengthens their connection.

Acknowledging that successful coordination of the whole range of civilian and military crisis management tools continues to be the crucial challenge surrounding efforts toward ensuring a CA at all stages of the EU crisis response cycle, some additional crisis management coordination mechanisms have been established within the EEAS. The Crisis Response and Operational Coordination Department is responsible to assist the HR/VP in her task to ensure the coherence and coordination of EU crisis management and to bring together all EEAS

tributed to a widely negative perception of the EEAS from EU staff perspective. On this issue see also Rettman (2011) who reports on difficult working conditions in the EEAS. The issue is also picked up in chapter 5.3, which *inter alia* suggests an organization-psychological path for further research. Yet, at the time of research for this dissertation, officials assessed a great future potential to the EEAS but not at present or the near future.

services dealing with that particular crisis (European External Action Service n.d., Crisis Response). Aiming at managing the complexity of crisis management issues, the EEAS Crisis Management Board, the EEAS Crisis Platform and the EU Situation Room are additional new EEAS bodies in this regard. The Crisis Management Board which is chaired by the HR/VP coordinates the actions of EU capabilities to crises of all types (European External Action Service 2011). The concept of a Crisis Platform has been introduced to the EU's crisis management structures to respond to specific conflicts or natural disasters and is of particular relevance from CA perspective. According to interviewees, the Crisis Platform effectively materializes as a roundtable of all stakeholders such as the CMPD, the EUMS and EUMC, the CPCC, the relevant geographic desks and Commission departments on the specific issue at hand and it can be considered a major CA tool. From CA perspective, the EU Situation Room is an important tool for developing cooperation and information exchange arrangements with relevant crisis management structures of EU member states and other EU institutions (European External Action Service 2011). Another inter-institutional unit set up is the Conflict Prevention Group which includes various EEAS departments, the chairs of CIVCOM, a Foreign Policy Instruments Service, the Directorate General for Development and Cooperation – EuropeAid (DG DEVCO) and the Commission's Directorate General for Humanitarian Aid and Civil Protection (DG ECHO) to review early warning and develop early response (European Peacebuilding Liaison Office 2012a, p. 5).

As assessed by respondents, coordination of EU crisis management activities was the whole idea behind the creation of these new departments within the EEAS. However, at the same time, questions of responsibilities and an overlap of competencies with other departments have complicated the fulfillment of this objective. Within the new structures, conflicts particularly occur between the CSDP crisis management structures consisting of the CPCC and the CMPD on the one side and the geographic and thematic desks on the other. In addition, the CPCC and the CMPD compete with each other over responsibilities on operational questions on civilian crisis management as a clear line between the civilian crisis management planning taking place in the CMPD and actual implementation by the CPCC is difficult to draw (Major/Bail 2011, p. 30).[263]

263 Major and Bail (2011, p. 32) also point to the debate on military dominance of EU crisis management structures. Contributors to this debate list staff background and numbers, structures and processes and well as strategic developments as criteria for a possible militarization of EU crisis management. From this perspective, civilian institutions have indeed fewer supporting structures at their disposal and more CMPD staff has a military background than a civilian one. Furthermore, the construction of CPCC was guided by military processes and structures. Proponents of the militarization effect argue that this development leads to a shaping of civilian

The Commission's external action and crisis management areas underwent some restructuring after the entry into force of the Lisbon Treaty as well. The DGs most involved in the EU civilian crisis management before the entry into force of the Lisbon Treaty were the Directorate General for External Relations (DG RELEX), the DG for Development and Relations with ACP States (DG Dev) and DG ECHO for Humanitarian Aid.[264] From an organizational perspective, the Commission is a strictly vertically organized system that like government systems is specialized mainly by purpose. As a result, organizational capabilities for everyday horizontal coordination are rather weak while horizontal coordination at the top through 'presidential' command is comparably strong (Trondal 2009, p. 241). The splitting of the Commission's different DGs contributed largely to the internal fragmentation of EU crisis management instruments. The new arrangements of the Lisbon Treaty aimed at resolving the separation of the DGs relevant for crisis management by transferring DG RELEX entirely and DG Dev partly to the EEAS. In this regard, the Commission's external cooperation programmes remain under its responsibility and the EEAS should contribute to the programming and management cycle and has responsibility for the preparation of the decisions regarding strategic, multiannual steps (Helwig/Ivan/Kostanyan 2013, p. 38). At the same time DG DEVCO incorporated the former Development and EuropeAid DGs. DG ECHO integrated the field of civil protection and humanitarian aid for a better coordination and disaster response which were merged under the new Commissioner for International

crisis management missions by a military mindset and patterns. While it is heavily discussed how this military dominance shows in practice, the issue has not been systematically examined yet. Major and Bail (2011, p. 32) resume in this regard that "[t]he military's considerable planning expertise should be acknowledged at some point."

264 From 1999 until 2009, the European Commission implemented its external relations responsibilities through the DG RELEX, DG Dev, the DG for Trade, the DG for Enlargement, and the DG for Humanitarian Aid. As analysed by Johannsen (2011, pp. 207/340), the growing external relations responsibility of the Commission resulted in regular reorganization and restructuring of portfolios in the last decade. The Commissioner heading DG Dev was also responsible for the DG Humanitarian Aid. Tasked with international commercial, investment and competition policy DG Trade is concerned with crisis management due to the obligation to include human rights and democracy clauses in trade agreements with third countries. Similarly, DG Enlargement's crisis management role is limited to accession negotiations with possible future EU member states. For an elaboration of the institutional architecture of the Commission for civilian crisis management and the distribution of responsibilities and competencies between DG RELEX, DG Dev and DG Humanitarian Aid until the implementation of the Lisbon Treaty see Johannsen (2011, pp. 208-210). Institutionally, already in 1999 a CFSP Directorate was established in DG RELEX to coordinate the Commission's input to CFSP. For more details on the European Commission's institutional adaptations to CFSP developments see Johannsen (2011, pp. 167-170).

Cooperation, Humanitarian Aid and Crisis Response. In an information note issued by the Commission President, the HR/VP is tasked to chair the group of Commissioners responsible for external relations. In addition to the Commissioner for Development and the Commissioner for International Cooperation, Humanitarian Aid and Crisis Response the group includes the Commissioner for Economic and Monetary Affairs, the Commissioner for Enlargement and European Neighborhood Policy and the Commissioner for Trade (Blockmans 2011, p. 11).[265] However, "the coordination potential of the entire external relations group of Commissioners has not been fully used" (Helwig/Ivan/Kostanyan 2013, p. 34) as it only met very few times.

In this process of structural modification and transfer of competencies to the EEAS, the Commission continued to vigilantly observe the maintenance of its competencies which led to power struggles particularly with member states. In the end, these disputes were decided at the expense of the EEAS. In spite of their importance for EU external action and civilian crisis management, areas such as development cooperation stayed with the Commission.[266] Conflict about competences also continued in domains formally ascribed to either DG RELEX, which is now with the EEAS and DG Dev, in particular in the area of humanitarian aid or disaster relief. It is precisely the field of crisis response, however, where the HR/VP sees a clear CSDP responsibility. As both sides claim responsibility, coordination is poor, resulting, for instance, in an insufficient sharing of information (Major/Bail 2011, p. 33). Controversy also occurred on the role of development assistance within the EU's new crisis management structures. This became evident in a background interview during which concerns were raised that by attaching policies of humanitarian assistance to the EEAS and consequently bringing them together with CFSP, these policies run the risk of becoming too politicized. However, it was also underlined in background interviews that the Commission has grown autonomous on its own terms and with a self-image of being independent from other actors. This development, according to a respond-

265 The Lisbon Treaty states that the President of the Commission shall "decide on the internal organisation of the Commission, ensuring that it acts consistently, efficiently and as a collegiate body" (Office Journal of the European Union 2010, p. 25). For a detailed analysis of the reconfiguration of crisis management competencies in the Commission see Van Vooren (2011).

266 As analysed by Blockmans (2011, p. 11) "[t]he risk of classic turf wars rearing their ugly heads is also the consequence of the pre-Lisbon manoeuvring by then Commission President designate Jose Manuel Barroso" whose actions shortened the HR/VP's responsibilities as entrusted to the post by the treaty. Removing the responsibility for the European Neighborhood Policy from the portfolio of External Relations to that of Enlargement, for instance, has effectively curtailed the HR/VP role who thus, in conflict with the post-Lisbon coherence driven process, represents fewer portfolios (Blockmans 2011, p. 11).

ent, explains the high level of interest by the DGs to remain institutionally independent.[267] Moreover, the initiatives taken by the Commission to improve coordination on various levels rather consisted of enhanced exchange of information than increasing complementarities and coherence with other EU actors (Aide à la Décision Économique 2011, p. iii).

EU funding mechanisms and financial arrangements are a highly debated topic also with regard to crisis management coherence as "[c]oordination both between the CSDP budget and commission funds and between their respective planning is a precondition for success" (Major/Bail 2011, p. 34).[268] Seen from the coherence perspective, EU crisis management funding is separated along the lines of resourcing the civilian and military side of CSDP, the Commission and member states. While civilian missions are mainly funded under Commission budget, military operations are financed by member state contributions. EU crisis management funding thus differs in functioning and structure. The variety of budget holders, different conditions and disbursement mechanisms as well as a fragmentation of funds that can be set aside for crisis response impede coherence of instruments to deal with crises (Drent/Zandee 2010, p. 43). Though this does not automatically lead to opposing decisions, it makes coordination essential and leads to loosing synergy effects and duplication. While the "fragmentation of financial instruments is undesirable in terms of coherence, effectiveness and clarity (…), and with regard to democratic legitimacy, speed of implementation and decision making" (Drent/Zandee 2010, p. 50), the pooling of funding, as considered and discussed on the national level has been argued to make little institutional sense at EU level as the Community budget can already be under-

267 For a highly detailed analysis of coherence and coordination between the EEAS and the Commission and its individual DGs see Helwig, Ivan and Kostanyan (2013).

268 According to Rieker (2007, p. 15) "[f]or civilian missions, there are at least four ways of funding" of which some are funded through the European Development Fund or by national contributions directly which is the case for most *ad hoc* missions. CSDP operations with a military component may be financed through the Athena mechanism or through national contributions based on a 'costs lie where they fall principle' (Rieker 2007, p. 15). The Athena mechanism which finances military CSDP operations "is a permanent administrative and cost-sharing mechanism which covers certain pre-identified common costs of an operation and is financed by all 26 member states (except Denmark)" (Drent/Zandee 2010, p. 50). For a detailed elaboration of the Athena mechanism see Drent and Zandee (2010, pp. 50-52). A detailed unraveling of the "financial conundrum" (Drent/Zandee 2010, p. 3) of EU external action in general and crisis management in particular is beyond the scope this dissertation. For a detailed elaboration see Rieker (2007, pp. 14-22) as well as Drent and Zandee (2010, pp. 43-56).

stood as a pooled budget managed by the Commission as a single actor.[269] The possibility of pooling of the Community with the CFSP budget

> "does not merit serious consideration, given the comparatively small amounts in the CFSP budget, the rare and exceptional circumstances in which it is used, and, perhaps even more importantly, the legal barriers that make it impossible within the existing institutional structures" (Banim 2009, p. 342).

Consequently, rather than the pooling of funding, proposals for complementary funding are discussed in the EU. In this context, the *Instrument for Stability* (IfS), which became operational in 2007, is a key element aiming at closing the gaps between CSDP emergency financing and long-term assistance and has been described by a respondent as a sign toward joint resources. The instrument which is shared between the Commission and the EEAS provides flexible decision-making and budget allocation measures on a case-by-case basis of rapid crisis reaction (Banim 2009, p. 343). In doing so, it finances both short-term as well as long-term projects. The IfS allows the Commission to provide short-notice funding in projects complementary to CSDP missions and permits the financing of actions when other means of funding are not available in a timely manner (Hanssen 2010, p. 18).[270]

In response to the creation of the EEAS, a *Foreign Policy Instruments Service* has been established as a new department within the Commission but under the responsibility of the HR/VP. The service has been created to handle foreign policy matters under Commission mandate which were not transferred to the EEAS but at the same time fall outside the mandate of DG ECHO and DG DEVCO.[271] The department manages the CFSP budget, the IfS and other actions under the authority of the HR/VP in her capacity as the Commission Vice-President. As a consequence, the Commission, by means of the Service, partly also deals with the CFSP budget to fund civilian crisis management missions. However, while the Commission manages the budget, it is the member states which settle its size per mission (Major/Bail 2011, p. 33).[272] As estimated by a

269 See respective parts on funding arrangements to increase crisis management coherence in country case studies.

270 The IfS succeeded the Rapid Reaction Mechanism. From 2007 to 2012 the IfS had a budget of €2 billion. Two thirds of this budged were assigned to short-term disaster relief aid and reconstruction projects and one third to long-term crisis management projects. However, as found out by Major and Bail (2011, p. 34) actual coordination of the two rarely takes place.

271 In this regard, the Foreign Policy Instruments Service is responsible for implementing the CFSP budget and short-term measures under the IfS (European External Action Service 2011).

272 A first evaluation finds that, "the cooperation between this service and the relevant EEAS departments is satisfactory, but there remain challenges in ensuring sufficient flexibility and re-

respondent, though the Foreign Policy Instruments Service should function as intermediary, a 'glass wall' between the Service and the EEAS remains. Nonetheless, interviewees underline the service's value as a catalyser and multiplier which fills an important gap due to its high flexibility. In general, the joint programming of funding increasingly involves both the EEAS and the Community side.[273] Supporting this development, the Council Decision establishing the EEAS states that the first three phases of the programming cycle have to be prepared jointly by the EEAS and DG DEVCO. Furthermore, an inter-service agreement was completed which determined how Community external action instruments cooperate with the EEAS regarding their programming (European Peacebuilding Liaison Office, p. 8).

The Lisbon Treaty provides supplementary potential solutions by passing bureaucratic hurdles concerning funding in the area of short-notice civilian or military missions (Drent/Zandee 2010, p. 53). The treaty creates a start-up fund for the HR/VP for crisis management operations made up of contributions from member states who decide on the use of the fund by unanimity after a proposal from the HR/VP. The fund thus provides the possibility to set-up funding that allows creating a budget for crisis management operations outside the general EU-budget provisions. This start-up fund can in theory be used for civilian and military crisis management purposes or a combination of both (Drent/Zandee 2010, pp. 53/54). Due to its institutional flexibility interviewees evaluate the fund as a promising provision regarding the EU crisis management coherence.[274]

In terms of vertical crisis management coherence between the EU and member states, the Lisbon Treaty does not introduce any major functional modifications or essential shifts of competencies and power from the member states to the EU (Kaski 2011, p. 2).[275] Rather than aiming at improving coherence in crisis man-

sponsiveness in implementing urgent measures in crisis situations given the constraints of the financial regulation" (European External Action Service 2011). However, Major and Bail (2011, p. 33) argue that "cooperation has thus far been inadequate". The Foreign Policy Instruments Service is planned to be located with the EEAS when it moves into the new headquarters building (European External Action Service 2011).

273 In EU terminology, "the programming of an instrument refers to the process of deciding how to spend the budget that has been allocated to it" (European Peacebuilding Liaison Office, p. 8).

274 Drent and Zandee (2010, p. 54), however, observe that the fund further increases the democratic deficit due to *ex-post* accountability and legitimism by the parliament. In general, however, the Lisbon Treaty increases formal parliamentary oversight powers on CFSP, in particular in the area of information exchange with the Commission and the Council. In addition, the High Representative in the function of the Vice-president of the Commission is responsible to the parliament.

275 However, the Lisbon Treaty entails new possibilities for cooperation between member states. In doing so, the treaty aims at enhancing the unity of member states' foreign policies with provisions such as the Permanent Structured Cooperation in the area of defence and military mat-

agement between the EU level and the national level, the treaty's approach was on "how to extend the range of possibilities for the Member States to avoid being disruptive by taking a conflictual stance in a key policy matter" (Gebhard 2011, p. 23). Even though the Lisbon Treaty maintains the obligations formulated in previous treaties for the member states to "support the Union's external and security policy actively and unreservedly in a spirit of loyalty and mutual solidarity and shall comply with the Union's action in this area" (Office Journal of the European Union 2010, p. 30) it does not contain enforcement mechanisms to ensure that these obligations are complied with. Though the HR/VP shall ensure compliance with these principles (Office Journal of the European Union 2010, p. 31), the post itself is highly dependent on the willingness of member states to support and agree on common positions. It is clearly stated that nothing in the Treaty will

> "affect the responsibilities of the Member States, as they currently exist, for the formulation and conduct of their foreign policy nor of their national representation in third countries and international organisations" (Office Journal of the European Union 2010, p. 343).

Not just do member states continue to hold their crisis management domains, at the same time, the Lisbon Treaty reinforces the role of the Council as a foreign policy forum. The treaty's structural reforms thus strengthen the intergovernmental dimension of comprehensive European crisis management by reinforcing the Council and thus the member states.

On the contrary, however, member states' access to relevant structures and opportunities for visible action and agenda-setting in the area of crisis management has been reduced due to the abolition of the rotating six-month presidency. While aiming at reducing potential inconsistencies between national agendas and the on-going priorities of the EU external action, the end of the rotating presidency comes at the cost of a certain drive of EU foreign policy by committed member states. Holding the presidency required member states to take on a more European approach to crisis management matters as they had to play a mediating role. Indeed, since the entry into force of the Lisbon Treaty, the member states seem less committed to the EU and showed a reluctant attitude toward any crisis

---

ters, a clause for solidarity and a clause for mutual assistance as well as the possibility to hand over a crisis management missions to a group of member states and conditions for an Enhanced Cooperation. Dealing with cooperation between member states, these provisions are not of relevance in the context of this dissertation. On the provisions and clauses see Kaski (2011) and the report by the European Peacebuilding Liaison Office (2012a, pp. 15/16).

management engagement (Major/Bail 2011, pp. 27-29).[276] Consequently, the abolition of the rotating presidency holds the risk of reducing the scope for Europeanization of member state crisis management approaches.

3.3 Synopsis: Comprehensive EU Crisis Management

As an actor without a security political framework from the Cold War period, the EU did not have to undergo difficult transformation processes. Rather its crisis management provisions as a whole were essentially planned to cover different dimensions of complex crisis management. Its policy coherence, however, has been clearly affected by the growing number of actors, institutions and posts in the EU's crisis management system and the lack of an existing crisis management legacy has structurally not facilitated the implementation of a CA and the development of a coherent crisis management architecture.[277] Though it presented itself in a picture of permanent institutional evolution, paradoxically, it was difficult for the EU to firstly avoid and later overcome crisis management fragmentation.

Though policy coherence has been a concern for decades and takes centre-stage as a political term and concept, a common understanding has not been agreed on between the EU's diverse actors. The EU has hesitated to define its CA officially in greater detail. The notion is not used coherently and "a deliberate terminological discussion has not yet taken place" (Johannsen 2011, p. 21).[278] When it comes to the strategic framework of the CA, rather than being adopted as a clearly defined concept and formal policy, the EU views the CA in terms of "an inspiration guiding the work towards a more integrated operational capacity" (Hanssen 2010, p. 9). Accordingly, analyzed documents outline objectives for EU's comprehensive crisis management activities in terms of general principles and often remain vague. Moreover, as CA issues are spread over so many policy

276 Major and Bail (2011, pp. 23-27) also observe that, though member states fulfilled a lot of the targets identified in the EU civilian headline goals, they continuously fail to satisfy EU demands for mission personnel.

277 Rieker (2006a, p. 512; 2007, p. 7) argues differently that the EU's lack of an institutionalized security policy from the Cold War period makes it easier for the Union to create innovative security approaches, facilitates CA implementation and opens up new paths to achieve crisis management coherence.

278 At a hearing at the European Parliament the Head of Integrated Strategic Planning Unit at the CMPD declared that there are as various definitions to the EU's CA as actors involved (Didier Lenoir, as paraphrased in Johannsen 2011, p. 48).

documents by different actors, the relevance of the concept itself looks rather rudderless in the EU.[279]

Nonetheless, the EU's crisis management expansion promoted a strong conceptual CA commitment. Crisis management coherence takes up a major part of the EU's larger conceptual development in this area and the CA has been repeatedly mentioned in numerous official EU documents as well as written and public statements of high level officials. Documents heavily signal the importance attached to it and the EU has formulated various concepts and strategies for particularly fostering coherence and coordination among its actors. Drivers of CA developments in the EU have been the increased demands of its crisis management engagement and ongoing lessons learned as a result of specific failures in missions.[280] Barriers, on the contrary, derived from competitive behavior on the part of the EU's diverse crisis management actors (Gross 2008, p. 35).

Along with the set-up of EU crisis management, the CA was reflected in the EU's Headline Goal Process. However, despite the parallel closure of the two headline goals, the civilian side of EU crisis management in this process came about as a "counterbalancing afterthought" (Drent 2011, p. 7). Concerns connected to the coherence of EU crisis management efforts initially derived from the motivation to develop a civilian ESDP dimension in addition to the military one. Moreover, though in parallel, civilian and military headline goals were defined and revised largely separately. The level of EU crisis management ambitions which acknowledges the character of complex conflict scenarios is thus not mirrored in the capability-development process which treats civil and military demands separately. At these early stages, efforts to improve crisis management coherence also took place with regard to the UN and with other donors in the development sector. As of 2003, it has been the ESS with its objective to prepare the EU to "share in the responsibility for global security and in building a better world" (Council of the European Union 2003a, p. 1) and its broad security background and the ESS' review in 2008 which have served as the overall framework concepts and a broad strategic roadmap for the EU's CA. Under this agenda, the CMCO doctrine and its Crisis Management Procedures can be regarded as the most direct reference of the CA in the EU's crisis management agenda. In the CMCO context, the aim to ensure a culture of coordination has been central to the EU's conceptual efforts to increase crisis management coherence. As such, it

279 Johannsen (2011, p. 48) suggests that this lack of terminological precision of the CA is closely related to the lack of clarity of the term crisis management in the EU.

280 Then High Representative for CFSP declared in 2005 already that the EU has "been developing a coherent and comprehensive EU approach to crises, more through the conduct of actual operations than by conceptual drafting" (Solana 2005).

is not limited to a certain set of procedures. The CMCO doctrine rather emphasizes the more general need of a cultural adaptation between the civil and the military level and across pillars. To this end, CMCO is concerned with the effective coordination of the whole range of EU instruments, external actors at the planning as well as the implementation phase. However, due to a lack of CMCO specifics as to what the envisaged culture of coherence entails, its effect on actual policy and CA implementation is rather limited.[281]

There are several features characteristic of the EU's approach to crisis management coherence which can be drawn from the conceptual analysis. First of all, the EU consistently repeats its own special capacity to activate a broad range of crisis management instruments within both the CFSP and the Community framework which should contribute to the EU's credibility as a crisis management actor. In this context, the EU's civilian oriented approach is also mirrored in the CA. All documents commonly underline the limited role of military instruments in international crisis management while at the same time emphasizing the civilian component. The CA for the EU also includes coherent crisis management in the most preventive way. Furthermore, the debate is institutionally focused. Generally, coherence is understood as the *sine qua non* for the EU to further develop itself as an effective and visible actor in international affairs. As such, the CA constitutes an alternative set of qualitative benchmarks for measuring its own institutional crisis management and offers an alternative perspective on the capability-expectations gap issue for the EU. The different EU actors appear to have an interest in an EU that draws upon its comparative strengths and utilizes is inherent potential toward the CA (Norheim-Martinsen 2009, p. 29).

As the EU's CA originated from the Union's first ESDP missions it largely developed against the background of respective experiences. While the term also entered the Commission's terminology and all EU institutions agreed to work toward the CA on paper, references to the CA in main documents mostly refer to the CFSP framework and appear to leave Community instruments aside. Considerations on comprehensive crisis management mainly centered on the civil-military interface within the CFSP/ESDP framework.. This reflects that in the first years, the EU and its member states largely concentrated on building up structures and capabilities for making ESDP operational (Johannsen 2011, p.

281 As Johannsen (2011, pp. 48/62) observed, the key period of the formulation and configuration of the EU's CA occurred between 1999 and 2004 in parallel with the emerging role of the EU as a civil-military crisis management actor. As of 2005, new EU documents and concepts dealing with the EU crisis management coherence contributed only little to further defining and developing an EU CA and rarely entail additional information on the EU's efforts to implement the CA, merely reconfirming its insertion in EU terminology without further explanation. They confirm the existence and importance of the approach in a general manner.

341). Though noting the Commission's contribution to most Council documents, interviewees found it remarkable that the Commission did not seek to contribute more to the EU debate on a common ground. This lack of contribution has been largely ascribed to the Commission's propensity to protect its competencies in background talks. As such, the conceptual development of the EU's CA took place largely separately within the Council or the Commission. Conceptual developments show that the two pillars each devised their own concepts for enhancing coherence. The analysis of EU documents revealed that the Council assumed a more active role in pushing toward more coherence in EU crisis management than the Commission. Moreover, while the Council follows an approach aiming at positive coordination, the Commission limits its aspiration by merely focusing on strategies mostly based on negative coordination. The conceptual analysis of Commission documents also demonstrated that as part of adopting a CA perspective, security has become an essential dimension of the EU's development policy and cooperation. However, as the Council and the Commission developed their conceptual crisis management frameworks on separate paths since the creation of the CFSP/CSDP framework, integration of these fairly independent tracks under a common approach to comprehensive crisis management has become particularly complicated. The strong focus on civil-military synergies as a key feature of the EU's CA consequently ignored issues of civil-civil coordination as an essential part of crisis management coherence for a long time. This became obvious in the development of the CMCO concept which was long treated as the "conceptual cornerstone of EU efforts towards implementing a comprehensive approach" (Gross 2008, p. 11).

As showed, EU crisis management policies developed conceptually disconnected from each other, following their own logic and with minor attempts to place them in a single integrated strategy. While awareness for this situation was raised with the ESS, it was even more challenged with the Lisbon Treaty and its institutional modifications. Indeed, coherence is a key objective of the treaty and its provisions confirm the centrality of coherence. Crisis management coherence is presented both as a concern and aim at the same time in this context. With the Lisbon Treaty, the EU has commenced to apply a crisis management approach that goes beyond CFSP and also includes community crisis management policies. Intrinsic to this new strategic approach is that the treaty conceptually and terminologically merges the different dimensions of EU foreign policy under the headline of External Action. Thus, it terminologically overcomes the strict division of its diverse actors. This formulation also demonstrates awareness on EU CA issues among member states (Gebhard 2011, p.7).

The Lisbon Treaty follows a coordination oriented approach which largely focuses on formal structural provisions to improve inter-institutional coherence.

The innovations of the Lisbon Treaty demonstrate that thinking on the issue has been taken on that goes beyond the CMCO's culture of coordination to a rethinking of the EU's crisis management architecture and structures as a whole. In this context, the EU's organizational crisis management form has been reorganized merging organizational entities and by moving them both horizontally and vertically to improve coordination. Structural changes introduced for many observers provide a large potential for actual changes to increase EU crisis management coherence and CA implementation as the treaty aims at bridging the EU's supranational and intergovernmental dimension and at bringing together different institutions, actors and policy areas to overcome the previous pillar structure. According to respondents, much will depend on how these newly created institutions work in practice. The reorganization and set-up of new crisis management structures on the scale of the EEAS are long-term developments which can ultimately only be assessed over a period of some years. Recognizing that it is too soon to make definitive judgments, nonetheless, some observation can be made and preliminary conclusions can be drawn also on the latest EU efforts to organizationally implement the CA.

Although the consolidation of EU crisis management structures after Lisbon at the time of writing is still pending, the process of modifying and establishing the structures through which the treaty is supposed to be implemented has proved difficult already. So far, the picture that has emerged on potentially increased crisis management coherence is rather mixed. According to interviewees, the structural innovations run the general risk of creating new tensions without solving existing ones. Even though institutions set up by the Lisbon Treaty are designed to help generate more coherence, they can also result in new conflicts of interest. Related to this assessment, respondents pointed out that some of the new institutional provisions may lead to further bureaucratization of EU crisis management. As outlined in a background interview, a major issue also is that different actors have different interpretations on their responsibilities, roles and leadership status. Indeed, at the time of writing, a major downside of the Lisbon Treaty reforms were the numerous uncertainties surrounding competencies of newly created bodies and posts. According to a respondent, existing division of crisis management policies might even be felt more strongly after the implementation of the Lisbon Treaty as it invigorated old institutional struggles and institutions fell back to protective and distrustful positions to safeguard control over competences. In particular, disagreements concerning the EEAS establishment demonstrate that such newly created institutions can become the focus of struggles for power that in the end go beyond the original matter of securing crisis management coherence.[282]

Organizationally, former first and second pillar bodies remain divided and while the new provisions build a bridge between the two spheres of EU crisis management action, they neither solved the dualism between the Council and the Commission nor changed the fundamental principles of the EU's CFSP. Thus, the substance of the fragmentation impeding EU crisis management coherence remains and the Lisbon Treaty's institutional changes fail to bring the two closer together. From CA perspective, the characteristic double quest for coherence remains. The wide range of EU civil and military crisis management instruments are not organized under one umbrella in a single organizational structure with respective decision-making power. In the same vein, the distinction between military and civilian crisis management operations carried out largely separated, without being civil-militarily integrated, mirror "outdated institutional patterns rather than the reality on the ground" (De Vasconcelos 2010, p. 23). Notably, the divide between security and development policies continues to be structurally strong. While a clear line between development and security can still suggest complementarity, this strict division of labor cannot be drawn anymore against the background of complex crises the EU faces (Major/Bail 2011, p. 33). EU crisis management fragmentation has thus only partly been corrected by creating new coordination mechanisms but not by actual organizational change. Despite the official 'depillarization', the particular procedural system which continues to be applicable in the area of CFSP means that challenges of delimitation continue to exist.

Nonetheless, "for the first time the EU could potentially have the desired capacity to wield the entire range of instruments (…) in unison" (Weiss/Spanger/Van Meurs 2009, p. 40), as the HR/VP is theoretically envisioned to represent the EU in all foreign policy issues, no matter whether the respective competencies eventually lie with the Commission or the Council (Hynek 2011, p. 83). The post pulls together the two formerly different strands and aims to be an institutional bridge

282 The structural implementation of the Lisbon Treaty's provisions, in particular of the EEAS, has been closely followed by scholars. Several publications evaluate and give insights into progress to date, criticize diverse aspects and make specific organisational recommendations. See, for instance, the written evidence for the UK's House of Lords' European Union Sub-Committee for External Affairs (House of Lords 2012), Burke (2012b) and Major (2012). See also Quille (2010), Blockmans (2011), Burke (2012b), the report by the European Peacebuilding Liaison Office (2012b) as well as Helwig, Ivan and Kostanyan (2013) for an assessment of the EEAS performance. A first EEAS review has also been undertaken by the HR/VP in 2011 which points to the challenging political and economic context of the service's launch. A review of the organization and the functioning of the EEAS will be provided by the HR/VP in mid-2013 and its mandate will be reviewed in 2014 (Helwig/Ivan/Kostanyan 2013, p. 1).

which links and coordinates the intergovernmental and the community dimension of EU crisis management (Hynek 2011, p. 83). The position unites the Commission's policies relevant to international crisis management with the instruments employed by the Council's CFSP and CSDP and brings the EU closer to its aspiration for policy coherence. Moreover, the formerly strictly intergovernmental CFSP gained communitarian traits and the HR/VP makes it possible to bring together intergovernmental crisis management instruments and communitarian tools (Drent/Zandee 2010, p. 15). It is this crisis management co-ownership of the HR/VP which promises greater coherence in a dual-key manner (Quille 2010, p. 70). Operationally expressed, the HR/VP serves as the key coordinator of all civilian and military EU crisis management instruments by overseeing them and "can be considered the preliminary institutional climax of forging a comprehensive approach" (Ehrhart/Petretto 2012, p. 10). As it becomes evident from the quote, the post is clearly framed to enhance EU crisis management coherence and has the potential to increase both horizontal and vertical crisis management coherence as well as the ability of the EU to speak with one voice. Indeed, the HR/VP represents the personification of the crisis management coherence debate as the former co-existence of the External Relations Commissioner and the High Representative for CFSP embodied the EU's crisis management incoherence. Merging the two positions symbolically is a strong political sign of increasing coherence. While the creation of such a position in the EU can be considered an achievement itself confirming the centrality of coherence with its mandate, questions about the post's effectiveness with regard to crisis management coherence remain. These mainly deal with the HR/VP's positioning in EU hierarchy and question, whether it is sufficient enough to result in a unity of command.

Looking at the concrete results so far, a mixed picture needs to be painted and the impression of the centrality of coherence needs to be mitigated. First of all, relations of the HR/VP with other institutions have been characterized by tensions and turf wars over competencies and structures which weaken the aim of coherence. Furthermore, the enormous range of tasks assigned to the HR/VP has been viewed skeptically. Being anchored in various institutions, the HR/VP does not 'own' one institution which weakens her position to enhance institutional coherence. It has been argued in several reports that the various 'hats' worn by the HR/VP could lead to "institutional schizophrenia" (Blockmans 2011, p. 8) and conflicting loyalties.[283] A point of criticism in this regard also is that despite all structural provisions the Lisbon Treaty assigned competences to

283 As described by Blockmans (2011, p. 8) "[i]ndeed, the Member States of the EU may well have created an even more impossible job than that of the United Nations Secretary-General – a post often called the most difficult in the world."

the HR/VP without providing the post with the necessary political support structures. Consequently, while the HR/VP is expected to drive initiatives forward and to lead, the post's mandate does not equip it to do so. This results in a high dependency upon member states support and leads to a "contradiction between supranational leadership tasks for the HR on the one hand, and unvaryingly intergovernmental control by member states over resources on the other" (Major/Bail 2011, p. 28). In practice, the HR/VP's coordination authority merely is horizontally while at the same time not supported by an enforcement authority. As stated by a respondent, it is thus within the HR/VP's mandate to effectively only deal with issues of unimportance.

Yet, the creation of the EEAS as an administrative service for the HR/VP further operationalises political commitment toward coherence and aims at completing the EU crisis management coherence framework. With the creation of the position of the HR/VP and the EEAS a double tracked approach was chosen which included the creation of a cross-institutional high-level position and a supportive administration. From organization theory perspective, the creation of the HR/VP represents a hierarchical approach. Though weak in terms of authority, the HR/VP represents the introduction of a hierarchical element as it represents the centralization of the coherence problem. The post also increases vertical and horizontal differentiation as it further distinguishes according to functional tasks and to command and control attributes. This hierarchical approach is supported by network efforts with the EEAS at administrative level which aim at bringing formerly different competencies and working cultures together. With the EEAS Crisis Management Board, the EEAS Crisis Platform and the EU Situation Room additional specific units were created to ensure crisis management coordination. Furthermore, the EEAS integrated CFSP and CSDP planning and strategic bodies such as the CMPD and the CPCC.

The EEAS is expected to foster coherence of EU crisis management action and to supply impetus to pursue this goal as a functionally autonomous *sui generis* institution. In doing so, it acts as a common source of expertise for any EU institution dealing with crisis management matters and provides an overview of activities of the different institutions. However, specific coordination tasks most likely consist of information sharing rather than the formulation of shared crisis management principles. It remains to be seen whether the EEAS will focus on enhanced exchange of information in its institutional development or use its potential to become a decision-shaping body on its own. Though the service did not work effectively at the time of writing and the suggestion that the EEAS has difficulities reducing inter-organizational coordination problems and improving coherence, its set-up is a remarkable accomplishment in such a short period of time. From an organizational perspective, however, as a newly created institution itself, the EEAS may have attached another layer of complexity to the existing ones. While the EEAS organizational set-up is appropriate to maintain flexibility of the EU crisis management network and in this regard reflects the characteristics and requirements of complex international crisis management, it may need more authority and influence over key actors. Additionally, early investigations show that the EEAS internalizes inter-institutional conflicts, rather than solving them.[284] Its performance to large parts is decided

> "by an invisible yet genuine distinction between two cultures: a Communitarian-like culture inherited from DG Relex and a political culture inherited from the Council policy unit and crisis management structures, deemed to retain a certain autonomy within the Service" (Blockmans 2011, p. 11).

Within the institution, "the cultivation of an esprit de corps (…) remains a challenge" (Helwig/Ivan/Kostanyan 2013, p. 9).[285] This mirrors theoretical assumptions that in processes of reorganization, the role of informal norms begins to show which can lead resistance and opposition to reform. At the same time, however, a lot of former DG RELEX officials kept their portfolios resulting in a kind of continuity in EEAS relations with the other institutions, particularly with the Commission. "Effectively, many of the institutional tensions related to the establishment of the EEAS were ironed out by these already existing working and personal relationships between officials" (Helwig/Ivan/Kostanyan 2013, p. 32). Yet, a structural division remaining is that some areas forming important parts of EU crisis management including development and humanitarian assistance stay outside the remit of the EEAS. They thus continue to be an obstacle jeopardizing EU ambitions for increased crisis management coherence and continue to pose a coordination problem in the future. It depends on inter-institutional coordination as well as effective management by the HR/VP whether this division can be overcome (European Peacebuilding Liaison Office 2012a, p. 1).

Seen from the organizational perspective, EU crisis management funding is separated along the lines of resourcing the civilian and military side of CSDP, the Commission and member states. As the pooling of funding makes little institutional sense at EU level, proposals deal with mechanisms for complementary funding. As it is supposed to be complementary to other EU instruments within the framework of broader crisis management strategies, the IfS provides a bridg-

284 Major and Bail (2011, p. 27) also observe that crisis management within the EEAS as well as between the EEAS and the Commission is coined "by disparate assumptions about their respective responsibilities and priorities in the realm of crisis management."

285 As stated by a respondent, diplomats want to rule, the military wants its chain of command, and the development side does not want to have to do with any of these.

ing function for the deployment of the Commission instruments (Drent/Zandee 2010, p. 47). As such, the IfS adds to the EU's CA in that it aims at linking Community and CFSP/ESDP crisis management instruments and also particularly contributes to an advanced exchange of information concerning conflict management issues (Aide à la Décision Économique 2011, p. 50). Also the Foreign Policy Instruments Service has the potential to function as an institutional link and coordination catalyser that keeps the Commission and the EEAS constructively engaged with each other. Also, the newly created start up fund is a promising provision regarding the EU crisis management coherence.

With regard to vertical coherence, member states continue to hold their crisis management domains. Moreover, as the strict structural and functional border between the Council and the Commission has been distorted, the role of the European Council as a foreign policy forum and the intergovernmental dimension of EU crisis management have been strengthened. On the contrary, the abolition of the rotating six-month presidency reduces member state access to relevant structures and opportunities for visible action and agenda-setting in the area of crisis management.[286] With regard to vertical coherence, respondents also point out that the EU's crisis management architecture is highly formed by considerations of what is considered politically legitimate. While it was opportune to create high-level and publicly visible coordinating structures, the coordinator was kept a weak player. The intergovernmental dimension of EU crisis management and weight of national perspectives also results in the development of symbolic organizational solutions to problems of crisis management coherence and coordination rather than considerations of effectiveness.

From the perspective of theoretic factors which contribute to facilitate CA reforms, the signaling and framing of these clearly increased on EU level with the Lisbon Treaty reforms. However, though the main objective of the Lisbon Treaty and the concrete solutions it introduces clearly is to provide the EU with a better capacity to implement the CA, many solutions offered only appear as such on the surface. Nonetheless, iteration and exchange to facilitate CA implementation and encourage coordination have also increased. Effectively, the coherence problem has been embarked upon by strengthening and reinforcing the intergovernmental nature of the EU's external action at the expense of the community

286 It can also be argued that the removal of the rotating presidency, as intended, supports action along lines of continuity, circumvents spoilers and agendas driven by member state interest. However, it can be observed that member states actually seem to loose interest in EU crisis management instruments. See, for instance, assessment by Major and Bail (2011, p. 35) in the dimension of civilian crisis management or Möller and Rapphold (2012, p. 13) who observe a German reluctance in this regard. This observation also manifested itself by the lack of unity and delayed action in the wake of the Arab revolutions.

and supranational dimension. In addition, the dissolution of the pillar structure has in fact not taken place on deeper levels. Structurally, the EU's crisis management architecture still presents various deficiencies and falls short of ambitious objectives declared in official documents. Against the background of this analysis, even after the reforms of the Lisbon Treaty, the EU's conceptual and structural framework does not serve the Union to fulfill the level of ambition set by it regarding a comprehensive European crisis management approach. Particularly, with regard to offering new prospects to strengthen inter-institutional coherence, the Lisbon Treaty has not suspended existing problems. In some aspects, the reform even created new ones. Also in the new crisis management environment created with the Lisbon Treaty, the EU thus maintains its routine of heavily demanding a CA in crisis management while, at the same time, relevant EU institutions follow individual tracks. This approach of CA implementation appears incompatible in itself.

4 Member State Comprehensive European Crisis Management

Also EU member states try to improve the coherence of their policies and actions in managing crises. Numerous governments carried out reforms that aimed at advancing the management of respective national crisis management efforts and the coordination between their own ministries.[287] After having identified the appropriate analytical framework for this dissertation and having discussed efforts toward a CA in international crisis management at EU level, the following chapters empirically analyze what the CA means in different EU member states.[288] National member state approaches to the CA and their means of trying to improve crisis management coherence are examined by looking at efforts in the UK, Germany and Sweden. In doing so, it is covered what has been elaborated as vital elements of the CA in the previous chapters: the crisis management approach, the conceptual dimension, the organizational dimension and the European dimension.

In accordance with these analytical dimensions, the country case studies firstly provide the backdrop of crisis management and security thinking in the particular country to examine national CAs in the light of the broader context of the country's crisis management approach. After these principles of crisis management commitments and role have been outlined, secondly, the country's conceptual and organizational efforts toward coherence of its crisis management policy are examined to explain the work that is ongoing in different national contexts. Thirdly, the country case studies deal with the role of the EU in this

287 Central governmental bodies are formally referred to as departments in the UK, as ministries in Germany and are terminologically known as Government Offices in Sweden. In line with these denotations, the central government bodies are named respectively in the country case studies on the UK and Sweden while being generally referred to as ministries in the remaining analysis in this dissertation.

288 Though these efforts collectively fall in the category of WoG approaches as they take place on national level, for the sake of terminological clarity and uniformity, respective initiatives are continued to be referred to under the label of the Comprehensive Approach. The dissertation does not go into detail on any particular crisis management engagements by the analyzed states. Details on particular operations are only included if relevant the specific context.

regard and the extent to which CA aims and crisis management activities are linked to the EU is assessed. After the analysis of the EU's role in the particular country's crisis management and policy coherence efforts, preliminary conclusions summarize the findings of the analysis in the broader theoretical context.

4.1 Case Study UK

4.1.1 The British Crisis Management Approach

The British approach to conflict is coined by several security political traditions and foreign policy factors that determine its policy. A historically important development influencing its policy until today is the decline from its hegemonic position as one of the world's leading nations to the status of a moderately large second class power.[289] For a long time, the UK took up a special position in world diplomacy which can be viewed to be at the intersection of three circles of influence.[290] Firstly, it was the dominant power in European affairs. The UK's position in this regard was that of "'a separate, closely and specially related ally and friend' rather than a direct participant in continental efforts at closer integration" (Self 2010, p. 2). The second circle of influence dealt with the country's role of a former colonial power and relations with its empire and former colonial territories. The third circle was the 'special relationship' with the United States of America (USA/ US).[291] British influence in each circle was seen to be mutually enforced by membership in the other two (Self 2010, pp. 1-3). According to respondents, another important aspect in this geographical context is the UK's island mentality and separation from continental Europe. This mindset is considered to be the foundation of the nation's self-perception of its own greatness (Hadfield-Amkhan 2010, p. 165). At the same time, the UK fears to lose its great power status to "a disgruntled second-rank island anchored off the European continent" (Self 2010, p. 7).

Due to this legacy, policymakers until today are reluctant to challenge the British self-image of the continuing great power status and a major bridge-builder in international affairs (Self 2010, p. 3). Despite material constraints, the UK has maintained this self-perception of itself as a global actor and its status as a leading

289 Until the end of the Second World War the UK was included in the superpower category alongside the US and the Soviet Union as part of the 'Big Three' of the Allied conferences and until the 1950s the UK produced almost one third of Western European industrial output (Self 2010, p. 2).

290 As strategically conceptualized by Winston Churchill in 1948.

291 The 'special relationship' between the UK and US is discussed fully later this chapter.

power is specifically held onto by political leaders (Rees 2011, p. 31).[292] Considering its perceived position in the world, the UK's foreign policy approach has frequently been expressed with the phrase that the country 'punches above its weight in the world' to ensure its influence.[293] At the same time, this aspiration demands "vision, alertness, a quick wit, capabilities and the willingness to accept a high(er) risk level" (Houben 2005, p. 119). The persistence of the belief of the UK's special and great power position in world affairs is all the more notable in face of the country's relative economic decline. As it is necessary to underline diplomatic ambitions with significant military forces making the country capable of intervening around the world, the UK's ability to perform internationally as an independent actor has increasingly been questioned (Self 2010, p. 5).

Equaling strong defence with good security is a major feature of British foreign and security policy orientation (Houben 2005, p. 128). A major issue in its defence policy since 1945 has been the aim to extend military activism in different global conflict areas while the resources devoted to this objective were reduced at the same time (Self 2010, p. 152). Adjustment to economic, diplomatic and military developments proved to be "a slow and painful process – not least because it has challenged many previously cherished images about Britain's place in the world" (Self 2010, p. 6).[294] This also shows in the UK's defence expenditure in proportion to its gross domestic product (GDP) which continues to be larger than that of any other European NATO member.[295] In particular, the possession of an independent strategic nuclear deterrent adds to the country's position in the world. It can be considered an international status symbol and the British ownership of a

292 According to David Cameron, British prime minister at the time of writing, the UK has "a position of authority that greatly exceeds our size. It is in our national interest to preserve and extend this influence, because the size of the problems and the scale of the opportunities we face demand our continued active engagement in the world" (Cameron 2010a, p. 9). For an overview of statements with a continuing highly positive rhetoric on the British position in international affairs as declared by policymakers see Self (2010, pp. 3-5).

293 This phrase is referred to by Houben (2005, p. 119), Mannin (2010, p. 311), Self (2010, p. 5), Smith (2010, p. 86) and Rees (2011, p. 31). Smith (2010, p. 86), however, observed that it has been reformulated by officials to the expression that the UK punches *at* its weight in world affairs instead of *above* (Smith 2010, p. 86).

294 Some scholars argue that the UK failed to adjust its international policy to a rapidly changing world. From this perspective, the end of the Second World War and the Cold War present two missed opportunities "symptomatic of a more fundamental failure as successive governments gradually adjusted to a changing global environment 'without thinking it necessary to re-define national goals or to launch an organized debate about history and identity'"(Self 2010, pp. 6/7).

295 In 2010, the UK reversed the long-term military expenditure trend. Until then, it has consistently spend an average 5-6% of GDP on defence which decreased to a 2,7% share of GDP in 2010 of $59,6 billion. However, the UK to retains third position in the SIPRI ranking of countries with the highest military expenditure (Perlo-Freeman et al. 2011, pp. 182/183).

nuclear deterrent also provides insight into the country's commitment to continuing world power status (Rees 2011, p. 31).[296] Hence, an important element of UK foreign policy is that security is viewed in essentially military terms. Security is strictly considered as the domain of the military, "reflecting the tenacity of traditional notion, even in an era when the UK's armed forces have increasingly been working alongside various agencies on operations" (Smith 2010, p. 87). This mindset is also reflected in the UK's global commitments. Its "legacy of empire has produced what has been perceived to be continuing interests outside Europe as well" (Houben 2005, p. 120). The UK consistently engages internationally taking on a leading role in respective activities. Its armed forces are seen to play an important role as a political government tool to enhance UK influence. Due to the colonial heritage, participation in international crisis management all over the world has traditionally been high (Houben 2005, pp. 119-121). As pointed out by an interviewee, the UK's status as a former colonial empire resulted in a vanguard attitude playing a strong role in its diplomacy which is focused on calming things down, while the military has a strong interventionist side.

In this context, the UK developed a sense of responsibility, in particular for UK dependent territories and internationalism has been a continuous theme of UK governments' conception of Britain's role in the world. British commitment to internationalism and justified interventions was revisited at the end of the 1990s by the government of then Prime Minister Tony Blair as part of the *Third Way* which viewed the UK's overseas interventionary role as one that resonated with, and reflected, the country's values (Rees 2011, p. 31). With the Third Way thinking, the Labour government presented a normative explanation for international crisis interventions based on logic of mutual interdependence of rights and responsibilities.[297] Placing great emphasis on the role of values, human rights, moral responsibility, dialogue and partnership added an ethical dimension to UK foreign policy

---

296 As Rees (2011, p. 32) points out, the UK's indebtedness and resulting fiscal stringency and defence cuts raise the fundamental question whether a global UK defense posture could still be afforded. Nonetheless, UK governments continue to restate the global orientation of UK foreign and security policy. The 2010 National Security Strategy (NSS) (Cabinet Office 2010) and Strategic Defence and Security Review (SDSR) (UK Ministry of Defence 2010a) of the same year have affirmed the UK as a military actor on a global scale. Both documents, the UK's 2010 NSS and SDSR, are analysed in detail in the following chapter.

297 The title of Blair's speech *The Rights We Enjoy Reflect The Duties We Owe* (Blair 1995) exemplifies this logic. Blair's argument was that state sovereignty has to be scrutinized in situations where states violate their peoples' human rights. Influenced by the crisis unfolding in Kosovo, he argued that the international community cannot remain inactive in such situations due to a moral obligation to intervene (Rees 2011, p. 34). The idea of the UK as a force for the good in the world has been reinforced in particular with the 1997 change of government and the Labour-Party coming into power (Smith 2010, p. 97).

which signified “a qualitative shift away from the narrowly pragmatic realist orientation of the past” (Self 2010, p. 217).[298] From British perspective, this justifies interference and when necessary the use of force in sovereign affairs of nation states, even without the support of the international community (Houben 2005, pp. 121/122). Above all, this ethical dimension of UK foreign policy became also associated with British willingness and responsibility to take unilateral action and to militarily intervene where necessary. Emphasizing the strength of the UK’s armed forces and its readiness to deploy them in combination with Third Way thinking and a moral imperative of doing the right thing did not just result in strong internationalist expressions and an active and internationally engaged role of the UK (Houben 2005, pp. 128/129). This also led to a heavy mission overstretch and caused enormous pressures on resources (Self 2010, pp. 181/243).[299] Reflecting disillusionment and wish for solid and tangible foreign policy foundations, the “‘ethical dimension’ was thus quietly abandoned” (Self 2010, p. 254) in the last couple of years. In particular, the lengthy operations in Iraq and Afghanistan with its high casualties challenged Third Way thinking. “If there was ever a discernible ‘Blair Doctrine’ of foreign intervention, it died in Iraq and Afghanistan” (Rees 2011, p. 31). Yet, it is still possible to observe certain preferences of UK international crisis management engagement. A key feature of UK operations is the country’s willingness to take the lead, and in doing so, to influence and form multinational operations. It follows a “risk-seeking strategy” (Houben 2005, p. 183) by being willing to take rather uncertain crisis management decisions and actions. On the contrary, the “British preference for workable solutions rather than grand plans or blueprints is well known” (Houben 2005, p. 183).

A key factor of the UK’s approach to international affairs is the above-mentioned bilateral ‘special relationship’ between the UK and the US which is highly important to the UK’s self-image and accounts for the position the country senses for itself in Europe (Houben 2005, p. 122).[300] The specialness multi-

298 However, as Self (2010, p. 222) pointed out, in practice, the UK record on human rights is rather inconsistent, particularly in the ‘global war on terror’ context and where British commercial interests are at stake.

299 For a detailed chronological overview of UK defence, MoD activities and use of resources see the UK MoD Annual Reports and Accounts available on the UK MoDs website.

300 The subject of US-UK relations and the extent to whether it is ‘special’, is highly controversial and today can “hardly appear in public unless wrapped in inverted commas and accompanied by a question mark” (EU HR/VP Catherine Ashton as cited in Self 2010, p. 74). One side argues that the shorthand notion is hardly more than an agreeable myth and sentimental language combined with self-interest and a congruence of international interests. Alternative views emphasize the close connection between the countries created by a shared language, history, culture and value system, which strengthen and deepen the relationship (Self 2010, pp. 103/104). For a detailed chronically analysis of the history of US-UK relations see Self (2010, pp. 74-103).

dimensionally differs between different policy sectors being closest and most comprehensive in the field of defence and surveillance where close collaboration became deeply rooted and institutionalized (Self 2010, pp. 105/106).[301] UK foreign policy focuses on working closely with the US and a central tenet of its security policy is to act as the main partner of the US with a close proximity of British and American views. The Anglo-American partnership is considered to be mutually beneficial on both sides of the Atlantic. For the UK, benefits of particularly close ties to a global superpower are self-evident. In this context, the UK-US relationship and respective policy is viewed to be a field in which the UK can exert expertise and influence and, in so doing, benefit from the prestige resulting from being treated as a 'special' partner (Rees 2011, p. 33). At the same time, the UK was often obliged to conform to US positions and actions. This is driven further by the burden of political will to act when Washington decides to become involved. A consequence of this special relationship "is a reflexive fear of ever being out of line with the USA on any issue" (Houben 2005, p. 122). Consequently, the Atlanticism resulted in UK strategies to a great extent being shaped by the US agenda and the UK's Atlanticist identity presents the key feature of its security and strategy thinking (Porter 2010, p. 6).[302] Also, striving to share US foreign policy burdens entails heavy demands upon British resources (Rees 2011, p. 33). The ambition "to be the next largest contributor to a mission has imposed considerable strains for the UK" (Rees 2011, p. 37).[303]

The priority the UK gives to its partnership with the US is closely linked to its NATO membership.[304] The Alliance is a central piece in British security policy and the UK plays an important role in its military and political structure as the next most important actor after the US (Rees 2011, p. 38). The UK's NATO commitment is also the area were above-mentioned features of the country's security and

301 Special UK-US relations become most apparent in the area of nuclear deterrence which underlines the centrality of the two countries' relationship and British dependency in this regard. As the UK procures the backbone of its nuclear component from the USA, "cooperation in this area has therefore come to represent one of the most high-profile and significant aspects of Anglo-American relations" (Smith 2010, p. 99). But also in the fields of intelligence, pre-warning and high-level political interaction the UK benefits from the close relation (Houben 2005, p. 122).

302 Dunne (2004) asserts that this Atlanticism bears the risk that UK strategic objectives and priorities are undermined and that Atlanticism and Internationalism are essentially incompatible.

303 For a comprehensive cost-benefit analysis of the Anglo-American partnership see Self (2010, pp. 107-111) who observes an "increasingly unequal and asymmetrical relationship" (Self 2010, p. 109).

304 The foundation of the UK thinking is encapsulated by the UK MoD stating that "most demanding expeditionary operations, involving intervention against state adversaries, can only plausibly be conducted if US forces are engaged, either leading a coalition or in NATO" (UK Ministry of Defence 2003, p . 8).

foreign policy such as the emphasis on military instruments and the eagerness to 'punch' internationally at a significant level meet and become most obvious. In military operational terms, the UK's policy is focused on either NATO or bilateral alliances and most major operations the UK conducted since the end of the Second World War involved the use of armed forces through NATO or as part of a US-led coalition (Smith 2010, p. 98).[305] Thus, the US and NATO are interdependent key priorities for the UK and the country's NATO membership can be considered a strengthening element of Atlanticism in British security identity. Interviewees emphasized that NATO is the UK's multilateral crisis management institution of choice and is seen as the cornerstone of security. Apart from the clear and outspoken preference for transatlantic cooperation and for the strengthening of NATO, the country's permanent membership of the UN Security Council poses another foreign policy obligation which gives the country a leading role in international crisis management and opportunity to exert influence.

However, in spite of the UK internationalism, a key legal principal and central political practice of the UK is its sovereignty which is perceived in a very traditional way that strictly separates nation state from supranational paradigms. Sovereignty in British policy "has been essentially a notion of undivided sovereignty, a zero-sum conception" (Bache 2008, p. 89). The importance of sovereignty is a key concept to understand both the country's development as a centralized state and its importance to external relations.

> "Within Britain, sovereignty remains a distinct 'legal concept, a status [that, S.P.] cannot be surrendered unless the units which form the political community, whether individuals or groups, abdicate their political rights.' Correspondingly, supranational initiatives are seen to embody a clearly illegitimate attempt to undermine political autonomy by renouncing the 'associative' elements of the sovereignty, including territorial integrity and monetary exclusivity" (Hadfield-Amkhan 2010, p. 181).

There are numerous reasons for the British emphasis on its sovereignty, most notably among them are the country's former world-power status and the continuity of British institutions since the end of the English Civil War. Yet, in practice, the British conception of sovereignty has been handled more flexible seeing internationalized policies as a measure to increase British influence. More than the erosion of sovereignty *per se*, it is its formal transfer which is perceived as highly sensitive (Bache 2008, p. 90).

305 In a 'wars of choice'-setting where the survival of the nation is not at risk it is the government of the day which decides on the deployment of armed forces and the scale of the commitment. Famously, in the case of Iraq the UK opted for the 'special' relationship' with the US instead of dealing with the issue in an international governance framework with other partners.

4.1.2 UK Comprehensive Crisis Management

4.1.2.1 Conceptual Dimension

Mirroring an early recognition of the complexity of the changing security environment, the UK was arguably among the first nations striving for crisis management coherence and explicit CA implementation.[306] Acknowledging the need for cross-departmental coordination and the awareness that policy issues cannot be addressed by one government department alone rose throughout the 1980s and early 1990s already and the country made a range of strategic and conceptual adaptations (Hull 2011, p. 6). Seeking to align cultures and structures to bridge organizational boundaries, early reforms were mainly domestically focused and dealt with policy issues such as drug abuse (House of Commons 2010a, p. 12).

When the Labour Party came into power 1997, first reforms toward policy coherence were introduced under the label of *Joined-Up Government*. The Joined-Up Government program and cross-departmental coordination became the administrative and organizational focus of the Labour government and received the explicit commitment by then Prime Minister Blair (House of Commons 2010a, Ev 141). Presented as the opposite of departmentalism, it was a central objective of this public sector reform in the government's modernization program (Christensen/Lægreid 2006, p. 8). The subjects of Joined-Up Government and holistic governance ran through the 1999 White Paper *Modernising Government* (Cabinet Office 1999). The document sketched an agenda of reform for the public sector which broadly covered the horizontal coordination between government departments as well as vertically between other relevant actors involved. The key argument behind the report was to ensure that public bodies have to focus on results to avoid that bureaucratic borders get in the way of sensible cooperation. The initial conceptual development of the British CA thus in large parts took place in the context of wider debate and efforts to increase cross-departmental collaboration of the UK government as a whole and not only in the crisis management context.[307]

306 See, for instance, assessments by Drent and Zandee (2010, p. 21), Hull (2011, p. 6), Major and Schöndorf (2011, p. 2) as well as Neal (2011, p. 6).

307 Christensen and Lægreid (2006, p. 7) view Joined-Up and WoG initiatives as a reaction to the negative effects of NPM reforms such as fragmentation and a lack of coordination that resulted from specialization in the name of efficiency. See also Verhoest, Bouckaert and Peters (2007, p. 326) on this issue. Unlike the prior reforms in the UK, which concentrated on government performance and efficiency, the Joined-Up Government initiatives focused on horizontal and vertical integration of policies in order "to align incentives, cultures and structures of authority to fit critical tasks that cut across organizational boundaries" (Geoff Mulgan, as cited in House of Commons 2010, Ev 140). See Bogdanor (2008, pp. 6-18) for an overview of UK efforts to deal with the overall problem of government fragmentation. Structural consequences and organisational innovations of Joined-Up Government are discussed in the next chapter.

As part of the Joined-Up Government reforms, the 1998 Comprehensive Spending Review introduced so-called *Public Service Agreements* (PSAs) in order to enforce inter-departmental cooperation and to make different departments work toward shared targets (House of Commons 2010a, Ev 141). With the Treasury as the central department responsible, the PSAs aimed at promoting joined-up working through the identification of cross-cutting goals. Initially, PSAs concentrated on single departments and the effective achievement of departmental outputs but developed into a framework of PSAs involving more than one department (Ling 2002, p. 623). While also focusing on outputs of individual departments at the beginning, the Foreign and Commonwealth Office (FCO), the Ministry of Defence (MoD) and the Department for International Development (DFID) shared a common PSA for conflict prevention and management from 2003 until 2006 as well as for 2005 until 2008. These cross-departmental PSAs required joined-up working on conflict issues and reflected the emphasis on inter-departmental working. A further step was taken with the 2007 Comprehensive Spending Review process leading to the *PSA Service Delivery Agreement 30* (Her Majesty's Government 2007).[308] The PSA 30 dealt with departmental use of key instruments and resources including tri-departmental funding for discretionary conflict activities. The delivery of the PSA 30 conflict agreement has been led by the FCO, which is supported by the MoD, DFID as well as other relevant departments (House of Commons 2010a, Ev 83). In the end, departmental contributions to PSAs were a comparably small part of the overall work (Teuten/Korski 2010, pp. 11/12). Yet, by signing up to inter-departmental PSAs, the departments set clear targets to achieve their common objectives and to improve inter-departmental performance. As departments are required to report for their performance against the published targets, the establishment of PSAs tightened accountability and embody a key change in the way government delivers and plans. Cross-governmental PSAs identify what needs to be achieved through joint action and thus encourage departments to work jointly on shared issues (Wall/West 2009, p. 218).

After these conceptual efforts involving the whole government, UK military deployments in Bosnia and Herzegovina renewed attempts for cross-departmental coordination in crisis management which were further pressed on by the UK engagement in Kosovo and the bilateral mission to Sierra Leone.[309] CA thinking

308 Service Delivery Agreements set out how departments deliver improvement defined in the PSAs (Wall/West 2009, p. 219). On specific targets and an assessment of progress between 2007 and 2008 see report by the Foreign and Commonwealth Office (2008, pp. 149-154).

309 The military deployment in Bosnia and Herzegovina took place under UN and later NATO command. In Kosovo UK military was deployed as a NATO force and with its bilateral mission to Sierra Leone the UK supported a UN mission to the country (Teuten/Korski 2010, pp. 1/2).

increased from the UK's crisis management efforts in these conflicts onwards as the British Armed Forces were employed in operations

> "where development activities, humanitarian activities and political activities were integral to the kind of objectives and end state they were trying to achieve, where they appreciated that they had to work more closely with non-governmental, humanitarian agencies" (House of Commons 2010a, Ev 1).

The complex interplay of different actors became increasingly apparent and departments evolved their understanding of conflict management challenges by drawing on international proceedings in political, development and military thinking. A series of department specific policies followed and the Joined-Up Government view started to inform crisis management policy thinking. Internationally oriented initiatives were developed and the need for cross-governmental coherence was acknowledged in several official documents (Patrick/Brown 2007a, p. 9).

In 2003, the government published its first White Paper entitled *UK International Priorities – A Strategy for the FCO* (Foreign and Commonwealth Office 2003) with the objective to outline the UK's strategic priorities reflecting the changing international system and the FCO's role in pursuing them. Its 2006 revision *Active Diplomacy in a Changing World* (Foreign and Commonwealth Office 2006) continued to describe foreign policy challenges and priorities for the whole UK government and adjusted objectives to changed international circumstances. The FCO concepts were again reviewed in the 2008 version *Better World, Better Britain* (Foreign and Commonwealth Office 2008) which renewed the UK's strategic priorities. The document again presents an effort to clarify the FCO's role in the interdependent environment it works in and to create a modern FCO which allows a focus on effort where it makes most difference (Foreign and Commonwealth Office 2008, p. 8). The review notes that "the FCO would reduce its involvement in those issues where other government departments in London are better placed to direct the Government's international priorities" (Foreign and Commonwealth Office 2008, p. 86). The documents positively assess that the FCO works closely with other departments and highlights the unique partnership with the DFID and the MoD to tackle causes and effects of conflict. The FCO appreciates that

> "[e]ach department brings a wealth of knowledge, skills and experiences and their own perspectives and, by working together, we can deliver a more effective response to complex problems than we could by working separately" (Foreign and Commonwealth Office 2008, p. 87).

On the development side, the UK's first White Paper on international development *Eliminating World Poverty: A Challenge for the 21st Century* (Department for International Development 1997) was composed right after the 1997 creation of the DFID to set out the new department's position. It clearly commits the country's development efforts to the elimination of poverty.[310] The document recognizes that "nationally and internationally, there is a complex web of environmental, trade, investment, agricultural, political, defence, security and financial issues which affect relations with developing countries" (Department for International Development 1997, p. 52). To effectively reduce poverty, the UK government must thus ensure consistency between different policies and all government efforts affecting development countries should take account of the UK's development objectives (Department for International Development 1997, p. 7). In doing so, the UK's first White Paper on international development delivered groundwork for the CA by highlighting links between UK national interests, international development and security.

The second White Paper *Eliminating World Poverty: Making Globalization Work for the Poor* (Department for International Development 2000) reaffirms the UK's development targets. Defining a policy agenda for the management of globalization to systematically reduce poverty, it focuses on the promotion of equitable and sustainable economic growth (Department for International Development 2000).[311] In subsequent 2006 and 2009 White Papers the UK government abandoned this "economist focus" (Green 2009) of the previous documents and renewed its position and commitment on good governance issues, fragile states and post-conflict situations. Moreover, while the 1997 White Paper stated that "conflict prevention is (...) crucial to combat poverty and reduce suffering" (Department for International Development 1997, p. 68) the subsequent White Papers successively stressed the importance of conflict management. In particular, the 2009 White Paper *Eliminating World Poverty: Building our Common Future* (Department for International Development 2009) changed the emphasis of the department's activities and focuses on security and justice in fragile states at the expense of social and economic development. In its chapter 'Working Across Government' the document gives specific attention to the CA which

310 Since 1975 no UK development policy had been formally codified (Teuten/Korski 2010, pp. 8/9). The 1997 establishment of the UK's Department for International Development as a fully fledged ministry is dealt with in detail the next chapter.

311 Also DFID's 2005 *Security and Development Strategy* sets out the department's intent to deal with conflict and security issues as an integral part of development and to view them as a necessary part of an enabling environment for poverty reduction. To facilitate cross-governmental working, the role of pooled expertise and analysis is highlighted (Department for International Development 2005, pp. 23-26).

"British diplomats, development and defence colleagues now adopt (…), working closely together in distinct but complementary ways" (Department for International Development 2009, p. 78). It demands that efforts to strengthen the UK's CA should include more joint government strategies to ensure "that cross-government funding arrangements continue to support conflict prevention activities" (Department for International Development 2009, p. 78). The White Paper also commits DFID to develop joint strategies with the FCO and the MoD in fragile states and emphasizes the integration of civilian and military efforts, while, at the same time, viewing more civilian expertise in conflict situations as necessary (Department for International Development 2009, pp. 78-83).

Also the UK MoD early evaluated the British Armed Forces new role and reviewed its policy, priorities and tasks. The 1998 and the 2002 *Strategic Defence Reviews* (UK Ministry of Defence 1998, 2002) reconsidered roles, mission and capabilities and respectively adjusted the British Armed Forces to match new strategic realities and to reflect the changing nature of security challenges facing the UK (UK Ministry of Defence 1998). Explicitly emphasizing that the review was foreign policy led and not merely driven by military or financial considerations, it pointed to the exceptional cooperation between the MoD and the FCO to ensure that defence was one instrument of a wider policy (Houben 2005, p. 130). The inclusive process in which the Strategic Defence Reviews were developed by drawing on a broad range of working groups and other consultations that informed its content, resulted in broad appreciation of the document (Drent/Volten 2008, p. 23).[312] Also MoD White Papers presented further revisions to bring national security into line with realities and to refine the department's policy approach in the light of evolving events and growing operational experiences. In particular since the 9/11 terrorist attacks, the set of crisis management concerns and foreign policy, security and development considerations have broadened and more widely been perceived to pose numerous dangers to the UK. As stated in background interviews, experiences in Afghanistan and Iraq further reinforced the necessity of a CA. They underlined the need to better coordinate the crisis management activities that the British government was

312 The document also contains several British security traditions such as an emphasis on NATO, nuclear weapons and the country's international responsibility. By stating that the UK does "not want to stand idly by and watch humanitarian disasters or the aggression of dictators go unchecked. We want to give a lead, we want to be a force for good" (UK Ministry of Defence 1998, p. 7), a proactive element and ethical dimension shows which aimed at combining realist and ethical assumptions by seeing the military as a 'force for good' (Drent/Volten 2008, p. 24).

undertaking and acted as a catalyst and quick impulse for conceptual and organizational developments.[313]

However, despite the large conceptual reflection of CA ideas in departmental documents, as shown above, according to respondents, departments in practice continued to view crisis management issues from their individual perspective and concentrated on their own role in setting objectives. In spite of conceptual adaptation, differences in conflict perceptions and observations were preserved. Regarding state fragility, for instance, the FCO continued to focus on regional instability while the MoD emphasized the potential role of failed states as sources of international terrorism.[314] While from DFID perspective crisis management policy coherence aimed at ensuring sufficient physical and human security for local individuals and communities, the FCO and MoD approached the issue from the security perspective of UK citizens, its implications for national interests and the benefits for the UK's national security. Another factor in this context has been that the FCO's leadership role in foreign affairs is traditionally strong in the UK. Worried about interferences on its role, the FCO for its part was persuaded of the value of the CA to a much lesser extent. Also, DFID was concerned that its development mandate and resources could be utilized to fulfil security objectives. Thus, departmental resistance to increase civil-military cooperation remained strong and in particular DFID was reluctant to engage together with the military. The UK's MoD, on the contrary, viewed the CA and respective inter-departmental efforts highly positive as a way that allows the military to focus on its core mission. It was frequently frustrated by the UK's lack of a coherent vision of national security priorities and worried that the military might increasingly be preoccupied with non-military activities (Patrick/Brown 2007a, pp. 12-15). Thus, "efforts to signal a change in culture have sometimes been undermined by actions inconsistent with rhetoric" (Teuten/ Korski 2010, p. 95) resulting in persistently strong departmental loyalties (Teuten/Korski 2010, pp. 95/96).

These diverse departmental crisis management concerns were addressed in 2005 in a comprehensive report prepared by the *Prime Minister's Strategy Unit*

313 In particular, the UK's efforts in the Helmand province in Afghanistan betrayed a lack of interdepartmental coordination across the security and development sectors. For an in-depth review of UK efforts and CA problems in Afghanistan see the report of the House of Commons (2010a, pp. 49-54).

314 As an indication of divergent perspectives, Patrick and Brown (2007a, p. 13) give the example of DFID being pressured by the MoD to consider Saudi Arabia as a fragile state and to develop programs for the country. On the grounds of Saudi Arabia's adequate domestic resources to promote their own internal development, DFID resisted and did not devote its resources to a country deemed critical to UK security interests.

(PMSU).[315] The report *Investing in Prevention: An International Strategy to Manage Risks of Instability and Improve Crisis Response* (Cabinet Office 2005) demanded a newly coherence focused strategy in particular with regard to the UK's engagement in fragile states and drew onto a strategic planning process to construct comprehensive interventions involving all relevant UK and international actors (Patrick/Brown 2007a, pp. 10/11). Going beyond department-specific strategy development, the PMSU's *Countries at Risk of Instability Initiative* in 2005 "provided the first definite guidance on cross-government country strategy formulation" (Teuten/Korski 2010, p. 20). As outlined in the report *Investing in Prevention* (Cabinet Office 2005) the UK government aimed at an institutionalization of a cross-departmental strategy with regard to failed states by proposing the initiative. Led by the PMSU, the initiative "envisioned the Cabinet Office secretariat convening cross-Whitehall teams to formulate a common UK strategy for countries in crisis" (Patrick/Brown 2007a, p. 16), to create a common understanding of UK crisis management policies.[316]

However, DFID and FCO participation in developing strategies remained limited while the MoD started to "search for other ways to compel integration" (House of Commons 2010a, Ev 142). By mid-2000, the focus on Joined-Up Government had notably shaped

> "the way in which key military officials thought the mistakes made during the initial US-led Iraq invasion could be avoided. In particular, senior military officials saw the absence of civilian departments (…) as key reason for the worsening security situation" (House of Common 2010, Ev 141).

Due to the resistance and passivity by the other departments, the UK's CA evolution thus was increasingly developed from a military point of departure. In the MoD's 2003 White Paper *Delivering Security in a Changing World* (UK Ministry of Defence 2003), the department "began laying out its argument that only by adopting a comprehensive approach to security policy, of which defence was but a

315 Kiso (2009, p. 126) refers to the role of reports as a central feature of UK policy initiation. Reports are formulated in project teams consisting of experts of the different departments, academia and the private sector.

316 The initiative was also supposed to create a win-win situation for relevant departments by promoting "the buy-in of each department by highlighting the implications of instability for its own interests, for instance by threatening the achievement of the Millennium Development Goals (DFID), undermining regional stability (FCO), endangering energy security (Department for International Trade), providing a safe haven for terrorists (MoD), or threatening uncontrolled immigration or facilitating organized crime (Home Office)" (Patrick/Brown 2007a, pp. 16/17). The Countries at Risk of Instability Initiative including the project's results and summary of the report is elaborated in Detail in Kiso (2009, pp. 127-132).

part, would British interests be best served" (House of Commons 2010a, Ev 142). The document reoriented the country's defence posture and revised key defence areas with an emphasis on the need for capabilities to meet a "much wider range of expeditionary tasks, at a greater range from the UK and at an ever-increasing strategic, operational and tactical tempo" (UK Ministry of Defence 2003, p. 11).

In 2005, the MoD produced its own departmental CA initiative in a *Joint Discussion Note on the Comprehensive Approach* (UK Ministry of Defence 2006) making the CA a topic itself and driving the concept forward.[317] In field experiences, it evaluated that departmental activities were rarely coordinated or concentrated on common goals while progress was made when efforts were coordinated (UK Ministry of Defence 2006, p. 1–1). The document describes the department's perspective on the rationale for, and the implementation of the CA in conflict situations and declares that

> "the realisation of national strategic objectives inevitably relies on a combination of diplomatic, military and economic instruments of power, together with an independent package of developmental and humanitarian activity and a customised, agile and sensitive influence and information effort" (UK Ministry of Defence 2006, p. 1–2).

The Joint Discussion Note's purpose "to codify emerging best practice and provide initial guidance to the MOD on the principles that should guide incorporation of CA thinking into the planning and execution of UK operations" (UK Ministry of Defence 2006, p. iii) also aimed at informing and supporting CA development across the whole of Whitehall.[318] In doing so, the UK MoD defined the CA as "commonly understood principles and collaborative processes that enhance the likelihood of favourable and enduring outcomes within a particular situation" (UK Ministry of Defence 2006, p. 1–5). This definition stresses the idea of the CA as a philosophy and frame for assessment and planning of objectives valid in each situation. Four guiding principles of the CA conceptual framework are underlined in the document, namely a proactive cross-Whitehall approach, shared understanding, outcome-based thinking as well as collaborative working (UK Ministry of Defence 2006, pp. 1–6/1–7). Notably, the CA is not seen as a determined method of doing things nor as a way of developing homogeneity across government departments and systems (House of Commons 2010a,

---

317 It has been pointed out that with the definition of a term in a *Joint Discussion Note* is very important as the word discussion implies that the MoD wanted to indicate to the other government departments that this was not a *Joint Doctrine Note*, but that it was open for discussion on the CA (House of Commons 2010a, p. 14).

318 The expression *Whitehall Model* or simply the term *Whitehall* describes the British core executive and central bureaucracy.

Ev 82). However, besides the cross-governmental CA view, the document considers it also in terms of a

> "natural extension of the UK Military Effects-Based Approach (EBA) and should be developed in a complementary manner within the MOD. The potential benefits of the CA can only be fully realised if absorbed within the culture of individual government departments" (UK Ministry of Defence 2006, p. iii).

Though the document was intended to set a cross-governmental approach, the notion of the CA has been explicitly linked to the military idea of the EBAO and moved the debate toward the implications of the CA for the military by emphasizing advantages and challenges of a CA adoption from military perspective. Even though cross-departmental consultation took place during the Discussion Note's development and it was not intended for military but cross-governmental audience (House of Common 2010, Ev 79), the MoD did not manage to secure agreement on its Joint Discussion Note and was unable to achieve cross-governmental agreement on its CA doctrine (Teuten/Korski 2010, p. 149).[319]

The CA was also covered in *British Defence Doctrines*.[320] In its third edition, the 2008 British Defence Doctrine (UK Ministry of Defence 2008) explores the topic of the CA with particular references, explicitly reflects on the necessity for cross-governmental and inter-departmental cooperation and aims at encouraging collaborative working. It does so by basing the CA on four guiding principles which mirror the guidelines developed in the Joint Discussion Note.[321] As mentioned above, the wars in Afghanistan and Iraq exposed deficiencies in the UK's crisis management capacities that became particularly clear during active

319 In 2007, the document was followed by the additional Joint Doctrine Note *Countering Irregular Activity within a Comprehensive Approach*. It conceptualizes the CA as a shared vision of a situation with exchanges allowing for a better comprehension of crisis management challenges individual actors face. The Joint Doctrine Note underlines the need of a sense of serving common objectives and the idea to build a "climate of trust through close exchange and networking" (Wendling 2010, p. 68).

320 The British Defence Doctrine "provides the broad philosophy and principles underpinning the employment of the British Armed Forces. It is the basis upon which all other subordinate national doctrine, describing in more detail how operations are directed, mounted, commanded, executed, sustained and recovered, is derived" (UK Ministry of Defence 2008, p. V).

321 Respectively, coordinated engagement in complex conflict situation firstly requires proactive engagement. Effectiveness of the various actors' capabilities is further optimized by a shared understanding that should be created through collaborative working. Furthermore, involved participants should focus on outcome-based thinking in crisis resolution. The fourth guiding CA principle is collaborative working which stresses the importance of institutional familiarity, generated through personal contact and human networking (UK Ministry of Defence 2008, pp. 1–8/1–9).

insurgency efforts to stabilize and reconstruct the countries. According to interviewees, this explains that the UK's CA concentrates very much on the insurgency environment. SSR experiences in these countries also served as an important vehicle for increased civil-military cooperation in the UK. Another aspect coining the British CA is that the conceptual framework in which the MoD began thinking about the CA stems from the development of EBAO doctrines and respective conceptions coming from the US which have been picked up by the British military. These served as a catalyst for doctrinal reforms and supported the MoD's intend for changes in Whitehall (House of Commons 2010a, pp. 13/14, Ev 143). Respondents pointed out that the CA is thus broadly perceived as a military phrased and led concept as it terminologically has originally been developed and distributed in UK discourse by the MoD. Not wanting to give credibility to a military concept, the CA was rejected and viewed highly skeptical by other departments. Fearing a loss of independence and autonomy, the FCO, MoD and DFID obstructed a common cross-departmental concept.

Against the background of budget cuts and the pooling of financial resources, the CA and effective crisis management coordination gained renewed focus in the UK with the financial crisis. The concept was terminologically reframed by the UK government under then Prime Minister David Cameron speaking about an *Integrated Approach.*[322] Different departmental concerns and the UK's increasing CA thinking in strategic and conceptual terms became most apparent with the government's decision to prepare an overarching *National Security Strategy* (NSS) to strengthen the UK's security policy-making framework.[323] With this decision, the government departed from the practice to formulate security and crisis management concerns in fragmented departmental strategies. The NSS follows an all-encompassing approach with the objective to connect different government strands to deal with various foreign and security policy challenges (Drent/Zandee 2010, p. 21). Echoing previous crisis management sentiments, the first NSS (Cabinet Office 2008) was formulated and published in 2008. It provided "the first systematic and comprehensive assessment of Britain's security challenges and responses" (Teuten/Korski 2010, p. 10) and is explicitly committed to a greater integrated civil-military approach which is a major theme of strategy. Acknowledging its own role in efforts toward policy coherence it states that

322 For matters of clarity and to avoid confusion with the UN Integrated Mission concept, the dissertation continues to use the term *Comprehensive Approach* in this country case study on the UK.

323 Pressure arising from outside and the Conservative Party pushed the government of Prime Minister Gordon Brown to issue the NSS which picked up on Blair's 2007 paper *Britain in the World* serving as a prototype (House of Common 2010, Ev. 146).

> "the major security challenges require an integrated response that cuts across departmental lines and traditional policy boundaries. The National Security Strategy outlines some of the ways in which our approach has become more integrated, but also identifies where we need to go further – indeed its existence is a reflection of the Government's ambition to work more closely together at home and overseas to achieve our security objectives" (Cabinet Office 2008, p. 8).

The document recognizes the significance of cross-governmental working by stating that "our response to global instability, conflict and failed and fragile states brings together a wide range of government activity, from diplomacy to development to overseas military operations" (Cabinet Office 2008, p. 33). It sketches concrete future steps aiming at the integration of UK military, civilian, security and development capabilities and intends to encourage UK partners to do the same to enable more timely integrated responses to conflict (Cabinet Office 2008, p. 40). These suggestions and key ideas formed the seeds for later structural changes.[324] Another major theme of the NSS is the changing role of the British military to providing long-term support for stabilization rather than preparing large-scale wars (Cabinet Office 2008, p. 53). However, little of the analysis of the changing security environment frequently referred to in the NSS is new to the UK. Thus, the significance of the document is its actual existence as a cross-governmental strategy as well as the key steps it points out to strengthen coherence in the UK's crisis management structure.

The UK's NSS has since been reviewed on a yearly basis and was up-dated in 2009, under the heading *Security for the Next Generation* (Cabinet Office 2009). The up-dated NSS again stresses the need for the CA and restated the necessity for a cross-government approach.

> "Our approach to conflict is to work across government bringing together the full range of development, diplomatic and military tools aligned to wider Government (and multinational) priorities, working together towards common goals. The FCO, MoD and DFID jointly deliver the Government's key public service agreement (PSA): our commitment to reduce the impact of conflict through enhanced UK and international efforts" (Cabinet Office 2009, p. 69).

It acknowledged that delivering the strategy requires cross-governmental efforts and further developed the CA's conceptual basis by not just taking into account the nature of UK interests but additionally considering the role of individual departments and crisis management capabilities (Cabinet Office 2009, pp. 15/33-36). With the third NSS revision in 2010, the strategy bedded down in the struc-

324 These are analysed in the next chapter.

tures of UK governmental thinking and the national strategic review process became institutionalized (Cornish/Dorman 2012, p. 221). In a written statement to Parliament announcing the publication of the 2010 NSS *A Strong Britain in an Age of Uncertainty* (Cabinet Office 2010), then Prime Minister Cameron reemphasized that the UK

> "will draw together and use all the instruments of national power to tackle these risks [as outlined in the UK NSS 2010, S.P.], including the armed forces, diplomats, intelligence and development professionals, the police, the private sector and the British people themselves" (Cameron 2010b).

The NSS again tries to meet pressures for reform and illustrates yet another attempt to fundamentally realign British crisis management policy. Need for action did not just increase due to experiences in Iraq and Afghanistan but even more because of heavy budgetary pressures caused by the economic and financial crisis. Another important factor was the broadly diagnosed strategic deficit of the UK's foreign and security policy which made the review necessary (Rotte/Schwarz 2011, p. 57). At the same time, the NSS and its up-dates are criticized for not being "really a strategy as such, but a methodology for a strategy" (Clarke 2010). Due to its unspecific nature, it is questionable, whether it "has yet gained enough political weight to inform, still less to shape [British politics, S.P.]" (Clarke 2010).[325]

Still, the NSS can be considered a "key component" (House of Commons 2010a, p. 14) in the CA development. Its establishment mirrors a CA both in document creation and in their conceptual alignment and as such can be considered "post-modern security strategy" (Clarke 2010). The document not just reflects CA ideas but itself serves as an example of CA implementation. The 2010 NSS explicitly sets out a WoG approach based on an extended concept of security which is defined as the principle of strategic implementation (Cabinet Office 2010, p.

325 With the revision of its NSS in 2010, the UK was arguably "the first Western state to undertake a complete defence and security review in the 'age of austerity'" (Cornish/Dorman 2012, p. 213). In justifying the country's defence spending cuts, the UK government clearly notes that the main threat to the UK is of economic nature. Defence cuts were smaller than expected and did not reach the scale of cuts for other Whitehall departments (Cornish/Dorman 2012, p. 215) but critics still point to the UK's reduced room for maneuver due to financial pressures not reflected in the NSS as the document still displays the UK's objective to shape a stable world (Cabinet Office 2010, p. 22). On the diagnosis of a perceived strategic deficit in British foreign and security policy see Gow (2009) and Porter (2010). Cornish and Dorman (2012, pp. 219/220) argue differently, however, considering it an unavoidable precautionary aspect of national security strategy to be based on hedging or on deliberate decisions due to the fundamentally unknowable, uncertain and possibly volatile security future.

10/34).[326] The strategies outline where responsibility across UK government lies for different crisis management aspects. Most importantly and a radical step with regard to CA implementation are the structural reform decisions pushed for. Furthermore, the NSS assessment processes require and lead to cross-governmental consultations. In its preparation, the documents drew on a wide range of expertise and activities across departments which contributed to the formulation of the strategy as a single overarching document. Conceptually, crisis management and security aspects are not strategically dealt with in fragmented manner anymore but in a comprehensive strategic framework. In adopting a broad and comprehensive perspective on security, the document allows a coherent picture on the UK government's point of view on wider crisis management questions.

In addition to the revision of the NSS, the UK government published the Strategic Defence and Security Review (SDSR) *Securing Britain in an Age of Uncertainty* (UK Ministry of Defence 2010a) in 2010. As both documents were initially planned to be published as one single document, the NSS and SDSR are coherently constructed and match each other.[327] With the SDSR, the British government set the parameters for fundamental defence and security reforms (Giegerich/Jonas 2011, p. 1). Conceptually, the formulation of the SDSR appears like a model of a thoroughly constructed strategic discourse which entailed inter-departmental collaboration, the institutionalization of new inter-departmental authorities as well as regular strategic revisions off key documents. Like the NSS, the SDSR formulation drew on a wide range of expertise from across departments, agencies and the armed forces which contributed to its development. It is the result of a broad cross-departmental process (Giegerich/Jonas 2011, pp. 3/5). Reflecting the CA, the SDSR positions itself as part of a wider and comprehensive security review and crisis management operations which require the use of the CA. Considerations in the SDSR are largely foreign policy led, perceiving defence as one instrument of wider policy.[328] The review also aims at ensuring that CA considerations are embedded in future government policy and that the armed forces are prepared to per-

326 This is also stated by then Prime Minister David Cameron and Foreign Secretary Nick Clegg who declare in the document's introduction that the NSS "mobilize[s] the whole of Government" (Cabinet Office 2010, p. 6).

327 While the NSS determines the overall context of UK security policy, the SDSR makes precise choices and decisions of required capabilities and resources and considers consequences and how to deliver NSS security priorities.

328 This approach became already obvious in the relabeling of the SDSR, which originally was supposed to be named *Strategic Defence Review* and reflects the aim to place defence issues under the broader foreign policy and security narrative. Thus, the added word *security* symbolizes an holistic approach that broadly involves all relevant issues, departments and actors in a consensual way (Giegerich/Jonas 2011, p. 9).

form the respective role (House of Commons 2010a, p. 66). On the downside, however, both the case of NSS as well as SDSR development also indicate that this planning process can easily be undermined by budgetary constraints as well as time pressure which allow departmental interests to come to the fore. According to interviewees, the formulation of both documents was conflict-ridden and the consensus based strategic development was largely challenged by financial pressures which damaged inter-departmental collaboration. As particularly the SDSR has heavy financial implications, after a while departmental budget safeguarding began to dominate the inter-departmental coordination and turf wars dominated the agenda (Giegerich/Jonas 2011, pp. 3/10).[329]

Stemming from and largely line with the NSS, the 2011 *Building Stability Overseas Strategy* (BSOS) (DFID/FCO/MoD 2011) is the first integrated cross-government strategy to address conflict issues put in place by the British government. Jointly developed by the FCO, DFID and the MoD, the strategy provides an operational framework for UK activities to address instability and conflict overseas through a coherent framework. It explicitly states that the departments "can improve the effectiveness of our efforts by strengthening our whole of government approach and refining our prioritisation at a time when resources are being squeezed" (DFID/FCO/MoD 2011, p. 4). The ability of the UK to take a whole of government approach is seen as its particular advantage (DFID/FCO/MoD 2011, p. 22). To this end, the strategy defines triggers and drivers of conflict and instability, identifies approaches to prevent conflict and sets out early warning, rapid crisis prevention and response as well as investing in upstream prevention as the three key dimensions of UK strategic response to conflict.[330] The BSOS outlines how different departments work alongside to bring development, diplomacy and defence together. In doing so, the strategy provides a comprehensive overarching framework for the UK's approach to conflict.

The British government also took a number of steps to improve the quality of strategy development toward specific conflict countries. Regarding their own department-specific country planning, both the DFID and the MoD regularly adjust their strategic guidance and doctrine. The FCO also generated internal

329 In addition, the SDSR was evaluated rather critically in particular over the mismatch between ends, ways, and means. See Dorman (2011, p. 84), as well as Giegerich and Jonas (2011, pp. 5, 10-13). A detailed analysis of the formulation, implementation and long-term implications of the 2010 strategy review of both the NSS and the SDSR can be found in Cornish and Dorman (2012).

330 Upstream prevention in this context means tackling the underlying drivers of instability before crisis occurs (DFID/FCO/MoD 2011, p. 2).

guidance on its own strategy development.[331] Other specific campaigns have produced additional frameworks for better cross-governmental strategic planning and civil-military cooperation. The UK strategy for Afghanistan, announced in December 2007, is an example of this as it "provided a specific enough framework to marry civilian and military activities" (Teuten 2010). Newly formulated strategies for Somalia, Sudan and Yemen also took on a more genuinely cross-government perspective (Teuten 2010). To avoid that strategies only deal with a specific sector in a conflict country or relate to only one source of funding, DFID's 2009 White Paper demands the UK government to "develop joint strategies in all fragile countries where they do not already exist and where the UK has a substantial development program, agreeing joint objectives within the framework of individual departmental mandates and capacities" (DFID 2009, p. 78). During the development of the 2009 UK government strategy toward Afghanistan and Pakistan, Cabinet Office was responsible for the overall coordination while the FCO led the development of the country strategy in close consultation with DFID and the MoD. The UK government agreed country strategies for other countries and regions thus showing a greater willingness to develop joint strategies. However, mutual governmental consultation has remained insufficient as the development of joint strategies often turned into an accumulation of what departments already do. While being useful to summarize cross-governmental intentions, another weakness of the current strategic approach is that it fails to prioritize conflicting objectives (Teuten/Korski 2010, pp. 22-23).[332] Furthermore, Cabinet Office established so-called *Departmental Capability Reviews* in 2006 which biannually evaluate cross-departmental collaboration across Whitehall. The 2009 review which covered both inter-departmental policy formulation and operational delivery states the improvement of relationships and inter-departmental coordination for all three departments (House of Common 2010a, Ev 83).

Consulting between departments thus increased with regard to doctrinal development. However, in its latest on the CA in the UK, the House of Commons comes to the conclusion that the "MoD and the Armed Forces, the FCO and DFID all recognise that engagement in future conflicts is likely to require the use of the Comprehensive Approach" (House of Commons 2010a, p. 3) and that it is "therefore essential that a shared understanding exists across Government and, in

331 For further details on department-specific country strategy development see Teuten and Korski (2010, pp. 19/20).

332 As Teuten and Korski (2010, p. 22) point out, the benefits such joint strategy development also depend on the amount of UK interests at stake and departments involved. In cases of insufficient involvement, transaction costs of developing a joint strategy may not be justified and would not generate "sufficient benefits in greater coherence to warrant the effort required to produce a cross-government strategy" (Teuten/Korski 2010, p. 22).

particular, within the MoD, the FCO and DFID about what the Comprehensive Approach is" (House of Commons 2010a, p. 3). While it sees that many components for such a policy and doctrine already exist, the report continues to demand a CA policy and doctrine to be further developed by the FCO, DFID and the MoD in close cooperation (House of Commons 2010a, pp. 3/4).

4.1.2.2 Organizational Dimension

The UK has made a range of structural adaptations to improve crisis management coherence and the centre of government changed considerably through the development of coordinating mechanisms but also through expansion (Teuten/Korski 2010, p. 31). The UK core executive enjoys wide foreign and security policy influence and with power of the Prime Minister's position being traditionally comparably strong. The Prime Minister in the UK's politico-administrative system is responsible to provide overall guidance and delegate respective matters to departments as needed (Teuten/Korski 2010, p. 28). The cabinet, as the top executive committee, is fairly large and does thus not operate as one collegial body. Instead, smaller committees are created to deal with specific issues of matter for higher-level decision. Being placed at the centre of government, Cabinet Office influentially supports the Prime Minister to deliver his main political programmes. Covering policy issues which cut across government with specialized units, it is also responsible that the UK government's foreign policy objectives and strategies are implemented in a coordinated way and serve the government as a whole (De Coning et al. 2009, p. 16). Cabinet Office plays a key role in cross-departmental working and in bringing different departments together to a certain end. While having a lot of weight forming the UK's long-term crisis management policy, it has less influence on the actual implementation (House of Commons 2010a, Ev 82). As part of the Joined-Up Government reforms at the end of the 1990s, prime ministerial influence increased and was institutionalized by the establishment of numerous task forces, inter-departmental committees and units at Cabinet Office level to achieve coherence across government. Particularly the *Strategy Unit*, created in 2002, can be regarded important as its responsibility is to analyze cross-departmental problems and to evaluate measures dealing with them (Bogdanor 2008, pp. 12/13). According to respondents, the Strategy Unit brings together teams from different departments and has developed a comprehensive web-based portal which provides material on comprehensive policy-making and project management.

However, only few government decisions are that controversial and important to be taken to Cabinet level but are taken by ministers and departments contributing to their political and administrative importance. Functional departments are the major organizational bodies for crisis management policy and implementation (Wendling 2010, p. 69).[333] The UK is characterized by a decentralized structure that gives a lot of autonomy to the individual departments which from structural perspective does not encourage cross-governmental policy coherence (Patrick/Brown 2007a, p. 15). Strong individual departments with a self-aware organizational culture and well-established policies, strategies and concepts result in "a strong culture of departmentalism in Whitehall" (Bogdanor 2008, p. 13). The organizational set-up of the UK executive is dominated by sector and function principles (Egeberg 2003, p. 122).

The central government machinery to execute UK crisis management responsibilities is made up by the FCO, the MoD and DFID. All three departments have carried out institutional reforms to increase crisis management coherence. DFID as an independent ministry and a full cabinet department dealing with development policy has been established in 1997. Its institutionalization is the outcome of one of the first initial reform decisions in the CA context, presenting development with a separate institution and an own set of objectives and an independent voice. With the creation of a new department, the government fulfilled its commitment to international aid and development as outlined in the first White Paper on eliminating world poverty. Moved from being a part of the FCO to a separate department, development issues became more influential and disconnected from other foreign policy goals (Teuten/Korski 2010, p. 30). In this context, the creation of DFID modified the agenda of the development debate away from aid as charity to a focus on economic development and a strategic conception of aid aiming at poverty reduction (Self 2010, p. 235). A major objective that drove the establishment of a new government department was to show that the UK's development policy is not subordinated to broad foreign policy priorities and to disconnect issues of poverty reduction from other FCO activities and political objectives (Self 2010, p. 235).[334] However, critics argue

333 Departments are accountable to department-specific parliamentary committees. Generally, the British House of Commons has little influence regarding security political decisions. The main role of its Select Committee on Defence, for instance, is to assess specific topics in special reports which are publicly available. In doing so, the committee monitors expenditure, policy and administration of the MoD and associated agencies. It launches several inquiries per year which increasingly become more influential (Houben 2005, p. 136). As already mentioned in the literature review, the Defence Committee report on the UK's CA (House of Commons 2010a) provided a valuable source for this country case study.

334 The 2002 *International Development Act* mandates that the department's funds are spend in low-income countries to support sustainable development. This should protect DFID to resist aid relocations to for security purposes and secures that the department's mandate focus is explicitly limited to be spent on poverty.

that the focus on poverty alleviation DFID is bound to, created an organizational culture within the department which impacts adversely on joint projects and "militates against spending resources in countries at risk of instability as well as alongside the military" (House of Commons 2010, p. 28). The creation of DFID thus partly resulted in ring fencing aid and set hurdles to cooperation with the other departments (House of Commons 2010a, Ev 141).

In the context of inter-departmental relations, it was indicated in background talks that, as DFID has been newly created, the department's position has been rather weak compared to the strong history and tradition of the FCO and the MoD. In particular, relations with the FCO, which fears to lose its traditionally strong crisis management role, are occasionally tense. It took some time for the FCO to adapt to the loss of its development policy authority and foreign aid resources (Patrick/Brown 2007a, p. 14). After its creation, the DFID had to work for its independence from the FCO and the expansion of its role (Macrae/Leader 2000, p. 3). Particularly in the beginning, critics argued that "DFID's humanitarian focus and reduced access to FCO 'political' expertise produced a rather naive approach" (Self 2010, p. 235). However, since its creation, the department adjusted the picture of an "over-promising and under-delivering" (Evans/Steven 2010, p. 13) actor to one very much "running its own foreign policy" (Evans/Steven 2010, p. 13). Still, both FCO and MoD officials are skeptical about DFID's contribution to UK and wider government objectives, despite a broad political consensus in favor (Evans/Steven 2010, p. 13), even more so as DFID follows a different management style that separates it from other departments. The diverse background of the department's staff and the strong focus on transparency and accountability provides the basis of this management style (Kiso 2009, p. 113). UK efforts to advance policy coherence have been pushed strongly by DFID and show the strength and dynamism of the department "as a fully fledged cabinet agency possessing significant resources and the liberty to pursue innovative approaches to the linkages between development, governance, and security" (Patrick/Brown 2007a, p. 9).

Though practically shared with the Prime Minister, the formal constitutional responsibility for foreign policy in the UK lies with the FCO which traditionally holds a central role in the departmental Whitehall structure (Self 2010, p. 258). This position has been challenged and decreased due to the decline of the UK as an international actor and by reactions to shifts in British foreign policy since the end of the Cold War. While the FCO remains highly important for the coordination of different aspects of British external policy system, when it comes to cross-cutting issues, it is just one actor among others (Oliver/Allen 2006, p. 54). While the FCO plays a major role in the British system of coordination, collective responsibility of the Cabinet ensures that the department is not the sole determinant of UK policy. Policy coordination and information exchange is ensured by the use of existing as well as *ad hoc* committees in an informal, reactive manner. In crisis management operations with force deployment the degree of inter-departmental coordination depends on the extent to which issues arise that need to be resolved at a high-level or inter-departmentally. Inter-departmental coordination is realized by routine and *ad hoc* meetings, correspondence and telephone contacts. In meetings concerning the security domain, Cabinet Office, the FCO and the MoD meet to discuss policy as well as practical matters. DFID in such cases sits in on the meetings (Houben 2005, pp. 124-127). However, the FCO still has a deeply rooted "culture of treating policymaking as an insular activity" (Self 2010, p. 258) and fears that the inclusion of other actors makes the policy process less controllable. Increasingly questioning its role and purpose in the challenging international environment also resulted in FCO recognition to broaden its expertise. It employed more specialists in individual subject areas to increase its value to other Whitehall departments "by providing an overarching perspective beyond the capability of more narrowly focused departments" (Self 2010, p. 274). As a result, the number, importance and influence of functional units dealing with cross-governmental issues increased at the expense of traditional narrowly focused departments (Self 2010, p. 274).

The MoD's role regarding the British CA focuses on military input and has led the conceptual and doctrinal CA developments. The British system and political leadership to a large extent underlines the importance of joint civil-military deliberations and advice as well as the functional imperative of the armed forces (Egnell 2009a, p. 7). In an attempt to further integrate civilian and military aspects in the planning process, the MoD has involved humanitarian, stabilization and development advisors in the process (Stepputat 2009a, p. 40).[335] Thus, the MoD's administrative body is comparably large with a high input of civilian officials, who dominate various defence policy areas. The defence minister has one military and one civilian principal advisor with none of them being subordinate to the other (Houben 2005, p. 127). Aiming at a more systematic approach to military support to international crisis management the new MoD task of *Military Assistance to Stabilisation and Development* was created which acknowledges that the military conducts operations to help stabilize situations where the security situation is too difficult to allow for independent civilian measures (Teu-

335 As Egnell (2009a, p. 106) observes, civil-military relations in the UK are facilitated by the historical tradition of such cooperation but also because both political and military leadership have historically been drawn from the same social group with same schooling and values. For a detailed analysis of the history and patterns of civil-military relations in the UK see Egnell (2009a, pp. 95-113).

ten/Korski 2010, pp. 106/107). Key feature of this task is the military assistance to civilians to enhance civilian crisis management effects such as the improvement of governance and non-violent political dialogue. To further develop this capability, the *Military Stabilisation Support Group* (MSSG) was established within the MoD in 2009 to offer training and to deliver and deploy stabilization planners and functional specialists. With its activities the MSSG also supports other government departments. To additionally enhance civil-military integration, the MoD has revised its career training by focusing more on stabilization and developed courses open civilians (Teuten/Korski 2010, pp.107/108).

To structurally ensure a nationally coherent approach to international crisis management, the UK government created some specific administrative and financial structures to encourage inter-departmental exchange and joint crisis management projects. Following the recommendations of the *Wiring it Up* report (Cabinet Office 2000, pp. 47-51) on improving funding for cross-governmental policies and a series of cross-cutting reviews on conflict prevention activities, the British Government partly bundled financial resources and jurisdiction of the FCO, the MoD and DFID to encourage common crisis management projects. The original push originated in the prime minister's office in line with efforts to increase coherence of the overall political system, and according to respondents, structural innovation were largely promoted by DFID and the FCO respectively.

The creation of the so called *Conflict Prevention Pools* (CPPs)[336] as part of the Joined-Up Government initiative in 2001 sought to increase UK conflict engagements by bringing together the FCO, the MoD and DFID as the three departments traditionally involved in crisis management (Foreign and Commonwealth Office 2011b). There were several reasons for the UK government for establishing such joint funding arrangements, starting with an overarching consensus in the government that the reduction of conflict demands a broad spectrum of measures from the different departments to improve the delivery of effective crisis management. Pooling resources as a mechanism was expected to improve effectiveness of UK activities through joint analysis, the development of shared priorities and common strategic approaches to conflict management. The reduction of resources spent in conflict management efforts and possible financial savings represented another incentive. Interviewees confirmed that the allocation for funding for conflict issues indeed has traditionally been a constraint in achieving common policy. Before the CPPs set-up, crisis management was

336 The term 'conflict prevention' as used by the UK government also covers peacebuilding activities (Lawry-White 2004, p. 3). Practically, the CPPs rather dealt with post-conflict situations and not conflict prevention indicated by the name (Stepputat 2009a, p. 36).

coined by inter-departmental tension due to budget issues.[337] The CPPs' establishment was encouraged particularly by DFID to ease the funding and inter-departmental coordination on crisis management activities that fall outside traditional *Official Development Assistance* (ODA),[338] such as SSR or *Disarmament, Demobilization, and Reintegration* (DDR) projects (Patrick/Brown 2007a, pp. 10/19). Furthermore, the creation of the CPPs increased DFID's involvement in overall policy-making which included interaction with the FCO and MoD. With the creation of the CPPs the government established a formal indicator for the departments to work together (Stepputat 2009a, p. 36).

To serve as an incentive for greater joint collaboration and as a capability for rapid response, ultimately two funding pools were created among the FCO, the MoD and DFID: the *Africa Conflict Prevention Pool* (ACPP) and the *Global Conflict Prevention Pool* (GCPP). While top-level organizational structures of both pools are similar, difference exist due to the GCPP's formal and standardized procedures. In this regard, the ACPP worked on a much more flexible base without fixed strategies and formal procedures (Debiel 2004, p. 27). Furthermore, the GCPP was managed by the FCO and funded programmes under different regional and thematic strategies. Having been managed by DFID with a slightly smaller budget, the ACPP as the common pool for Africa concentrated on conflict prevention and post-conflict activities in various regional and conflict priority areas such as pan-African SSR initiatives (Patrick/Brown 2007a, pp. 20/21).[339]

---

337 Resource allocation in the UK is traditionally rather decentralized with departments discussing and agreeing their budget individually with the Treasury. As regards conflict issues, the MoD's funding responsibility entailed the cost of deploying UK peacebuilding and stabilization operations as well as the application of defence assets for longer-term diplomacy and conflict prevention. The FCO is responsible for UK contributions to multilateral peacekeeping missions of the UN, EU and OSCE. DFID's funding responsibility covers development and humanitarian programmes (Teuten/Korski 2010, pp. 48/49). See Teuten and Korski (2010, pp. 48-65) for a detailed discussion of UK crisis management funding.

338 The OECD (2008) defines ODA as those flows to countries, territories and multilateral development institutions which are "provided by official agencies, including state and local governments, or by their executive agencies" and transactions which are "administered with the promotion of the economic development and welfare of developing countries as its main objective; and (…) concessional in character and conveys a grant element of at least 25 per cent."

339 A respondent stated that the first major difficulty in establishing the joint conflict management funding mechanism was to find a definition of activities covered by conflict prevention and to identify measures and projects qualified for funding by the CPPs. Furthermore, the identification and setting of regional and thematic priorities posed another fundamental question. Several reasons led to the creation of a separate pool for Africa. First of all, the continent is particularly affected by conflict and the UK has been frequently engaged in African conflicts. However, the creation of two pools also is a result of the bureaucratic context and DFID securing that its departmental regional priority is adequately taken into account (Debiel 2004, p. 279).

The pools' resources consisted of a share of each department's budget to be used for common initiatives. Funding for the ACPP and GCPP was additional to DFID, FCO and MoD allocations and the three departments had to reach agreement on standards for accessing pooled resources (Patrick/Brown 2007a, p. 20).[340] Thus, while the individual departments remained responsible for expenditures, they were forced to engage in policy discussions (Stepputat 2009a, p, 36). Funding from the CPPs was "allocated by consensus (the so-called 'triple keys') between regions, countries and themes, and then between projects sponsored by each of the three departments" (Teuten/Korski 2010, p. 49). Strategies for both pools were selected in "a competitive bidding process" (Patrick/Brown 2007a, p. 20) between departments which had to develop potential bids in a tri-departmental process. Proposals were then passed to a tri-departmental steering group, which sent recommendations and options to the cabinet ministers for approval. Importantly from CA perspective, funding by the pools was restricted to projects in which joint inter-departmental efforts have powerful synergies and promote inter-departmental cooperation (DFID/FCO/MoD 2003, p. 8).[341] In 2008, a separate *Stabilisation Aid Fund* was created which focused on financing UK stabilization activities in Afghanistan and Iraq and supported joint stabilization planning in the two countries. Similar to the CPPs, the FCO, MoD and DFID were key holders. However, the new fund was primarily managed by the *Stabilisation Unit* (SU) and established to protect the GCCP from funding pressures arising from the two conflicts and to "pay for civil conflict stabilization activities in volatile or hostile areas where the security situation does not (yet) permit implementing programmes that the Conflict Prevention Pool has traditionally funded" (Drent/Zandee 2010, p. 22).

In 2009, the Stabilisation Aid Fund and the CPPs were merged to a single pool of funds with the objective to further join expertise in development, diplomacy and defence across the three departments.[342] Dedicated to discretionary

340 To achieve their objectives the CPPs also shared a joint PSA target, which states that "[i]mproved effectiveness of the UK contribution to conflict prevention and management as demonstrated by a reduction in the number of people whose lives are affected by violent conflict and a reduction in potential sources of future conflict, where the UK can make a significant contribution" (Austin et al. 2004, p. iii).

341 The three other criteria for funding by the pools were, firstly, the significance of the conflict to the UK government; secondly, whether a UK conflict management activity makes a difference to the conflict and thirdly, if the approach allows greater UK influence on the international community.

342 Teuten and Korski (2010, p. 50) point out that a key factor for the merging of the Stabilization Aid Fund with the CPPs was the financial pressure resulting from the removal of Treasury compensating mechanisms for the effect of sterling fluctuations on overseas expenditure and the sterling's fall. Also substantial overlaps between the GCPP and the ACPP called for a restructuring of the joint funding arrangements (Debiel 2004, p. 292).

activities that support worldwide conflict prevention, stabilization and peacekeeping the newly created *Conflict Pool* is managed by the SU. Management structures across the three departments include a *Building Stability Overseas Board* which is made up of the conflict directors of each department and which is responsible for setting the overarching strategy for the Conflict Pool and directing its reform. The board is supported by a tri-departmental secretariat. Different programmes of the pool are managed by separate Programme Boards which are also responsible for strategic decisions (Independent Commission for Aid Impact 2011, p. 1). The SU reports to the MoD, the FCO and DFID which are on the board of the pool (Stepputat 2009a, p. 41). Activities supported by the Conflict Pool focus on dealing with threats in risk, fragile and conflict-affected regions.[343] Decisions and activities delivered through the Conflict Pool remain to be based on tri-departmental joint analysis. Funding of the Conflict Pool comes from a joint Treasury settlement, which is separate and complementary to FCO, DFID and MOD departmental budgets. The UK government's 2011 BSOS as the operational strategy for the Conflict Pool guides the allocation of Conflict Pool resources from 2012/13 onwards (Foreign and Commonwealth Office 2011b).[344]

UK experience with pooled funding varies widely and assessments differ.[345] Undisputedly, the sharing of pooled resources leads to genuine cooperation and

343 The Conflict Pool's activities are divided into the five regional programmes Afghanistan, South Asia, Africa, the Middle East and North Africa and Wider Europe and one thematic program, the so-called Strategic Support for International Organisations/Strengthening Alliances and Partnerships program, which seeks to support the conflict prevention capacity of IOs. The largest program by expenditure was Afghanistan with more than 40% of the total Conflict Pool spending (Independent Commission for Aid Impact, 2011, p. 2).

344 See also United Kingdom Parliament (2011). In the coming years, the Conflict Pool's budget is supposed to increase to allow the UK "to deliver more cross-government support to long-term conflict prevention and stabilisation programmes" (Foreign and Commonwealth Office 2011a). Separate arrangements are in place for additional cost of military operations in Afghanistan which are not included. Another joint funding instrument also managed tri-departmentally is the UK's Peacekeeping Budget, which pays for the UK's legally binding commitments to UN, OSCE and EU peacekeeping missions (Foreign and Commonwealth Office 2011a). Other crisis management activities or operations are funded by individual department funding resources from within existing allocations or by drawing on the Treasury's Reserve to fund urgent operational requirements for the military (House of Commons 2010a, Ev 86).

345 The CPPs have been assessed several times in different evaluations. Most notably, an extensive evaluation by DFID in 2004 (Austin et al. 2004) found evidence of the overall value of the CPPs and several improving steps were taken as recommended by the report. Analyzing four country cases and two thematic subjects the study concluded that numerous valuable activities have been started. However, it also comes to the result that the three departments lacked a "consistent set of ideas" (Austin et al. 2004, pp. 4/5) and that a mismatch between the amount of resources and the level of ambition existed. Kiso (2009, p. 120), Teuten and Korski (2010, p. 62) and Hull (2011, p. 6) come to similar conclusions and view the UK's pooled funding positively.

labour division "on the basis of joint analysis and strategizing, thus contributing to a mutual understanding and gradual convergence of institutional cultures" (Weiss/Spanger/Van Meurs 2009, p. 28). Respondents evaluate the joint funding mechanisms as a very useful instrument concerning CA objectives that actually leads to integrated planning of crisis management projects. The pool successfully encourages more coherent analysis across departments and ensures coherence by not just forcing departments to work together but also having them to seek consent and prioritize. In doing so, pooled funding has raised understanding of other departments and their crisis management projects, which is a particular value of the funds. The establishment of joint funding mechanisms for crisis management activities in the UK had the effect that the three participating departments and major crisis management actors started to look for agreement on their resource allocation as well as setting joint targets for crisis management efforts. As projects are evaluated together and experiences shared, working together in the CPPs can be considered as an 'educative process', according to a respondent, which has led to a socialization process with improved dialogue and coherence (Patrick/Brown 2007a, p. 20). Evaluated as "the most innovative donor instruments to promote whole of government action" (Patrick/Brown 2007b, p. 4), the set-up of pooled resources thus represents a consensus building approach that has been a success facilitating cross-departmental coordination by offering an important networking opportunity resulting in long-term cooperation, increased interaction and mutual understanding of each other's work (Kiso 2009, p. 120).

Though being evaluated mostly positively as a success from CA perspective, the pools are also assessed dysfunctional in some respects. In this regard, respondents questioned the effects of the pools by referring to their comparably small financial volume. Questioning, what impact the CPPs actually have on crisis management coherence at such modest levels of funding, according to interviewees, the pools' budget is not substantial enough to excite any of the departments involved. The FCO, DFID and the MoD have their own funds and the pools are not seen to have a significant impact compared to other crisis management initiatives managed by the departments individually. Compared to the financial resources of each department, the pools have been poorly equipped and major budgets remain in departmental silos driven by departmental considerations and objectives. Nonetheless, according to other respondents, the pools can be considered a truly tri-departmental CA implementation. In spite of their moderate means, they stimulate inter-departmental collaboration and allow for a faster provision of resources for integrated conflict management measures (Major/Schöndorf 2011, p. 2). Another criticized aspect is the pools' complex decision-making machinery which is out of proportion considering the resources it manages. Transaction costs have been significant by devoting a disproportionate amount of coordination effort to managing relatively small sums of money (Patrick/Brown 2007a, p. 20). Furthermore, the pools have also been utilized by the departments to finance non-integrated individual projects (Patrick/Brown 2007b, p. 4). Pooled funding instruments were thus misused as pots of additional money for dealing with regular projects that would have received funding from either of the departments (OECD 2006, p. 27).[346] Another criticism of the UK's pooled funding arrangements has been the lack of an overarching strategy which led to a tendency of treating the pools as a pot of funds for projects without a clear strategic focus and "piecemeal funding decisions" (Independent Commission for Aid Impact 2011, pp. 3/4). A weakness of pooled funding in this regard has been "a tendency to succumb to a lowest common denominator approach with each department negotiating protection of its top priorities in the absence of any agreement on an overarching strategy" (Teuten/Korski 2010, p. 58). The underlying problem when crisis strikes is that most department budgets are already programmed. Thus "the UK government faces a constant balancing act between insulating these modest resources for conflict prevention purposes and drawing on them for urgent needs" (Patrick/Brown 2007a, p. 23). Departments lean to the conflict pools as contingency funds and risk that the pools are misused for short term crisis rather than long-term conflict prevention (Patrick/Brown 2007a, p. 18).[347]

CA implementation in the UK's policy further became apparent in the establishment of cross-departmental policy units formed on specific international issues or particular countries to encourage more coherent policy-making and implementation. These cross-cutting units aim at facilitating policy coherence at the inter-departmental level. While some of them were genuinely cross-departmental working for a common aim with shared staff and funding from participating departments others were cross-departmental in form but origin departments still conducted parallel policies. Strategy groups such as the *Afghanistan Strategy Group* have allowed for a thematic focus on a conflict due to a joined-up process involving Cabinet Office, the FCO, the MoD and DFID. Furthermore, *ad hoc* cross-departmental committees and working groups were established, yet, with uneven

346 This means that pools have been one-sidedly utilized by individual departments to enforce already planned projects and departmental aims instead of treating the pools strategically to ensure a joint conflict prevention approach.

347 Another aspect criticized in UK funding on conflict in general deals with the implications of the MoD's preferential access to the Treasury's Reserve to fund urgent operational requirements. As DFID and the FCO have to handle unforeseen expenditures by reallocating existing budget plans, "the relative ease of access to funds by the military resulted in occasions when a military solution was chosen, or the military used as the implementing agency, even though a civilian approach might have been preferred" (Teuten/Korski 2010, pp. 57/58).

participation of the three departments (Stepputat 2009a, pp. 39/40).[348] A major difficulty in view of comprehensive crisis management in the UK was posed by the problem of distinct cultures of each department and the limited experience of working together. Interviewees stated that officials with different departmental backgrounds from the FCO, DFID or the MoD, including the armed forces, spoke 'different languages'. They barely understood each other which led to misunderstandings on conflict management policy issues.[349]

According to an interviewee, these experiences provided the final impetus for actual organizational change that would not have been possible after the British engagement in the Balkans and which led to the construction of an innovative coordination body and the institutionalization of a center for post-conflict reconstruction work within the central UK government machinery. In 2004, the UK government established the tri-departmental and multi-disciplinary *Post Conflict Reconstruction Unit* (PCRU) as a coordinating unit to be jointly owned, financed and staffed by the FCO, the MoD and DFID to harmonize and support cross-government crisis management assessment, planning and execution. An interviewee described that the initial conception and construction of the new func-

348 Other examples for the establishment of cross-departmental units include the Sudan Unit jointly ran by the FCO and DFID which was created in 2002 due to the high level of both departments' engagement in Darfur and the 2004 Afghan Drugs Inter-Departmental Unit established by the FCO, DFID, Home Office and the Serious Organized Crime Agency (Teuten/Korski 2010, p. 35). In particular the Sudan Unit is considered an example of successfully promoting cross-government coherence according to an interview. See also Teuten and Korski (2010, p. 40). Also with regard to UK engagement in Iraq, post-conflict planning was centralized in the FCO's Iraq Planning Unit which worked closely together with Cabinet Office and which included FCO and DFID officials, MoD civilian and military staff (Houben 2005, p. 142). However, despite the set-up of a joint Iraq-Unit by the FCO and the MoD, it proved highly difficult to carry out cross-departmental planning and to ensure the availability of financial resources and staff (Teuten/Korski 2010, pp. 34/35, 40).

349 A respondent pointed out that the main danger of these inter-departmental misunderstandings is that actors involved are not aware of them and think, they agreed on the same issue which increases the damage of trust even further. This was revealed in particular in the UK's engagement in Afghanistan and Iraq which created "intense interdepartmental disputes over the desirable relative roles of the FCO, the MoD and DFID in designing and implementing such operations" (Patrick/Brown 2007a, p. 11). As the Defence Committees report states with regard to the UK engagement in Afghanistan, "[t]he MoD provided advice on the size and scale of proposed British forces and other military issues. The FCO briefed on the composition and personalities of Afghan and other relevant political figures, and DFID produced papers on local agriculture and other development issues. (…) What seemed to be lacking was a detailed, holistic and comprehensive analysis of how the different factors that effected stabilization were related to each other" (House of Commons 2010a, Ev 150).

tional unit proved difficult.[350] Due to inter-departmental turf wars, the PCRU's mandate was downsized to an entity of strategic planning and capacity building in support of the other government departments. The FCO, DFID and the MoD rejected the PCRU's ambitions for a stronger mandate by arguing that existing crisis management mechanisms would be adequate (Patrick/Brown 2007a, p. 28). As indicated in background interviews, this lengthy creation process based on the good will of the three departments instead of a clear demand by the Prime Minister left the PCRU ending up more modest than it could have been.

In the end, it was decided for the PCRU to report to to all three departments while formally both physically and administratively being located within DFID. The PCRU was supposed to coordinate civilian and military specialists and to improve the strategic and operational capacity and coherence of UK stabilization efforts in particular at the aftermath of conflict. By providing operational support and facilitating communication between DFID, the FCO and the MoD, the PCRU was expected to align cross-departmental planning processes (OECD 2006, p. 52). The unit's tasks additionally included the provision of civilians to work in hostile stabilization environments, to identify lessons learned and to share best practices. In doing so, the PCRU was also supposed to deploy civilian force's expertise and skill sets to complement those of UK military troops (Patrick/Brown 2007a, p. 11).[351] To do so, the PCRU drew staff from different parts of Whitehall, the military as well as the private sector (Teuten/Korski 2010, p. 35). The unit acted in response to a board of directors composed by the FCO, MoD and DFID as well as Cabinet Office as its main stakeholders. The board reported to the Conflict Prevention and Reconstruction Committee in Cabinet, which was also responsible for governing the pools. Being co-owned and co-staffed by the FCO, DFID and the MoD, the PCRU was intended to deal with crisis management responses in the broadest sense, combining military and civilian components (Gordon 2006, pp. 344/355).[352]

350 After the government decided to create the PCRU, each department appointed one medium level officer to work on the conditions. As the officials with different departmental backgrounds were not used to working with each other and moreover did not use a 'common language' to defuse tensions, inter-departmental infighting characterized the PCRU's creation. Another initial question proving difficult was where in the existent structure and hierarchy the PCRU could be linked easiest. Issues were settled in the end by bringing in a Cabinet Office official to mediate disagreements.

351 It has done so on tactical level for instance by deploying increasing numbers of Civil-Military Humanitarian Advisors.

352 For a detailed elaboration of the PCRU's early years and functions see Patrick and Brown (2007a, pp. 27-29).

As stated in interviews it took some time until the PCRU found its feet within the UK' crisis management structures and its role provoked opposition from traditional units within the respective departments. A respondent explained that the PCRU worked more like a think tank after being created and only slowly took on a serious role after 2006. In 2007, the body was renamed into Stabilisation Unit and redefined its complementary and coherence facilitating role (Weiss/Spanger/Van Meurs 2009, p. 28).[353] Organizationally, the SU is given instructions and directions from a Board of Secretaries of State of the FCO, the MoD and DFID as well as Cabinet Office. Being dependent on DFID for nearly all its funding until 2009, the SU's funding then moved to the Conflict Pool (Major/Schöndorf 2011, p. 2).[354] Nonetheless, of the three departments involved in the SU, DFID can still be considered to play a dominant role as the head of the unit is accountable to the DFID director, who also chairs the tri-departmental board. With respect to the management of resources, the unit is held responsible by DFID's permanent secretary. However, the SU also works particularly close with the MoD to sustain the department's thinking on the role of the military in stabilization efforts. Doing so led to the SU being a large contributor to the development of a Civil-Military Coordination Group as part of the Military Support to Stabilisation Group in the MoD, and to advancing stabilization doctrine and improving training courses (House of Commons 2010a, pp. 37/38). As of 2008, the SU doubled its size "to become a 'one stop shop' for stabilization planning, deployments of civilians and cross-government conflict lessons" (Teuten/Korski 2010, p. 35).

The concept of stabilization behind the unit conceptually is closely linked to the CA as it covers the humanitarian, political, military and development aspects (Wendling 2010, p. 69). The SU today concentrates on three key areas.[355] The unit's first major role is the facilitation of cross-governmental assessment, planning and execution of UK efforts to achieve stability in countries emerging from conflict. By involving civilian and military actors from the beginning, the UK's ability to plan more effectively across government should be strengthened. This

353 According to a respondent, the development from the PCRU into today's Stabilisation Unit was only possible because a different generation of officials was involved.

354 As stated in a Written Ministerial Statement, "[f]rom 2011-12 the conflict pool will meet the running costs of the tri-departmental Stabilisation Unit, which supports UK civilian deployments." The statement continues affirming that "[t]he Stabilisation Unit is the Government's front-line delivery service for stabilisation in countries affected by conflict, most notably working alongside the UK military in Helmand province. The allocation of £12 million to the unit will support the strengthening of civilian-military stabilisation co-operation and the deployment globally of civilian stabilisation experts from the unit's thousand-strong pool. It will also enhance the unit's integrated planning expertise in priority countries" (UK Parliament 2011).

355 As identified by Hogan-Hern (2011, p. 44). See also House of Commons (2010, p. 30).

element also aims at preventing frictions between civilians and the military by involving both from the out-set. In doing so, the SU aims at developing a common understanding and has been established to ensure that all three departments develop, agree and ultimately operationalise a common policy on crisis management issues and to integrate cross-governmental crisis management planning (Stepputat 2009a, p. 37). The SU's second key area of focus deals with the identification and dissemination of lessons learned across government and to embed relevant experiences in the institutional memory of relevant actors (Hogan-Hern 2011, p. 44). It is the lead actor which identifies and shares best practices in crisis management both at national and international level. The SU thus presents a cross-governmental standing capacity to learn lessons on conflict, which already resulted in an uptake in the quality of deployed civilian staff (House of Commons 2010a, Ev. 28). The SU's third responsibility concerns the development of civilian expertise and the deployment of experts. By recruiting, training and deployment of civilian crisis management experts, the provision of civilians which have the right skills and can be deployed alongside the military is ensured. In this context, the SU supports joint civil-military trainings (Drent/Zandee 2010, p. 22). In doing so, the SU seeks to seal the gaps between emergency responses to conflict and military interventions on the one hand and long-term civilian conflict management instruments on the other (Stepputat 2009a, p. 35). To provide suitably experienced civilian personnel, in 2008 then Prime Minister Gordon Brown (as cited in Hogan-Hern 2011, p. 44) proposed that "in the same way as we have military forces ready to respond to conflict, we must have civilian experts and professionals ready (…) to help rebuild countries emerging from conflict." Acting in response, the SU created the *Civilian Stabilisation Group* (CSG), which was formally launched in February 2010 and pools over 1000 civilian experts and police willing to work in conflict-affected states and war zones. The CSG is managed by the SU as a database of deployable civilian experts. Individuals deployed by the CSG come from both the public sector being seconded by their departments and from outside the government (Teuten/Korski 2010, p. 104).[356]

As a House of Commons report assesses, the SU developed itself "as a repository of expertise and experience on stabilisation. It is coordinating cross-

356 Half of the CSG civilians were deployed to bilateral missions such as Haiti, Southern Sudan and in the Helmand PRT in Afghanistan and the other half to multinational operations (Teuten/Korski 2010, p. 104). However, the SU does not have the sole responsibility to deploy civilians in hostile environments. A large number of posts abroad in UK Embassies and DFID offices are filled by the respective departments using their standard recruitment processes (House of Commons 2010a, pp. 30/31). For more detailed information on CSG composition and training see Teuten and Korski (2010, pp. 104/105).

Whitehall work on improving joint assessment and planning at the strategic and operational levels" (House of Commons 2010a, p. 30). In doing so, the SU has also taken pressure from FCO and DFID shoulders and allows that these focus on the strategic level (Stepputat 2009a, p. 56). Its tri-departmental nature has a number of benefits. It attracts staff from all three departments and thus enables an increased understanding of the differing cultures and systems (Teuten/Korski 2010, p. 124). On the contrary, precisely the SU's tri-departmental character also poses the disadvantage that "it has not been 'fully owned'" (Teuten/Korski 2010, pp. 124) by one of the departments and consequently did not have full high level backing by either one of them. Effectively, no single government body has the authority and capacity to take the overall lead, which has impeded interdepartmental coordination (Stepputat 2009a, p. 37). Disagreement between the parent departments also constrained the SU's work and prevented it to reach its full potential (Teuten/Korski 2010, p. 125). Range and scope of SU crisis management is highly dependent on this interdepartmental board and its capability to agree on a common denominator. While the SU effectively does fill a gap between relevant UK crisis management departments, ar major challenge are the limits of the unit's power (Stepputat 2009a, p. 37).

Nonetheless, the SU's efforts in recruiting and maintaining this cadre of deployable civilians for assignments of various kinds are evaluated "significant in the UK's capacity to implement the Comprehensive Approach" (House of Commons 2010a, p. 33). This is mainly because the CSG gives the SU its own staff resources and capacity of civilian experts capable of being deployed. On the contrary, however, in oral evidence given by DFID to the House of Commons, the department emphasized that it does not "anticipate 'that the Stabilisation Unit would be the principal driver' of the work that DFID is taking forward jointly with other government departments" (House of Commons 2010b, p. 36). While drawing such a clear distinction between the departmental working lines, the role the unit is anticipated to play remains unclear. This lack of clarification underlines concerns pointed out in background interviews. Respondents indicated that while the creation of the SU has indeed neutralized interdepartmental tensions, the unit never actually worked as a coordination instrument in the CA sense.[357] As the development, planning and formulation of the crisis management policy remains departmental, more than a policy unit the SU has to be viewed as a delivery unit and in doing so mainly fulfils a logistics function. It provides exper-

357 In the state fragility context, a Chatham House report (Evans/Steven 2010, p. 12) comes to a similar conclusion stating that while the SU "can play a valuable role in providing 'surge' capacity, it has not emerged as the cross-governmental coordinating apex on fragile states that it was initially intended to be." An interviewee noted that the SU does not even serve as a mediator, as mediators do have some power, which the SU does not have.

tise and functions like a facilitator of comprehensive crisis management. Additionally, respondents pointed out that the cross-governmental background of SU also causes conflict and the level of frustration when working together can be high. As an interviewee indicated, the mixed backgrounds of the departments SU officials come from is obvious and bears the risk for potential for tribalism. Different departmental backgrounds and cultures become particularly apparent when it comes to the tackling of concrete issues. On the other hand, mutual understanding in the SU constantly improves and bears the additional value that officials carry the awareness into their former departments when going back. Staff working in the SU undergoes a kind of 'trilingual training'. The SU leads to the development of strong inter-departmental networks on the personal level.

In spite of steps taken and improved cross-departmental coordination, departmental traditions and policy structures have continued to be strong in the UK, according to interviewees.[358] A particular weakness of the system is that "loyalty remains to departments, rather than the government as a whole" (Teuten/Korski 2010, p. 5). Decision-making continues to stay in the individual departments which are free to choose whether they decide in a coherent way or not (Teuten/Korski 2010, p. 6). In spite of the CA and the need for improved cross-governmental coordination being clearly supported by the highest levels of government, senior officials in the FCO and DFID have remained skeptical, which in effect led to "a process of stop-start reforms" (House of Common 2010, Ev 143) with support for the CA in Whitehall growing again from 2007 (House of Common 2010, Ev 143). Traditionally, the formal coordination mechanism for international policies on the cabinet level has been the Defence and Overseas Policy cabinet committee. In 2008, the committee was restructured into the *Ministerial Committee on National Security, International Relations and Development* (NSID). NSID members included relevant ministers, police chiefs, the heads of the intelligence agencies, and the Chief of Defence Staff to oversee the government's work on national security. The NSID met on a regular basis with regard to Afghanistan and Pakistan.[359] Separate meetings took place on other issues. Also in 2008, an inter-departmentally staffed *National Security Secretari-*

358 A number of studies draw a similar conclusion. Teuten/Korski (2010) note that because of the UK's failings to successfully implement the CA, "Britain is not yet delivering on its full potential to engage in fragile states" (Teuten/Korski 2010, p. 5). In the wider context of the British central government system Mulgan (2008, p. 187) concludes that barriers towards government thinking in a joined-up way remain substantial.

359 According to the House of Commons report, however, the NSID only met every couple of month and not the NSID but the three Secretaries of State meetings were the principal vehicle overseeing the case of Afghanistan. This tri-departmental state secretary meeting, however, is not a formal Cabinet Sub-Committee and lacks a Cabinet Office secretariat (House of Commons 2010a, pp. 23/24).

at was created in the Cabinet Office to support the NSID, to formulate the NSS and to coordinate and administer national security policy across government (Teuten/Korski 2010, p. 33). Furthermore, the prime minister created an Interim National Security Forum in 2009 which consisted of experts providing substantive independent advice to the government. The National Security Forum's propositions were submitted to the NSID to contribute to strategy development. With the Defence Advisory Forum, a similar instrument was established to advise the MoD (De Coning et al. 2009, p. 16).

In addition to the Conflict Pool and the SU these bodies presented important elements of the overarching structure that directed the work of different departments until 2010 when the NSID was transformed into the National Security Council (NSC). The council is headed by the prime minister and includes relevant ministers and high-level civil servants.[360] Since its creation, it met on weekly basis. The NSC explicitly pursues the objective to achieve a cross-governmental CA in UK foreign and security policy through the establishment of an institutional base which should ensure coherent and coordinated decision-making. One of its major responsibilities is to oversee implementation and to evaluate and manage NSS revisions and the development of the SDSR. NSC members "work with all departments with a stake in the issue" (UK Ministry of Defence 2010a, p. 68). The NSC thus brings together formerly disparate foreign policies into a coherent whole and ensures that the whole of government operates in a joint manner toward this end. In doing so, UK government initiated a comprehensive process of strategy formulation and oversight that broadly includes all relevant departments in the NSC. This already played a central role in the NSS and SDSR development (Giegerich/Jonas 2011, p. 9). The NSC has three ministerial sub-committees which consider the topics threats, hazards, resilience and contingencies, nuclear deterrence and security and UK relations with emerging international powers.[361] In addition to these thematic sub-committees, ministers, who are members to the NSC, are responsible for coordinating priority areas of work to deliver NSS priorities by working with all departments concerned. These so called lead ministers are supported by officials who direct cross-governmental work and manage collaboration with external actors including the private sector, NGOs and international partners (UK Ministry of Defence 2010a, pp. 68-70). Additionally, the NSC is assisted by associated cross-government

360 Permanent members of the NSC are the Prime Minister, the Deputy Prime Minister, the Foreign Secretary, the Chancellor of the Exchequer, the Secretary of State for International Development, the Home Secretary, the Defence Secretary, the Secretary of State for Energy and Climate Change, Chief Secretary to the Treasury, Cabinet Office Minister of State and Minister of Government Policy (Cabinet Office n.d.).

361 See also Cabinet Office (n.d.).

high-level official groups. The creation of inter-departmental sub-structures to support the NSC also included the new appointment of a National Security Adviser who chairs the Permanent Secretaries Group, heads the Cabinet Office National Security Secretariat and acts as secretary to the NSC. The advisor is expected to set the agenda and to provide leadership to other government departments (Cabinet Office n.d.).[362]

According to respondents, the latest NSS and SDSR formulation already showed that also these new structures do not succeed to totally dominate individual department interests. As noted in background interviews, the establishment of the NSC gives greater influence to the cabinet process in crisis management with Cabinet Office playing an increasingly powerful coordination role and its lead in international crisis management grew stronger in an evolutionary process. However, as pointed out by respondents, conflict and crisis management issues have to be sufficiently important enough for Cabinet Office to actually get involved which proves particularly problematic regarding long-term structural conflicts. Furthermore, the effect of the NSC is seen to be highly dependent on the cooperation between its chairman and the prime minister (Major/Schöndorf 2011, p. 3).

Above structural changes have been substituted by an enhanced prominence on inter-departmental liaison mechanisms. To improve coherence in the British crisis management approach, numerous other initiatives on the interpersonal level aim at creating an advanced culture of collaborative working. Apart from these bodies particularly established to improve and encourage cross-governmental coordination and coherence, the UK utilized internal non-institutionalized inter-departmental collaboration (Giegerich/Jonas 2011, p. 9). The structural changes to increase crisis management coherence in the UK were supplemented by an enlarged emphasis on inter-departmental coordination at the director level in the three departments and the establishment of civil-military liaison posts (Gordon 2006, p. 345). In 2006, the *Top 200* initiative was introduced across Whitehall. This group of permanent secretaries and director-generals meets twice a year to exchange best practices and deal with cross-departmental issues. The initiative was supplemented by the creation of the *Whitehall Strategy Programme* in 2009. The initiative is a network of senior officials across Whitehall that considers strategic issues facing government through discussion (Teuten/Koski 2010, p. 92). Considering comprehensive crisis management efforts, specific steps taken by the FCO, the MoD and DFID also include joint visits by the departments' permanent secretaries to conflict regions to reinforce common efforts and working. The three departments also increasingly seconded staff to each other which has resulted in a growing number

362 Reforms such as the NSS and the NSC are noticeably along the lines of the US national security state (Porter 2010, p. 6).

of officials with cross-departmental experiences at Whitehall. In particular, the FCO has started to distribute all senior civil servant posts across government (Teuten/Korski 2010, pp. 92/93) and opened its positions to staff from other departments. Additionally, the FCO has established a team specifically dealing with inter-departmental relations in the civil-military area. Though small in terms of manpower, the unit aims at providing CA platforms such as inter-departmental lunches. According to a respondent, pressure has been renewed with the aim to further promoting the concept and putting the network idea to the next level. Also the MoD employed a number of diplomats. Joint civil-military training has also increased with civilian participation in military exercises as mentioned above.[363] However, such training initiatives have not been coordinated across government and are still individually managed by departments. No single overarching guide exists for the growing number of inter-departmentally relevant crisis management courses. Notwithstanding these steps toward a greater culture of collaborative working, progress also remained limited due to the *ad hoc* nature of these informal initiatives that were limited in scale. Though numbers of posting between departments increased, the overall extent of joint training and cross-departmental secondments of officials is in practice rather limited, not only with regard to the number of staff exchanged. Interchanges are also limited to few narrow areas. Though formerly separate departmental training programmes are increasingly brought together and opened up for other departments,[364] most departmental training is still undertaken in separate institutions.[365] Even though some cross-departmental training opportunities exist, actual take-up and use is limited (Teuten/Korski 2010, p. 162).

4.1.3 European Dimension and EU Interface in British Crisis Management

The British engagement in the EU and participation in European integration has been loaded with obstacles and ambivalences. Generally, the UK has acquired a reputation for being a 'reluctant European' and an 'awkward partner' when it

363 The Permanent Joint Headquarters' Joint Venture series is a case in point with representatives of the FCO, DFID, the SU and other parts of Government as well as representatives from NGOs participating in a predominantly military exercise, "unthinkable for non-MoD ministers a couple of years ago" (House of Common 2010, Ev 146). The Joint Venture series in discussed in detail in the House of Commons Defence Select Committee report (House of Commons 2010a, pp. 39/40).

364 The FCO and the SU, for instance, increasingly provided civilian speakers and participants to military training programmes and there is also a growing cross-governmental involvement in pre-deployment training for military units (Rintakoski/Autti 2008, p. 196).

365 Individual departmental training institutions are the Ashridge Management School for the FCO, the Defence Academy for the MoD and the National School of Government for DFID.

comes to the country's relation with the EU. British responses to the EU are often described as Eurosceptic and the country's management of EU issues as functional and pragmatic.[366] The financial crisis has underlined that the UK maintains largely ambivalent toward the EU.[367] While it has been a forerunner of efforts both to improve its own crisis management capacities and to strengthen the capability of multilateral institutions to manage conflict, a consistent feature of British foreign and security policy is the low priority accorded to the EU and UK skepticism in particular shows in relation to a European dimension in its international crisis management policy where the country strictly guards its independent position (Forster 2000, p. 57). Signs that British objectives were transformed by the country's EU membership are rare and British crisis management approaches and security identity largely remain the same.

Ongoing concerns are largely related to the country's sovereignty, relations to the US, independent capabilities and the national policies of other EU member states. Historically, UK engagement with the EU is coined by the traditional British image as a great nation and the country's attempts to maintain a global role ignoring Europe as a possible field for active involvement (Mannin 2010, p. 333). This self-perception fits uneasily into an EU engagement and the UK's attachment to sovereignty plays an important role as this fundamental principle finds its main expression in an independent foreign policy. Engaging too close with the EU is thus viewed as a threat to British political autonomy. The idea of undivided British sovereignty makes it highly difficult for the UK to deal with the EU and a transfer of sovereignty to Brussels is perceived as a loss of power (Bache 2008, pp. 89/90). The framework of the UK's EU discourse is thus very much concentrated along the line of the "'widening vs. deepening' debate" (Hadfield-Amkhan 2010, p. 165). In this context, the UK preferred EU enlargement over the Union's deepening in terms of increasing institutionalization, supranational policy-making and the transfer of competencies to EU institutions (Had-

366 See, for instance, assessments by Self (2010, p. 112), Hasfield-Amkhan (2010, pp. 165/166), Jokela (2011, pp. 82/83) and Aktipis and Oliver (2012, p. 72). Mannin (2010, pp. 11-36) and Aktipis and Oliver (2012, pp. 73/74) give detailed overviews on British history and statehood within Europe. For a chronological overview with a focus on prime ministers' EU relationships see Self (2010, pp. 112-151). Aktipis and Oliver (2012, p. 87) point out that the characterization of the UK as an awkward partner particularly resulted from a negative view on European integration by the political elite and an extremely unfriendly media, deepening British EU-skepticism.

367 The issue of the adoption of the Euro as a key symbol of EU deepening showed the British self-other attitude with regard to continental Europe. However, this case study does not look at the British rejection of the EU's 'fiscal compact' and the campaign in the British conservative party to 'repatriate' powers from Brussels. With the effects of Lisbon treaty being still uncertain and the EU's difficulties to respond to the 'Eurozone-questions' doubts on the EU's ability to act further increased on the British side.

field-Amkhan 2010, p. 165). Rooted in the Atlanticist camp, the UK also opposes an Europeanization of its foreign policy because it might influence "the continuance of the special relationship with the Americans" (Oliver/Allen 2006, p. 190). The 'special relationship' accounts as a distinctive obstacle to the Europeanization of its foreign policy and EU debates in the UK centre on the issue of a possible weakening of the transatlantic link (Miskimmon 2007, p. 108). Consequently, the Europeanization of British foreign and security policy toward a prioritization and attitudinal change toward an increasing EU role largely depends on US support for a more independent EU policy in this regard. At the same time, a major aim of British policy is to bring Europe and America together by holding the two sides of the Atlantic in its role of a transatlantic bridge, an interlocutor between the Atlantic and European spheres and a broker between US and European perspectives (Mannin 2010, pp. 318-320).[368]

Respondents emphasized that the UK is highly suspicious toward Brussels and views the EU mainly as an opportunity to strengthen the UK and to maintain its status. On the domestic strategic level, the EU did not play an important role in both NSS and SDSR preparation and formulation. This largely reflects British assessments that the EU ultimately cannot add anything to improve its crisis management capabilities (Giegerich/Jonas 2011, p. 17). However, the SDSR views the EU as "a key part of our international engagement and means of promoting security and prosperity in the European neighbourhood" (UK Ministry of Defence 2010a, p. 62) and assesses that "[t]he EU's ability to integrate civilian and military responses coherently will become increasingly important" (UK Ministry of Defence 2010a, p. 62). At the same time, the document subordinates EU CFSP developments to NATO. In this context, a major SDSR aim is to "persuade other member states to direct effort and resources towards improved national military and civilian capabilities, rather than institution building and bureaucracy" (UK Ministry of Defence 2010a, p. 63).

Despite its sketicism, British governments aimed at shaping EU foreign and security policy and the emergence of EU crisis management policy and EU CFSP has been largely driven by the UK, starting with the British-French St. Malo Initiative (Aktipis/Oliver 2012, p. 72). The step to support the development of a strong and coherent autonomous EU military capacity in this regard signified a departure from the traditional British NATO centric approach. The St.

368 Regarding the issue, whether the UK's orientation was either European or Atlanticist, UK governments mostly deny the existence of conflict between the two identities by arguing that if Britain is stronger in Europe, it also is stronger in the rest of the world (Self 2010, p. 216). Particularly the Blair government adopted a position viewing British and European interests not as opposed but to be supported by future collaboration while at the same time reassuring the British commitment to the transatlantic alliance (Self 2010, p. 137).

Malo initiative indicated a British willingness to accept an autonomous EU military crisis management capacity and points to the British desire to be leading in Europe, despite its doubts. At the same time, the UK government viewed the initiative and creation of ESDP as a means to strengthen the European role within NATO. Furthermore, in particular the MoD hoped that the establishment of a common EU policy on security and defence would lead to military reforms in other EU member states reflecting the transformation of the British military (Dover 2007, pp. 6/7).[369] Also conceptual UK input into the development and formulation of the ESS was particularly strong and the FCO published its own strategy documents at the same time. The ESS largely reflects UK concerns (Oliver/Allen 2006, p. 63) and the country also viewed the ESS as a bridging device in transatlantic security issues. In its 2005 EU presidency, the UK pushed for stronger civil-military coordination and made the topic a centerpiece of its CFSP activities. Following the UK presidency initiative, the EU advanced its central CMCO coordination instrument and developed its Concept for Comprehensive Planning (Drent/Zandee 2010, p. 12).

The UK government continued to play an instrumental role throughout the process of ESDP development. It strengthened emerging CFSP with the establishment of the Political and Security Committee which was initiated by a British proposal. Traditionally one of the EU member states most opposed to supranational EU integration and in favor of intergovernmental decision-making, the British goal in this context was based on the will to keep the control of the CFSP in the hands of national foreign ministries and to keep external policy in the scope of the Council of Ministers and the Council Secretariat, not the Commission. It was a major British objective to keep CFSP as an intergovernmental process distinct from other policy areas (Mannin 2010, p. 331). In the context of wider EU crisis management structures, the UK early supported the creation of a more permanent President of the Council. Again, the key aim in this regard was that the new institutions come under firm control of the European Council and not the Commission (Aktipis/Oliver 2012, p. 77). The UK also played a crucial political and military support role for the Union's takeover of NATO's mission in Bosnia in 2004 making it one of the EU's most demanding crisis management missions. The increasing operationalization of both civilian and military ESDP missions resulted in an increasing recognition of the EU as a capable crisis management actor in UK ministries which goes beyond viewing CSDP merely as an instrument to reinforce NATO (Aktipis/Oliver 2012, p. 83). Yet, close involvement on the EU level in this context continues to aim at the preservation of a

369 For a detailed elaboration of UK positions and debates in the St. Malo process see Dover (2007, pp. 22-38).

leadership status, to pursue certain interests and to avoid the destabilization and weakening of the UK's Atlanticist predisposition (Mannin 2010, p. 331). The UK also continued to aim at avoiding EU institutional crisis management innovations, such as the *European Union Battle Group* (EUBG) initiative, which could threaten NATO's security prerogative in Europe (Aktipis/Oliver 2012, p. 77). At the same time, the country has been aware of the necessity of reform and practical change to increase EU crisis management effectiveness and capabilities. However, as explained by a respondent, the UK's attempts to expand the EU's role in international affairs and the extent to which this position is accepted at domestic level are different matters.

With regard to capability development, the UK was positive and active toward civilian aspects of EU crisis management and particularly reluctant to provide the EU with any significant military capability (Jakobsen 2009, p. 97). This focus on the building and the projection of civilian crisis management capacities aimed at avoiding a duplication of NATO's traditional defence responsibilities. As indicated by an interviewee, the UK wants to limit EU competence and independent EU capabilities to a certain extent. The country is most sensible when it comes to topics such as a European Army or an EU Headquarter as it strictly wants to avoid the undermining of NATO. From UK perspective, EU military crisis management first and foremost has to be complementary to NATO and the EU has to work closely with NATO before it can develop its own CA.[370] Its support for EU civilian crisis management is based in the thinking that the topic is less controversial than the military dimension and the UK has been vocal and active in support of the civilian side to keep military side low. At the same time, the UK is a strong advocate of more integrated civilian-military capabilities to balance and to complement the traditional defence role of NATO. The EU is rather seen as an economic organization with an emerging foreign and security policy but largely lacking capability and political will in this field (Jokela 2011, p. 87). British skepticism toward EU crisis management is also pushed by a lack of confidence that EU member states can find common interests and perceptions on crisis management issues and a lack of military capabilities to project force (Rees 2011, p. 38).

The UK's resistance to any integration of foreign and defence policy into the Community resulted in the enhancement of the pillar model in the first place, where the country succeeded in avoiding a perceived Europeanization of foreign policy (Oliver/Allen 2006, p. 191). Nonetheless, regarding the EU's ability to adopt a CA, the UK critically viewed the pre-Lisbon pillar structure of the EU

370 In this regard an interviewee pointed towards the importance of the right rhetorical framing of EU issues to convince EU-skeptics and the persuasive power of the argument, that certain action will also benefit NATO.

institutions to result in a fragmented approach to crisis management (House of Commons 2010a, Ev 87). Thus, in parallel with the CA development in the UK, the concept's development at EU level was largely promoted by the UK and the CA development in the UK provided fundamental guidance for the EU in this area (Hanssen 2010, p. 9). As indicated by interviewees the UK tried to upload its CA structures and CA experiences, such as the CPPs, as a solution on the EU level and to run EU crisis management according to UK principles. In this regard, the UK supported and assisted the EU via the FCO, the MoD and DFID to influence and step up thinking and awareness on EU crisis management fragmentation through the active participation of diverse departmental units in debates on EU level (House of Commons 2010, p. 46).[371] As stated by the three departments,

> "[t]he UK fully supports and helps to influence and accelerate EU thinking on this subject [the CA, S.P.], through the active participation of the Development, Concepts and Doctrine Centre, the Joint Force Headquarters and the Stabilisation Unit (House of Commons 2010a, Ev 88).

In this context, the UK strongly influenced and pushed for a CA in particular in the early years, starting with the term Comprehensive Approach entering the EU terminology via the British delegation and the usage in the UK MoD (Johannsen 2011, p. 58). As noted above, the British MoD's push toward the CA met the resistance of the FCO and DFID on Whitehall level which also drove British military officials to reach out internationally. The UK's work on the CA inside the EU led to an EU crisis management policy framework that similarly underlined the need for close civil-military cooperation as a prerequisite for effective crisis responses (House of Commons 2010a, Ev 144). In particular during the UK's 2005 EU presidency, the MoD accomplished to "put the idea of a 'comprehensive approach' into EU documents and processes – before it had even become official British policy" (House of Commons 2010a, Ev 144). It is of major importance in UK policy toward the EU to advance the ability of the EU to plan and conduct operations that include both civilian and military elements (House of Commons 2010, Ev 88).

Officially, reforms proposed in the Lisbon Treaty were welcomed offering potential solutions to CA issues in the EU. In particular the creation of the post of the HR/VP is seen as a chance to ensure consistency of the EU's external action. The UK supported the consolidation of the EU's role as a civilian crisis management actor and the HR/VP is viewed to better integrate planning for gen-

371 Indeed, the UK has been "one of the strongest advocates of the Comprehensive Approach across a wide range of international organisations, in order better to combine civil and military measures and coordination within any given operational environment" (House of Commons 2010a, p. 42).

eral crisis response. In this regard, the UK government underlines that the post-Lisbon EU is much more effective in ensuring UK interests (Burke 2012a, p. 14). As mentioned by a respondent, the UK, while not being in favor of the creation of the EEAS, sees the EU's organizational structure as an EU version of the SU. The Lisbon Treaty also raised concerns dealing with the negative impact it would have on the UK's position in international affairs. Declarations of the UK government emphasized the importance to preserve the notion that the Lisbon Treaty does not expand the EU's responsibilities on foreign policy to the disadvantage of the member states. Particularly with regard to the EEAS, the UK emphasizes that the service does not impact on the British prerogative to make its own political decisions. Furthermore, the substitution of the title 'High Representative' over 'Foreign Minister' was a gesture toward the UK. UK ambitions for the EEAS are comparably low and while the Conservative Party was against the Lisbon Treaty, the government now accepts the positive role it can play in achieving both British and EU policy objectives.[372] According to a respondent, the EEAS is seen negatively as a costly new bureaucratic body. Downgrading the role of the EEAS, however, the UK does not reject the services purpose itself, but emphasizes the primacy of member states in EU foreign policy formulation and guidance (Burke 2012a, p. 14). This approach mirrors UK preference for intergovernmental decision-making also for the provisions of the Lisbon Treaty on international crisis management. While officially announcing a future focus on spelling out a more coherent vision of EEAS priorities by setting clear objectives for the service and actively working toward ensuring its success, the UK in practice blocked a lot of EEAS work and the articulation of common EU positions which would allow it to speak with one voice (Burke 2012a, p. 15).[373] Reflecting the UK's wider vision of inter-governmental EU without strong and

372 As Giegerich and Jonas (2011, p. 4) observe, chances for active UK support for EU crisis management and foreign and security policy further decreased with the 2010 elections. Under the conservative-liberal government in power at the time of writing, support for multilateralism further diminished to a form of 'multiple Bilateralism', which makes pragmatic international partnerships driven by interest a key feature of British policy. However, at the same time, an interviewee pointed out that the UK's approach to the EU remains pragmatic to a certain degree, no matter what government. Even when the conservative government at the time of writing came into power, EU policy did not change as much as expected by civil servants and the working practices behind the scenes did not change much.

373 As Burke (2012, p. 15) further points out, the EU regularly spoke with one voice on multilateral level already before the adoption of the Lisbon Treaty, even on issues where full competence was with the member states. It came thus unexpected to other EU member states when the UK did not follow established procedures anymore and blocked EEAS officials from speaking at international institutions. The Commission warned to take the UK to the European Court of Justice if it continues to block a proactive role of the EEAS at multilateral institutions where it had partial competence, for instance in UN bodies dealing with development issues.

centralized institutions, the country wants the EEAS to play a restricted and strictly subordinated role to its own diplomacy. The UK favored the pre-Lisbon structure with clear separation of the intergovernmental CFSP.

The British ambivalence toward the EU and the country's struggle with respective change is particularly notable as relations between government departments and EU institutions are strong and policy coordination and administrative exchange is high (Forster 2000, p. 57).[374] The FCO "has worked hard to identify quality diplomats from among its ranks for secondment to the EEAS" (Burke 2012a, p. 14).[375] Despite the country's political ambivalences and difficulties toward the emerging EU crisis management policy, "the institutional links between the 'core executive' and Brussels are often held up as a model for other states to follow" (Aktipis/Oliver 2012, p. 78). UK EU policy-making consists of the Prime Minister's Office, the European Secretariat within Cabinet Office, the FCO and the UK Permanent Representation. The Cabinet Office's European Secretariat arranges chairs and records the results of interdepartmental discussions at all levels. Cabinet Office and the FCO are the main EU policy agents at UK level. While not having an EU ministry, the UK does have a minister for European affairs who assists the foreign secretary in his overall responsibilities (Mannin 2010, p. 238). While departments have considerable autonomy, the growing involvement of the Prime Minister and Cabinet Office points to the generally increasing importance and need to address EU issues. Particularly in the FCO, EU membership resulted in internal restructuring and the creation of new posts. Doing so, the FCO's position shifted from that of a gatekeeper of the lines of communication between Whitehall and Brussels and a manager of UK EU policy, to that of an EU actor and interest party itself (Bache 2008, p. 103). The FCO particularly acknowledges EU importance structurally with the establishment of the post of Director-General Europe (Self 2010, p. 257).

At the same time, the FCO experienced a decline of departmental power over the policy-making toward the EU due to involvement of other domestic departments challenging its dominance (Oliver/Allen 2006, pp. 52/53).[376] Specif-

374 All departments at central government level have EU units and sections. Furthermore, the UK has a European staffing unit to advance the employment of British nationals to the EU (Mannin 2010, p. 242).

375 In addition, it can be argued that the placing of Catherine Ashton in the newly created position of the HR/VP allowd the Eurosceptic country to make a significant contribution regarding the most challenging aspect of further EU crisis management developments.

376 Most departments in the UK deal directly with the EU through their own specialized EU units thus further reducing FCO influence within Whitehall. Most of the staff in the Brussels' UK Permanent Representation is seconded from other departments (Self 2010, p. 257). The UK Permanent Representation thus is "akin to a mini-Whitehall in Brussels" (Mannin 2010, p. 326). Due to the British Cabinet government and the principle of collective responsibility, min-

ically the CFSP of the EU "has given the FCO a particular area of competence over which it alone has expertise, and that has compensated to some extent for the loss of influence over general domestic and European issues" (Oliver/Allen 2006, p. 53). Thus, the FCO maintained a significant role within managing EU relations while on the contrary "Europeanization has been the most obvious sign that the FCO cannot play a traditional role of gatekeeper through which all contact with the outside world must flow" (Oliver/Allen 2006, p. 63). While the FCO's role regarding EU matters has declined, the CFSP development gave the department an area of individual competence. These developments also resulted in a transformation of its structure and the creation of several new posts (Aktipis/Oliver 2012, pp. 78/79). Furthermore, an inter-departmental committee headed by the Europe minister aimed at creating EU-policy awareness across departments (Mannin 2010, pp. 232/233, 243). According to respondents, the UK government actively supported FCO officials and diplomats to work in the EU to improve British contributions and influence but also its worth to UK departments. While in this context also EU military crisis management led to an adaptation of structures, at the same time, the UK government was reluctant to establish structures and institutions completely independent from NATO structures. For instance, the MoD officials in charge to support the EU's Headline Goals were the same responsible for NATO's capability development process (Aktipis/Oliver 2012, pp. 80/81). While EU impact on the British central government can be observed, rather than as processes of Europeanization, these can be termed as a "'quiet revolution' in the way Whitehall manages Britain's dealings with the EU" (Bache 2008, pp. 102/103).

In the UK, the EU is used as an arena for the pursuit of national objectives, however, "[t]he choice of which institution through which to pursue foreign policy objectives is based on a robust cost/benefit analysis of the competing advantages and disadvantages of different institutions and the extent of their legal competences" (Forster 2000, p. 57). An interviewee described the UK EU efforts additionally as largely bilaterally forced. As other EU member states are much more focused on

isters and the Permanent Representative work closely together. However, from the beginning of British EU membership, Cabinet Office was recognized as the main institution in coordinating British government policy, rather than the FCO. This was caused by worries that other Whitehall departments might deem the FCO to merely precede diplomatic political solution while disregarding interests of other departments. Cabinet Office, on the contrary is expected to provide consensus view. Though being left with the formal task to communicate issues to the Permanent Representation, this understanding has the advantage that the FCO was indeed allowed to argue for its departmental view (Forster 2000, p. 49). While these formal FCO responsibilities also include the preparation EU General Affairs Councils and Presidencies, the FCO was thus able to play a particular role in CFSP development.

the EU, it is of highest interest for the UK to maintain close relations to these states. In order to have good working bilateral relationships, UK policy needs to have an EU perspective. The UK focuses its international crisis management policies on bilateral relations and also sees the EU as a forum to advance bilateral relationships. UK efforts on the EU level and toward the development of European crisis management are mainly motivated by the wish to see the UK as a leading actor in European international relations. As pointed out during background interviews, the UK evaluates EU crisis management through the lens of the policies effectiveness which is only worth to be supported, if it proves more effective and a benefit for UK. An aspect contributing to the UK's critical perspective is the country's view on the EU as a highly inefficient bureaucracy. EU issues always experience an 'extra test' from UK-side.

4.1.4 Synopsis: The UK and Comprehensive European Crisis Management

4.1.4.1 Crisis Management Approach and Conceptual Dimension

In order to deliver a CA in crisis management, the British government developed a number of key policies that influenced the UK's approach to international conflict. Attempts to adapt to the changing international environment are a major theme of UK policy and the CA plays a large role in this reorientation and readjustment. The country clearly is at the forefront of showing willingness and flexibility to experiment with new ideas in its so-called Integrated Approach efforts. The CA as a concept and the objective toward improved crisis management coherence to build stability enjoys clear political commitment on the strategic level. References to the need to establish a CA are frequently repeated in UK government documents and illustrate the importance placed on crisis management coherence. UK documents dealing with specific crisis management issues acknowledge the necessity for more coherence and coordination in British crisis management. Concepts and statements illustrate that the UK government openly embraces the CA concept and its strategic intent. The UK's approach in this regard is framed by the country's economic and great power decline. This baseline has been an important factor shaping much of the UK's crisis management policy and also coined the implementation of the CA in a pragmatic way.

Conceptually, the government's emphasis on an ethical dimension to foreign policy helped defining common objectives between different departments. The UK's Third Way politics thus facilitated joint objectives in the CA sense and provided an umbrella for different actors (Macrae/Leader 2000, p. 3). British CA developments have to be viewed as a part of a more general move of the UK policies toward Joined-Up Government and of the wider debate and efforts to increase cross-departmental collaboration of the UK government as a whole. Since then Prime Minister Blair introduced the Joined-Up Government, reforms to also increase crisis management coherence began. Facilitating security and development coordination can be understood as a continuation and extension of these attempts to join-up all of UK government. An important factor in this regard was the strong support and effort by government leadership who stressed the need for greater cross-departmental coherence and the development of shared concepts. Crisis management coherence was made a high-profile priority. In the wider government context, the PSAs represent a conceptually important instrument to improve inter-departmental discussion and coordination and provide a strategic framework. The development of an outcome-focused performance culture and respective tightening of accountability encouraged working together toward shared targets.[377] A setback of the PSAs is that while fostering inter-departmental coherence they also resulted in a bureaucratized system and agreements which do not always correspond to the departments' particular strategic priorities (Teuten/Korski 2010, pp. 13/14).

The individual departments' White Papers, the MoD's Joint Discussion Note and Strategic Defence Reviews, the NSS and its up-dates, the SDSR and the BSOS demonstrate the increasing focus on the topic and response to cross-governmental interdependencies. Also, the development of country strategies increasingly took place in a cross-governmental approach. Cross-governmental planning and coordination is evaluated in capability reviews which assess that documents are increasingly developed in an inclusive cross-governmental process. Thus, relevant overarching strategies are complemented by specific policies, and regular reviews. By clarifying departmental crisis management roles, and recognizing interdependency and the importance of coherence, these documents have contributed to conceptually clarifying the CA and to conceptual complementarity of the different departments. The UK's CA on the conceptual level is focused on inter-departmental consultation, practical measures and operational issues concerned with doctrinal CA implementation. It is conceptually based on the key principles of a shared understanding of problems and objectives of UK crisis management and a proactive engagement in the planning process. In doing so, the UK broadly interprets the CA as a concept concerning all levels and dimensions of crisis management as well as all actors. At the same time, the UK's CA underlines individual departmental strengths and emphasizes the competencies and resources of each department. CA implementation in the UK does

377 See respective analysis in DFID (2009, p. 78), by Teuten and Korski (2010, p. 12) and by the House of Commons (2010).

not develop cross-governmental guidelines for specific modes of crisis management action or homogenizes operating modes of different actors. Crisis management coherence is rather aimed at in a joint departmental effort to achieve coherence as a whole.

Notwithstanding these conceptual steps forward toward comprehensive UK crisis management, a downside of the conceptual CA development is that the concept's evolution for a long time was largely MoD-driven rather than resulting from of an actual cross-departmental deliberative process. CA understandings in the UK have been operationalised in military doctrine. Furthermore, the UK's CA conceptualization has developed around the issues of NATO or coalition-led military operations, EBAO doctrines and SSR. In consequence, ownership of CA operationalization was not shared by the FCO and DFID for long and, in addition, broadly perceived as military concept without true comprehensive crisis management credibility. As such, the UK's focal point of CA implementation and aim toward comprehensive crisis management in a pragmatic way focuses on coherence to enhance national security. This framing and the intensification of civil-military interaction is central to the materialization of the UK's CA. At the same time, development policy and civilian crisis management issues were reflected on with pragmatism.

In this context, the establishment of the NSS can be considered a key development and important step to implement the CA on the conceptual level as the strategy represents the first comprehensive cross-departmental statement on the UK's international security environment and respective interdependencies. The NSS' establishment renewed the process of setting out the CA in the UK. Pushed by heavy budgetary pressures and a perceived strategic deficit of the UK's foreign and security policy, the NSS has been established as a comprehensive strategic framework, explicitly committed to the CA in its background, planning process and content. The establishment of the NSS as a grand strategy together with the SDSR in a series of successive documents developed in a cross-governmental formulation process which in itself improves inter-departmental planning on conflict issues. Also, the BSOS presents a genuinely cross-departmental strategy and sets another example of a coherent government approach to crisis management issues. With the BSOS, the UK has a conflict management strategy that has been developed in a joint effort by the FCO, the MoD and DFID and which represents a major step forward considering comprehensive crisis management and the CA implementation on the conceptual level. In doing so, the documents bring together the key issues that need to be addressed in a structured and coherent manner. With this approach to strategy improvement, the British government attaches particular importance to the development and formulation process as an effort to build consensus and broadly coordinate and secure interests between departments

(Giegerich/Jonas 2011, p. 9). Though the documents have been criticised on other aspects and the interdepartmental planning process was not entirely unproblematic, it ultimately produced widely respected and agreed on inter-departmental documents.

4.1.4.2 Organizational Dimension

In line with conceptual developments, the UK experienced a shift in emphasis toward inter-departmental coordination concerning all of government in the late 1990s. Organizational steps taken by the UK underscore that crisis management challenges are coherently seen at once as security, development and foreign policy challenges. Due to the central role of the Prime Minister and Cabinet Office, a rather strong departmentalism has been illustrated as being characteristic of the British political system leading to tensions between departments. Due to the large departmental autonomy and inevitably different departmental mission statements, strategic approaches and working cultures, inter-departmental coordination has been hampered by deep frictions among the departments. The UK government introduced a number of steps to signal the importance of the CA and cross-departmental coordination and the country is widely seen as a leading example of the emergence of CA at national level.[378] Institutional changes reflect the continuing awareness already shown on the conceptual CA side and the UK's crisis management apparatus has responded to the requirements of complex crisis management. The aftermath of the wars in Iraq and Afghanistan largely triggered structural reforms in an effort supported by the whole government. Respective reforms concentrated on the horizontal coherence of both policy and implementation.

Though being acclaimed an innovative coordination mechanism, the overall assessment of a joined FCO, MoD and DFID budget turns out to be more nuanced after more than a decade of the pooled funding mechanisms being in action. On the one hand, the pooling of resources is considered to have improved the coherence of UK crisis management policy particularly by promoting dialogue and information sharing. The process of joined planning ensures that tensions in departmental priorities and approaches are exposed. The UK's pooled resources show that such arrangements can lead to compromises and also reduce

378 Stepputat (2009a, p. 35), De Coning (2009, p. 118), De Coning et al. (2009, p. 16), Weiss, Spanger and Van Meurs (2009, pp. 20/44), Drent and Zandee (2010, p. 21), Major and Schöndorf (2011, p. 2) as well as Hull (2011, p. 6) also view the UK as a major promoter of the CA agenda and a pioneer of CA development and implementation.

time lags for addressing urgent needs. Yet, though the instrument is certainly worthy and from international perspective is highly innovative, there are also limitations as the pools revealed that common resources do not automatically lead to common understandings and result in joint implementation. Furthermore, the pool's role as an incentive for coordination between the departments has been questioned due to its small financial volume compared to overall departmental resources. These issues show that pooled funding instruments cannot always be perceived as resources leading to increased coordination and coherence of activities. They can also generate inter-departmental competition for limited financial resources and thus discourage CA ends. However, the various reforms and transformations of the Conflict Pool into a more comprehensive funding instrument showed a learning-by-doing attitude of the UK government. Despite some dysfunction, the joined funding mechanism increased the density of communication and led to increased interaction between the three departments. Also, with the creation of the CPPs the UK government made inter-departmental crisis management coordination more formal.

The SU has been the most significant organizational innovation and "departure from the uni-departmental approach" (Teuten/Korski 2010, p. 6) to implement the CA in the UK. It illustrates a clear effort to create a single government organization that takes in the multi-departmental crisis management expertise to deal with complex conflict. The idea of the SU as a one-stop-shop for UK comprehensive crisis management can be considered as a hybrid between hierarchical and network structures. Yet, the creation of SU as a new functional unit has to be viewed two folded. As an exemplary model of CA implementation in the organizational dimension it sheds some lights on potential benefits and drawbacks of the creation of a permanent cross-governmental body to improve coherence. Experience indicates that standing cross-governmental units advance the institutionalization of procedures and improve inter-departmental awareness. Furthermore, they facilitate the inter-departmental exchange. The unit's strength is that it is a standing capacity which promotes common understanding and shared objectives. While it has proved that cross-governmental crisis management planning is possible, the department nonetheless "has often been constrained by the reluctance of its parent departments to give it responsibility and by a lack of a fully effective mechanism across government to respond to crisis" (Teuten/Korski 2010, p. 6).

While it is particularly valuable in carrying inter-departmental awareness back into relevant departments, several unintended consequences of this institutional model became apparent. Firstly, it does not serve as a coordination instrument in the CA sense but rather as a delivery unit fulfilling a logistic function. In this regard, the SU as a newly established institution appears to increasingly become a fraction of the coordination problem. While its staff works well together and the inter-departmental awareness in the SU is high, it increasingly detaches itself from its parent departments. Having most crisis management expertise within the unit, feeding back to the relevant departments becomes insufficiently, according to respondents, and the SU becomes part of the bureaucratic turf wars and inter-departmental struggles for influence, funds and prestige with existing relevant departments it is supposed to coordinate. Lacking political weight to declare a lead role in inter-departmental coordination, it increasingly guards its own institutional survival acting as a provider of expertise to create further significance within the UK government system.[379]

Scholars proposed several alternatives to address the above-outlined weaknesses of the SU.[380] Most prominently among them, the call for greater ministerial championing and full ownership by one department that would help catalyze the unit's goals. This, however, cannot be realized without losing the sense of tri-departmental ownership. It is questionable, if the unit can maintain its neutral character and can still draw on such a diversity of staff with one department being the dominant leader. Another suggestion is to link the SU to the National Security Secretariat. With regard to lessons learned, it is suggested that the SU "should make it a priority to encourage those involved in the Comprehensive Approach to learn lessons from each situation and to disseminate the lessons as appropriate" (House of Commons 2010, p. 34). To carry out this critical function, however, the unit has to be given sufficient resources which calls for a significant increase in the Conflict Pool's budget.

The latest attempts to institutionalize the CA are the establishment of the NSC and a National Security Advisor. With the creation of the NSC, the UK government system has an additional high-ranking forum for bringing about cross-departmental coordination. It remains to be seen how the NSC impacts on British efforts toward crisis management coherence. Critics argue that the NSC's role is rather that of a clearinghouse than working as an institution that critically evaluates information and decision-making (Rotte/Schwarz, p. 58). However, with the NSC, an organizational capacity at managerial level has been developed at the core of the UK government, which can provide focus and overall strategic direction ensuring a coordinated approach across the whole of UK government. The main goal behind the NSC establishment is to create a network of decision-makers, which is declaredly cross-governmental in contrast to a coordinating

379 See also skeptical comments by De Coning et al. (2010, p. 2).

380 See the House of Commons Defence Committee which suggested some practical steps the SU could take (House of Commons 2010, p. 34). See also Teuten (2010) as well as Teuten and Korski (2010, pp. 147-177).

body merely collecting contributions by other departments. Thus, a major advance made with the establishment of the NSC, the National Security Advisor and the supporting structures is that it created a clear point of responsibility and accountability for strategy and to make sense to questions of comprehensive security. According to interviewees, the SU, in contrast to this, rather than offering conceptual leadership, constitutes a forum where different departments meet and are confronted with each other's position. With the creation of the NSC and the National Security Advisor with central coordinating roles, the CA can make its presence felt more throughout Whitehall.

Conceptual efforts toward improved crisis management coherence in the UK resulted in relatively extensive structural innovations and institutional transformations to improve inter-departmental crisis management coherence and coordination. These included the establishment of joint funding mechanisms as central financial oversight mechanism across departments and the creation of the tri-departmental SU. Reforms of Cabinet sub-committees, working groups and inter-departmental units supported these structural innovations. The UK has clear hierarchical components in its style of CA implementation. Political leadership is a major driver of the UK's CA. Mechanisms to improve inter-departmental coordination have been developed as overarching authority mechanisms to coordinate different parts of government thereby strengthening the role of central government. While this illustrates the seriousness with which the CA is being taken, it also shows that the UK approach of CA implementation is largely based on a centralizing way of approaching CA issues, hierarchical coordination mechanisms and authority as a means for implementing the CA. In this regard, policy networks in the traditional sense are being pressed for in a top-down style and hierarchical direction of the chief executive. With the creation of inter-departmental units, the expansion of the SU and funds, the UK government directly aims eliminating gaps in crisis management activities no department can fill independently on its own by advancing coordination of the three key departments (Stepputat 2009a, p. 6). However, the top-down institutionalization of cross-departmental coordination at the top decision-making level also entailed instrumental aspects. The reforms largely took place in terms of horizontal differentiation and specialization of existing mechanisms horizontally according to functional tasks. Efforts were oriented toward positive coordination in terms of sustaining effort which aims at assuring that all parts of UK government actively engage with each other to maximize effectiveness and efficiency of crisis management policy. From the organization theoretical perspective of aspects facilitating policy implementation, in the UK most factors have been present to a certain degree. Signaling in terms of clarity of relevant CA facts to all involved was provided by the political leadership. Also, key points of common interest were framed in the CA context and the two factors of iteration and exchange of UK actors encouraged coordination and strengthened coherent action over time, for instance due to the pooled funding. Also commitment by all actors increased over time.

Nonetheless, UK efforts and experience also leave a mixed impression on efforts to CA implementation with the set-up of extra bodies. In spite of the improved formal cross-departmental coordination, old departmental traditions, loyalties and remaining policy structures continue to be strong, dominating cross-departmental working initiatives and hampering the UK's CA to international crisis management and efforts to coordinate its policies. There are still insufficient incentives on the level of informal coordination, for instance, for staff to train together or to gain experience in other departments. With the rationale of cross-departmental coordination at the heart of the UK's CA implementation process, the British approach to facilitate crisis management coherence has so far clearly favored formal structures. In doing so, it has been predominantly built around departments and departmental objectives. This structural and formal focus revealed the importance of input and participation by all relevant actors.

In the UK, the greatest CA success is the clear identification of the problem with strong institutional solutions backed up by resources. According to one respondent, conceptual as well as structural emphasis on the CA in UK crisis management resulted in a high degree of shared inter-departmental responsibility and in sum, the analysis of the CA in the UK makes clear that a great deal has been achieved. In this regard, the case provides a classic case of dynamics of change in government institutions observed by scholars of organization theory. The country experienced the failure of its policy in its Afghanistan and Iraq engagements and post-war planning while at the same time, Joined-Up Government and international CA efforts provided an alternative set of beliefs about how to better organize international crisis management. Furthermore, it was the government leadership itself that served as the political champion willing to advocate CA efforts, at the beginning in the broader context for the whole government machinery and then specifically for the area of crisis management giving a top-level cover to the reforms. The overall CA discourse in the UK can be characterized as a result of perceived necessary adaptation to external changes.

4.1.4.3 European Dimension and EU-Interface

UK membership in the EU has been deeply controversial and the country's approach and involvement to EU crisis management appears largely ambivalent. The analysis showed that the country both confirmed significance to the role of the EU and a degree of Europeanization of its crisis management policy and, on the contrary, also demonstrated the British desire to retain the capability to operate as an independent actor. On the one hand, the country showed commitment such as the St. Malo initiative leading to the development of the ESDP. On the other hand, the UK's political accounts emphasize the country's independence, national sovereignty and a general unease about any EU appearance and moving beyond the intergovernmental framework. The majority of British objectives remain remarkably unchanged by EU membership and its security identity and crisis management approach has not changed. The contradictory British approach reflects the UK's strong US-support and aim to ensure transatlantic security relations while also being actively engaged in the construction of the EU crisis management. In security matters, the UK remains largely committed to NATO and a dominant role for the Atlantic Alliance in European security.

But UK Europeanization is also restricted by the limited scope the UK intends for EU crisis management as well as by the limited scope it actually has from UK perspective. Furthermore, UK patterns of adaptation to the EU are characterized by an emphasis on adjustment of existing procedures and attempts to keep changes at a minimum. The analysis showed that changes which occurred do not mean a break with the past. Foreign and security policy traditions and crisis management approaches have rather been rearticulated. However, the EU and EU crisis management became important in this re-articulation of ideas of British leadership in international affairs and great power identity. Europeanization of the UK takes place in terms of tactical shifts rather than a shift in key crisis management preferences and traditions. Furthermore, British crisis management policy has become Europeanized only as long as traditional perspectives could be maintained.[381] In this context, it largely took place with a greater leaning toward the uploading than the downloading dimension with intergovernmentalism being at the core of moves to develop EU institutions. Policy promoted on the EU level has been inspired by nation state priorities. The UK does not develop a proactive approach if it does not serve some specific interests. Europeanization only takes place when issues are not sufficiently important (Oliver/Allen 2006, p. 65)

Structurally, however, efforts have been made to closer link the UK's core executive and the EU's crisis management machinery. This organizational adaptation is underlined by the attention given to the EU which attempts to provide political direction across controversial and complex issues. In this regard, Europeanization of British policy making took place also in the downloading dimen-

381 Bache (2008, p. 103) describes British attitude with the sentence "Europeanization yes, but on our terms."

sion within traditional Whitehall arrangements. The UK's institutional engagement with the EU grew significantly, in particular in the area of the emerging ESDP and the high degree of coordination with the EU became a part of the organizational outlook of British foreign policy. British reforms toward comprehensive crisis management both formed and have been formed by international developments. UK CA efforts thereby not only aimed at joining up UK crisis management activities, but also focused "on improving the effectiveness of the wider international community" (House of Commons 2010a, Ev 82).

While EU crisis management did not have a transformative impact on British policies the crucial role of the UK was vital for the development of EU crisis management. The UK's engagement in this area offered a way to provide the country with a greater leadership role. These initiatives represent a successful uploading of UK priorities to EU crisis management developments. In this regard, the country instrumentally uses the European arena. Though acknowledging the broad range of EU crisis management instruments, the UK is very EU-skeptic in these high policies and little space is given for the development of EU military crisis management capabilities. As the military side is largely neglected, the UK does ultimately not acknowledge the EU as a comprehensive security actor. While it sees value in a civilian crisis management role for the EU, it is very skeptical about the potential of the EU to develop into a 'hard security' actor and doubts that the majority of European countries will become credible military partners. The UK has thus largely promoted the development of the EU's civilian crisis management, but rather than being a core-promoter with the aim to maximize capacity and size of EU civilian crisis management, the UK focused on functional aspects and particularly stresses the importance of improving civil-military to assure coherence in crisis management. In doing so, the country largely contributed to pushing the CA debate forward on EU level. At the same time as the CA was adopted in the EU and it became a growing concern internationally, the concept, which has been strongly pushed for by the UK very early on the level of IOs, came back to Whitehall.

4.2 Case Study Germany

4.2.1 The German Crisis Management Approach

German foreign policy and crisis management engagement is largely affected by its history. Debated in the framework of its past, the country's foreign and security policy culture is coined by the experiences of the Second World War and National Socialism (Pradetto 2006, p. 21). As a consequence, two dimensions play against each other: "the Nazi preoccupation with power against Bonn's [today Berlin's, S.P.] reluctance to admit its power" (Andrei Markovits and Simon Reich, as cited in Houben 2005, p. 185). German national identity is defined in opposition to the Nazi regime, with ideas of nationalism being largely absent from political vocabulary, and a political culture that stresses multilateralism and cooperative institution-building over the pursuit of national interest (Aggestam 2000, p. 66). Historical, moral and strategic dimensions are tightly interwoven. Political discourses tend to be moralized and the country's participation in international crisis management operations is symptomatic of its broader role and history, identity and collective memory (Houben 2005, pp. 185/186).[382]

This resulted in a German foreign and security policy which has been described to be reluctant and a traditional leaning of German governments to a very cautious and balancing foreign policy approach (Houben 2005, p. 194).[383] The country seeks to "generally 'keeping a low profile' avoiding claims of leadership and the impression that Germany was pursuing narrowly defined national interests" (Pradetto 2006, p. 22). As the German security culture is coined by a rejection of its military past, civil-military relations in Germany are founded on civilian-oriented principles such as the soldier as a citizen in uniform (Giegerich 2009, pp. 66/67). However, Germany also feels deeply obliged to contribute to peace and stability in proportion with its political weight and a moral imperative of having a duty toward peace. The operationalization of such a duty, again, is deeply conditioned by historical experiences leading to a German political culture which is portrayed to be non-confrontational with a high degree of continuity, guided by a culture of restraint (Houben 2005, pp. 187/188/191).

In this context, the German foreign policy identity has frequently been labeled with the description of the *Zivilmacht,* or civilian power identity.[384] Being a civilian power implies a focus on non-military means to secure national objectives with military power in the role of merely a side instrument, the promotion of human rights and peaceful conflict resolution. In line with the two fundamental 'Civilian Power Germany' principles of "never alone and never again" (Wilke 2007, p. 49), violence as a way to solve conflict is strictly neglected. Complemented by 'politics before force' as another defining characteristic of this civilian power paradigm, other characteristics entail the acknowledgement of the need of international cooperation in the pursuit of common objectives and the willingness to develop respective supranational structures to deal with critical issues in international affairs. A

382 This also becomes obvious with certain notions such as national interest being unmentionable in German political discourse for a long time (Houben 2005, p. 187).

383 As also observed by Giegerich (2009, p. 66) as well as Harnisch and Wolf (2010, p. 43). Houben (2005, p. 194) finds that apart from the German history as a reason, this foreign policy reluctance partly also originates from a "fear of a sudden decline in domestic prosperity."

384 See, for instance, Harnisch and Maull (2001), Houben (2005) and Kirchner (2010b).

civil power thus emphasizes multilateralism and engagement through international institutions. Another key element is the strong normative policy dimension and trust in the validity and viability of international law (Houben 2005, pp. 186/187). In the particular German context, the civilian power identity also largely serves the objective to reassure its European neighbors that they do not have to fear a potential return of German militarism or hegemonic dominance (Dettke 2009, p. 28). German crisis management is based on a concept that assigns priority to civilian conflict resolution and moreover preventing conflict and crisis to arise in the first place. However, changes in German foreign and security policy that include the use of the military led to increased questioning whether Germany still is a civilian power. These are also reflected in notions describing Germany as a 'normal civilian power' or a 'civilized power plus' which modify the civilian power paradigm (Kirchner 2010b, p. 142).[385]

Apart from strongly favoring non-military crisis management, the German self-perception as a civilian power also demonstrates a great preference for multilateralism which showed early in the policy of *Westbindung* (Aggestam 2000, p. 64).[386] The German commitment to the West and deep rooting in the Western community at the same time aimed at avoiding any German *Sonderweg* (Pradetto 2006, p. 21). Germany indeed has mostly acted in broad international terms and through supranational institutions in international affairs. Consistent with "the multilateral thrust of German foreign policy" (Patrick/Brown 2007a, p. 105), Germany framed its international policy largely in the context of broader EU, UN, NATO and OECD crisis management initiatives which constitute the pathways through which German conceptions of security are pursued. The strong drive toward multilateralism and international cooperation from the German perspective, however, can also be explained by "the softening or even 'comforting' effect [of multilateral institutions] in that they can take the sharp edge of a difficult decision" (Houben 2005, p. 194). German multilateral institutionalism thus served to legitimize its actions and characterizes a major thrust of German foreign policy. In doing so, Germany sees itself primarily as a European power with only regional reach. The Transfer of stability is a key factor in German foreign policy in this regard. Its multilateralism is also reflected by channeling resources through IOs

385 Dettke (2009, pp. 71-114) deals with the issue of the German civilian power paradigm and the country's use of force since the 1990 Gulf War. On the question whether Germany still is a civilian power see Aggestam (2000, pp. 79/80), various contributions in the volume edited by Harnisch and Maull (2001), and Overhaus (2006). Also, see chapter 3.1 which analyses the EU's civilian power context.

386 The *Westbindung* policy of the first German chancellor Konrad Adenauer attempted to embed Germany in Western institutional structures viewing NATO and the EU as the respective anchors (Aggestam 2000, p. 64).

(Unger 2009, pp. 144/147). As Dettke (2009, pp. 30/31) summed above elaborated characteristics of German foreign and security policy up,

> "Germany's willingness to adopt the role of a civilian power with a 'culture of restraint' (…) – characterized by multilateral decision-making and a rejection of the use of force – was a rational choice both for its protection and to reduce its military vulnerability, as well as to guard against a return of nationalism and militarism."

Considering German international engagement, key features contain a concern for human rights, a moderateness of approach, extensive consultation with partners, renunciation of unilateral action and a willingness to seek compromise (Aggestam 2000, p. 80).

Furthermore, Germany has a strong legalistic state tradition which reflects in the German *Rechtsstaat* conception. The German Rechtsstaat idea signifies that "the sovereign was to be bound by laws and rules (…), which were to be equally and fairly applied to all state subjects, and that judges and administrators were to be neutral" (Kickert 2008, p. 6). Legislation thus forms the basis of the state and the "[c]onstitution, laws and regulations became the exclusive source of administrative action. Therefore, the law gained the monopoly of the only relevant expertise for the effective functioning of the state" (Kickert 2008, p. 6). Against this background, Germany shows a particular kind of constitutional patriotism (*Verfassungspatriotismus*) which stems from the above-mentioned close link of the German identity to domestic institutions and to the federal constitution (*Grundgesetz*). Specific norms of German foreign policy are coined by those preserved in the Grundgesetz. In the foreign and security policy context, Article 26 of the Grundgesetz requires both the German state and the citizens to work for the continuance of international peace. Also, the constitution obligates the country's international policy to "cooperative internationalism; to this end, the constitution explicitly provides for the possible transfer of sovereign powers to intergovernmental institutions" (Pradetto 2006, p. 16). In consequence, Houben (2005, p. 191) observes a German orientation along a "formalized regulatory or legal culture" in contrast to the flexible bargaining culture of other countries and a commitment to institutionalize international relations on a legal footing. Moreover, the German federal constitution and Rechtsstaat idea are deemed to be internationally promotable norms (Aggestam 2000, p. 66). Germany regards its foreign policy as a *Weltinnenpolitik* which is oriented toward the global rule of law and supported by the normative policy premises in the Grundgesetz (Houben 2005, p. 196).

As elaborated above, the German security political approach is characterized by a 'pacifistic impulse', an aversion to exercise military power and, moreover, large skepticism about the utility and effectiveness of military force.[387] Based on the the concern of overcoming its history and establishing itself as a peaceful European partner, "Germany has pursued a consistent national security policy of deemphasizing military power as a foreign policy tool" (Johannsen 2011, p. 45). However, during the 1990s a change and emancipation regarding its "deep-seated anti-militarist sentiments combined with pacifist inclinations" (Kirchner 2010b, p. 139) has been observed. This new foreign policy departure, "displayed a new robustness in security affairs" (Harnisch/Wolf 2010, p. 43) which partly replaced the dominant civilian power paradigm. This new robustness is explained by the unification of the country and increased material power leading to a broader spectrum of foreign policy choices (Harnisch/Wolf 2010, p. 44). Indeed, in particular regarding international conflict resolution, the German reunification led to a discussion on whether a reunified and stronger Germany should also take on more responsibility.[388] Germany's interest in reacting to conflict and instability was further pushed by the Balkan experiences and events in conflict prone countries in Central Africa. The major impulse for the increased German crisis management activity during the 1990s thus was humanitarian and largely focused on human security and poverty alleviation (Patrick/Brown 2007a, p. 105). In this context, the German involvement in crisis management, stabilization and reconstruction efforts are a rather recent phenomenon, which began to change the country´s global posture, as also the German military began to reform itself from a force of traditional defence to more of an expeditionary force (Patrick/Brown 2007a, p. 112). Though decisions to deploy German forces are unavoidably controversial and involve explicit parliamentary approval, the idea of the use of military force as an instrument of German policy in this humanitarian context increasingly became accepted (Houben 2005, p. 196).[389]

387 The skepticism against the use of military force is also reflected in a highly negative attitude of the German public toward any use of military power (Dettke 2009, p. 30).

388 Political and academic debate after the German unification circled around the issue of German power and the question, what the unification means in the context of European power equation. After the German reunification, also the country's approach to civil-military cooperation changed from the host nation perspective it was discussed under during the Cold War, which mostly focused on cooperation between German national actors and the friendly forces, to the adaption of a CIMIC conception that takes the changing security environment into account (Voget 2008, p. 144). For a detailed analysis of the German CIMIC doctrine see Voget (2008).

389 German involvement in international missions indeed developed from military abstinence (Gulf War in 1991) to robust involvement in UN peacekeeping operations (UNOSOM) to military engagement not mandated by the UN (NATO Operation Allied force in Kosovo). The Balkan conflicts for Germany represented a big political issue high on the agenda due to the regions geographical and cultural proximity to Europe as well as the flow of refugees coming into the country. The German engagement in Kosovo in this regard meant a turning point for German crisis management policy and was of major symbolic importance (Miskimmon 2007, p. 100).

Since the 2001 terrorist attacks, more sober national security considerations were added to the normative German approach, with the country gradually undergoing a 'normalization' of its foreign policy approach which was linked to questions of Germany as a nation state and discussions of national interests and responsibility.[390] In doing so, Germany recognized the threatening spillover effects of instability and violent conflict for the country itself as well as its European neighbors (Patrick/Brown 2007a, p. 105) and the country developed a "new 'culture of realpolitik'" (Dettke 2009, p. 3), which takes a perspective into account that acknowledges that Germany has to play a leadership role in international diplomatic and security affairs that is commensurate with its economic strength, population size, geographic centrality and political power. This, however, meant for Germany to acknowledge its responsibility to become more active to ensure peace and stability (Aggestam 2000, p. 72). Lately, however, history seems to repeat itself and a renewed strictly anti-interventionist behavior becomes noticeable. The German polical considerations during the uprisings in the Arab World suggest that the German government developed into an adversary of any military engagement again. This inconsistency of responses and resulting ambivalence has been a source of debate on the credibility and reliability of German crisis management policy both within and outside Germany. These developments furthermore indicate that the issue of how to use its power in the conflict management context will remain a problem in the future as well. In addition, the habit of German governments to simply follow public opinion points to a somehow weak foreign policy foundation.[391]

390 The so called 'normalization' of German policy is a highly contested and raises several questions beginning the problematization of what can considered to be the standard and nomal. For a detailed review of this discussion regarding the use of force see Wilke (2007, pp. 70-93). Dettke (2009, pp. 18-48) provides an in depth chronologic overview and the historic background of German crisis management responses. For a deeper elaboration focusing on Germany as a nation state see Aggestam (2000, pp. 71-72).

391 This analysis is shared in diverse newspaper examinations on German foreign policy such as the articles by Geis (2012) and Ulrich (2012). These recent press pieces also point to the resulting ambivalence in German foreign and security policy in the last decade that has been broadly criticized at home and abroad. Often cited examples of contradictory German responses to conflict and will to set multilateralism aside include the unilateralist recognition of Croatian independence 1992, the absence of participation in any military action against Iraq even if a UN Security Council resolution were taken in 2003 or the 2008 rejection to expand German NATO forces in Afghanistan (Kirchner 2010b, p. 139). The German rejection of a no-flight zone and a military intervention in Libya are in line with these events. Also The Economist (2012) noted that despite past talks about 'normalization', "yet little about Germany's army seems normal", in particular as German public still is highly skeptical of the military and the political class acts according to that sentiment.

4.2.2 German Comprehensive Crisis Management

4.2.2.1 Conceptual Dimension

Several guiding documents and conceptualizations impact on German efforts to implement the CA on the strategic level and coin respective debates in Germany. Initial discussions about the need to deal with the complexity of crisis management challenges arose in the end of the 1990s after the increasing German involvement in international crisis management operations. These conceptually concretized under the then Red-Green Coalition government in 2000, with the adoption of the *Comprehensive Concept for Civilian Crisis Prevention, Conflict Resolution, and Post-Conflict Peacebuilding* (*Gesamtkonzept der Bundesregierung: Zivile Krisenprävention, Konfliktlösung und Friedenskonsolidierung*) (Die Bundesregierung 2000) which was largely driven by the Balkan experiences and primarily aimed at pursuing civil crisis management goals. Calling for a nationally coordinated comprehensive strategy, acting together, also with international partners, is seen as a means to ensure strategic and conceptual coherence. The document takes on a wide definition of security which involves political, economic, ecological and social stability and which is largely civilian oriented. Conflict prevention is viewed as the decisive pillar of German policy and it is explicitly mentioned that the objective of the German development efforts is to lay the basis for avoiding violent conflict for which poverty alleviation is regarded the key. With regard to precise measures toward coherence, the Comprehensive Concept mentions the elaboration of a strategy and the implementation of country specific discussion groups (Die Bundesregierung 2000). Though only one page long, and only skeletal containing vague principles and statements regarding their implementation, the concept represents the first formal policy document expressing efforts toward the CA by using the multiplicity of German instruments and involving a wide range of actors both from government and from outside (Ehrhart 2011, p. 68). Outlining initial goals and pointing to the interconnectedness of various crisis management dimensions, it can be considered a first effort to bring different crisis management dimensions together and attempt to give coherence policy a formal base in Germany.

The debates to advance the German crisis management capacity and to establish a government-wide set of objectives and goals for engaging in conflict-prone countries further advanced at the cross-governmental level early in 2004 when the Federal Chancellery commissioned an internal report which suggested the establishment of inter-ministerial task-forces, conflict pools and working groups to improve government performance in fragile states (Patrick/Brown 2007a, p. 106). While there was no direct follow-up proposal to this report, the German government approved the *Action Plan for Civilian Crisis Prevention, Conflict Resolution, and Post-Conflict Peacebuilding* (*Aktionsplan Zivile Krisenprävention, Konfliktlösung und Friedenskonsolidierung*) (Die Bundesregierung 2004) with the above-mentioned Comprehensive Concept forming the base of the new plan (Die Bundesregierung 2004, p. 11). The topic has been debated in the 2002 coalition negotiations already, and in 2004 the document was formulated under the lead of the German Federal Foreign Office (AA) but also involved and consulted experts from other relevant ministries, civil society and academia. The formulation of the document also was the result of the interest, particularly on AA-side, of Germany to be part of the group of countries that were active on this issue and to display German CA engagement (Unger 2009, pp. 151/154).[392]

The Action Plan presents principles of German policy in the context of the new security environment and establishes civil crisis prevention as the core area of German foreign and security policy. In doing so, the document clearly states that "[c]risis prevention requires coherent and coordinated action on the part of all state and non-state actors" (Die Bundesregierung 2004, p. 1). As it is directed by the German emphasis on multilateralism, it notes that no actor has all strategies and instruments in its toolbox. The plan in particular aims at strengthening German capabilities in the civilian conflict prevention and post-conflict peacebuilding field. In doing so, the document has a heavy long-term and civilian crisis management impetus. To guarantee a coherent strategy, it highlights the need for a greater involvement of civil society actors and NGOs (Die Bundesregierung 2004, p. 4). Furthermore, it recommends concrete actions to improve the German civilian crisis management infrastructure (Die Bundesregierung 2004, pp. 5/13).[393] Calling for policy coherence, the Action Plan places the subject as a cross-cutting issue of the German government and defines conflict-related activities as cross-ministerial tasks. Comprehensively describing and summarizing crisis management challenges of various fields, the document sketches 161 concrete initiatives at the national and international level that Germany should start dealing with conflict at all stages. It also obliges the government to review the

392 According to an interviewee, the Brahimi-report was of particular influence in this context. The development of the Action Plan is elaborated in greater detail in Weller (2012, pp. 8-11).

393 The Action Plan's concrete initiatives to create the preconditions for an effective civilian crisis management infrastructure entail the appointment of cross-sectoral contact persons for civilian crisis prevention, the creation of an Inter-ministerial Steering Group for Civilian Crisis Prevention, cross-ministerial training, the availability of sufficient funding and numbers of personnel for crisis prevention tasks, the elaboration of inter-ministerial country and regional strategies as well as the possibility of establishing a jointly administered crisis prevention pool (Die Bundesregierung 2004, pp. 90-93). Initiatives that were structurally realized are analysed in detail in the following chapter.

document every two years and to monitor its implementation.[394] The Action Plan presents the conceptual framework in which CA innovations are arranged in and can be considered the central framework of German crisis management by constituting the most precise conceptual policy agenda for the CA in Germany and an important political highlighting of the topic. The German government for the first time carved out its CA idea and drafted concrete inter-departmental guidelines and objectives. Moreover, the plan represents the first inter-ministerial taking of some form of inventory of German crisis management measures.

However, the Action Plan has been largely criticized for not prioritizing proposed initiatives and that it rather reads like a long wish-list of wide-ranging initiatives from non-proliferation to democracy promotion and environmental protection.[395] Also respondents regarded the lack of priority setting and weighting as a major downside leading to a diffusion of the subject. It was furthermore noted that the document lacks clear guidance how to ensure the implementation of its initiatives and can thus hardly be considered a strategy. Even more so, as it provides neither financial nor human resources to sustain its implementation (Patrick/Brown 2007a, p. 108). It thus lacks the structure capable of giving it political authority which is the Action Plan's major limitation. Missing functionality, the plan rarely is a point of reference in crisis management policy action and thus not actually operational (Wittkowsky et al. 2011, p. 3).[396] Due to its complexity and lack of priority setting it remained in a practical niche existence. In addition, the piecemeal structure of the plan made implementation difficult as actual capacities were not discussed beforehand. This resulted in a kind of unsorted conceptual perception of the document in the different ministries. Another major downside is that the Action Plan lacks political support and leadership which has been considered a highly important factor by interviewees. Furthermore, the document does not reflect upon ministerial self-interests and neglects aspects of bureaucratic behavior with a strong belief in ministerial self-coordination. The major question, how to achieve crisis management coherence is not addressed (Debiel 2004, p. 297). Another issue from the coherence perspective is that, though the document also refers to military aspects and emphasizes the need to take the civil-military interface into account, it mainly and deliberately focuses on civilian activities. Largely emphasising the civilian side of crisis prevention, it explicitly states that military crisis prevention

394 Responsibility for further execution of the Action Plan has been placed in the hands of an Inter-ministerial Steering Group accompanied by an Advisory Board (Die Bundesregierung 2004, p. 99). For an analysis of both see the following chapter. For a summary of the Action Plan's different chapters and structure see Wenger et al. (2006, pp. 29/30) and Weller (2007, pp. 11-13).

395 See, for instance, Debiel (2004, pp. 296/297).

396 See Patrick and Brown (2007a), Rintakoski and Autti (2008) and Nachtwei (2011) who assess the document in similar ways.

instruments "are not subject of this Action Plan" (Die Bundesregierung 2004, p. 2). This approach, however, actually blanks civil-military coordination as part of the CA idea out.

A major positive effect of the Action Plan, is that it circulated and established notions of civilian crisis prevention, conflict resolution and peacebuilding across the German government and relevant ministries. Thus, the document has provided these policy areas with a higher profile and "increasingly established itself as a part of security policy" (Rintakoski/Autti 2008, p. 149). Similarly, a respondent evaluated the strengthening of the overall policy field with the entity of its instruments itself as a major benefit of the document. The plan's mere existence in terms of a new foreign and security agenda is seen to have supported inter-ministerial awareness. Domestic ministries such as the Ministry of Justice became aware of their role in foreign policy and the importance of their international capacities (Patrick/Brown 2007a, p. 109). Still, this did not result in a broad debate and the Action Plan as well as subsequent Implementation Reports are only known in small circles (Unger 2009, p. 155). Moreover, while it drew a lot of attention from the expert community, it was rarely discussed in parliament (Nachtwei 2011, p. 3). Consequently, some observers assess the Action Plan and its capabilities as largely overrated viewing the document as product of his age only, but out-dated today. Implementation reports of the Action Plan mostly assessed that many of the initiatives and structures envisioned in the document had not been implemented or developed any relevance declaring numerous remedies and trying to gain more commitment for the Action Plan (Unger, p. 167). Just as its predecessors in 2006 and 2008, the latest 2010 Implementation Report renders an account of the period under review by addressing a range of topics, however, without working out particular details or clarifying progress or obstacles for its implementation. The Implementation Reports thus underline the impression that ministries lack orientation and direction for a consistent implementation of the Action Plan's demands.[397]

In 2006, the German government released its White Paper on Security Policy and the Future of the Bundeswehr (Weißbuch 2006 zu Sicherheitspolitik Deutschlands und zur Zukunft der Bundeswehr) (Federal Ministry of Defence 2006) for the first time after twelve years. The White Paper represents the second main document largely influencing the German debate on the CA. It was drafted

397 See The Federal Government (2006, 2008, 2010). See also the assessment by the Plattform Zivile Konfliktbearbeitung and the Forum Menschenrechte (2010, pp. 5/6). In 2011, it was decided to formulate the formerly bi-annual implementation reports every four years only to allow the parliament to follow the long-term developments. Interim reports with a focus on a specific topic are intended to be written regularly (Deutscher Bundestag 2011, p. 5).

under guidance of the Federal Ministry of Defence (BMVg) and lays out the country's security policy within the new strategic framework and international security conditions as well as German security political values and objectives.[398] According to the White Paper "German security policy is based on a comprehensive concept of security; it is forward-looking and multilateral" (Federal Ministry of Defence 2006, p. 6). The paper explicitly draws focus to new threats and challenges and calls onto Germany to take on an integrated, comprehensive and multilateral approach to deal with them. This approach "requires the application of a wide range of foreign, security, defence, and development policy instruments in order to identify, prevent, and resolve conflicts at an early stage" (Federal Ministry of Defence 2006, p. 5). The White Paper includes various dimensions of comprehensive crisis management, and inter-ministerial coordination and policy coherence are heavily emphasized. However, like previous documents, the White Paper does not clarify how to actually implement the suggested objectives while reaffirming basic ideas of policy coherence as developed in the Comprehensive Concept and the Action Plan. Though the White Paper refers to comprehensive security, it abstains from moving beyond the military sphere and does not provide precise information on the possible CA and ways of implementation potential. Discussion of the CA remained on a declamatory, general and abstract level with a vagueness that opened doors for different interpretations.

Yet, the White Paper notably refers to the 2004 Action Plan as an example and reference point of German inter-ministerial and comprehensive security provision. In this context, however, it largely neglects challenges of defence policy and civilian crisis prevention by neither discussing nor linking the issues. By downgrading the Action Plan to merely being an example of inter-ministerial crisis management coordination, the White Paper does not view the concept as an integral part of German policy (Weller 2012, p. 27). Furthermore, though mentioning the Action Plan's Inter-ministerial Steering Group as the body responsible for its implementation, the White Paper does not refer to the Steering Group as an interministerial instrument of the CA, but to other structural examples as steps taken to improve CA structure (Federal Ministry of Defence 2006, p. 23).

While the White Paper does "not reflect its [the Action Plan's, S.P.] spirit of civilian approaches" (Unger 2009, p. 157), it is also based on an essentially preventive political approach (Rintakoski/Autti 2008, p. 144). In this regard, the document demands a stronger inter-ministerial approach and closer integration of

398 Based on an analysis of the strategy's framework, the first chapter 'German Security Policy' aims at defining German security political interests and names important institutional partners. The White Paper's second part 'The Bundeswehr – An Instrument of German Security Policy' reports on structural as well as conceptual innovations.

political, military, development policy, economic, humanitarian, policing, and intelligence instruments as a foundation and prerequisite for conflict prevention and crisis management (Federal Ministry of Defence 2006, pp. 10/11). The emphasis on policy coherence is comparably larger in the White Paper, in particular, as it stresses the importance of organizational matters. Being a defence political document, it draws particular attention to the role of the Federal Armed Forces. According to the White Paper, armed forces undergo a continual process of transformation with the objective to improve operational readiness across the entire mission spectrum. It notes that

> "the chief determinants of future security policy development are not military, but social, economic, ecological and cultural conditions which can be influenced only through multinational cooperation. It is therefore not possible to guarantee security by going it alone, or with armed forces only. What is called for, rather, is an all-embracing approach that can only be developed in networked security structures based on a comprehensive national and global security rationale" (Federal Ministry of Defence 2006, p. 22).

The changing role of the military in conflict management is thus reflected in the White Paper. The Federal Armed Forces' major task cannot be found in the area of traditional territorial defence anymore, but includes tasks of international crisis prevention and resolution which need a strong inter-ministerial approach (Federal Ministry of Defence 2006, pp. 10/11).

Importantly, the White Book specifies the German CA conceptualization under the notion of *Vernetzte Sichterheit*.[399] It establishes this concept as the new leitmotif and approach to build up 'networked' security structures and improve coordination on the national level but also in relevant IOs. The concept of Vernetzte Sicherheit emphasizes the importance of an all-round picture of the situation, a shared situational understanding of all relevant actors and the need of an inter-ministerial approach that brings together the different crisis management aspects to this end (Federal Ministry of Defence 2006, p. 22). However, the clarifications on the concept of Vernetzte Sicherheit are limited to the instrumental dimension and largely neglect reasons for CA necessity and ignore conflicting goals of actors.

399 To avoid conceptual and terminological confusion a translation of the German term *Vernetzte Sicherheit* is not undertaken even though Vernetzte Sicherheit has been changed to Networked Security in the official White Paper translation. Attempts to translate the term, however, are heavily debated on academic as well as political level in Germany. Furthermore, the English translation of Vernetze Sicherheit into *Networked Security* runs the risk of being confused with the term *Network-Centric Warfare* in international discourse which adds to terminological confusion on international CA level.

As explained in background interviews, conceptual discussions surrounding the CA in Germany are heavily coined by this terminological and conceptual vagueness as well as both deliberate and unintentional misunderstandings. On the one hand, the White Paper's concept of Vernetzte Sicherheit has established itself as the German terminological conception of crisis management coherence efforts and interdepartmental coordination but, on the other hand, the approach is strictly rejected by German civil society for being military coined due to its first mentioning in the White Paper. Also, the White Paper conceptually subordinates the concept of *Civilian Crisis Prevention* as emphasized by the Action Plan. Civilian crisis management actors in Germany commonly view Vernetzte Sicherheit as a strategy to enforce 'hard' security objectives and the concept is largely perceived as an attempt of governmental security policy to enforce security and defence political interests which securitize and take over development policy and humanitarian aid. Not perceiving it as a neutral concept (Thiele 2012, p. 81), many civilian actors sense the establishment of a hierarchy with a subordination of civilian instruments under the military and a militarization of civilian crisis management policy areas (Plattform Zivile Konfliktbearbeitung/Forum Menschenrechte 2010, p. 2). Understanding Vernetzte Sicherheit in such terms makes it not just for humanitarian aid and development NGOs but also for the Federal Ministry for Economic Cooperation and Development (BMZ) difficult to accept the concept as a foundation for further coordination with other crisis management actors. It has also been argued in background talks that by focusing on this perception, crisis management actors in Germany have largely neglected the civilian component as part of the civil-military coordination in the concept.

German civilian actors thus often refer to the notion of Civilian Crisis Prevention as used in the Action Plan and which clearly underlines civil and preventive engagement, when approaching the CA subject. Concequently, the concept of Civilian Crisis Prevention developed into a second notion to describe institutional preconditions of crisis management coherence in Germany. Just as the two documents, their key concepts, Civilian Crisis Prevention and Vernetzte Sicherheit, are left unclarified to each other. Terminological confusion is added as the Action Plan itself uses the term in an unclear manner as a synonym for measures taken in the whole conflict cycle ranging from prevention, crisis management and post-conflict situations (Major/Schöndorf 2011, p. 3). Furthermore, as pointed out by a respondent, from CA perspective, the use of the Civilian Crisis Prevention concept excludes any military involvement and is thus debated similarly on military side as the concept of Vernetzte Sicherheit is on civilian side which further increases conflict. The notion of Civilian Conflict Prevention from this perspective implies a deliberate demarcation of actors which questions the CA and *Vernetzung* from the beginning.[400]

The framing of the CA in the Vernetzte Sicherheit context thus resulted in a concept which is politically contaminated. Moreover, with the term Vernetzte Sicherheit the whole idea became ideologically connoted. According to background interviews, the White Paper's development in the BMVg with the other ministries merely being involved via formal notification was a key strategic fault which mostly contributed to the perceived military dominance of the concept. Apart from the concept being a matter of heavy domestic disputes, in particular between the BMZ and the BMVg, it is not used in a uniform way by different ministries which also work with varying degrees of intensity, effort and participation regarding the concept's implementation. Furthermore, interviewees describe that the quantitative affliction and interest in the concept's underlying matters largely varies depending on the particular ministry. In this regard, the BMVg is mostly affected when it comes to issues of comprehensive crisis management in the field, the AA is less concerned already and the BMZ is least directly affected in most of its concrete crisis management efforts. In addition to the lack of a coherent concept on the CA in Germany, interviewees assess that the clarification of tasks and objectives from the political level are missing. Another issue perceived is that the concept is pursued for its own sake and an end in itself. This leads to endless discussions about responsibilities and competencies while conflict specific issues are lost sight of. As noted by respondents, the disagreement and conflicting interpretations between the German parties and their respective parliamentary groups within the German *Bundestag* on the actual meaning of Vernetzte Sicherheit further contributes to the insufficient political legitimacy of the concept. Notably, however, consensus exists on the general meaning and overall objective of comprehensive crisis management. Disagreement only and particularly relates to the conceptions Vernetzte Sicherheit and Civilian Conflict Prevention.

There are several other documents relevant for the further development of the CA in Germany. As the country prefers tangible concepts of civilian conflict prevention and crisis management, which clearly showed in the activities represented in the Action Plan CA discussion, conceptual developments also took place in this arena with an active engagement and input of sector-spanning pro-

400 An interviewee pointed out that the public debate on the CA and the notion of Vernetzte Sicherheit is largely dominated by only a few NGOs. While the discussion itself damaged the CA in Germany it does not reflect the mainstream thinking of relevant actors in Germany, according to the interviewee. On the contrary, Wittkowsky et al. (2011, p. 2) constitute opposing policy opinions even within ministries in their study which methodological largely relies on interviews.

motional concepts. In particular, the BMZ's 2005 *Sector Strategy for Crisis Prevention, Conflict Transformation and Peace-Building in German Development Cooperation* (*Übersektorales Konzept zur Krisenprävention, Konfliktbearbeitung und Friedensförderung in der deutschen Entwicklungszusammenarbeit*) (Bundesministerium für Wirtschaftliche Zusammenarbeit und Entwicklung 2005) elaborates principles established in the Action Plan and includes binding directives for the planning, implementation and steering of development projects. With this document, the BMZ developed its own broad cross-sector strategy which underlined the security-development nexus, highlighted the linkage between peacebuilding and German development policy and emphasized that the strengthening of peace and security is an inter-ministerial task (Bundesministerium für Wirtschaftliche Zusammenarbeit und Entwicklung 2005, p. 5).

The widened notion of security and the CA also entered the 2003 *Defence Policy Guidelines* (*Verteidigungspolitische Richtlinien*) (Federal Ministry of Defence 2003). Referring to the changed security environment, the guidelines entail a geographically and conceptually broadened and far reaching concept of defence and point to the need of synergy of all security relevant capabilities and instruments (Federal Ministry of Defence 2003, pp. 3/18). In consequence, the Defence Policy Guidelines explicitly lay down new tasks for the German Federal Armed Forces which derive from past missions they have been assigned to. The document calls for the Federal Armed Forces' further transformation toward a professional and flexible army (Federal Ministry of Defence 2003, pp. 16-18). Also the subsequent Defence Policy Guidelines published in 2011 (Federal Ministry of Defence 2011) reflect the changing international and security environment, the broader security-political background and point to an inter-departmentally connected and wide spectrum of tasks. The guidelines further develop the German CA from MoD side by explicitly calling for a German WoG approach which demands "a national, comprehensive and coordinated security policy that includes political and diplomatic initiatives as well as economic, development policy, police, humanitarian, social and military measures" (Federal Ministry of Defence 2011, p. 5). To this end, all relevant and competent German institutions and forces have to "take far-sighted, whole-of-government action while taking into account their responsibilities and capabilities laid down in the constitution" (Federal Ministry of Defence 2011, p. 5). To implement this approach it is necessary to improve "the goal-oriented interaction of the foreign service, development cooperation, police, armed forces, civil protection, disaster control, and the intelligence service" (Federal Ministry of Defence 2011, p. 5). The German Armed Forces in this context merely serve as a tool of comprehensive security and defence policy (Federal Ministry of Defence 2011, p. 9). Some observers, however, criticize the lack of strategic supplementation of these reforms.

Also coalition agreements between government parties, in which the CA increasingly started to be mentioned since the end of the 1990s, are of relevance. In the 1998 coalition agreement between the Social Democratic Party and the Greens, the official wording explicitly referred to a coherent coordination of policies in the field of foreign affairs, security and development. Development and aid policy was considered an overarching governmental task. In contrast to the 2005-2009 grand coalition, where the Vernetze Sicherheit approach was only promoted by the then Minister of Defence while the concept was almost ignored by the AA and the BMZ (Nachtwei 2011, p. 1), governing parties as of 2009 have been clearly committed to the concept.[401] The coalition agreement underlines that the new government is "committed to the cross-linked security policy approach. It requires modern and efficient armed forces and suitable civil instruments for international conflict prevention and management, and for closer integration and co-ordination" (CDU/CSU/FDP 2009, p. 175). The Action Plan, however, is not mentioned explicitly in the document pointing to its low political significance. In the context of German development cooperation structures, CA ideas are reflected in the most concrete way. Organizational reforms are viewed to increase the effectiveness of German development policy and to eliminate the duplication of structures between the government's implementing organizations (CDU/CSU/FDP 2009, pp. 183).

Additionally, the German CA debate is largely framed in the context of the country's engagement in Afghanistan.[402] CA ideas in the 2009 coalition agreement are particularly emphasized in the context of Afghanistan, where the government stresses the importance of cross-linked security and inter-departmental efforts for the country's sustainable reconstruction (CDU/CSU/FDP 2009, pp. 174/175). While the German Afghanistan engagement raised the militarization

401 The 2009 election resulted in a change of coalition from the grand coalition of conservative Christian Democrats (CDU/CSU) and the Social Democratic Party to the coalition between the CDU/CSU and the liberal Free Democratic Party (FDP). Weiss, Spanger and Van Meurs (2009, pp. 30/31) point out that in German government coalition building usually each party claim a different CA relevant ministry. The fact that CA relevant ministries mostly go to different parties of coalition partners and never belong to the same political party, a traditional feature of German government, can also be accounted as a complicating CA dynamic in Germany. This factor, which gives the different parties the opportunity to perform different policies, however, is not investigated in more detail in this dissertation.

402 In this context, Gauster (2012, p. 77) points to a Afghanization ("Afghanisierung") of German security policy. Nachtwei (2011, p. 2) also observed that the Vernetzte Sicherheit conception was mentioned in German parliamentary debates mostly in connection with Afghanistan. According to a policy briefing by the German Center for International Peace Operations, in particular coordination on the ground in operations such as the German Afghanistan engagement helped to improve *ad hoc* coordination between actors. Coordination on the ground, however, is compromised by existing inter-ministerial boundaries (Wittkowsky et al. 2011, p. 1). This is discussed more fully in the next chapter.

fears of civilian crisis management actors even further, at the same time, the 2010 Afghanistan Strategy has been developed as a basis of agreement by all relevant ministries and has been mentioned as an example for a comprehensive country strategy. Generally, however, formulation of country and regional strategies in conflict zones are guided by the AA and a standing government-wide process is missing (Patrick/Brown 2007a, p. 110). In particular in the area of development cooperation, country strategies are developed independently within the individual ministries. The BMZ's plans, for instance, though informing the AA of what it plans to spend resources on, usually are at an elaborated stage already with the AA not having capacities or the mandate to change the proposed aid allocations. Also, conflict monitoring and early warning is maintained by each ministry from its individual perspective with the AA observing from a national German interest related perspective, the BMVg from a military one and the BMZ monitors countries in crisis with view to poverty and the impact on its habitants (Patrick/Brown 2007a, pp. 111/112). Also, sector strategies adopted by ministries on conflict, governance, and poverty reduction matters have a propensity to mirror the preoccupations of particular ministries, instead of presenting a unified, coherent German approach (Patrick/Brown 2007a, p. 106).

Taking up coalition agreement elements, one of the latest efforts to increase coherence of German foreign and development policy is a 2012 inter-ministerial agreement between the AA and the BMZ which aims at improving cooperation and reallocated responsibilities. As stated by the Federal Foreign Minister,

> "[w]ith the greater dovetailing of the activities of the two ministries we are acting upon a central concern of this Federal Government, namely the implementation of a fully coordinated foreign and development policy guided by values and interests" (Federal Foreign Office 2012b).

The agreement regulates that ODA is coordinated in its entirety by the BMZ while humanitarian aid, a field previously separated between the two ministries, is steered entirely by the AA.[403] The most recent effort in the context of concep-

403 For this purpose, the BMZ transfers 95 million Euros from its budget to the AA. Long-term transitional aid, however, remains with the BMZ (Federal Foreign Office 2012b). While the agreement implements a goal as laid down in the coalition agreement and deletes redundant structures and confusion about responsibilities, critics point out that humanitarian aid is an integral part of German development policy and needs to be closely connected with BMZ activities. Furthermore, the lack of a coherent overall concept and continuing transfer of competences back and forth between ministries was criticized (Weiland 2012). Notably, the BMZ was in charge for about 60 percent of German development assistance. The AA as the second largest ODA ministry is accountable for 8,7 percent. The Länder are responsible for some 7 percent of German ODA and approximately 14 percent is channeled through the EU budget (Federal Foreign Office 2012b).

tual CA implementation in Germany took place in September 2012, when Germany's ministers of defense, foreign affairs and development presented their ideas about the country's strategy toward fragile states in a common press conference introducing the country's *Interministerial Guidelines for Coherent Federal Government Policy Towards Fragile States* (*Ressortübergreifende Leitlinien für eine kohärente Politik der Bundesregierung gegenüber fragilen Staaten*) (Auswärtiges Amt/Bundesministerium der Verteidigung/Bundesministerium für wirtschaftliche Zusammenarbeit und Entwicklung 2012). The guidelines, which stem from the 2009 coalition agreement, were developed under AA leadership and with CA ideas under the notion of Vernetzte Sicherheit at its core. Both, the BMVg and the BMZ, as well as members of parliament and academics were closely involved in the consultation and formulation which also publicly showed in the common press conference by the three ministries. The ministers acknowledged during the press conference that, despite the repeated declaration of the concept of Vernetzte Sicherheit, the ministries have remained unknown to each other. As a consequence, according to the guidelines, relevant German crisis management ministries intend to create country-specific or regional task forces to analyze crisis situations and to ensure faster and better coordinated crisis management activities.[404] This entails the distribution of information to all relevant departments, the clarification of German goals, the coordination of ministerial efforts as well as international cooperation. The new guidelines aim at strengthening the German government's commitment to a consistent policy position in particular with regard to fragile and failed states. With the development of such an inter-ministerial strategy, the three ministries came together to reach common objectives with the result of a unified German conception of a strategy for fragile states. This constitutes a major step toward more coherent crisis management thinking in Germany.

Despite these recent efforts, however, respondents note that the conceptual debate on the CA in Germany does not take reality into account and largely deals with questions which are of secondary relevance in practice. Thinking on comprehensive crisis management in Germany takes place in a shortened and one-dimensional manner of traditional security policy thinking with a focus on either development or security. This increasingly leads to exhaustion and frustration between relevant actors.

404 These task forces will be guided by AA regional directors to coordinate the German response to a crisis instead of responding with each ministry separately. As the AA points out, such inter-ministerial task forces have already been established in the past for Sudan, Syria and the Sahelian zone which successfully pooled relevant expertise (Auswärtiges Amt 2012b).

4.2.2.2 Organizational Dimension

The German political-administrative system has demonstrated a high degree of continuity over the last decades and only gradually adjusted structures and procedures to new conditions and challenges (Reichard 2008, p. 43). This observation also applies to organizational attempts to improve crisis management coherence in Germany. Formally, the chancellor as the head of government and chairman of the cabinet is in a rather strong position. It is the chancellor's prerogative and responsibility to lay down and determine the general guidelines of government policy, the so called principle of chancellor authority (*Richtlinienkompetenz* or *Kanzlerprinzip*) leaving it to the ministries to concentrate on the details. The Federal Chancellery does not have a large decision-making authority and limits itself to a coordinating and mediating role in case of overlapping issues where it has to intervene, interpret, explain and control. Structurally mirroring government ministries, the Chancellery's main responsibility is to guard the principle of chancellor authority. The Federal Foreign Minister is not the only foreign policy advisor, as the Chancellery has an additional Security Counselor for advices on foreign and security policy. Another advisory body tasked with the coordination of German security policy is the Federal Security Council (*Bundessicherheitsrat*). However, this cabinet committee mainly serves as a forum for discussion, to share information and to debate urgent issues in order to build inter-ministerial consensus. However, the Federal Security Council mainly engages in a reactive manner and is largely limited and restricted to issues such as arms exports (Houben 2005, p. 193).

Within the general guidelines provided by the chancellor, federal ministers in Germany run their departments independently and on their own responsibility. The *Ressortprinzip*,[405] which is protected by the German constitution, grants wide autonomy and room for maneuver to each ministry in formulating its policies and implementing cabinet decisions. It thus creates strong ministerial organizations along strict hierarchical lines which are highly focused on their own institution (Hustedt/Tiessen 2006, p. 25). Ministers are responsible for successes and failures in their own ministries, and not for policies that cut across ministerial boundaries. Respectively, staff is concerned to ensure success for their minister, with their skills judged according to managing their ministry. In the context of crisis management coherence, German government actors thus act largely independent to each other (Ehrhart 2011, p. 72). This ministerial independence and sovereignty implies that key crisis management ministries are not obliged to

405 The Ressortprinzip has been translated with the terms 'departmental rule' or 'departmental principle' to describe this ministerial responsibility. See, for instance, Hustedt and Tiessen (2006, p. 24).

cooperating and working together. In the case of inter-ministerial disagreement there is hardly a way of settlement other than going up to the cabinet level which rarely occurs (Patrick/Brown 2007a, p. 110). Particular interests of ministries are dominant and prone to lead to disjoint policy formulation and execution. Consequently, German crisis management engagement is very likely to be fragmented, with the AA, BMZ, and BMVg each pursuing autonomous courses of action "that reflect their unique agenda and concerns, whether these be diplomatic, developmental, or defense-related" (Patrick/Brown 2007a, p. 110).[406] As evaluated by an interviewee, the Federal Chancellery in this context merely engages in a monitoring role.[407] The Ressortprinzip presumes that one ministry is in the lead and sets the policy coherence agenda to deal with the individual case at hand (Rintakoski/Autti 2008, p. 145). From CA perspective, the Ressortprinzip thus makes pro-active cross-ministerial coordination all the more necessary.

This is supposed to be enabled through inter-ministerial collaboration via both *ad hoc* and regular meetings on different departmental levels. On priority themes and under pressure to act, ministries meet in different committees and working groups to discuss strategies and to ensure successful case by case coordination. Several *ad hoc* cross-ministerial forums on specific issues exist on different echelons of ministerial hierarchy, from the working level to the state secretary level. As of 2005 long called for inter-ministerial country discussion groups (*Ländergesprächskreise*) materialized. However, the results were mixed, mostly because "the purpose and relevance of the exercise never became clear to the participants" (Unger 2009, pp. 169/170) and the discussions took place on an abstract level. A recent structural innovation in this context is the cross-ministerial collaboration of the policy planning staffs of the AA, BMZ and BMVg to develop joint policy guidelines (Wittkowsky et al. 2011, p. 2).[408] An important form of *ad hoc* coordination in German international crisis management are case related task forces such as the inter-ministerial Task Force Syria

406 For instance, Nachtwei (2010, p. 32) observed with regard to the separate engagement of the ministries in Afghanistan that each does so from strictly individual perspectives: the AA aims at the implementation of political objectives such as support for the central Afghan government, the BMZ focuses its work on medium- and long-term development projects, the Federal Ministry of the Interior deals with training of the police and the priority of the BMVg as the military component is to ensure local security.

407 Contrasting the Chancellor's Richtlinienkompetenz and the ministries' Ressortprinzip, the later rule is given greater priority. The Chancellor only provides guidance under extraordinary circumstances.

408 The Interministerial Guidelines for Coherent Federal Government Policy Towards Fragile States (Auswärtiges Amt/Bundesministerium der Verteidigung/Bundesministerium für wirtschaftliche Zusammenarbeit und Entwicklung 2012) are a result of these joint planning staff meetings.

established by the AA in August 2012 to ensure the coordination of all measures undertaken by the German Government concerning Syria.[409] Such task forces present a new unit, most times within the AA with greater human resources particularly to coordinate efforts.[410] In particular, in the context of the German engagement in Afghanistan different coordinating mechanisms were created which resulted in consultation and coordination at various levels.[411] Generally, however, with regard to specific conflict or country cases, the German government undertook only limited attempts to improve cross-governmental coherence. The different ministries cooperate with each other on different projects and in different compositions (Patrick/Brown 2007a, p. 108). Most *ad hoc* coordination mechanisms and actions have been largely reactive (Unger 2009, p. 165).

In addition to these informal and *ad hoc* forms of coordination, also formal requirements for horizontal coordination exist which are regulated by the *Joint Procedural Code of Federal Ministerial Departments* (*Gemeinsame Geschäftsordnung der Bundesministerien*). In particular, the concept of the 'lead ministry' (*Federführendes Ministerium*) is relevant with regard to inter-ministerial coordination as it assigns the responsibility for comprehensive involvement of other ministries to the leading one. The lead ministry which is decided on according to the issue at hand drafts a first proposal and sends the outline out to affected ministries. According to the 'rule of co-signature' affected ministries are supposed to signal either agreement or disagreement with the in-

409 According to the German Foreign minister at the time of writing, the Syria Task Force has been created to coordinate the German Government's wide-ranging tasks to a greater extent (Auswärtiges Amt 2012a). Foreign Minister Westerwelle stated verbatim: "Mit der Gründung der Task Force Syrien geht es uns darum, die umfassenden Aufgaben in der Bundesregierung noch stärker zu bündeln" (Auswärtiges Amt 2012a).

410 The first task-force in the CA context specifically aiming at inter-ministerial coordination was the 1995 Bosnia Task Force in which the Heads of Division of the AA, the BMZ, the MoD, the Federal Ministry of the Interior and the Federal Ministry of Economics met once a week (Konrad-Adenauer-Stiftung 2009, p. 4).

411 According to the OECD (2010, p. 44) "[r]egular meetings involving the state secretaries of the ministries involved (…) serve the political management of Germany's involvement in Afghanistan. In addition, the respective ministers meet regularly to discuss Germany's Afghanistan policy (…). At the working level, weekly video conferences involving the ministries concerned and an intensive process of consultation at working level are helping to establish a dense network that allows continuous and close co-ordination within the policy framework." Furthermore, in 2009, the AA conducted the first inter-ministerial seminar preparing for Afghanistan (Nachtwei 2011, p. 5). In addition to regular and frequent meetings between ministry officials responsible, the Federal Government Commissioner for Afghanistan functions as a coordinator and is connected to the highest government management level (Major/Schöndorf 2011, p. 3). The OECD's 2010 DAC Peer Review of Germany notes the success of a German CA in developing a crisis management policy in Afghanistan and recommends the case as a starting point for future learning in the area (OECD 2010, p. 44).

coming proposal by co-signing it. To jointly prepare drafts to solve inter-ministerial conflict from the beginning, the government can also set-up above-mentioned *ad hoc* working groups (Hustedt/Tiessen 2006, pp. 27/28).

However, while procedures of information sharing and commentating are regulated, these do not automatically result in open dialogue. According to respondents, inter-ministerial coordination has remained difficult in situations where different perspectives and working cultures became apparent. A formal mechanism to improve inter-ministerial coordination and awareness in this regard is the mutual secondment of officials which are delegated between relevant ministries. The posting of such liaison officers has been increased in Germany. Though considered valuable in general, interviewees pointed out that the mechanism highly depends on the integration of liaison officers in the ministry and how both ministries utilize the officer's insights on the respectively other ministry. The coordination benefit drawn from liaison officers thus varies. In addition, the mechanism affects only a limited number of officials and is viewed with some skepticism due to the negative impact such a secondment can have on prospects for promotion and compliance with staff regulation. While liaison officers in their function as a coordination mechanism are evaluated rather poor, their role to improve inter-institutional awareness is appreciated and they are seen to add a channel of communication. Interviewees underlined that the liaison officers are an important instrument to bring in different perspectives. Inter-ministerial coordination, however, is not facilitated by the secondment of additional officials.

As evaluated by interviewees, in general, ministries are open to informal exchange. Respondents assessed above *ad hoc* coordination mechanisms an important, flexible and uncomplicated coordination measure. Sharing of information and networking between ministries has greatly improved and responsible ministries exchange information in an issue-oriented manner via phone and e-mail on a daily basis. Informal collaboration is viewed as a coordination instrument which is successfully employed in a lot of cases, though in a rather invisible manner. According to respondents, interaction both via *ad hoc* and regular structures works smoothly between ministries and yielded tangible results.[412] Inter-ministerial specialized groups established between the AA and the BMZ have frequently been mentioned as examples for successful inter-ministerial coordination on specific topics. As some interviewees note, due to these mechanisms the CA is embedded in their everyday work. Their quality, however, depends on personal relations between officials.

412 See also evaluations in Deutscher Bundestag (2011, p. 3) and by Wittkowsky (2012).

Institutionally, the main ministries involved in German crisis management engagement are the AA, the BMZ and the BMVg and efforts toward inter-ministerial crisis management coordination mainly involve these ministries.[413] The role of the AA as the leading ministry for foreign relations is to deal with the overall political crisis management framework (Houben 2005, p. 193). Specific crisis management cases are mainly dealt with by the AA country desks. The BMVg is involved in the military arrangements and tasked with the overall planning of military defence, the conception of basic principles of military policy as well as operational planning and mission conduct (Urban 2009, p. 160). According to background interviews, inter-ministerial coordination between the AA and the Ministry of the Interior as well as the BMVg are very good on the working level while relations with the BMZ remain on a conceptual meta-level. Indeed, due to the urgent operational pressures to succeed in international crisis management, in particular the BMVg has pressed for coordination of civil-military activities and changes toward CA implementation (Gauster 2012, p. 77). This also became obvious with the formulation of the concept of Vernetzte Sicherheit in its 2006 White Paper which resulted in the development of the German CA model in different directions with largely differing degrees of ministerial acceptance. On the contrary, the BMVg and the Federal Armed Forces are a highly ambivalent factor when it comes to the German approach to comprehensive crisis management. With regard to the CA, the German defence establishment is caught between the necessity and expectation to find a new role to meet post-Cold War Challenges and its in Germany often as inappropriate perceived resources to do so (Unger 2009, p. 149).

The extension of missions of the Federal Armed Forces challenged traditional parameter of Germany as a civilian power and yet is fully compatible with the same viewing the German forces as a political instrument rather than a military tool (Pradetto 2006, pp. 19/20).[414] The Federal Armed Forces have undergone considerable transformation due to the emerging new forms of conflict which challenge its traditional principles and increase the German role also in military crisis management.[415] However, in particular the latest reforms serve as

413 In case of rule of law issues and policing, the Federal Ministry of the Interior also plays a major role.

414 The scope of the German Federal Armed Forces was widened in 1994 by the Federal Constitutional Court to include robust peacekeeping and peace-enforcement operations within the collective security system framework of the UN and NATO. The court's decision made an employment of the German Federal Armed Forces possible as long as the systems of collective security were bound to the preservation of peace. Another condition the court enforced was that the German Parliament has to endorse such a decision.

415 The transformation of the German Federal Armed Forces is in full swing at the time of writing.

an example that ministerial reform plans and projects have largely been developed separately by the ministry (Wittkowsky et al. 2011, p. 1). The transformation of the Federal Armed Forces takes place solely as a ministerial matter of the BMVg. Though being declared on the rhetoric level, in such an insular approach, the CA is practically blanked out. Observers thus fear that the reform, while making armed forces more efficient, will not result in more coherence on side of German security policy (Nachtwei 2011, p. 4).[416]

Since its creation in 1961, political competencies dealing with development were transferred from other departments to the BMZ which became a member of the Federal Security Council in 1998. As a consequence to its institutional development, some rivalry and frictions over competencies arose due to the reassignment of responsibility between the AA and the BMZ. The ministerial divide was latent at first but then became obvious in diverging policy conceptions. This, however, showed that the inter-ministerial conflict was not just caused by competence rivalry but also by differences in policy content and implementation. At the same time, this situation of rivalry with the BMZ, from AA perspective "seems to have given issues of cooperation with the military side a more prominent position" (Weiss/Spanger/Van Meurs 2009, p. 31). Coherence from the BMZ's side is further hampered by the ministry's desire to be present in a very large amount of countries rather than concentrating efforts on a lower number of aid recipients. Furthermore, despite repeated requests for broader buy-in particularly from the AA and the BMVg, the BMZ itself has complicated efforts by insisting on a narrow development focus rather than taking into account the broad range of German approaches in a country (Patrick/Brown 2007a, pp. 107/108).

The Action Plan provides the conceptual framework for organizational CA innovations in Germany and some specific bodies where established to create an institutional framework for the implementation of the plan, to particularly support action at a cross-ministerial level and to promote crisis management coherence. In the regard, in 2004 the Inter-Ministerial Steering Group for Civilian Crisis Prevention (*Ressortkreis Zivile Krisenprävention*) was established to monitor and implement the Action Plan. The Steering Group directly aims at over-

416 It has to be pointed out in this context that the so-called Vernetze Operationsführung (network-enabled operations and command) which is a key instrument of the transformation of the German military cannot be ranked as an input by the reform on CA level. Though terms such as network and coordination take center stage in the reform context, this type of interconnectedness should not be confused with the concept of Vernetzte Sicherheit in the CA sense as it aims at enabling command and conduct of operation "auf der Grundlage eines streitkräftegemeinsamen, Führungsebenen übergreifenden und interoperablen Kommunikations- und Informationsverbundes" (Thiele 2012, p. 85).

coming interdepartmental blockages.[417] Under AA guidance and headed by the ministry's Commissioner responsible for Civilian Crisis Prevention on ambassadorial level, the Inter-ministerial Steering Group meets on an average of every six weeks. It consists of the Civilian Crisis Prevention Representatives and contact points from the AA, the BMZ, the BMVg, the Federal Ministry of the Interior, the Federal Ministry of Justice and the Federal Ministry of the Environment which each delegate ministry representatives (Wittkowsky et al. 2011, p. 2). The Steering Group's primary task is to mutually exchange information and to provide a platform for inter-ministerial coordination. In doing so, it aims at creating structural pre-conditions for improving German crisis management coherence (Federal Foreign Office 2012a). It also seeks to enable inter-ministerial continuity, consensus and to facilitate crisis management agenda setting. Notably, it was not created to duplicate existing inter-ministerial coordination mechanisms and its efforts should not overlap with inter-ministerial taskforces created in cases of acute crisis. Rather, the Steering Group is charged with proposing and overseeing the development of mechanisms for coordination among the ministries and to compensate for current weaknesses rather than responding to specific crises. In doing so, it focusses on civilian crisis prevention streamlining and the formulation of general strategies for inter-ministerial coordination (Unger 2009, p. 161).

Almost a decade after its creation, the Steering Group is widely perceived as a disappointment when it comes to evaluating its effect on increasing crisis management coherence.[418] As it lacks authority and actual steering capacities it is considered an unsuccessful mean to bridge the inter-ministerial fragmentation created by the Ressortprinzip (Major/Schöndorf 2011, p. 3). Only little has happened in terms of the departments coming together to pursue a more coherent approach to crisis management and the Steering Group's function is reduced to a symbolic role. This perception is further intensified as it remains unclear, how issues are discussed in the group, and fed back and regenerated to ministries. This also depends on the individual departmental representative (Wittkowsky et al. 2011, p. 2). The Steering Group is perceived "[o]verly bureaucratic and unable to come up with real, prioritized, and actionable strategies to accomplish a mandate that is massive and unrealistic" (Patrick/Brown 2007a, p. 109). It is largely detached from the top levels of government in practice, and consequently

417 The Steering Group chose four specific projects from the overall 161 initiatives of the Action Plan to prove its relevance and start work in practical terms. Its work focused on the fields of security sector reform, joint financial instruments, refugee issues, and country pilot programs (Patrick/Brown 2007a, pp. 108/109). See Unger (2009, pp. 166/167) for more details on the projects.

418 See, for instance, evaluations by Patrick and Brown (2007a), Ehrhart (2011), Major and Schöndorf (2011), Nachtwei (2011), Wittkowsky et al. (2011) and Wittkowsky (2012).

also downgraded within the bureaucracies of relevant ministries. As a body without any operational capacity due to a lack of human and financial resources as well as a lack of steering authority, it has not been able to go beyond a non-binding exchange of information between ministries (Nachtwei 2011, p. 4). The lack of initiatives by the Steering Group can also be ascribed to the frequent change of officials on the heading AA position (Plattform Zivile Konfliktbearbeitung/Forum Menschenrechte 2010, p. 7). However, despite its weak mandate, the Steering Group in some cases has successfully served as a catalyst for specific topics in a broad manner (Wittkowsky et al. 2011, p. 2). Furthermore, it sensitized actors within ministries (Unger 2009, p. 162). Interviewees assessed that the Steering Group created greater understanding and appreciation on a general level. Another positive effect of the Steering Group is that it has increased the binding character of coordination on a regular basis. The inter-ministerial exchange in this context is also valuable when it comes to the early identification of conflict developments. Due to the unbinding character of exchange it is also easier to identify possible inter-ministerial compromises.

Most respondents were keen to point out several deficits of the Steering Group, pointing to the body's unsatisfactory existence and an inability to fulfill expectations due to a lack of political weight and detachment of an appropriate ministerial level. This results in problems when it comes to decision-making, starting with the setting of priorities to the determination of resources which, in sum, lead to a 'shadowy' existence of the body. Due to its establishment on the ministerial working level, the Steering Group does not have the authority to take decisions, and has not worked in a decision-oriented manner from the beginning. Neither does it have a precise work plan or other terms of reference giving it strategic direction. Interviewees also pointed out that, apart from officials being actively involved, the Steering Group is hardly recognized in the operational work and practice of ministries and barely influences these. In this regard, it is rather perceived as another institutional entity which needs to be coordinated. The feedback of the Steering Group's work largely depends on its ministerial representatives who mostly work according to the interest of their respective ministry which makes the group to a politically charged and detached forum at the same time. In addition, these officials can only devote minor parts of their working time to the Steering Group as they do this in addition to their daily work portfolio. Even though its deficits were pointed out in Action Plan Implementation Reports, the Steering Group did not gain much practical clout and generated only limited interest across ministerial portfolios. In addition, it was criticized that other relevant ministries such as the Federal Ministry of Economics and Technology are not represented in the group. Due to its 'discussion group character' it has been assessed an unnecessary institution by some respondents who point to existing coordination mechanism. The Steering Group did not succeed in identifying cross-cutting issues or even increase the cross-governmental visibility of CA issues or the concept itself.

To advice and inform the Inter-ministerial Steering Group an Advisory Board for Civilian Crisis Prevention was established in 2005. With the creation of the Advisory Board as an intermediary between the Inter-ministerial Steering Group and relevant civil society actors, the German Government sought to assure and institutionalize the involvement of non-state actors in the cross-sectoral decision-making processes of civilian crisis prevention and peace-building in accordance with the Action Plan.[419] The Advisory Board is supposed to accompany the work of the Inter-ministerial Steering Group and give expert advice (Federal Foreign Office 2012a). With the board, the German Government recognized civil society actors as key partners of its crisis management policy, established a forum to regularly engage on respective issues and created the space for a critical analysis of crisis management policy and CA efforts. As the major interface between German civil society and government, the Advisory Board can provide both the government and civil society with information about each other's policies, actions and perspectives and thus offers an opportunity to harmonize and align their policies accordingly.[420]

However, though being a valuable forum for collective discussion and an instrument for transparency the Advisory Board largely operates "in a grey zone between communication, consultation, and supervision as regards the Steering Group" (Unger 2009, p. 162). Due to the institutional weakness of the Inter-Ministerial Steering Group, it is highly difficult for the Advisory Board to develop impact and to use its potential, in particular, as the board only meets twice a year due to a lack of resources. Thus, so far common initiatives of the board members as well as slightest attempts to set an agenda for the Inter-ministerial Steering Group are missing (Wittkowsky et al. 2011, p. 2). In this regard, the Advisory Board largely mirrors the Steering Group and is mostly evaluated in a similar pessimistic way. Respondents criticize that the board's expertise is not utilized by the Steering Group but that it is rather informed with hindsight.[421] However, also the

419 Its nineteen members are drawn from academia including experts on security policy, development policy, human rights and humanitarian issues as well as the environment, the church, industry and political foundations, or they are policy advisors or individuals with particular expertise in crisis prevention.

420 Unger (2009, p. 162) points out that the Advisory Board is also met with suspicion in the civil society side.

421 It is characteristic in this context, how the board's comments to the implementation reports are dealt with as advice included in these is rarely taken up but published separately (Plattform Zivile Konfliktbearbeitung/ Forum Menschenrechte 2010, p. 7).

different civil society actors participating as board members do not form a homogeneous group rarely agreeing on issues or speaking with one voice. As such, the board has not found its role yet and tends to lose itself in discussions on its self-perception and what it should give advice on precisely, accompanying the Steering Group as some form of 'academic jester'. Similar to the Steering Group, it does not have a work agenda to generate impetus, initiatives or agenda-setting into the debate. Consequently, consultation output so far has been limited.

The set-up of inter-ministerial structures with the Steering Group and an Advisory Board marked the first institutionalization of the CA with regard to the policy coherence question in the German crisis management infrastructure.[422] Though not illustrating an institutional mechanism to directly increase government crisis management coherence, the German Bundestag's *Subcommittee on Civilian Crisis Prevention and Networked Security* (*Unterausschuss Zivile Krisenprävention und Vernetzte Sicherheit*) was created to contribute "significantly to making the topic an integral part of the parliament's work" (Federal Foreign Office 2012a) and provides an opportunity for parliamentary initiatives.[423] Established in 2010, the subcommittee has been created within the parliamentary Committee on Foreign Affairs to specifically deal with civilian crisis management with a focus on giving strategic recommendations. It regularly conducts expert hearings in the CA context and maintains dialogue with relevant stakeholders on the national and international level. In doing so, the subcommottee addresses both specific conflict cases as well as the development of CA ideas in Germany on the conceptual level, in particular in the context of the debate around the concept of Vernetzte Sicherheit, with the aim to clarify debated issues. It closely collaborates with the Steering Group and the Advisory Board (Federal Foreign Office 2012a).

In background interviews, the parliamentary subcommittee was evaluated as the only 'engine' further promoting the CA in Germany by encouraging interaction between various actors, succeeding to place the subject in the spotlight and by fostering exchange between experts and practitioners on the issue at an informal level in its hearings. Furthermore, it drafts suggestions in broad consensus

422 Also, including officials from other ministries as well as an officer from the General Staff, a *Unit for Civilian Crisis Prevention, Conflict Resolution and Post-Conflict Peacebuilding* was established at the AA in 2005. However, officials are not expected to ensure the consideration of their home ministries (Unger 2009, p. 162).

423 Party-political tactical reasons were a decisive factor behind the Subcommittee's creation as the Greens, as the party requesting the subcommittee, anticipated that the establishment of a Subcommittee on civilian crisis prevention was only possible when combined with the Vernetzte Sicherheit concept (Nachtwei 2011, pp. 1/2). Inter-ministerial exchange also takes place in meetings of the parliamentary Committee on Foreign Affairs and the Defence Committee as representatives of the respectively other ministry participate (Nachtwei 2011, p. 7).

between parliamentary groups and the topic of Civilian Crisis Prevention increasingly occurred on the agenda of the German parliament since the creation of the subcommittee (Plattform Zivile Konfliktbearbeitung 2012, p. 3).[424] As such, the sub-committee is a useful instrument to focus the CA debate in Germany on specific topics and to increase overall attention. Practically, it only plays a minor role with regard to organizing the space for political action to improve crisis management coherence. But though its influence with regard to government control has been assessed marginal it has the potential to increase pressure on the German government to act according to CA principles. The creation of such a parliamentary working structure also holds the chance to increase transparency and openness of debates. However, the German conceptual CA conflicts are reflected in the subcommittee's discussions and it is difficult to uncouple political intentions behind debates in the subcommittee.

International crisis management funding is not integrated in one funding instrument or line item in Germany.[425] With regard to crisis management funding, tension in particular exist between the AA and the BMZ on allocation and use of the latter's significant resources, which the former would frequently like to deploy. In this regard, "the AA would prefer to see greater selectivity in BMZ's global engagement, including greater targeting of aid resources to German diplomatic priorities" (Patrick/Brown 2007a, p. 114). With the Action Plan the German government has committed itself to working toward the consolidation of the budgetary funds for crisis management. Considerations on pooled financial resources ended without any long-term results in Germany. An examination of this subject by the Steering Group arrived at the conclusion that pooled crisis management funding is a mechanism which is not applicable in the case of German crisis management due to national budgetary regulations and the constitutional setting, particularly the high degree of ministerial independence (Die Bundesregierung 2006, pp. 85/86). Even the Inter-Ministerial Steering Group, as the only inter-ministerial body, does not have an own budget at its disposal. One exception, however, occurred in 2005 when the Steering Group received a budget of ten million Euros, the so called *Nachtwei-Millionen.*[426] The sum was created after "some political horse trading"

424 See also assessment by Nachtwei (2011, p. 8) and the Plattform Zivile Konfliktbearbeitung (2012).

425 The AA has a budget line for crisis prevention, peace maintenance and conflict management (Förderung von Projekten zur Unterstützung von internationalen Maßnahmen auf den Gebieten der Krisenprävention, des Friedenserhalts und der Konfliktbewältigung). However, in contrast to the BMZ, the AA's administrative and financial management capacity is limited.

426 The fund was named after Winfried Nachtwei, a former Member of Parliament for the Green Party, who is particularly active in the field of civilian crisis prevention and crisis management coherence in Germany.

(Unger 2009, p. 173) in which the Green parliamentary group only approved the development of a missile defence system in the NATO framework on the condition that a budget of ten million Euros from the defence budget was allocated for civilian crisis management activities between 2006 and 2008 to be spent by the Inter-Ministerial Steering Group. While the budget was modest in scope and the process in itself illustrates the systematic shortcomings of German crisis management, the allocation of the amount nonetheless led to a new quality of inter-ministerial coordination (Nachtwei 2011, p. 4). However, while the money in this joint funding mechanism was primarily used for small reconstruction projects, at the same time, the BMVg ensured that the budget was spent according to its own ministerial priorities and as a supplementary funding of ongoing activities (Unger 2009, p. 173).[427] This aspect largely questions the original purpose of pooled funding to increase inter-departmental coherence of its spending. On a more general level, however, respondents appreciated the pooled funding mechanism as a flexible short-term instrument which has promoted common inter-ministerial decision-making and activities. At the same time, interviewees expressed the view that the Nachtwei-Millionen were rather counterproductive. While leading to increase of coordination, the funding instrument at the same time resulted in great bureaucratic efforts and transaction costs.

Several other institutions have been created in Germany to increase coordination between both governmental and non-governmental actors of crisis management on the individual level. These were established independently from the Action Plan but nonetheless contribute to an overall CA implementation. In this context, two institutions particularly promote inter-departmental coordination by fostering a culture of cooperation on the working level and by serving as brokers for inter-ministerial networking and the CA: The *Federal Academy for Security Policy* (BAKS) and the *Center for International Peace Operations* (ZIF). While the work of the BAKS in the German CA framework focuses on its strategic dimension, the ZIF deals with its operational dimension.

Founded in 1992, the BAKS is "Germany's highest-ranking interministerial institution for advanced education and training in the field of security policy" (Bundesakademie für Sicherheitspolitik 2008). Funded by the BMVg, the BAKS is mandated by a Board of Trustees which is composed by the members of the Federal Security Council. The academy aims at disseminating the comprehensive security concept via courses and seminars for officials from different ministries of rele-

427 Such as West African peacekeeping, the Provincial Development Fund in Afghanistan or NATO's Trust Fund Project for Bosnia and Herzegovina (Die Bundesregierung 2006, p. 84). Patrick and Brown (2007a, pp. 113/114) mention minor initiatives mainly by the AA in their study which served as some sort of limited funding experiments.

vance in security policy.[428] Officials working for the BAKS for practical reasons continue to have access to the intranet of their ministry of origin. The BAKS thus largely focuses on CA issues in terms of both subject and practice. From CA perspective, the BAKS is important on three levels. Firstly, the academy promotes inter-ministerial networks between relevant officials and decision-makers through its events which provide a platform for informal exchange and a neutral meeting point. In doing so, the BAKS views itself as "an elevated forum of discussion, promoting the proactive exchange of ideas on national as well as international solutions" (Bundesakademie für Sicherheitspolitik 2008). Secondly, it improves strategic and practical crisis management competence and "a comprehensive, interministerial understanding of the long-term security interests of the Federal Republic of Germany" (Bundesakademie für Sicherheitspolitik 2008). It does so by considering diplomatic, military, economic, development, police and other aspects. Thirdly, the academy provides impetus for discussions on the topic in a broad sense. According to interviewees, the value of the BAKS is that it is perceived as an independent and neutral actor which promotes the ability to interact interministerial. In this context, respondents emphasized the importance of joint interdepartmental training. Though it increasingly develops into a sort of information and knowledge center on its own, the BAKS' true benefit is that 'it knows who knows what'. A problematic aspect with regard to BAKS seminars is that human resources development concepts greatly differ depending on the respective ministry. Interviewees pointed out that the academy's seminars too much act on the assumption of the armed forces human and financial resources planning. The academy could thus play a more important role if seminars would consider other ministries' every day work plans and resource allocations.

CA thinking in Germany also gave birth to the ZIF as a non-profit entity financed by the AA in 2002. The centre "is the implementing organization of the Federal Foreign Office for recruiting, training, deploying, preparing, supervising and qualifying German specialists and managers for peace operations" (Federal Foreign Office 2012a). The center professionalized recruitment and training of the German civilian peacekeeping personnel. In doing so, the ZIF's mandate is to improve Germany's civilian crisis response capacity by recruiting and training civilian staff for deployment to multinational peace support efforts under the auspices of the EU, UN, or OSCE. It also carries out academic analysis and evaluation in the context of international operations, crisis management and on implementing the CA in Germany. As such, the ZIF hosts a working group on the CA. The working group's coordinator works on behalf of the BMZ and addi-

428 According to a respondent, participants increasingly also come from ministries which are not presented in the Federal Security Council.

tional representatives are assigned to the working group by the relevant ministries. The working group's task is to provide analytical input for CA implementation by preparing policy briefings, working papers and expert reports and by giving expert advice to the ministries as well as parliament and implementing organizations. It also gives recommendations on CA implementation in the center's own human resources and training activities (Zentrum für Internationale Friedenseinsätze 2012). From CA perspective, the center's benefit with regard to German CA implementation efforts lie in the training of military and civilian personnel alike in preparation for crisis management operations as well as in its contribution for analysis and for drawing inter-ministerial lessons-learned.[429]

4.2.3 European Dimension and EU-Interface in German Crisis Management

Germany has been portrayed as a strong supporter of European integration and the German view on Europe is coined by an emphasis on notions of a shared European destiny (*Schicksalsgemeinschaft*). This is also mirrored in the widely held view that the country´s reunification had to take place together within the wider European unification.[430] The German commitment to the EU largely follows an idealistic vision and is granted high symbolic importance (Aggestam 2000, pp. 67/75). In this regard, active EU involvement is also viewed as a part of the German historical legacy in expressing the clear ambition to prevent a repetition of history. Also the EU's crisis management policy and particularly its CFSP are seen to prevent potential renationalization of member state policy. In these terms, the EU crisis management framework provides "a useful channel to exert an international influence without raising historical fears [of a German hegemony, S.P.]" (Aggestam 2000, p. 70). Interviewees underline the positive connotation of a common European policy in Germany which is perceived as "replacing the discrete conception of national interest as the guiding principle of foreign policy" (Aggestam 2000, p. 64). As a respondent observed, by seeking consensus, the German position in the EU is very much coined by the different centers of gravity within the Union and respective dynamics. At the same time, the country strongly believes in institu-

429 Other organisations constituting important pillars of the German civilian crisis management profile are the Working Group on Peace and Development (*Arbeitsgemeinschaft Frieden und Entwicklung*/FriEnt) and the Civil Peace Service (*Ziviler Friedensdienst*/ZFD). On these institutions see Weller (2007, pp. 16/17).

430 On the impact of the German unification on the country's international role see Dettke (2009, pp. 49-70).

tions, with a particular faith in the institutions of the EU which also coins German approaches to EU crisis management.[431]

However, at the beginning, the role of the EU in German crisis management policy was limited due to the primacy of NATO during the Cold War. But while Germany has continued to view NATO as the organization for its core security in form of collective defence, "the broader notion of security in terms of *Friedenspolitik* (peace policy) has (…) been channeled through membership of the EU" (Aggestam 2000, p. 64). The German conceptualization of EU crisis management sees the EU's key contribution in the area of 'soft security', where the Union can exert influence through its economic strength. As such, German and EU crisis management conceptions harmonize very well due to the common emphasis on their civilian normative power identity. The German self-perception and the EU's image as a civilian power provided both with an essentially similar starting point for development. Both the uploading and downloading dimension of Europeanization blend well on the normative level of crisis management. Against this background, Germany has made strong declarations toward the EU since the beginning of the 1990s, emphasizing the Europeanization of its foreign and security policy. The delegation of responsibilities to Community institutions provided an instrument for Germany to strengthen its voice in crisis management policy questions where the country did not have means comparable to that of other member states. Historically, European institutions shaped German domestic institutions from the very beginning. The institutional congruence "paved the way for Germany gradually to acquire normative power and shape the contours of the European integration process in an active manner" (Daehnhardt 2012, p. 37).

In doing so, the German government adopted a proactive role in the first phase of ESDP by striving for an integration of the bilateral British-French St. Malo initiative into the multilateral EU framework, in particular during its 1999 Council Presidency (Kirchner 2010b, p. 143). At the same time, the conflict in Kosovo revealed European crisis management deficiencies and increased the pressures on German policy and the necessity to address the EU's lack of relevant crisis management capacities also in the military field. But the presidency also involved an increased opportunity to upload German preferences onto the level of EU crisis management.[432] This was of particular importance "[g]iven Germany's

431 See, for instance, Dyson and Goetz (2003, pp. 4/5) as well as Bulmer and Jeffery (2010). Miskimmon (2007, p. 63) states that "Germany has consistently sought to play the role of Musterknabe [in the EU, S.P.]." This analysis has been questioned lately by some scholars who observe a de-Europeanization of Germany. See, for instance, Daehnhardt (2012).

432 For a detailed elaboration of the German role at the initial ESDP phase and the 1999 German Council Presidency see Miskimmon (2007, pp. 100-144) and Jonson (2006, pp. 71/72).

traditional weakness in shaping security policy within the EU" (Miskimmon 2007, p. 101). While the German engagement at the beginning of its 1999 presidency was rather reactive to events such as the Kosovo conflict and the St. Malo initiative, it started to shape events by coordinating efforts toward an agreed basis for discussions.[433] In these discussions, Germany argued that the combination of the civilian and military crisis management dimension would best equip EU crisis management giving it an operational profile congruent with the framework applied by the country itself. Germany took advantage of the opportunity of the Council Presidency to bring in own thoughts and was keen to push for extending the civilian ESDP dimension and for the establishment of linkages between the community pillar and the emerging CFSP/ESDP (Johannsen 2011, p. 156). The importance of conflict prevention and the necessity for civil-military coordination as an integral part of EU crisis management was underlined. Working toward strengthening of the civilian components of CFSP, Germany was particularly active in supporting idealist conceptions and the model of the EU as a normative civilian crisis management actor (Daehnhardt 2012, p. 39).

Even though Germany conceptually strongly worked to increase the importance of civilian crisis management, the country's normative basis was challenged due to the creation of military capabilities with effects both in the uploading as well as the downloading dimension. The emerging ESDP and Germany's position at the center of these developments since its initial phase put some German foreign policy and crisis management traditions under adaptational pressure. The creation of military EU crisis management caused some ambivalence and seemed largely incompatible with the German orientation as a civilian power (Overhaus 2006, p. 69). Germany thus was keen to point out that the strengthening of the EU's military capabilities would not mean a militarization of the EU and focused on institutional matters, rather than on capabilities (Miskimmon 2007, pp. 108/109). Instead of engaging with regard to military aspects of ESDP, the German government largely contributed to the establishment of civilian structures where "the German voice could be heard most effectively" (Johannsen 2011, p. 154).

Nonetheless, the central issue in the downloading dimension was "the acceptance of new circumstances in which Berlin would have to play a more active role on military operations as part of an EU force" (Miskimmon 2007, p. 111). Acknowledging these developments, Germany increased and showed activism

433 As Jonson (2006, p. 71) observed "[i]t was clear that the presidency did not have a clear road map of where it wanted to go with the ESDP process, (…) it also seemed as if it lacked a certain degree of self-confidence in this new area of European integration, where traditionally Germany has had a weak role because of its size, in terms of proportionally limited capabilities, and obvious historical reasons. The German Presidency's paper raised considerably more questions than providing guidelines for the development of the ESDP process."

also in EU military crisis management. In doing so, Germany began to view its military engagement as a normal part of its firm commitment to multilateralism. The German activities in the context of the developing EU crisis management capabilities were characterized by a reinterpretation of the German historic responsibility "as a spur to action rather than standing in the way of German intervention in crisis management" (Miskimmon 2007, p. 149). As such, the EU's ESDP inputs were downloaded as part of the process of 'normalization' of German foreign policy and participation in both civilian and military EU missions pushed German domestic limitations. As elaborated in the chapter on the German crisis management approach, the country's efforts have become more robust and partly aimed at overcoming existing pacifist and anti-militarist principles. The EU and the development of the military dimension of EU crisis management acted as a facilitator which supported the German developments in this direction. In addition, the country's efforts to shape the EU's crisis management policy have been largely supported by a domestic political and societal consensus regarding the important role of the EU as a crisis management actor (Daehnhardt 2012, p. 46).[434]

On the level of declarations, Germany positioned itself with several conceptual initiatives in the crisis management field.[435] The country played a recognizable role in developing the expanded notion of security enshrined in the ESS which "has a German imprint in Berlin's preference for effective multilateralism, civilian crisis management instruments and the use of military force as a mechanism of last resort" (Daehnhardt 2012, p. 47). Additionally, the country pushed toward the wording of 'preventive engagement' instead of the 'preemptive' denotation. As the ESS is largely in line with its security political thinking, Germany developed its own documents on the strategic level on the basis of compatibility with the ESS (Daehnhardt 2012, p. 47). From the German perspective, the

434 The country, for instance, took over the military command of EUFOR Althea in Bosnia and Herzegovina and participates in the EU's naval mission Atalanta with up to 1400 military personnel (Daehnhardt 2012, p. 40).The German debate regarding the country's role in the EU's mission to support the UN mission in the Democratic Republic of Congo highlights many of the German reservations concerning CFSP participation. See, Miskmmon (2007, p. 150).

435 Miskimmon (2007, pp. 127-137) views the Stability Pact for South-Eastern Europe as a major example and success of uploading German conceptual preferences onto EU level by "presenting an all-encompassing view on security and stability on the Western Balkan region" (Miskimmon 2007, p. 128). The plan can also serve as an example of broad inter-ministerial involvement in an issue on EU level as several ministries contributed to the framing of the proposal which resulted in a comprehensive strategy for conflict prevention in the region (Miskimmon 2007, p. 132). While the Stability Pact was designed to create a better EU response to developments in the region, it also indicates German willingness to play a kind of leadership role.

ESS also meant an order to develop and formulate its own comprehensive crisis management. The Action Plan picked up the debate and pictured the EU as a model of success in terms of promoting stability and structural crisis prevention. The document argues specifically in favor of the EU as a civilian power (Die Bundesregierung 2004, p. 38), and sketches several initiatives of particular German engagement to improve the coherence of EU policy in crisis management and to strengthen its capacity for action. According to the Action Plan, the German government specifically aims at improving the EU's operational civilian capabilities. On the EU level, the enhancement of the Council Secretariat's civilian crisis management capabilities was to be focused on. In addition,

> "all relevant fields of work of the Commission (trade, development, external relations, humanitarian aid, etc.) and of the Council/Council Secretariat [shall be reviewed, S.P.] with regard to the coherent application of crisis prevention principles in order to identify and address weak points" (Die Bundesregierung 2004, p. 40).

In the context of civilian and military ESDP missions, the Action Plan points to the German contribution to police missions and calls for a significantly larger pool of qualified personnel. To this end, the government wants to establish "a civilian division for the adequate recruitment and training of personnel for civilian crisis prevention missions is advocated in this context" (Die Bundesregierung 2004, p. 40). Also, the Union's military capabilities should be reinforced through new forms of cooperation to conduct military Petersberg operations (Die Bundesregierung 2004, pp. 40/41). Importantly from CA perspective, the plan also demands the "intensified exchange of expertise with the relevant ministries of other EU Member States concerning concepts for and experience in civilian crisis prevention" (Die Bundesregierung 2004, p. 41).

However, the adoption of the German Action Plan took place after developments at the EU level. The country was not in a position to point to German concepts and developments in these EU processes. Thus, Germany approved the strengthening of inter-institutional crisis management on the EU level even before working on the issue on the national level. The Action Plan, the concept of civilian crisis prevention and demands to coordinate diverse crisis management instruments in a CA thus largely follow the ESS' logic. In addition to the Action Plan, the ESS' release also resulted in the expansion the BMVg's definition of security and is largely reflected in the 2006 Defence White Paper which underlines that the ESS "is in accord with the goals and interests of German security policy" (Federal Ministry of Defence 2006, p. 34). It supports the EU's ambitions in the area of military crisis management and at the same time pushes toward an expanded spectrum of tasks and responsibilities of the EU in the crisis management arena (Fischer 2004, pp. 314/315).

In general, Germany's EU policy is structurally based on a decentralized policy-making system and institutional structures in which each ministry participates. This results in a "plurality of competing EU policy-coordinating ministerial departments" (Daehnhardt 2012, p. 46) which complicates the coordination of crisis management preferences and ultimately weakens the German position in Brussels. Regarding EU policy-making processes, most assessments point to this highly sectorized nature and key actor's opposition to centralize coordination in Germany. As regards bureaucratic reorganization, only minor bureaucratic changes or reallocation of resources took place in consequence of the developing EU crisis management machinery (Miskimmon 2007, p. 176). EU policy making remained embedded with a central role of the AA. However, the Chancellery has taken greater interest in pursuing policy objectives in this area. With the growing importance and profile of EU crisis management, former policy autonomy of the AA in this area was partly reduced by the Chancellery which strengthened its position by providing more firm EU policy guidelines and with the creation of a department dealing with EU issues in the Chancellery (Miskimmon 2007, p. 14). Nonetheless, in particular with regard to CFSP development, the AA continues to perform a central role and still can be considered the "'sponsoring ministry' of the CFSP process" (Miskimmon 2007, p. 139). The number of officials in the AA focusing on coordinating CFSP has been increased which points to the growing role of EU crisis management. Reflecting the German support for civilian crisis management, the German government created structures to support and embed EU developments on national level with the establishment of the ZIF to organize the recruitment on training and personnel (Major/Bail 2011, pp. 21/22).[436]

The development of EU crisis management with the inclusion of military crisis management also broadened the role of the BMVg. In this regard, EU crisis management policy again is less and less exclusively the domain of the AA. While the role of the BMVg in CFSP augmented, the ministry continues to take on a rather Atlanticist perspective due to the deep integration of the German Federal Armed Forces into NATO structures (Miskimmon 2007, pp. 13/14). Nonetheless, the incremental steps Germany has taken to normalize the use of its armed forces conceptually played an important role with regard to the Europeanization of its crisis management policy. As mentioned in the previous chapter, the Federal Armed Forces have undergone significant transformations. These also occurred due to demands from EU crisis management developments. EU pressures developed in connection with a domestic reexamination of the role of the armed forces on the international stage and the EU served as a cover or 'fa-

436 For a detailed analysis in German bureaucratic reorganization as a result of emerging CFSP in the 1999-2007 period see Miskimmon (2007, pp. 177-179).

cilitating milieu' for reform. According to interviewees, the debate in this context has largely focused on the concern that the country could be involved in too many missions without the adequate funding or a clear strategy of German crisis management. This discussion is heavily emphasized on the CSDP context as the EU presents Germany with the challenge to deploy forces in regions where the country has only little experience and no interest of getting involved. However, EU missions are also very much in line with German crisis management tradition as they address humanitarian crises rather that strategic interests and deployments have often been motivated by civilian power norms (Harnisch/Wolf 2010, p. 57). This allows German contributions to the EU military efforts which also show in the high commitment and strong contribution to the Battle Group concept (Miskimmon 2007, p. 151).

However, Germany did not always follow the conduct of EU crisis management and has also polarized and split opinion within the EU over decisions in the context of the Iraq war or more recently an engagement in Libya. Issues in this regard mostly result from the country's ambivalent standpoint on military commitment and instruments despite being part of CSDP. Political constraints in providing the necessary resources add the scrutiny that each mandate undergoes lengthy debates in parliament (Daehnhardt 2012, p. 47). Though the reform of the Federal Armed Forces to participate effectively in crisis management as part of ESDP forced Germany to speed up the pace with which its reform was progressing (Miskimmon 2007, p. 140), the traditional German preference of a role limited to self-defence of the armed forces is still challenged by evolving expectations for a greater German involvement. In this regard, Germany leans toward national security traditions and is also prepared to put up with conflict with EU institutions and other member states to match German objectives and substance in delivering the respective contributions (Kirchner 2010b, p. 143).[437] However, the shortage of sufficient structural power, discrepancies between the potential for engagement and existing capabilities to deliver troops and material as well as between the country's ambition on the one hand and the lack resources on the other hand is characteristic for both, the EU and Germany. Interviewees as well see a difference between the country's commitment on the planning level and at the operational crisis management level as the major issue of German-EU crisis management problems.

The push toward more communitarization of EU crisis management was one of the German long-term objectives in the context of the developing EU

437 However, Kirchner (2010b, pp. 143-150) comes to a differentiated picture when examining the country contributions in to ESDP missions.

crisis management.[438] Germany was a strong advocate and active contributor with regard to the Lisbon Treaty reforms to improve EU crisis management coherence. In particular, the country's support and draft document for a 'European Foreign Service' during the European Convention in 2002 helped to materialize the EEAS. The German 2007 Council Presidency played a major role in achieving the breakthrough of the reform treaty. The country also focused on the implementation of the double majority voting system to strengthen its position in the EU system. However, since the entry into force of the Lisbon Treaty, the German government's approach developed into "one of modest reticence" (Möller/Rappold 2012, p. 13) and the country became less active in the area. The German restraint toward the new institutional provisions shows clearly in a lack of strategic vision how to deal and how to use them. "There seems to be a wait-and-see approach until the EEAS' structures have settled in, and the big institutional battles are over" (Möller/Rappold 2012, p. 13). Also in the German parliament, the EEAS was heavily disputed in its initial phase due to questions surrounding budgetary issues of parliamentary control, resources and structures for civil-military cooperation. Activism merely took place on symbolic side, for instance, with the then Foreign Minister Guido Westerwelle being among one of the signatories of a letter to the HR/VP which suggested improvements to make the EEAS more effective. Though such initiatives underline the German interest of a strong and efficient EEAS, the new institution is not prioritized in foreign and security policy debates in Germany. According to a respondent, German concerns regarding the EEAS come from a practical standpoint which doubts that the EEAS can match its own ambitions. The low ranking of the subject on the German government's agenda also shows with regard to staffing the EEAS and led to a German underrepresentation in the EEAS.

In this context, the AA has played an important role. While on the one hand struggling with its decreasing lack of relevance in German EU policy due to the increasing importance of the Chancellery in EU questions, the AA takes on a similar perspective on the international level and leans toward seeing the EEAS in terms of a competing rather than a complementary institution. Though the AA stated staffing as a priority for the EEAS, it merely aimed at securing few top-positions in the EEAS and could not reach levels of representation other member states did (Möller/Rappold 2012, pp. 10/11). According to interviewees, the AA

438 On structural level, the German government viewed the EU pillar structure critically and pushed towards more CFSP integration in the supranational structures of the first pillar to increase coordination and implementation of the EU's CFSP policy and instruments. In this context, Germany also strongly promoted European defence cooperation with explicit supranational elements and collective structures beyond the ESDP process and second-pillar intergovernmentalism (Aggestam 2000, p. 73).

has realized this shortcoming of its policy toward the EEAS and the importance of staffing and recruiting and reinforced its efforts in this regard. EU post-Lisbon structures increasingly become a constitutive part of Germany's foreign and security policy.

Europeanization of the German policy is also pursued with an emphasis on the country's own political and economic interests. As Germany continues to participate in the EU's further development, this increasingly takes objective and power-political interests into account and is not merely based on the objective of European integration anymore.[439] This also is the outcome from the fact that Germany is more comfortable with its economic power than with its role as an important actor in the areas of crisis management and foreign and security policy. The Euro crisis has further strengthened this attitude (Möller/Rappold 2012, pp. 9/10).[440] The German focus on civilian crisis management can also be seen as a reflection of the fact that the country has considerably more influence in this area than in the military sphere. However, though the German policies toward the Lisbon Treaty provisions and contributions have been reluctant lately, the observed 'normalization' and self-confidence in German crisis management remains anchored in the EU framework. Due to the country's traditionally strong preference for multilateral frameworks of its crisis management policy, the EU remains the ideal IO for Germany to pursue its policy, and the country, according to interviewees, is highly committed toward the EU and the EU playing a stronger role in crisis management. As such, the effective organization of EU crisis management gains particular importance from German perspective.

4.2.4 Synopsis: Germany and Comprehensive European Crisis Management

4.2.4.1 Crisis Management Approach and Conceptual Dimension

German international crisis management is deeply rooted in strong security norms and the country acts according to clear crisis management principles. In doing so, the German government largely focuses on the civilian crisis management side and aims at preventing crises rather than getting involved in existing conflicts. German crisis management engagements are preferably undertaken in

439 See also analysis by Rummel (2006).

440 At the time of writing, the Euro crisis has dominated the German EU policy for three years already. As a consequence, there is not just a growing skepticism not just towards the monetary union but also with regard to a joint foreign and security policy. Moreover, "[g]enerally speaking, EU foreign policy 'fell off the radar'" (Möller/Rappold 2012a, p. 9).

multilateral frameworks on a basis of broad legitimacy. While the country is conceptually and also with financial and human resources dedicated to international crisis management, Germany is in practice confronted with the problem of combining its principles and actual policies consistently. From CA perspective, the German international crisis management environment, with its ongoing aversion of military force and a policy of restraint, hampers reforms toward increased CA policy coherence by creating a highly sensible political environment.

As shown in the above analysis, German governments aimed at improving crisis management coherence with several documents which reflect CA ideas and show the importance Germany places on the subject. Rhetorically, crisis management activities are based on a wide security concept that embraces political, economic, ecological and social stability. German strategic analysis is based on CA thinking and political leaders employ a broad concept of security excessively. As the then Federal Minister of the Interior Thomas de Maizière declared in 2010, the CA represents the key element of German security policy (Lange 2010).[441] However, this did not result in a coherent conceptualization and conceptual agreement on the CA and its strategic direction is missing. Respective documents and concepts do not take any form of hierarchy or a regular series, but are developed and formulated as required. While the goal of comprehensive crisis management is highly present on the conceptual level, the logic of strategic CA thinking is far from being recognized and appreciated in Germany. German CA efforts take place in terms of a necessary adaptation rather than a profound change and respective debates are coined by an absence of pragmatism.[442]

This can largely be accredited to the conflict between the diverse CA models developed in Germany which were not able to find a remedy in this regard. Quite the contrary, the co-existence of the Action Plan and its Civilian Crisis Prevention CA conception and the White Paper's concept of Vernetzte Sicherheit largely exacerbate tensions that 'naturally' come with the CA idea, onto a level of deep conflict resulting in a 'political-cultural blockade' of the CA. As such, the intensity of the debate and conflict in Germany is rather unique in the international CA context. Although both initiatives can be considered as commendable and both documents reflect a shift in the crisis management perspective, the conflict they create is invariably normatively charged. In particular, the concept of *Vernetzte Sicherheit* takes on a highly negative stance and appreciation of the CA not only differs between ministries but also within them (Wittkowsky et al.

441 At the time of writing in the position of Federal Minister of Defence.
442 The Konrad-Adenauer-Stiftung (2009, pp. 6/7) comes to a similar conclusion.

2011, p. 1).[443] The CA debate in Germany has thus sustained cultural and bureaucratic cleavages. Generally, the whole subject of comprehensive crisis management and respective debates in Germany are very unpopular and not prioritized, particularly among practitioners. "[T]his – in official parlance – 'important policy area' was on its way to an 'irrelevant niche project of German foreign policy" (Unger 2009, p. 156).

As the debate around the two conceptions shows, efforts to react to CA developments challenge German foreign political culture particularly on the normative level which is reflected in the conceptual adaptations. Though Germany attempted to synchronize its policy traditions with the requirements of comprehensive crisis management, these adaptations largely maintained roots in the traditional German foreign policy culture of civilian power Germany. A major issue of the German discussion in this context is that it does not aim at implementing the CA by involving different crisis management dimension on an equal level but clearly claims a primacy of the civilian dimension. Civilian Crisis Prevention is assigned a special role in German crisis management not just on side of the peace research community, but also in public debate. Alamir (2012, p. 74) observes with regard to the security-development nexus that while half a decade ago debates focused on whether development policy should get involved into security topics at all, today the nexus is acknowledged and questions rather deal with the way and extent development should address security issues. This observation, however, does not seem to be valid in Germany. Due to the particular bias of its post-World War II foreign and security policy and the relative modesty of the military with a decisive role being assigned to civilian crisis management in the German constellation, issues of cooperation with the military require a strong civilian impetus. In this context, the German political sensitivity concerning military crisis management and involvement has 'softened' the country's CA model. The German CA model materializes as an instrument for the country to express its positioning in relation to its military commitment and has been undertaken as close as possible to its particular security political background.

443 A ZIF policy briefing (Wittkowsky et al. 2011, p. 1) denotes the issue 'a downright cacophony' ("*eine regelrechte Kakophonie*"). The survey identifies two opposite opinions on the German CA efforts present in ministries. One side sees Germany as a pioneer of CA implementation undertaking the necessary reforms. Reforms such as theInter-ministerial Steering Group and exchange of personnel from their perspective already resulted in coordinated inter-ministerial activities. The opposing side evaluates that Germany continues to fall short of its opportunities by implementing measure which do not lead to any outcome. This side observes a lack of both political will and resources (Wittkowsky et al. 2011, p. 2). This split of opinion is largely consistent with observations during interviews conducted for this dissertation.

The conceptual implementation of the CA in Germany and the lack of clarity in this regard also shows in the absence of an overall strategy which could contribute to the development of a common understanding and clear goal setting within the German CA model. The two guiding German CA documents, the Action Plan and the White Paper, stand rather uncoordinated side by side.[444] Overlaps, differences and characteristics of the two concepts remain largely unsolved, resulting in an uninformed use in the national language and also confusion of international crisis management partners. Thus, the challenge on the conceptual level is to ensure a clear understanding and consolidation of the concepts of Vernetzte Sicherheit and Civilian Crisis Prevention (Major et al. 2011, p. 136). In this regard, it was emphasized in background talks that the concept of Vernetzte Sicherheit cannot serve as an overall conceptual framework but rather as an idea. Interviewees demand a clarification and definition of the CA in Germany which makes clear that it does not mean a subordination of civilian actors under military guidance but the adjustment of all instruments toward one goal. It was suggested that the CA in Germany needs a complete re-naming of the idea away from any Vernetzte Sicherheit denotation toward a term that underlines actor equality in a broadly defined international environment and that eliminates the lack of a common vision in the German concept of Vernetzte Sicherheit.[445] The discussion around the development of Vernetzte Sicherheit underlines, what also interviewees observed, namely, that CA thinking in Germany has largely evolved along ministerial lines of bureaucracy. The conceptual enhancement of the German CA model is needed. In this context, an overall strategy which is built on both approaches could be established to provide guidance.[446] The process of strategically clarifying CA concepts and their relations would substantially add to German conceptual crisis management coherence. An inter-ministerial strategic discourse, which so far does not exist, seems important. In this context, efforts such as the 2012 *Interministerial Guidelines for Coherent Federal Gov-*

---

444 Formally, as it is a government document, the White Book is of much greater relevance than the Action Plan (Nachtwei 2011, p. 1).

445 A re-naming of the concept to *Vernetztes Handeln* (Networked Action instead of Security) is considered an improvement as the term security is avoided. Also the use of the international term Comprehensive Approach is regarded more meaningful and sensible as it is not obviously framed in the security context. Furthermore, a terminological reframing could replace the word 'concept' instead talking about the 'approach' of Vernetzte Sicherheit, which implies a less closed departure.

446 There is a continuing debate on the benefits of a national security strategy in Germany. In this regard, Ehrhart (2011, p. 79) points to the lack of an elaborated strategy in Germany due to the German structural inability to develop and plan on strategic level. See also Major et al. (2011, p. 134).

ernment Policy Towards Fragile States have send out an important signal of efforts to advance the CA in Germany.

4.2.4.2 Organizational Dimension

The lack of conceptual clarity and the co-existence of the concepts of Vernetzte Sicherheit and Civilian Crisis Prevention are also reflected in structural efforts to increase crisis management coherence in Germany and complicate inter-ministerial coordination. The organizational side to advance crisis management coherence is not in line with the heavily debated but nonetheless strong political declarations. Being highly in favor of the CA conceptually, traditional ministerial competencies in structural terms have been left untouched in Germany.

CA implementation in existing German crisis management structures has proved difficult for various reasons. The relatively high degree of ministerial autonomy represents a key aspect in these dynamics. There is a large degree of tension between the CA idea and the Ressortprinzip as a high-level constitutional convention and procedure which proves difficult to resolve.[447] From coherence perspective, the Ressortprinzip constitutes the initial structural barrier for inter-ministerial coordination and poses a "critical obstacle" (Patrick/Brown 2007a, p. 110) to horizontal crisis management coherence with major consequences for ministerial structures and culture. Institutionally, ministries pursue their own interests and priorities with a great focus on ministerial matters. In this regard, problems in Germany also lay in the diplomacy-development nexus due to the opposition between the AA and the BMZ which is a major stumbling block (Weiss/Spanger/Van Meurs 2009, p. 48). In Germany, the high self-confidence of the autonomous BMZ furthers inter-ministerial fragmentation. In inter-ministerial debates, development issues generally hold the upper hand, making for a lack of balance between the three dimensions security-governance-development and respective actors. Although rhetoric changed, ongoing inter-ministerial turf battles are a key constraint and ministries in practice very much adhere to the Ressortprinzip which dominates coordination efforts particularly on the working level (Wittkowsky et al. 2011, p. 2). Thus, structural CA reforms in Germany have been mainly triggered by sectoral developments, rather than cross-ministerial reform policies. Strong hierarchical characteristics and vertical

447 According to a ZIF-survey, ministry officials themselves view the Ressortprinzip largely diverse with views blaming the principle for departmental egoism and self-centeredness while other opinions stressing the importance of the strict functional division of work (Wittkowsky et al. 2011, p. 2).

differentiation were difficult to be overcome by introduced structures. At the same time, this hierarchy has lacked required leadership and political decision-making. In this regard, CA implementation on the organizational level can be viewed as a market type of organization in Germany without an organizing or directing centre and based on opposition and pragmatic exchange between actors. Factors to encourage interorganizational coordination such as signaling to ensure that everybody shares the CA commitment and a framing of CA matters to highlight a common interest have been largely absent in Germany. Exchange took place in an *ad hoc* and infrequent manner and repeated, predictiable interaction in the iteration factor sense was not observable. However, though alleviated procedurally by these mechanisms, due to the strength of the Ressortprinzip and other mechanisms such as co-signature, it is negative coordination which prevails in the structural dimension of German crisis management coherence. Coordination in the German CA context mainly deals with potential opposition and limits CA aspirations to avoiding confrontation.

While the functional division of work proves valuable, the strong legal basis for this division of labour needs to be softened with inter-departmental coordination mechanisms just as strong. This would include both standing inter-departmental bodies, as well as informal and *ad hoc* coordination but first and foremost the identification and creation of organizational interfaces. Indeed, rather than setting up an entirely new inter-ministerial entity, the German efforts to implement the CA have involved respective mechanisms such as *ad hoc* interministerial task-forces, working groups and the exchange of staff between ministries. Interviewees evaluated inter-ministerial coordination based on voluntary and informal involvement as high and being widely recognized. The value of liaison officers in this regard highly depends on their integration in the ministry and how both ministries utilize the officer's insights on the respectively other ministry. Their role as an interface between ministries can thus be further improved. It is also important to clarify the function and responsibility of liaison officers as well as their integration into the ministry's day-to-day work (Wittkowsky 2012, p. 5). In addition to informal forms of coordination as well as *ad hoc* working groups and task forces, stronger formal requirements for horizontal coordination might be needed.

As a result of such unfavorable institutional preconditions, the German government chose a rather cautious approach to built governmental substructures to organize crisis management in a cross-ministerial manner (Weiss/Spanger/Van Meurs 2009, p. 49). It gradually created structures, such as the Inter-ministerial Steering Group and the Advisory Board, to organize crisis management as a cross-ministerial measure. Evaluated as little effective to advance the CA, both from government officials as well as from academia, these institutions largely lack appropriate support structures in administrative, financial and staff terms.

Difficulties are also given account of by the German government's struggles to declare political priorities. The issue of coalitions and the fragmentation of competences between ministries further contribute to the constraints (Major et al. 2011, pp. 134/135). As a result, respective institutions miss anchorage on a higher level, institutional clout and controlling competence in form of incentives and sanctions to actually enforce inter-ministerial coordination. German structural changes and newly created bodies are rather in charge of the facilitation of communication and refining ministerial practices than being actual operational policy actors (Unger 2009, p. 145). The most notable accomplishment of organizational efforts toward crisis management coherence in Germany to date is the growing sensitivity for CA issues.[448]

Indeed, from an institutional policy perspective, no central actor driving the CA forward can be identified. In particular, the Steering Group is perceived as an inter-ministerial learning field without policy impact or an independent decision-making authority but with great difficulties reducing inter-ministerial differences of crisis management approaches. The Steering Group as the body to facilitate comprehensive crisis management lacks decision-making power and has struggled to find a proper role within the German crisis management architecture. Major impulses which drove the German CA agenda were triggered by either international processes or developments or by civil society rather than conscious, proactive assessments with an overall minor input from the German government. Generally, in its attempts to implement the CA, Germany heavily emphasizes the role of civil society organizations in the process toward coherence. Arguably, a lot of drivers of the CA and the crisis management coherence agenda are to be found outside the government. However, the newly created parliamentary subcommittee provides fresh impetus to the debate.

In sum, German efforts to implement the CA hardly provide a sufficient institutional basis for inter-ministerial coordination and policy coherence. As the lack of political weight and a high-level actor within governmental institutions are often referred to as a major obstacle, suggestions for an improvement of the institutional design of arrangements for coordination mostly demand respective changes. The spectrum of suggestions includes the possibility for a stronger role of the Chancellery as a driver of policy coordination and the German coherence agenda to an operational structure for the Federal Security Council in terms of a National Security Council to develop coherent political-strategic concepts. Most prominently, however, both academia as well as practitioners demand a strengthening and 'institutional up-grading' of the Steering Group onto the level of state-

448 This growing awareness, however, is partly negatively coined due to the Vernetzte Sicherheit debate.

secretaries with a permanent secretariat (Deutscher Bundestag 2011, p. 5). It has also been suggested to greater interlock the AA and the BMZ with the establishment of a state secretary for crisis management tied to both ministries to perform a sort of hinge function between them. Both the Steering Group as well as the Federal Security Council are obvious candidates to be enhanced to function as a central body in the discussion on how to increase coherence of crisis management structures. Introducing more substance behind the Steering Group by bringing in higher levels of executive and by providing appropriate staff and funding, however, also has some downsides. As interviewees assess, even after an institutional up-grading of the Steering Group, in most cases it will be the same people on working level ultimately dealing with issues and it will not change much in inter-ministerial every day work. Also from the perspective of organization theory an actual increase of inter-ministerial coordination and coherence due to a group of state secretaries working on an issue is questionable. While it is difficult to evaluate the impact of committees or working groups of state secretaries on inter-ministerial horizontal coordination due to a lack of systematic information, it should be noted that the existing regular gathering of state secretaries to prepare cabinet meetings already is a coordination process at the administrative-political level where inter-ministerial coordination takes place (Hustedt/ Tiessen 2006, p. 31). Meetings on the state-secretary level generally "can be considered a bottle-neck in the coordination process" (Hustedt/Tiessen 2006, p. 34) and effectiveness of such meetings can hardly serve to debate issues, mainly due to a lack of time. Nonetheless, since state secretaries are politically appointed, such an institutional anchorage would increase the importance of political considerations and also the pressure to solve divergences in order to avoid conflicts. Lifting the Steering Group up on the level of state secretaries can thus be viewed as "a counterbalancing structural technique to the effects of departmental egoism on a higher level of hierarchy" (Hustedt/Tiessen 2006, p. 34).[449] Consequently, though not directly resulting in increased inter-ministerial coordination, with the explicit up-lifting of the Steering Group on the level of state secretaries, the debate around the subject of crisis management coherence would gain political quality. As pointed out during a background talk, organization and structures

449 Hustedt and Tiessen (2006, p. 31) demonstrate this by referring to the so-called 'Green Cabinet', a group of state secretaries working on sustainable development, which was established in 2001 and brought together 11 state secretaries and the Chancellery with the objective to carry the idea of sustainability into their ministries and to develop cross-ministerial policies and strategies on sustainability. Hustedt and Tiessen (2006, p. 31) conclude that the group did actually not get involved in inter-ministerial coordination "but rather resembles a (symbolic) political signal or priority respectively." As such it mirrors "the usual way of shifting issues upwards through the layers of hierarchy as is typical in German federal government."

are the key element of the concept of Vernetzte Sicherheit and platforms to network are important to for inter-ministerial coordination. Germany has sufficient 'soft institutions' and misses a leading institution.

A major issue concerning CA implementation in Germany is not just the lack of interest in concepts such as the Action Plan and its objectives but apparently also a lack of efforts to increase interest in the issue from the beginning. The picture of inter-ministerial coordination in Germany is incompatible with the government ambition to pursue comprehensive conflict management. Due to the lack of operability of CA structures, efforts to improve crisis management coherence on the organizational level have to be evaluated rather symbolic than substantial. This can be assigned to the fact, that the CA debate in Germany is stuck on the conceptual side and has, apart from the discussion on existing structures, not reached the level of discussing structural issues yet. The steps taken in Germany on the structural level represent some cautious innovations which are impeded by a set of problems that largely lie on the strategic level as key conceptual questions on the CA and German inter-departmental *Vernetzung* are largely unclear. Respective debates are coined by an absence of pragmatism regarding CA implementation and the careful changes are rather undertaken in response to the need to show that a weakness in German crisis management activities is being addressed, without fully embracing a fundamental change and an actual surge to bridge divides. The CA debate in Germany largely serves to sustain the cultural and bureaucratic cleavages described above which collectively undermine the comprehensiveness pursued. The lack of any strategic and capstone formula leads to a troublesome duality of strategic independence and functional interdependence of individual actors. There is no strong protagonist or lobby for the CA and related issues, but an increasing lack of interest. As the CA in Germany is merely adopted in terms of reform symbols rather than problem-oriented reasons, the institutional implications of the objective to improve crisis management coherence is ignored by political decision-makers. Rather than prioritizing inter-ministerial coordination on the national level, Germany has been pushed to take up the debate mostly in multilateral contexts, not by an intrinsic motivation. In Germany, even more than implementing structural changes, the fundamental acceptance and commitment toward the idea itself should be focused on.

In the context of the outlined key ingredients organization theory views as relevant to trigger major organizational innovation, the country case study on Germany only shows one of the three. The country has not experienced a relevant disaster of its crisis management policy in its last two decades or at least none it felt directly responsible for with failures in its policy. There is no well-placed political champion in the German system which actively advocates changes to support CA efforts. However, the country could not elude from internationally emerging CA ideas, in particular at EU level.

4.2.4.3 European Dimension and EU-Interface

German crisis management is deeply embedded within a multilateral framework and, generally, the development of ideas and concepts in Germany are strongly internationally pushed. The country has been portrayed as a strong supporter of European integration and as being Europeanized, particularly due to the strong congruence between German and EU norms and principles in the field of foreign and security policy. German crisis management policy started to Europeanize after the Balkan conflicts in parallel with the increasing crisis management capabilities on the EU level. In particular in the beginning of the EU's ESDP development, Germany provided significant input to the EU developments. As a military introverted country, Germany was particularly committed, vocal and active to support the civilian side of ESDP as the topic was less controversial and sensitive compared to the military dimension. In addition to the civilian crisis management advancement, the conceptually driving force behind the German approach to EU crisis management was also to reinforce the primacy of conflict prevention. The country uploaded and largely promoted its traditional anti-militarist stance and contributed to the framing of the EU as a normative power (Daehnhardt 2012, p. 46). In this regard, the EU's crisis management framework has constituted an ideal framework for German political priorities in crisis management. The German crisis management approach and identity fitted in the emerging EU crisis management culture and its key contributions in the area of 'soft security' understanding security in broad terms.

At the same time, EU crisis management clearly became an important aspect of Germany's crisis management policy and reflected in several concepts. Also, improvements made by the EU in implementing policy coherence left a mark as Germany reacted with the conceptual development of its Action Plan. The country thus supported CA reforms on EU-level even before working on it on the national level. With its strong preference for multinational frameworks in crisis management, the German government was compelled to download the EU's agenda of policy coherence an to follow EU logic. Germany leans on the EU in devising its own crisis management objectives also in terms of identifying and developing policy for a government-wide approach as the Union provides legitimacy for German activities. EU crisis management also allowed Germany to strengthen its voice in crisis management most effectively without raising historical fears of its European neighbours. Most developments of EU crisis management are uncomplicated for Germany on the conceptual level and in the normative sense.

However, German principles come under pressure vis-à-vis the use of force and the demands to play a greater role in military crisis management which became a cental issue in the downloading dimension. In this regard, EU crisis management developments forced Germany to reconsider its reservations concerning scale and geographical scope of its crisis management policy as well as the responsibility it had to take in the EU. In this regard, the observed 'normalization' of the German security policy was closely accompanied by its Europeanization. The development of EU military crisis management served as a facilitator of this 'normalization'. Thus, the impact of the EU and respective Europeanization of German crisis management witnessed a two-way process in terms of uploading German civilian preferences with regard to CFSP and, at the same time, downloading in terms of redefining thinking on international military operations as a part of German crisis management. Most significant adaptational pressures have been faced by BMVg as a result of the increasing German role and domestic reassessment of the Federal Armed Forces, where Europeanization has been a facilitating setting for transformation and reform.

Yet, though Germany has faced severe pressures to engage more actively in EU military engagements, it maintained many of its views and distinctively historically coined principles on this issue. Despite the highly pro-European perspective, some obstacles remain with regard to the Europeanization of the contentious aspect of EU military crisis management. A mixed Europeanization in terms of adaptation due to the involvement in EU crisis management can be observed and conceptual change and word has not met operationalization. In terms of crisis management Europeanization, the German limitations become particularly evident concerning the operationalization of appropriate institutional structures. Generally being characterized as very sectorized, only minor reorganization of EU policy-making processes took place. In this regard, however, EU policy is less exclusively the domain of the AA as the Chancelery but also the BMVg have stenghtened their positions. Though the EU is a central pillar of Germany's crisis management policy, it has not led to a deep restructuring of crisis management within the country. According to a respondent, in terms of numbers of staff but also on a more general strategic level, Germany so far has not been able to find a way and the means to link German and EU external action and crisis management.

This observation also applies to the new crisis management framework after the Lisbon Treaty reforms. While heavily Europeanized and congruent in the normative dimension of crisis management, obstacles fall into structural factors relating to the country's institutional set-up. Germany has been an early and strong advocate of respective reforms with various initiatives. Yet, since its entry into force and the start of reform operationalization in the institutional sense, the German positioning and attitude is rather reserved which particularly showed with regard to the EEAS establishment.

Aggestam (2000, p. 65) differentiates between the normative and the instrumental dimension of German foreign policy. This separation also applies with regard to the Europeanization of German crisis management policy. EU's crisis management is highly important in the German normative and symbolic view on the EU as a political union and reflected in documents and initiatives. However, instrumentally, Germany hesitates to exercise German foreign policy through existing EU structures. The country can thus be understood as a 'contributory follower' which on the one hand is highly supportive concerning developments in particular on the conceptual level but which only slowly and limited transforms national capacities in a reactive manner. However, despite this observation and the late reluctance and 'de-normalization' of German crisis management, Germany remains deeply anchored and committed in the EU crisis management framework.

4.3 Case Study Sweden

4.3.1 The Swedish Crisis Management Approach

In general, Scandinavian political-administrative systems are described as being structurally and culturally comparably homogenous, emphasizing collective norms rather than individual rationality and efficiency (Christensen 2003, p. 113). This observation also applies to the Swedish case. Political tradition in Sweden is historically characterized by an atmosphere of mutual trust and collaboration (Calabrò 2011, p. 27). Its consensus-oriented nature is reflected in the country's political-administrative system and a tradition of negotiation and consultation (Wollmann 2003, p. 595). Solidarity and equality are other highly valued features of the Swedish political system. Decisions are intended to be taken with the agreement of virtually all actors. A key feature of the political culture is thus the ability of opposing factions to come together in reasonably harmonious discussions. Even if actors oppose to a decision, the consensus is not challenged "since groups could be in disagreement without being in conflict" (Arter 2008, p. 153).[450]

Apart from this "culture of technocratic consensus" (Levin 2009, p. 39), an important defining feature of the Swedish broader political system is an emphasis on openness, transparency and accountability which becomes apparent in two modes. In addition to the already mentioned first feature that interested and af-

450 While this consensual element is evident in the policy-making process itself, it is not apparent in the formation of the government due to a strong separation of political parties on the basis of a left-wing/right-wing spectrum (Arter 2008, p. 153).

fected actors are consulted before a decision is taken, demands on public insight into the Swedish system are very high. This becomes evident in the right to access official state documents and also in the freedom to communicate information.[451] Furthermore, Swedish policy-making is approached in a rationalistic and problem-oriented way that is characterized by a pragmatic and intellectual style that aims at taking decisions on a base of the maximum amount of information (Arter 2008, pp. 152/153). This aspect becomes apparent in the deliberative Swedish policy preparation process which is often dominated by experts on particular policy issues. In doing so, Swedish policy-making and institutions seem to have been "designed to avoid open conflict at all costs" (Arter 2008, p. 152).[452] The emphasis on seeking consensus in the domestic context also spills over into the making of foreign and security policy and the Swedish approach to international crisis management is shaped by some very particular traditions and guiding lines (Miles 2006, p. 77).[453]

Sweden takes on a heavily internationally minded and peace-driven approach to international relations and an "anti-militarist stance to international affairs" (Browning 2007, p. 32). The country's fundamental position in the global security scene aims at following an effective multilateralism by being embedded in IOs. As mentioned by respondents, international engagement in this regard is also considered a matter of improving security and reflects the belief that

451 More than by the legal dimension of accountability, the Swedish political system is coined by a culture of openness in support of accountability (Levin 2009, p. 44). It has been argued, however, that such high openness and transparency leads to an overflow of official information which can also be perceived as "a very effective way of concealing sensitive matters" (Larsson/Bäck 2008, p. 282). For further details on different forms of openness and transparency in the Swedish political system see Larsson and Bäck (2008, pp. 279-282).

452 This resulted in the labeling of Sweden with expressions of a 'negotiating democracy', a 'bargaining democracy', a 'consensual democracy' or a 'harmonious democracy' and to references to the 'Swedish model of the politics of compromise' (Arter 2008, p. 156). Other components of a Scandinavian model of governance often mentioned in scholarly literature are the Nordic welfare model and the strong Social Democratic Party dominance (Arter 2008, p. 161). At the time of writing, however, the Social Democrats were in opposition in Sweden. Also other Scandinavian countries experienced extended periods of non-socialist government, which questions if the Social Democrats can be regarded the dominant party any longer. Also the Nordic welfare system is increasingly challenged by pressures generated by forces of globalization and demographic change (Arter 2008, p. 189). For a detailed analysis of these components in the Swedish model of government and on the question of certain characteristics of a special Nordic governance model see Arter (2008, pp. 151-190).

453 The international dimension of the Swedish "spirit of political consensus" (Levin 2009, p. 39) as well as a solid faith in the key principles of equality and solidarity also shows in the idea of *folkhem* (people's home), namely "the vision of a government as a home that protects the nation's people as much as a family's home protects each of its members" (Erik Åsard and Lance Bennett 1997, as cited in Wunderlich 2010, p. 6).

Swedish security cannot be built in isolation but needs international cooperation. Against this background, the country very actively contributed to peace and security with a high commitment to the UN, which hold a strong legitimizing role. Sweden traditionally supports UN peacekeeping and actively participates in international peace support and humanitarian efforts. In addition, Swedish foreign policy is profoundly framed in the human security context. The development paradigm and a strong focus on humanitarian aid lie at the heart of Swedish foreign policy and crisis management (Patrick/Brown 2007a, p. 118).[454] The country's international engagement thereby particularly focuses on the reduction of global poverty and adopts a human rights-based approach to development (Government Offices of Sweden 2003, pp. 19-22). These two fundamental perspectives guide and design Swedish policy which is committed to the prevention of conflict and mediation efforts (Patrick/Brown 2007a, p. 116). Additionally, environmental and ecological issues are placed emphasis on in Swedish foreign policy (Hopkinson 2004, p. 19).

A highly dominant characteristic of Swedish foreign and security policy and integral part of its national identity is played by the *neutral* approach Sweden historically takes to its international engagement. Theoretically, neutrality either evolves in an attempt to improve national security or stems from an aim to fulfill an international role according to the country's self-conception. Thus, neutrality has a strategic dimension that is concerned with security and the avoidance of conflict; and a normative dimension concerned with national identity and the appropriateness of the country's role in international affairs. In practice, however, a mix of these two components impacts neutrality policy (Möller/Bjereld 2010, pp. 367-369). Neutrality is an essentially contested concept with different meanings depending on the particular political, societal and security emphasis and respective political practices. For an overview of the history, definitions and differing forms of the concept see Agius and Devine (2011). Basic principles of neutrality in Swedish policy were developed in the early 19th century. In Sweden, neutrality is highly positively coined and viewed to have saved the country from involvement into wars since 1814 and from having to take sides during the Cold War (Miles 2006, p. 90). This perception is reflected in the equation of the Swedish neutrality policy with peace, which is deeply rooted in Swedish society (Rieker 2002, p. 32). It

454 Sweden is among the most generous providers of ODA and one of the few countries that lives up to the UN aid target, devoting some one percent of its gross domestic product to this purpose (Government Offices of Sweden 2011a, p. 9).

becomes also evident in the continuingly strong majority support the policy enjoys in the Swedish public (Möller/Bjereld 2010, pp. 371/372).[455]

During the Cold War, the country's neutrality policy was officially referred to with the formulation "non-alignment in peace, aiming at neutrality in the event of war" (phrase as cited in Rieker 2006b, p. 66). The end of the Cold War and the bloc system, however, raised questions about the logic of the neutral status.[456] In consequence, the official description of Swedish neutrality was reframed with a stronger focus on the term *military non-alignment* and the official formulation was amended to "non-alignment in peacetime, in order to enable Sweden to remain neutral in the case of war in its vicinity" (phrase as cited in Rieker 2006b, p. 70). In 2002, the above formulation was extended by an assessment on the past benefits of the concept of neutrality.[457] In 2003, however, Swedish political parties and government decided to avoid mentioning neutrality or any reference to any kind of neutral policy option (Rockberger 2005, p. 7). While the disappearance of references to neutrality from the Swedish discourse has been considered as a "meaningful silence" (Devine 2011, p. 338) and military non-alignment became less relevant in Swedish international engagement, it continues to be valid.[458] At the same time, the modification toward the actually narrower meaning of non-alignment has allowed for a greater flexibility regarding international crisis management matters (Miles 2006, p. 80).

Since Swedish neutrality was domestically not codified in a legal act and never formalized in treaties, it is not anchored in international law and no international obligations can be derived from it (Rockberger 2005, p. 7). It is thus open to diverse interpretations and the actual implementation and consequently

455 Swedish neutrality policy is well covered in numerous studies. Ambivalences, inconsistencies and controversies of the Swedish neutrality and non-alignment policies with regard to the Second World War, NATO's Euro-Atlantic Partnership Council and Partnership for Peace, which Sweden joined 1994, are not discussed here. Though affecting public debates in Sweden, societal aspects are not taken into account either. For the historical background of Swedish neutrality see Rieker (2006b, pp. 64-69). For an overview of Swedish non-alignment policy with regard to NATO see Schlüngel (2005) and Ojanen (2002, pp. 178-184).

456 "Neutral from what?" (Bailes 2006, p. 8), since the rationale for a neutral policy disappeared with the end of bipolarity and the need to take sides in the Cold War bloc system.

457 The policy was officially formulated as follows: "The aims of Sweden's security policy are to preserve our country's peace and independence, contribute to stability and security in our vicinity and strengthen international peace and security. Sweden pursues a policy of non-participation in military alliances. This security policy, making it possible for our country to remain neutral in the event of conflicts in our vicinity, has served us well" (phrase as cited in Rieker 2006b, p. 74).

458 Internationally and by its own citizens, Sweden continues to be seen as a neutral country and neutrality has remained a point of reference despite non-alignment being the officially accurate term (Rieker 2006b, p. 73).

also the credibility of Swedish neutral policy in peacetime has always been the responsibility of the respective government. This involves pursuing a policy in times of peace that enables to stay neutral during war as well, and means a holistic adjustment of all foreign policy aspects toward the neutral policy. Neutrality consequently influences all foreign policy areas of a state, making it a foreign policy doctrine itself (Schlüngel 2005, pp. 6/7). In this regard, Sweden conceptually strictly draws a line between defence and crisis management, which makes it politically possible for Sweden to participate in peacekeeping operations (Wedin 2005, p. 12). Non-Alignment stands for Sweden's continued non-participation in any form of military alliance (Wedin 2006, p. 142), and strictly refers to defence issues in military terms. The status explicitly excludes mutual military security guarantees and bilateral defence alliances (Ojanen 2000, p. 6). This explains how Sweden's neutral stance can include a highly active foreign policy element that aims at contributing to stability and common security (Huldt 2003, p. 47). Historically, neither its neutral status nor the newer concept of non-alignment has deterred Sweden from active involvement in international affairs (Väyrynen 2006, p. 228). Quite the contrary, its highly active international role and emphasis as a mediator can be considered a characteristic of Sweden's dynamic policy of neutrality (Möller/Bjereld 2010, p. 376). Engaging in crisis management and the prevention of humanitarian crisis is regarded an important responsibility of the international community. "The specific relevance of military non-alignment as a limiting factor for Swedish international cooperation (…) became a non-issue – at least as long as there was some form of UN mandate or consent" (Huldt 2003, p. 48).

Another factor that has contributed to shaping Swedish foreign policy is the concept of *Norden,* which views Sweden, together with its Scandinavian partners, as a special member of the international community (Åselius 2005, pp. 26/27). The idea of Nordic exceptionalism has an international dimension and influenced Swedish foreign policy and international engagement.[459] It led to a peace-driven approach to international relations as it was the case during the Cold War "when the 'peace-loving and rational' Nordics differentiated themselves from the rest of 'warlike' or 'conflict-prone' Europe" (Browning 2007, p. 27). The foreign policy dimension of this perception becomes apparent in the

459 The conception of Nordic exceptionalism involves a geopolitical element that sees the Scandinavian countries at the periphery of Europe. This position at the edge of the continent as well as a shared history and cultural development enabled the perception of a homogeneous and closed region (Vogt 2005, p. 141). Another aspect of Nordic exceptionalism refers to the socio-economic organization of Nordic countries as a model welfare society and egalitarian social democracy (Browning 2007, p. 27). For a more detailed discussion on Nordic exceptionalism, the idea's development and different elements see Browning (2007, pp. 27-51).

perception of Nordic countries being international bridge builders. Even more so, the idea of Nordic exceptionalism is understood as a model that is better than other models and should be copied and applied elsewhere (Browning 2007, pp. 27/28). This translates into a commitment to a particular form of internationalism which seeks to export advantages on the global stage (Miles 2010a, p. 186).[460] In this regard, it also attained a heavy moral dimension that aimed at not just being a bridge between the East and the West but also to overcome the North-South divide by focusing on the Third World and international justice. In particular during the Cold War, the Nordic countries' international engagement was considered in terms of multilateral, moral and unselfishly oriented generous and concerned donors (Strömvik 2006, p. 199).

Though today this approach is rather seen as "enlightened self-interest" (Vogt 2005, p. 144), the concept promoted morality and solidarity to become key factors in Swedish foreign policy (Browning 2007, pp. 33-35).[461] Until today "a sense of global solidarity" (Patrick/Brown 2007a, p. 117) and moral responsibility remains a motivation for getting involved in international crisis management and is seen by the Swedish government as

> "[t]he primary justification and rationale for Sweden's policy for global development (…). This rationale reflects the basic values in accordance with which Swedish society has evolved and that are also expressed in the UN Charter and the Universal Declaration of Human Rights. (…) Solidarity is reinforced by the realization that security, equality and sustainable development are not an exclusively national concern. (…) Responsibility must also be shared" (Government Offices of Sweden 2003, p. 19).[462]

The perceived role of Sweden being the world's moral conscience and a moral superpower resulted in "an internationalist program designed to influence developments in the direction of peace and security" (Arter 2008, p. 332).[463] While potentially at odd with the pacifism that is deeply rooted in the country's self-image (Wunderlich 2010, p. 3), these features of its security approach character-

460 This refers even more to the Nordic socio-economic organization and ideas of egalitarianism and the social democratic model (Browning 2007, p. 27).

461 Both values are also attached to the above mentioned *folkhem* concept and are socially deeply embedded in a strong sense of collective identification (Lawler 1997, pp. 567/568).

462 Examining the self-images of Swedish peacekeepers, a study showed that this image is also reflected in the self-perception of the Swedish personnel deployed in peacekeeping operations. The Swedes described themselves as having very strong ambitions and higher ethical and moral standards than the peacekeepers from other nations. Swedes also claimed to take a more neutral approach than others (Hedlund/Soeters 2010, pp. 410).

463 For an in depth consideration of Sweden's role as a moral superpower and post-Cold War developments of this image see Dahl (2006).

ize Sweden "'as a country with disproportional wide-ranging internationalist ambitions, given its size and location, and with strong moralist and ideological convictions to match'" (Rieker 2006b, p. 81).[464]

Another doctrine that influences Swedish security policy is the country's former leading Cold War defence strategy called *Total Defence*. Rooted back to the Second World War, the notion of Total Defence presents a Cold War conception that was invented in the context of protecting homeland against an external threat, namely a full-scale Soviet invention, as several interviewees emphasized. The concept was initially constructed to the physical survival of the Swedish nation in the case of war and is based on territorial integrity. In the face of an external attack, every sector of society and all resources should be mobilized to ensure Sweden's security, to support the military and to have the most comprehensive risk assessments available. As the concept aimed at bringing together all non-military actors, state controlled institutions, private enterprises and societal functions to support the Swedish Armed Forces (SAF), it presents a largely military centred approach (Grönberg 2010, pp. 25/26). After the end of the Cold War, the Total Defence notion gradually ceased to exist and a transformation of the concept took place which recognizes that today's challenges are reversed in logic. "Instead of mobilizing civil society to support the military in the face of external attack, the military is now one element to be mobilized as part of an overall response to major societal disruptions, including terrorism" (Hamilton 2006, xxvii). The Total Defence approach that applied to interstate confrontations was not seen to meet the requirements of today's challenges anymore and has lost its relevance. The notion has been widened to embrace a broader approach to threats with the modern concept of *Societal Security*.

Total Defence stood for efforts taken by civilian and military actors to oppose war strains. The key method through which it was implemented is the close civil-military coordination of planning and operation at every societal level (Grönberg 2010, pp. 20/25). As it involved detailed and systematic coordination of civil and military administrations and competencies, it is attributed as a benchmark for successful civil-military cooperation (Björkman/Friman 2011, p. 233). Though being a highly sensitive matter, Sweden thus has considerable experience and a long tradition of close and extensive civil-military cooperation

464 Reflecting the human security and development paradigm in Swedish engagement, the scope and nature of Swedish international crisis management to a big extent depends on the context. The Swedish involvement in Liberia presents a good example as a poor country, a factor of regional instability and additionally an important arena for Swedish EU-partners (Patrick/Brown 2007a, p. 118). Also Afghanistan serves as a case in point for Swedish crisis management that reflects the Swedish paradigm quiet well (Patrick/Brown 2007, pp. 124/125).

(Hopkinson 2004, p. 33).[465] According to background interviews, due to this tradition, civil-military coordination has long been seen as a general good in Sweden. It was furthermore pointed out that due to the Total Defence tradition, cross-governmental coordination is traditionally strong and horizontal and vertical partnerships long existed between relevant actors. Civilian and military authorities historically worked together and plans for civil-military coordination, integrated planning and common exercises existed early.

Though the concept of Total Defence is strategically not applicable anymore, its core principle nonetheless is retained in the CA, namely, the need for comprehensive efforts by all levels of government and the society (Hamilton 2006, xxiii). The Swedish Total Defence concept can be considered as "an example of a successful Comprehensive Approach" (Björkman/Friman 2011, p. 229) Due to the concept of Total Defence, processes toward comprehensive crisis management have been viewed as already being integral parts of the national security discourse. Respective changes have been facilitated by the greater salience of Total Defence in the country „including the placing of both civilian and military defence under the authority of the defence ministries" (Rieker 2006c, p. 313). The Swedish Total Defence concept thus had a strong CA side and, as declared by interviewees, a wide crisis management concept was already inherent to Swedish policy. Interviewees further explained that though the Total Defence concept was dominated by military thinking, Swedish actors were used to deal with civil-military action and the concept provided a valuable point of departure for the CA. While the phrasing in the CA context was new, the idea and philosophy for effective engagement to combine civilian and military means and to work together was already there. Government officials were trained in this kind of thinking. Also, the existing correspondence between the ministries already matched doctrines about good organization. As interviewees assessed, CA acceptance was gained more readily, and it was easy to achieve CA legitimacy and endorsement. Though the international context and security environment

465 Involvement of the military in civil and domestic issues is highly contested and the Swedish military is not permitted to use their combat capability in the service of civil society at peacetime (Hopkinson 2004, p. 33). SAF are not allowed to militarily assist the police since 1931, when Swedish military troops called in to reinforce the police opened fire during a labor strike killing 5 participants in the so-called Ådalen riots (Ekengren 2006, p. 276). The legacy of this tragedy has been that strong that the deployment of armed forces within Sweden was unimaginable until the 9/11 terror attacks. A debate regarding military support has arisen since then (Grönvall/Lundberg 2005, p. 31).

changed since the times of the Total Defence concept, the CA was therefore nothing actually new for Sweden.[466]

4.3.2 Sweden's Comprehensive Crisis Management

4.3.2.1 Conceptual Dimension

As elaborated in the previous chapter, Swedish external policy is guided by some strong foreign policy norms which result in an emphasis on the importance of active international engagement while also sticking to its non-alignment policy and a tradition of territorial military defence conceptions. The Swedish policy of non-alignment as the country's principal foreign policy approach for a long time continued to be dominated by a focus on territorial defence and up-coming CA ideas did not immediately change these defining elements. Furthermore, both, the Swedish military non-alignment as well as the country's moral ambitions and international responsibility required a high level of military self-reliance that had to be covered by credible defence policy and strong military on which Swedish policy options could be based.[467] However, as sketched out with regard to the non-alignment reformulation in the previous chapter, Sweden's security discourse gradually started to change in the second half of the 1990s and a profound recognition of the complexity of the international security context took place (Rieker 2006b, pp. 66/69/80/81). Though Sweden had extensive crisis management experiences, especially through the UN, conflict issues have been considered largely in separate terms than those characterizing national security. Respective aspects have not been paid much attention to in documents elaborating the Swedish national security approach and have mostly been advanced by the

466 For a different opinion see Grönberg (2010, p. 27), who argues that "the Swedish Total Defence Concept has very little – if any – similarity with today's comprehensive approach to civil–military cooperation in international crisis resolution (…), as they are based on totally different conditions". Also, various authors in the volume edited by Brimmer (2006) view the concept from a different perspective in analyzing the Total Defence concept in the context of homeland security transformations.

467 Thus, Sweden maintained a considerable number of armed forces, defence capabilities and a large national defence industry sector coupled with significant expenditure (Åselius 2005, pp. 26/27). The SAF had to be strong enough to deter potential outside aggression. These strong military capacities were based on the assumption that they should have a Swedish profile "designed not to be interoperable with those of other countries, thereby underpinning the credibility of the 'policy of military non-alignment'" (Wedin 2006, p. 142). An effective force was viewed to maintain Swedish neutrality and furthermore support the ambition to be a moral super-power (Browning 2007, p. 33).

Swedish Ministry for Foreign Affairs (MFA) only. As Swedish security parameters became more comprehensive, also civil–military cooperation became a renewed matter of interest (Rieker 2006c, pp. 309/310). While this did not lead to the formulation of one single overarching concept or strategy specifically dealing with the CA itself under a certain label,[468] the need for coherence and interministerial coordination is reflected in various different Swedish crisis management policy documents and strategies.

The necessity for developing a CA to international crisis management has been given increased conceptual attention. In 1995, Sweden incorporated a widened concept of security into the Swedish security doctrine and the Swedish MoD started defining policy coherence as a guiding principle for its security and defence policy. A report by the Swedish Defence Commission in the same year emphasized the high priority Sweden gives to conflict prevention, crisis management and humanitarian efforts, calls for a focus on these aspects in Swedish international commitment rather than on the tradition of national territorial defence (Rieker 2006b, pp. 76/81).[469] As outlined above, on the part of Swedish defence policy, the Swedish Total Defence concept has been widened in several parliamentary acts since 1996 calling on to crisis management to be adaptive, holistic and international. The Swedish government increasingly started to set out principles for a reformed security policy and Total Defence objectives. Though stating that main objectives of Swedish defence persist, as of 1999, key documents described an integrated view of military and civilian crisis management aspects and a new demand for adaptability in short- and long-term responses to changes in the security situation. A wide definition of security in international peace support and humanitarian operations became a focal point. It was particularly emphasized that both civilian and military resources shall be able to assist in international missions and that only sending both civilian experts and armed forces abroad to support peacebuilding efforts advances stability and security (Grönvall/Lundberg 2005, pp. 21/22).[470]

Starting in 1999, major defence reforms transformed the Swedish military away from territorial defence toward more flexible forces consistent with the broad spectrum of challenges after the end of the Cold War, giving crisis management capabilities the priority (Rieker 2006b, p. 79). The CA was increasingly

468 Compared to the very clear labeling of the CA as a part of Joined-Up Government or Integrated Approach efforts in the UK or the it conceptualization in the Vernetzte Sicherheit framework in Germany.

469 The report is available in Swedish only. Thus, the dissertation relies on secondary sources on the document. For the sake of clarity, the original document is referenced to in the list of references under Government Offices of Sweden 1995.

470 See Government Offices of Sweden (1999a, 1999b, 1999c).

referred to in official documents produced by the MoD, and became a vital part of daily work. The active engagement in international peace promoting and humanitarian operations became a major task of the SAF (Grönvall/Lundberg 2005, p. 30). Efforts that followed the initial 1999 defence reforms stressed civilian security and an *ad hoc* Defence Commission on the issue was established by the MoD to examine and present suggestions concerning a more integrated approach to defence and emergency planning. The transformation of the Swedish military was further advanced by the Government Bill *Our Future Defence* (Swedish government Offices 2004) and the focus of the Swedish defence transformation was again renewed with the 2008 Government Bill *A Functional Defence.* Largely based on Defence Commission proposals, the bill aimed at providing Sweden with a significantly stronger defence capability. As a part of this, the document contains specific proposals in numerous areas which include measures to concretize and increase civil-military coordination specifically acknowledging that

> "security and development presuppose a holistic approach and collaboration between military, peace-building and development-related resources. (…) Establishing stronger links between security and development in a conflict area require greater coordination of national efforts" (Government Offices 2010b, p. 7).

Another early key document is the Swedish Action Plan *Preventing Violent Conflicts* (Government Offices of Sweden 1999a) which aims at advancing the promotion of a culture of prevention, the identification of structural risk factors, the development of an international system of norms and its implementation, and the strengthening of international conflict prevention activities in different policy areas. The action plan emphasizes the possibilities of prevention for Swedish and international politics and seeks to encourage new attitudes in diplomacy, in international peace-support activities and in development cooperation (Government Offices of Sweden 1999a, pp. 15-18). Pointing toward an overall search for innovative crisis management efforts, the objectives were framed as an integrated approach that included economic, political, legal and military instruments in the plan. Recognizing that the numerous players of different relevant sectors are in constant collaboration, it demanded that the organization of the MFA and interdisciplinary knowledge, understanding and coordination should be strengthened from a conflict prevention perspective.

In doing so, the Action Plan moves beyond a mere declaration of intent and includes practicable measures such as coordination in project form between relevant MFA units or the appointment of a special steering group to provide the MFA with a decision-making basis for a unified policy (Government Offices of Sweden 1999a, pp. 54-56). Furthermore, the plan points to the difficulty of a separation of conflict prevention from other initiatives taken to reduce conflict.

"In practice, a clear delimitation between conflict prevention measures and measures aimed at settling on-going armed conflicts is not always possible, or even desirable" (Government Offices of Sweden 1999a, p. 2). While conceptually acknowledging that conflict prevention has to be broadly based on all available instruments (Government Offices of Sweden 1999a, p. 52), the document, however, largely ignores questions of coherence and coordination between the different Government Offices. Proposals for reform and the mainstreaming of conflict prevention are limited to MFA structures only.

In parallel with these conceptual developments, which show the recognition of the CA, Sweden adopted an exceptionally strong CA in the area of development policy and the Swedish perspective on crisis management coherence is heavily dominated by a focus on policy coherence of its development policy. This focus on coherence is one of the most notable features of the 2003 government bill *Shared Responsibility – Sweden's Policy for Global Development* (Government Offices of Sweden 2003) laying the foundations for Sweden's Policy for Global Development (PGD). The document has been widely recognized internationally and acknowledged by other countries as it made Sweden the first country with an official coherence policy (CONCORD 2011, p. 1). In the bill, the Swedish government assesses that the country should follow a coherent policy for global development, which shall be included at all levels of political decision-making and policy areas. The overall aim of the PGD, to contribute to the achievement of equitable and sustainable development, should apply to all policy areas (Government Offices of Sweden 2003, p. 18).

Coherence as the key concept of Swedish PGD and its need in different policy areas is strongly emphasized in the document which states that it is needed due to "the fact that national, regional and global perspectives are increasingly interwoven" (Government Offices of Sweden 2003, p. 3, 31). Sweden's PGD demands the harmonization of the whole spectrum of official instruments with which the country engages in developing countries and calls on to establish linkages between them (Government Offices of Sweden 2003, pp. 3/32). It aims at doing so by strongly promoting inter-ministerial coherence and coordination to engage a wide range of actors. "The policy will involve all ministries and activities in a number of areas of state and government affairs and policy" (Government Offices of Sweden 2003, p. 18) and "the responsibility for achieving these goals will be divided among all the ministries" (Government Offices of Sweden 2003, p. 8). Stressing the importance of coordination between civil and military crisis management measures, humanitarian assistance and development cooperation, coherence is seen to improve effectiveness by linking different policy areas that reinforce instead of contradict each other (Government Offices of Sweden 2003, p. 30). If actions taken to realize objectives of one policy area oppose

Swedish efforts to achieve equitable and sustainable development, the policy area in question is responsible to search for alternatives as part of its standard policy preparation work (Odén 2009, p. 4). The document recognizes the importance of conflict prevention and management for global development and their close connection with development efforts, "in which an integrated approach and collaboration are now considered necessary" (Government Offices of Sweden 2003, pp. 22/23). This means that PGD coherence not only entails traditional development activities but involves other areas like security and defence policy, trade, migration, economic and financial policy, agriculture, cultural policy, environmental policy and public health.[471]

However, as the bill also notes, it "does not regulate future policy in detail" (Government Offices of Sweden 2003, p. 9) in terms of suggesting any concrete measures with regard to the organization of the general PGD conduct or organizational arrangements for implementation of the policy. It rather broadly demands that the organizational set-up for implementing the PGD needs to be coherent at all levels and generally points out, that results based management has to precisely define the objectives, description and analysis of the effect of the measures to be taken (Government Offices of Sweden 2003, p. 77). To monitor the realization of the policy, regular progress reports to parliament, so-called government communications, measure and summarize PGD efforts to enhance coherence in various policy areas (Government Offices of Sweden 2003, pp. 19/20).

These follow up communications to parliament continue to demand greater coherence between different policy areas and actors as crucial to make the fight against poverty effective.[472] By examining the first years of PGD implementation it is pointed out that the main coherence challenge of the PGD is not conflicting interests and objectives between different policy areas. Instead the biggest issue is formed by the question on how to identify and utilize potential synergy effects, and how to design actions that promote the realization of goals. Efforts to implement coherence are viewed to a large extent in terms of promoting knowledge of the development perspective (Government Offices of Sweden 2006, p. 10). After a change in 2006, the new Swedish government announced that the PGD has to improve its efficiency by a focus on fewer issues and a result based approach by establishing specific areas for stronger coordination, including minis-

471 Concerning this matter, the government bill also clearly illustrates the links between security and development cooperation and particularly emphasizes connections between poverty and armed conflict (Government Offices of Sweden 2003, p. 34).

472 Due to the lack of quantifiable targets in the 2003 document, it proved rather difficult to monitor achieved results or possible progress. As the 2003 PGD bill did not concretize how to implement the new policy and where the responsibility for the PGD lays, the first communications to parliament evaluating the PGD mainly dealt with organizational matters (Odén 2009, p. 3).

terial programmes to specify how coherence can be achieved and how the policy should be implemented (Odén 2009, pp. 4/5).

Thus, since 2006, the Swedish government has aimed at clarifying PGD features and at making them more transparent. In this regard, special emphasis is placed on an increased understanding for coherence in terms of not just aiming at avoiding that policies contradict each other but also to generate complementarities to ultimately benefit from each other's skills (Odén 2009, p. 14). The 2008 government communication *Global Challenges – Our Responsibility* (Government Offices of Sweden 2008a) further advanced the Swedish PGD conceptually. The 2008 communication aimed at laying the basis for improvement toward a more focused development policy that strengthens developmental effects created in other policy areas. Evaluating Swedish effort's to implement PGD since 2003 it claims that policy coherence implementation has been found to be incomplete in some respects. The communication again demands a results based management approach and results oriented initiatives that include "procedures and a work organisation that ensures coherence and cooperation across policy areas and between the Government Offices and relevant agencies" (Government Offices of Sweden 2008a, p. 55).[473] Though the MFA considers the 2008 communication "a re-launch and concretization of the Policy for Global Development" (Government Offices of Sweden 2011d, p. 10), the repeated call for coherence between different policy spheres is nothing conceptually innovative. While implementation objectives are more clearly defined in the 2008 communication (Government Offices of Sweden 2008a, pp. 52-57), numerous issues like management by result, monitoring and evaluations were already included in the 2003 document (Government Offices of Sweden 2003, p. 57).

In 2010 the Swedish government presented its most recent PGD communication *Meeting Global Challenges* (Government Offices of Sweden 2010a), which entails examples of results for each one of the 2008 identified six global challenges. The 2010 communication itself is considered an improvement regarding the reporting on PGD progress which was a weak aspect in previous communications (CONCORD 2011, p. 1). The communication re-emphasizes the need of coordination within the government, relevant government authorities and other actors and suggests the establishment of six inter-ministerial working groups for the implementation of the PGD in each of the six previously identified

473 In doing so, the PGD should focus on six global challenges: oppression, economic exclusion, climate change and environmental impact, migration flows, communicable diseases and other health threats and conflict and fragile situations (Government Offices of Sweden 2008a, pp. 5/6/9).

global challenges (Government Offices of Sweden 2010a, p. 88).[474] However, in spite of conceptual adaptations, PGD government communications continue to disregard a major CA aspect which is that an effective promotion of coherence needs "[s]trong political commitment to coherence on the part of Governments with leadership and clearly defined policy objectives, priorities and criteria for assessing progress" (Odén 2009, p. 9). In this context, the communications also only pay minor attention to institutional coordination, the Swedish institutional architecture and administrative culture to promote inter-ministerial cooperation in a systematic way. Analytical capacities to examine possibilities and resources to achieve the PGD objective also fall short in the communications. Without such analysis of resources and options, PGD coherence is difficult to accomplish (Odén 2009, pp. 9/10). A recent review by the OECD DAC, while also seeing improvements, criticizes the Swedish government's focus on the promotion of synergies between different policy areas, while disregarding how to deal with policy inconsistencies and thus calls for taking into account conflicting issues not only synergies and to further develop mechanisms to handle them (OECD 2011).

Sweden also unveiled a document named *Peace and Security for Development,* which clarifies and confirms fundamental principles and approaches of the country's engagement on development and security issues with the intention "to illustrate how foreign and security policy can contribute to promoting development goals, in accordance with Sweden's Policy for Global Development" (Government Offices of Sweden 2011d, p. 9). The policy particularly emphasizes the roles and responsibilities of Swedish actors and declares that it is essential to clarify objectives and limitations of the various policy areas and actors. In this context, exchange of information, consultation and close collaboration of instruments and actors is vital, according to the document. On the contrary, retaining the appropriate division of responsibilities is considered important (Government Offices of Sweden 2011d, pp. 14/28).

In addition to conceptual developments to improve policy coherence in the area of development, two other government documents are particularly relevant setting out Swedish policy practice and providing important prerequisites for crisis management coherence against demands for solving complex crisis situations. To enhance its crisis management engagement in an integrated way, Sweden adopted a *National Strategy for Swedish Participation in International Peace Support and Security Building Operations* (Government Offices of Sweden 2008b) in 2008. It re-emphasizes Swedish support for combined civil and military action and formulates numerous guidelines relevant for the development of a CA. In this regard, Swedish crisis management efforts should be developed

474 See the next chapter on the structural implementation of these working groups.

as an integral part of the overall policy toward a particular country or region. Nationally and internationally, complex crisis management of today requires the coordination of available instruments in both the planning and implementation phases, according to the document (Government Offices of Sweden 2008b, p. 16). By applying a long-term perspective, the strategy aims at providing support for combined civil and military action in areas of operations Sweden participates in. Furthermore, the strategy clearly recognizes that foreign, development, security and defence policy shall be linked more closely together in respective efforts and declares that only if the security-development nexus is realized, work on both crisis management and development can be successful (Government Offices of Sweden 2008b, pp. 1/2/5). The strategy presents a complete and comprehensive picture of CA aspirations and focuses Swedish efforts by reemphasizing important CA dimensions.

Secondly, the 2009 Afghanistan Strategy (Government Offices of Sweden 2010c) has been pointed out by interviewees as the most coherent document on the country strategy side which sets out clear ambitions for the Swedish contribution. As described by interviewees, all relevant Government Offices contributed to its development and the strategy covers the whole spectrum of the Swedish government's engagement from diplomatic and political efforts, military, police, development and humanitarian operations, development cooperation as well as the promotion of trade and migration issues. However, the military's role in Sweden's Afghanistan Strategy is rather limited and seen in a restrictive way (Grönberg 2010, pp. 28/29). The strategy underlines that the overall aim of the Swedish engagement in Afghanistan is to improve living conditions of the people, especially women and children, in a peaceful and democratic society by reducing poverty. While it is difficult to draw a conclusion regarding the effect of the strategy, it is noteworthy, that in the Afghanistan Strategy the Swedish government explicitly declares its willingness to apply a WoG approach when engaging in international crisis management. The strategy acknowledges that connecting different engagements in Afghanistan is a prerequisite for effective stabilization (Government Offices of Sweden 2010c, p. 18).

4.3.2.2 Organizational Dimension

Looking at the organizational dimension of Swedish attempts to improve crisis management coherence, it is necessary to put a comparably high focus on the wider Swedish political-administrative system which differs from that in other countries and which composition essentially influences Swedish CA efforts. The Swedish core executive is essentially built around the delegation of power (Levin 2009, p. 42). Sweden is a "politically decentralized and administratively deconcentrated country, with a remarkable degree of institutional continuity" (Wollmann 2003, p. 598). Its central administration is divided into Government Offices that are organized along functional lines, and government agencies.[475] The core executive is thus divided into the formulation of policies on the one hand and their implementation on the other. The decision-implementation is the responsibility of the government agencies while policy planning and articulation is the task of the Government Offices. Special-purpose agencies thus realize important central level executive functions, implement government decisions and do much of the work that in other countries are performed in ministries (Wollmann 2003, p. 598). Consequently, a lot of expertise and competence is located within the agency system (Larue 2006, p. 124). Each agency is subordinated to certain Government Offices that are relatively small in size compared to the number of officials employed in the agencies. Though being under the jurisdiction of a certain ministry, the agencies are responsible to the government as a collective and report to the whole government on measures taken and results achieved (Levin 2009, p. 41).[476]

This *administrative dualism* as the distinguishing characteristic of the Swedish core executive signifies a clear separation of the political ministries and the non-political agencies, whose independency and autonomy is protected by the constitution (Levin 2009, p. 42). The agencies are granted extensive freedom to manage and have considerable power to decide without Government Offices being involved. Yet, they do not have the legal status to formulate government policies. While the agencies are independent from single Government Offices they are not from the government as a whole which decides on the conditions

475 All Swedish ministries and the prime minister's office were united in 1997 into one single organisational form, officially called *Regeringskansliet*, Government Offices respectively in English. The objective of this reform was to reduce sector thinking and improve coordination as well as administrative efficiency (Hustedt/Tiessen 2006, p.39). On the history of government and central administration organisation in Sweden see Larsson 2002 (pp. 181-183). At the time of writing, Swedish Government Offices included the Office of the Prime Minister, eleven ministries, and the Office for Administrative Affairs. However, as one ministry may be divided between two or more ministers, the Swedish government at the time of writing consisted of 23 ministers and the Prime Minister. In contrast to the creation of agencies in other countries, the traditional Swedish separation of Government Offices and agencies has not been inspired by the agencification trend of NPM reforms. In Sweden, the agency model has a long tradition and the power of independent decision-making of agencies was repeatedly strengthened in 1986-1989 and again in 1991 (OECD 1997, p. 29). For a detailed description of the division of the Swedish administrative system into Government Offices and agencies see Larsson (2002, pp. 181-183).

476 The Prime Minister's Office along with the eleven Government Offices employs approximately 4500 officials (Government Offices of Sweden 2011c, p. 5). The 387 agencies on the contrary employ about 250000 civil servants (Arter 2008, p. 17).

under which the agencies have to operate by a set of government instructions. Annual appropriation directions establish both an economic framework for each agency as well as the objectives and focus of its activities. However, neither the government nor any individual minister is allowed to influence the way in which an agency deals with a specific matter.[477] The relationship between them can be described in terms of interdependence with the Government Offices depending on the semi-autonomous agencies' expertise and participation in the formulation and execution of policies and the agencies depending on the ministries' financial and political support (Larsson/Bäck 2008, p. 190).

From a coherence perspective, the Swedish system entails the risk that special-purpose agencies perform their tasks rather cut off from the ongoing processes at the government level and with little contact to other ministerial departments than their own. There has been persistent concern that the Government Offices lack the capacity to effectively supervise their agencies, in particular, as the agencies' relative autonomy results in fewer opportunities to influence policy-making. A mechanism established to advance coordination between respective agencies and also Government Offices is the creation of boards that link agencies in a broader political context and ensure a considerable degree of coordination (Wollmann 2003, p. 600). Furthermore, a great extent of agency management by the Government Offices involves regular dialogue on objectives and results between agencies and agency liaison officers at the ministry level (Government Offices of Sweden 2011c, p. 20). While varying in intensity, contact and information exchange between Government Offices and agencies are frequent (Larsson/Bäck 2008, p. 190).

The Swedish administrative dualism also has organizational implications on the way the country implements the CA as relevant agencies largely impact on the development and implementation of the Swedish international crisis management policy. Though not part of the ministerial structure, government agen-

477 It is considered a serious offense for a minister to act in what could look like ministerial rule manner. A Standing Committee of the Constitution in the Swedish parliament annually examines cases of suspected illegitimate interference in the work of the administrative agencies by government members. Yet, government ministers are allowed to exercise general control over the agencies under their jurisdiction; they are just not permitted to interfere in matters pertaining to individual cases or specific decisions (Levin 2009, p. 42). The agencies autonomy finds its limits in the fiscal instruments that can be used to govern administrative agencies (Levin 2009, p. 42). Additionally, the government can influence the agencies by appointing the personal to the high-level position of the agency (Larsson/Bäck 2008, p. 188). Nonetheless, "[w]hile there are thus several instruments available to Government ministers and their politically appointed staff to govern the general direction of the agencies' work, the average Swedish civil servant retains a degree autonomy that is unusual if not unique by international standards" (Levin 2009, p. 43).

cies are very active in participating in inter-ministerial working groups on conflict issues. However, though relevant agencies are in touch and exchange information, strictly speaking no joint and formalized analysis and planning between them exist. Without formalized structures among actors, coordination mainly occurs through informal groups and individual initiatives. In this regard, the divide between relevant agencies and the Government Offices is rather big and feeding back proves difficult in both directions and results in many discussions about responsibilities (OECD 2006).[478] As evaluated by a respondent, in particular in case of coordinating crisis management activities between relevant agencies, a high risk persists that they will adopt too narrow perspectives in accordance with their respective task and more guidance from the Government Offices in this regard would be useful. On the contrary, due to the guarantied independence of the agencies, giving them clear directions is problematic.

As the Government Offices are absorbed by day-to-day activities, the Swedish government frequently establishes *ad hoc* commissions or committees that include a mixture of ministers, civil servants and experts. Such a committee can be illustrated as a "public authority set up on a temporary basis to prepare the ground for a government policy decision. It prepares and presents proposals in its own name" (Government Offices of Sweden 2011c, p. 13). According to the Swedish government the benefit of an *ad hoc* committee is

> "that it can bring together the leading experts on a given policy matter for a limited period of time. This procedure also gives the political opposition and other public actors a chance to engage in the reform effort at an early stage" (Government Offices of Sweden 2011c, p. 13).

Structural reasons for the appointment of such pre-legislative commissions of inquiry are the small size of Government Offices and their lack of direct authority over policy implementation (Arter 2008, p. 153). However, Government Offices are responsible for bringing together and coordinating the work of the commissions. Though having less control over their resources, commissions have the legal standing of an agency and can be considered in terms of another functional extension of the core executive (Larsson 2002, p. 184). *Ad hoc* commissions are controlled by government directives and instructions that provide the framework in which it shall propose solutions. The government decides on the organization and composition of the commission as well as on its financial

---

478 Government Offices, though deciding on the general direction of agency activities, sometimes have the manner to refer to the agencies' autonomy in cases of policy failure (Larsson/Bäck 2008, p. 191).

resources. It also influences commissions in a more informal way by positioning own officials (Larsson/Bäck 2008, pp. 237/238).

The almost total absence of ministerial rule and the prevailing collegiality norm in the Swedish political system guide governmental decision making by demanding that "all decisions by the government are taken jointly by all ministers as a collective as opposed to the individual minister in charge of the relevant ministry" (Levin 2009, p. 41).[479] Policy-making in Sweden generally begins with debates inside the relevant ministry and other ministries are consulted after an opinion has been achieved inside the Government Office formally in charge. After submitting a proposal, all government agencies and organizations considered relevant have the opportunity to take part in the referral process and submit written comments. The coordination of inter-ministerial policy in form of this so-called *Joint Preparation Process* is an institutionalized process compulsory required by law on matters that involve more than one Government Office and seeks to create agreement across ministries (Patrick/Brown 2007a, p. 120). After agreement has been reached within the respective Government Office, the final proposals and drafts are circulated between all other ministries and the Office of the Prime Minister in a joint drafting procedure before they are addressed in Cabinet meetings (Larsson/Bäck 2008, pp. 238/239).[480] While the Joint Preparation Process takes place at an early stage of decision-making, the process of sharing is the last step before cabinet decision (Hustedt/Tiessen 2006, p. 41). Interviewees emphasized that single strategies are drafted either across Government Offices or within individual ministries strongly taking into account other ministry opinions. Also in the crisis management context, the Joint Preparation Process thus formalizes coordination processes "that need to be followed to ensure that a whole-of-government approach has been integrated into the policy decisions that lead to, and sustain, foreign interventions" (De Coning et al. 2009, p. 39).

This extensive policy-making process leads to the production of deeply and comprehensively documented arguments (Yates 2000, p. 154). With the mechanism of the Joint Preparation Process, Sweden has established a formal procedure at the governmental level to comprehensively coordinate issues before a decision is taken on a matter. The procedure in particular allows for an increased

479 There is no official English translation for the Swedish term *Ministerstyre*.

480 "Government decision-making is discharged on a collective basis. This means that, at their weekly meetings, the members of the Cabinet take joint decisions on all government business. Consequently, all the ministers have a say in government decisions and the Cabinet as a whole is collectively responsible for them" (Government Offices of Sweden 2011c, p. 5). The process continues until the government has reached a decision and the proposal proceeds on to the Parliament (Larsson/Bäck 2008, p. 239). For a detailed description of the sequences of the Joint Preparation Process regarding the PGD see Patrick and Brown (2007a, p. 121).

information exchange and discussion between Government offices and their ministers and officials. While one ministry is responsible, all relevant actors are involved in the policy-making. Inter-ministerial meetings that take place as part of the process aim at bringing out latent tensions, promoting a balanced and coherent agreement as well as clear division of responsibilities. The formal absence of ministerial rule further reinforces coordination and collective responsibility as ministers have to gain support of other government members for their legislative proposals (Mazey 2001, p. 261).

To deal with crisis management issues, networks exist in form of strong mechanisms of horizontal coordination with the involvement of all relevant Government Offices. However, mechanisms of vertical coordination with the respective agencies are relatively weak. As shown, promoting inter-ministerial coordination in such networks led to the development of informal arrangements on an *ad hoc* basis, some of which were then institutionalized on a more formal basis such as the regular lunch-meetings. This *ad hoc* coordination in Swedish policy-making plays an important role as a mechanism of conflict avoidance from the very beginning (Arter 2008, p. 153). At the same time, given the collective responsibility of the Swedish government, the comparably small number of formal structures of policy coordination places all the more demand on officials to coordinate informally.

"In many ways the Swedish administrative system can best be described in terms of informality and mutual interdependency instead of hierarchies and control" (Larsson 2002, p. 207). Being a key characteristic of this, the country's administrative structure is characterized by a high degree of openness and accountability as well as a "spirit of consensus" (Levin 2009, p. 39), as outlined above. In addition, the feature of agencification attaches much importance to professional and expert considerations. The Swedish consensus oriented administrative culture and the principle of collective responsibility structurally reflect in a strong tradition of both vertical and horizontal coordination with an emphasis on informal contacts cutting across different levels of administration. In addition to the formal Cabinet meetings, the actual negotiations within the government are held during informal sessions of Cabinet that usually follow a Cabinet meeting. These are complemented by other forms of collective informal negotiations such as daily, private lunches of Cabinet members that ministers usually have together every working day and where a lot of issues of common interest are freely debated and decided on (Larsson/Bäck 2008, pp. 178/179). The strong tendency toward consensus-oriented policy making and focus on expertise facilitates inter-ministerial coordination, and the search for pragmatic solutions and compromise politics (Arter 2008, p. 6). Recognizing each other's roles, Government Offices concerned with international crisis management pragmatically include other ministires and relevant actors in debates on common issues, as evaluated by respondents.

On the downside, as an interviewee pointed out, also if the relations between the Government Offices and the respective agency are good, this division between policy-making and technical expertise can be problematic, in particular with high staff rotations. Also, agreements that follow from consensus-oriented compromises are often "based on elastic formulations [resulting in, S.P.] long unprioritized wish-lists or 'catalogues' of good intentions that are open to interpretations" (Christensen et al. 2007, p. 151). This observation also applies to Swedish attempts to implement the CA. With no clear government guideline on how to coordinate, political direction on the CA is missing and in particular agencies demand more direction, clear goals what to achieve and how deal with CA demands. It was underlined by a respondent that the Swedish consultative structures and consensus oriented policy style has the downside of being rather time-consuming, cumbersome and an ineffective routine which can lead to frustration which in turn is not favourable for the CA as an idea itself. Also, the fact that expertise is not located within Government Offices but in the agencies makes the procedures to use this expertise lengthy.

Several governmental authorities are responsible in the Swedish crisis management organization. Sweden does not have a separate ministry responsible for development policy. It is the MFA, the MoD and the Ministry of Justice who are the main crisis management actors. While there is no single actor to position government-wide agendas or direct other actors, the MFA traditionally has a superior position and is largely responsible for the inter-ministerial coordination in international crisis management.[481] Horizontal inter-ministerial coordination thus needs to be seen in the light of the MFA's relationship with other relevant Government Offices (Johansson 2003, p. 376). In doing so, the MFA is described as a highly valuable information gathering body and source for other Government Offices by interviewees. Points of contact have increased enormously between actors, particularly between the MFA and the MoD and inter-ministerial coordination takes place on all levels of government, from ministerial to the desk officer level (Patrick/Brown 2007a, p. 119). In this context, the under-secretary of state is responsible for planning and establishing the coordination between activities of the different ministerial units (Yates 2000, pp. 154/155).

With big cuts in the defence budget, Sweden adopted the concept of a small and modern force for international deployment and does not have a territorial defence anymore. The country emphasizes the development of its armed forces to play non-traditional military roles with the objective "to contribute to stability in

481 The MFA, for example, has the privilege of formulating the Swedish non-alignment policy. With the decrease of this policy, however, meaning and importance of this privilege changed as well (Johansson 2003, p. 376).

the surrounding world" (Government Offices of Sweden 2009a). The military forces' restructuring took place to organize them to profoundly support crisis management tasks, including confidence building, conflict prevention and humanitarian engagement. Swedish military strategies are based on the increased deployment with civilian crisis management teams (Patrick/Brown 2007a, p. 124). In addition to the Swedish MoD's broad transformation to achieve a more functional and flexible defence, it has also aimed toward the development of close cooperation to facilitate inter-ministerial relations. According to a respondent, the ministry has acknowledged that its part in the new security environment becomes merely instrumental. The MoD particularly focused on the issue of civil-military coordination and established several commissions to analyze and submit proposals concerning a more integrated approach (Rieker 2002, p. 37). Since 1999, civil-military relations were not only emphasized in MoD strategies, the ministry also focused on practically advancing joint civil-military training activities (Rieker 2006b, p. 84). On MoD initiative the Swedish government established a standing forum on civil-military coordination under MFA lead. As noted by respondents, in Sweden, terminological differences between the military and civilian side created problems. However, over the time both sides became increasingly aware about the differences. In 2002, the MoD published a CIMIC-Handbook and at the time of writing, the SAF were in the process of creating a glossary of key terms that has been asked for from the civil side.[482]

As Sweden does not have a separate development ministry, departments responsible for development policy are located within the MFA, guided by the Minister for International Development Cooperation.[483] Most coordination of developmental aspects of Swedish crisis management processes thus takes place at the Department for Development Cooperation of the MFA, with the MFA taking the lead in shaping the overall developments strategies. The department is responsible to ensure that all Government Offices contribute to the formulation of development objectives. Coordination regarding the Swedish PGD takes place among the MFA, the MoD, the Ministry of Justice as well as relevant government agencies, which can contribute to formulating development goals and with a view to jointly generating strategies for countries that Sweden has a specific aid relationship with (De

482 In addition to the SAF, the *Swedish Civil Contingencies Agency* (MSB) is an agency of importance when it comes to international crisis management. Under the direction of the Ministry of Defence, the MSB is tasked to coordinate non-military actors operating in crises and emergencies in Sweden and abroad (Grönberg 2010, pp. 26/27). MSB is thus the agency responsible for Sweden's involvement in international humanitarian aid.

483 As diplomacy, foreign trade and development are all subsumed in the MfA, the Government Office was headed by the Minister for Foreign Affairs, the Minister for International Development Cooperation and the Minister for Trade at the time of writing.

Coning et al. 2009, p. 15). The *Swedish International Development Cooperation Agency* (Sida) is the government agency of the MFA which is responsible for most of Sweden's ODA to developing countries. By recognizing that 75 percent of the agency's programme countries are affected by violent conflict, Sida "begun to explore the links between security and development, including the potential of foreign aid to help prevent conflict and address the root causes of instability" (Patrick/Brown 2007a, p. 118). In this regard, Sida's *Shared Responsibility Unit* and the *Division for Peace and Security* are responsible for the mainstreaming of conflict issues into development assistance and the implementation of conflict management operations (Tannous 2008, p. 36). From structural comprehensive crisis management perspective, not having a separate ministry for development did not result in a disregard of the development dimension by other Government Offices. Quite contrarily, MFA country strategies are often heavily coined by Sida (Patrick/Brown 2007a, p. 120). In addition, not having a development ministry on the cabinet level brought senior development officials into the corridor of the MFA. As an OECD study found, Swedish cooperation strategies in the PGD framework led to improved inter-ministerial dialogue and joint planning (OECD 2006, p. 28). However, the working groups presented as the main instrument for PGD implementation in the 2010 PGD communication have not been active by the end of 2011. Thus, the intra-ministerial Joint Preparation Process remains the main institutionalized mechanism for inter-ministerial coordination. Another aspect of structural PGD implementation is a focal point on the policy in each Government Office that collaborates with relevant counterparts in the MFA. In this regard, "[t]he limited knowledge about the PGD and its implications within government offices outside of the Ministry of Foreign Affairs continues to be an impediment to effective implementation" (CONCORD 2011, p. 2).

According to background interviews, secondment of staff between different Government Offices and agencies frequently takes place in Sweden and contributes to the very strong informal networks that stretch across different institutions and which are considered very important in the crisis management arena. The overall organization of the Swedish core executive and the merging of ministries into Government Offices have facilitated the exchange of officials. The biggest flow of staff in this regard exists between the MFA and the MoD. Regular contact and exchange via telephone and e-mail between relevant officials and their counterparts in other Government Offices exist. It was emphasized by respondents that personal contacts rather than formal structures constitute the backbone of inter-ministerial coordination and cooperation in Sweden. Only very few formalized and ritualized mechanisms and formal structures ensuring crisis management coherence exist.

However, the Swedish government has also taken some specific decisions that aim at advancing CA implementation and several steps have been taken to strengthen coordination for instance by establishing weekly meetings of the MFA, the MoD and the Ministry of Justice (Rieker 2006b, p. 85). As part of the CA, the range of actors involved in debates on operationalizing crisis management has been broadened inter-ministerially, with relevant Government Offices present depending on the subject concerned (OECD 2006, p. 33). As mentioned above, usually, the MFA is having the lead in inter-ministerial working groups, but responsibility can also be designated to other ministries. Several *ad hoc* commissions were established by different ministries to analyze and suggest improvements for civil-military cooperation in crisis management. However, also such commissions form "deliberative rather than an official policy-making body" (Bailes 2009, p. 20) and a forum for consultations on defence and security issues.[484] As pointed out by an interviewee, also in the crisis management context, inter-ministerial working groups are set-up in an *ad hoc* matter depending on the issue. Furthermore, specialized consultation groups on project basis are created in an event driven manner. Especially since the Swedish engagement in Afghanistan, such coordination mechanisms as well as coordinating meetings increased on all levels between the MoD, the MFA and the Ministry of Justice.

Also the creation of the *Folke Bernadotte Academy* (FBA) presents an important example of CA awareness and its organizational implementation in the country's politico-administrative structures. The main objective that led to the creation of the FBA was to improve quality and effectiveness of Swedish international crisis management. Based on commission reports, the agency has been established under the authority of the MFA as a civilian-military cooperation centre for the recruitment and training of personnel for international crisis management operations (Rieker 2006b, p. 85). In doing so, the FBA functions as a platform between the ministries and government agencies. It also serves as a national point of contact with IOs and partners regarding crisis management engagements. As personnel for international operations is recruited by several government agencies,[485] the FBA also acts as a coordinating recruitment agency that provides a "coherent overview of the combined national civil expertise and

484 The Swedish Defence Commission, for instance, is not permanently established but periodically reappointed to produce reports on specific questions. Still, its reports play a decisive role in drafting and formulating Swedish security policy (Grönvall/Lundberg 2005, p. 18), in highlighting the broadened security agenda and also to advocate intra-governmental coordination links (Bailes 2009, p. 20). In the instrument of the Defence Commission, civilian and military state servants work together with parliamentarians and independent experts (Bailes 2009, p. 22). The role of commissions also shows in chapter 4.3.2.1 on the conceptual dimension the Swedish CA by the inclusion of various commission reports.

485 Other government agencies for recruiting civilian experts for international operations are the National Police Board, the Swedish Prosecution Authority, the National Courts Administration, the Swedish Prison and Probation Service, MSB and Sida.

categories of personnel the authority concerned can contribute, and already contributes, to ongoing operations" (Government Offices of Sweden 2008, p. 20). Apart from the recruitment of Swedish civilian personnel to international peace operations and its multifunctional education and training, the national cooperation and coordination of Swedish crisis management is the FBA's main area of responsibility. In doing so, the FBA carries out the training of both civilian and military crisis management personnel. Together with the SAF and other agencies it identifies best practices and lessons learned in the field of crisis management by closely cooperating with the SAF and the Swedish Armed Forces International Center (SWEDINT) (Rieker 2006b, p. 85).

As stated by interviewees, in terms of crisis management financial resources, Sweden has made no effort to create pooled funds to stimulate interministerial coordination, though it continues looking at the UK mechanism of pooled funding for inspiration. In clearly dividing governmental and departmental responsibility in the Swedish political system, Government Offices are given budget guidance once a year. However, as crisis management situations can occur rather short-term,

> "the MFA is frequently frustrated by its inability to obtain flexible funds for critical purposes – such as promoting political reconciliation or security sector reform – since the vast majority of resources are already tied up in Sida's longer-term, programmatic work aimed specifically at poverty reduction" (Patrick/Brown 2007a, pp. 122/123).

Even in a crisis situation, the MFA is not authorized to shift or relocate Sida's flexible contingency funds (Patrick/Brown 2007a, p. 155), which is a rather important aspect when considering the Swedish crisis management funding situation.

4.3.3 European Dimension and EU-Interface in Swedish Crisis Management

Swedish crisis management policy is based on a strong focus on multilateral partnerships. The UN has historically been the primary pillar of Swedish international engagement and the most important IO fulfilling Swedish peace and security objectives and traditions. Though the UN continues to hold the legitimizing responsibility on international security matters in Sweden, today, the EU presents the player with at least the same importance in the Swedish foreign policy arena.[486] According to background interviews, the EU is the international actor

486 Sweden maintains its principle UN policy, namely that Swedish participation in international military operation is considered a conceivable option under UN mandate only. However, scholars observe "an increasing willingness to deviate from this position in the case of humanitarian crisis situations" (Wunderlich 2010, p. 33). Sweden's UN policy thus "can be overruled by moral concerns in exceptional situations of humanitarian emergency" (Wunderlich 2010, p. 33).

Sweden seeks to work through and provides the main platform for Swedish foreign policy. The Union transformed "into the focal point for Swedish security and crisis management efforts" (Grönvall/Lundberg 2005, p. 29). As declared by respondents, it is Sweden's utmost strategic priority to strengthen the EU in this context.[487] EU foreign and security policy from the Swedish perspective as a small to medium sized member state more and more substitutes Swedish national policy, in particular in the context of budget cuts, new global challenges and in the field of defence cooperation.[488]

Yet, the development of the EU's crisis management capacity has presented Sweden with some challenges to its security policy traditions and own crisis management approach since joining the EU in 1995. An obstacle to EU membership on the part of Swedish policy perceptions was posed by the role of Nordic exceptionalism. From Swedish perspective, joining the EU clearly entailed the risk of ending Nordic exceptionalism (Lawler 1997, p. 582). At the same time, EU membership was seen as the only way to come to terms with global political and economic realities "whilst preserving, and hopefully transferring upwards, some essential features of exceptionalism" (Lawler 1997, p. 566). Almost two decades after joining the EU, Nordic exceptionalism can be considered to be increasingly undermined by the European model. Nordic crisis management conceptions are dominated by EU structures and a Nordic profile is hard to preserve. However, this does not imply that Sweden changed its international behavior, but rather questions whether it identifies certain values, attitudes and actions as European, rather than Nordic (Browning 2007, p. 28). Despite some sustained Nordic exceptionalism, Sweden increasingly considers the EU as the key outlet for a specific Swedish foreign policy and crisis management preferences. As assessed by an interviewee, though EU membership might have reduced Sweden's international visibility in terms of a Nordic presence, the EU is seen as a more effective way to accomplish substantial and concrete results.

487 At the same time, the strengthening of EU crisis management capacities is considered as a way to also reinforce UN capabilities and to promote EU-UN cooperation (Eriksson/Foghelin 2002, p. 94). As it had been pointed out during a background interview, EU engagement on the one hand does not exclude UN activities on the other.

488 This is also visible in government statements about the importance of the European crisis management policy area. As stated by the Swedish Minister for Foreign Affairs at the time of writing Carl Bildt, "European Foreign Policy is our Foreign Policy" (Government Offices of Sweden 2011, p. 1). Against the background of Sweden generally being a country with a comparatively high degree of EU-skepticism, the high levels of support with regard to EU foreign and security policy are particularly noteworthy (Bailes 2006, pp. 16/17). However, some scholars point to a prevailing gap between the population and the political elites. "The Swedish state has been Europeanized, while Swedish society has not" (Rieker 2006b, p. 75). See also Herolf (2007, p. 57) and Johansson (1999) on this matter. On the particular case of EU-skepticism in the Scandinavian countries see Arter (2008, pp. 323-332).

The most difficult factor, in particular during the first years of Swedish EU accession, was posed by the country's non-alignment policy which was considered incompatible with emerging EU crisis management efforts. At the time of Swedish accession to the EU, the then future security and defence dimension of the Union was problematic for Sweden and viewed rather skeptical. But also from EU side it was suspected that for Sweden as a non-aligned country it would be impossible to play an important role in EU crisis management and that the country would lack a commitment for the EU's CFSP from the beginning (Lee-Ohlsson 2009, p. 124).[489] The Swedish government's domestic argumentation in this regard was to underplay the significance of a future EU security policy, pointing toward its immaturity and to concentrate arguments for EU-accession on socio-economic issues (Rieker 2006b, p. 69). It also argued that due to the end of the military bloc confrontation in Europe, neutrality was not a barrier to Swedish EU-membership anymore.[490] Yet, the advancing CFSP made it increasingly difficult for Sweden to maintain its non-alignment policy, which on the one hand, had to be compatible with the CFSP and, on the other hand, if doing so, could lose its political effect with the whole Swedish security policy being perceived as unreliable and unpredictable (Miles 2000, p. 182). At the time of accession, the reformulation of the policy and the specification of non-alignment in military terms, "paved the way for Swedish membership in a multilateral framework not classified as a military alliance" (Rieker 2006b, p. 70). Sweden adjusted its non-alignment policy in order to bring it in line with EU membership. These modifications allowed Sweden to clearly express intentions to strengthen EU crisis management capabilities and to support the EU's goal to take greater responsibility at the global level, while the country at the same time was still able to stress the maintenance of its non-alignment policy (Rieker 2006b, p. 77).

489 The issue and consequences of accusations of Sweden not being fully committed to the objective of a common EU security policy is elaborated in detail in Ojanen (2000, pp. 1-3; 2002, pp. 155-157/172). Swedish non-alignment with regard to the developments of EU security and defence policy has been closely examined in numerous studies. See Bjurner (2003) and Wunderlich (2010). Several relevant articles have been published in an edition of *Cooperation and Conflict* (2011, Vol. 46, No. 3).

490 Swedish neutrality did in fact not dominate the debate about the country's EU accession. Instead, arguments related to sovereignty and to the issue of Sweden still being able to take its own decisions, to old Swedish ways against new Brussels dominance, did (Huldt 2003, p. 47). Rather than foreign policy and security reasons, the Swedish application for EU accession was justified by essentially economic and domestic factors that were at the forefront of the debate and economic necessity appears to have been the decisive factor of the Swedish EU accession ensuring the positive outcome of the referendum (Arter 2008, p. 322). On this issue see also Eriksson and Foghelin (2002, p. 88) and Miles (2000, p. 186). See Eriksson (2006, pp. 79-85) for details on the Swedish EU relationship before accession.

Additionally, in parallel with the continuing official declaration describing Sweden as militarily non-aligned, a declaration of solidarity has been formulated in the 2004 Parliamentary Defence Commission report which has been repeated since then in slightly different formulations. As part of its new functional defence doctrine, the government restates

> "that the security of our country is founded on community and cooperation with other countries. Sweden will not remain passive if another EU Member state or other Nordic country suffers a disaster or an attack. We expect these countries to act in the same way if Sweden were affected. We must be able to give and receive military support" (Government Offices of Sweden 2009a).

Solidarity in this regard, however, is interpreted in a strict political sense (Wunderlich 2010, p. 10), and the Swedish government, in line with its overall approach to engagement in international conflicts, clearly distinguishes crisis management from collective or common defence understood as territorial defence and mutual defence guarantees (Lee-Ohlsson 2009, pp. 127/128). Nonetheless, this illustrates Swedish reasoning that speaking about security and defence in the EU, actually means crisis management, not collective defence (Rieker 2006b, p. 74). This also allows Sweden to participate in the Lisbon Treaty's Solidarity Clause, which commits EU member states to collectively engage in response to emerging attack or disaster (Myrdal/Rhinard 2010, pp. 5-7). Furthermore, the Mutual Defence Clause on cooperation in case of an armed aggression accommodates the Swedish non-alignment policy and allows the protection of the Swedish traditional status by stating that the clause will not impact security and defence policy characteristics of individual member states. In sum, while the country holds on to non-alignment, the concept was stretched to great extents to assure Sweden's loyalty and reliability in the context of EU crisis management. In the wake of continued European integration, Swedish security traditions as outlined above were reviewed with the objective to bring them into line with the actual circumstances (Wunderlich 2010, p. 7).[491]

Overall, Sweden has been very outspoken on the strategic instrumentality of the EU as the government felt pressed "to relate the benefit of any given institution directly to the nation's own needs" (Bailes 2009, p. 30). As argued in con-

491 Some observers view Lisbon Treaty principles to be incompatible or competing with Swedish non-alignment policy. This argumentation particularly applies to innovations such as the Permanent Structured Cooperation on defence issues, the extension of the Petersberg Tasks, and the provision of combat units to be deployed outside of EU borders to undertake unlimited EU military action. It is argued that extending Swedish principles in such ways brings them close to meaninglessness and irrelevance (Devine 2011, pp. 355/360).

trast to Swedish NATO and UN engagement, in the EU, Sweden is directly involved in policy-making from the idea to the implementation.[492] Sweden's possibilities on the UN level are much more limited than in the EU-framework in which it can take up initiatives more easily. In particular with regard to strategic crisis management development, it can be argued that the lack of official guiding strategies in Sweden contributed to the development of Sweden leaning heavily on EU policy and echoing EU-strategies (Patrick/Brown 2007a, pp. 118/119).

While these developments are reflected in the above statement that Swedish foreign policy today is more the EU's than its own,[493] which has also been pointed out during background interviews, EU membership has not just one-sidedly influenced Swedish policy approaches. The country also informed the development of EU crisis management policy. As Swedish non-alignment policy is one of active participation, the country strove to impact on the creation of an European security architecture. An underlying rationale in this regard is that these efforts ultimately promote Swedish interests and strengthens its influence on international issues (Bjurner 2003, pp. 43/44). The Europeanization of Swedish crisis management is, though being heavily led by its foreign policy values, in big parts also driven by Sweden's objective to influence policy. It considers EU crisis management as a chance to consolidate the country's EU credentials and views the area of foreign policy as an opportunity to further the overall goal of internationalism and for the EU to play an active role in human rights and peacekeeping questions (Miles 2000, pp. 187/196). Respondents pointed out that emphasizing its influence and viewing the EU as an opportunity to promote own interests are directly linked to the Swedish eagerness for the EU to be strong.

In this regard, Sweden promoted several initiatives concerning the development of EU crisis management. As a non-aligned country, it was particularly concerned to avoid the development of European security policy toward a collective defence system and thus considered ESDP initially to be too one-sided on the military dimension. Sweden could only accept a defence element to the EU if it was combined with a credible non-military crisis response (Drent/Zandee 2010, p. 9). Therefore, together with Finland, Sweden initiated the incorporation of the Petersberg Tasks in the CFSP remit in the 1997 Amsterdam Treaty and

---

492 Though Sweden also increasingly engaged in NATO as of 1994, scholars argue that policy changes have to be seen in relation to the European integration process and the development of the EU's foreign and security policy as the timing of decisions indicates this and also the Swedish political leadership "explicitly referred to the ESDP process as the main reason for the changes" (Rieker 2006b, p. 76). This is also advocated by Wenger et al. (2006, pp. 110-112) and Lee-Ohlsson (2008, p. 18).

493 Ojanen (2002, p. 155) observed this development in 2002 already.

emphasized the importance of civilian crisis management.[494] The successful enhancement of the EU's civilian crisis management role and capabilities had the intention to demilitarize the EU's security dimension and was sensed a major diplomatic success. In doing so, EU crisis management became to be viewed comparable with UN peacekeeping and thus an activity Sweden has been traditionally active in (Ojanen 2000, p. 7). During its 2001 EU presidency, the Swedish government continued to be particularly active on non-military crisis management aspects and to put great emphasis on the civilian dimension of ESDP, while for the first time exploring a more proactive role of civilian instruments.[495] The adoption of the *EU Programme for the Prevention of Violent Conflicts* by the Göteborg Council in June 2001 was a major success in this regard (Rieker 2006b, p. 75). Furthermore, EU-UN dialogue was strengthened during the Gothenburg Council and CIVCOM was established to be responsible for inter-pillar coordination in the field of civilian crisis management.[496] During the Gothenburg Council "[t]here was an awareness, (…) that the aim was not so much to develop new capabilities, but rather to guarantee that the EU had more of a coherent and coordinated approach to conflict prevention" (Jonson 2006, p. 127). Reacting to EU military crisis management developments, Sweden aimed at balancing ESDP developments by proposing non-military counterparts such as the Civilian Headline Goals to the military Helsinki Headline Goal or later on the Civilian Response Teams as a counterpart to the concept of the EUBGs 2008 in order to counter respective developments on the military side (Drent/Zandee 2010, pp. 9/10). Also, with regard to the 2003 ESS the country has been highly cautious that the multifunctional coordination of European crisis management instruments does not result in subordinating the EU's civilian crisis management policy under the military rationale (Bailes 2006, pp. 19/20). Sweden strongly supported the establishment of the ESS and viewed it positively as an added value to the CFSP and contribution to fostering a common European strategic culture (Bjurner 2003, pp. 44). According to interviewees, Sweden also sought quite a prominent role in the 2008 process of reviewing the ESS.

With respect to institutional-structural questions of the CFSP set-up, Sweden has tried to avoid the establishment of purely military bodies, underlining the need to have both civil and military expertise. The country particularly emphasized the necessity for the EU to have comprehensive crisis management capabilities, proposed the development of consensus building mechanisms and the enhancement of

494 On the joint Swedish-Finish Petersburg initiative see Ojanen (2000, pp. 6-9).

495 Other priorities of the 2001 presidency laid on employment, enlargement and environment issues, the so called 3 E's (Miles 2010b, p. 81). For an overview of the Swedish 2001 EU presidency see Jonson (2006, pp. 122-128, 205-209).

496 CIVCOM can be considered a Swedish invention, according to an interviewee.

opportunities for civil-military cooperation (Ojanen 2002, pp. 14/169). In parallel with the institutional development of the EU's military dimension with the set-up of the EUMC and the EUMS, Sweden became eager to avoid the 'militarization' of EU crisis management mechanisms and structure. To this effect, the country also proposed the founding of the Civilian Crisis Management Committee. While it was also regarded "as being easier to sell the ESDP politically if it was not too military and thus not seen as calling into question the policy of non-alignment" (Lee-Ohlsson 2009, p. 128), the promotion of non-military aspects also resulted from a belief that a comprehensive security approach is necessary. From Swedish perspective, the EU is a suitable actor to encourage the civilian crisis management dimension, since the deployment of both civilian and military means is its actual strength (Lee-Ohlsson 2009, p. 128). The Swedish attempts met with success as the Helsinki Council Conclusions explicitly recognized non-military crisis management aspects (European Council 1999b, II/28). Sweden has also been at the forefront within the EU to create a joint and integrated civil-military planning capacity for example with an initiative approved by the EU in 2005 under the rubric of Effects-Based Approaches to Operations (Patrick/Brown 2007a, p. 124). Civil-military coordination as part of the CA idea thus is the major issue Sweden promoted on EU level. The success of the Swedish argument that EU crisis management has to be comprehensive "is also visible in that it quickly became the hallmark of the 'EU approach' to crisis management" (Jakobsen 2009, p. 92).

During the Swedish 2009 EU presidency, the country again focused on the aspects renowned for and declared that its policy priority is to "continue to work to strengthen the EU as a global actor with a clear agenda for peace, development, democracy and human rights" (Regeringen 2009, p. 9). Also, civil-military cooperation as a traditional Swedish priority was highlighted. In this regard, Sweden considered the EU's unique potential for comprehensive crisis management as not sufficiently explored and called for more civil-military cooperation (Sundberg/Nilsson 2009, p. 4). Another 2009 priority was the ratification of the Lisbon Treaty which the Swedish presidency successfully contributed to. Generally, the impact of the Lisbon Treaty on both Swedish and the EU's crisis management is evaluated highly positive by respondents. As stated, Sweden views the Lisbon Treaty and the latest developments of the EU's foreign and security policy affirmatively and has great hopes for its implementation. The Swedish government is in particular positive to the fact that the new treaty institutionalizes and strengthens the collective responsibility of EU member states for Europe's security. In this regard, the country initially supported a stronger mandate for the HR/VP to act on common grounds. Sweden is a strong supporter of the treaty's institutional innovations, in particular the EEAS which is viewed as an appropriate instrument to increase crisis management coherence and also transparency. The EEAS is seen as

> "a key tool for achieving an effective and coherent European foreign policy. Sweden will continue to contribute to the development of the External Action Service with a view to pursuing strategically important issues on the global arena" (Government Offices of Sweden 2011a, p. 2).

Consequently, the country sought to proactively insert Swedish points of view into the EEAS decision-making machinery and to ensure adequate staff representation. So far, however, though the EEAS is seen as a valuable step in the right direction, it does not fulfill Swedish expectations, according to background talks. The Lisbon Treaty developments are believed to provide for an inevitable momentum toward a unified EU foreign policy on all counts. However, according to interviewees, Sweden would have favored a closer integration of EU crisis management decision-making competencies and improved coordination also on the vertical level between the EU and member states. At the same time, Sweden came under criticism with regard to negotiations on the candidates for the Council President and the HR/VP during its presidential term for not being more up-front at ensuring the involvement of higher profile appointees (Miles 2010b, pp. 84/88-91). In sum, hosting the EU presidencies forced the whole government to think in EU terms and to manage EU issues on all levels of administration (Lee-Ohlsson 2009, p. 133). On the contrary, the EU presidencies allowed Sweden to prioritize a set of issues and orchestrate collective support for them. Work on EU issues has hence also become an extension of domestic policy-making and Government Offices developed their own bilateral relations with Brussels (Mazey 2001, p. 262).

However, also the other way around, the EU-agenda of the Swedish Ministry for Foreign Affairs circulated back into the Swedish MoD and resulted in a heightened focus on comprehensive security and led to a greater interest in civil-military cooperation and, to an explicit ambition to base national policy on a wide security concept (Rieker 2006c, p. 310). While the Swedish support in the area of coherence for civilian crisis management and global development comes as no surprise, the increasing Swedish engagement in the ESDP area is all the more remarkable, as Sweden was highly skeptical of its creation in the late 1990s, and even considered to block it (Lee-Ohlsson 2009, p. 124). Widely perceived as a success, ESDP became more uncontroversial in Sweden.[497] The importance attached to the EU's crisis management ability by the Swedish govern-

497 In this regard, the 2003 EU military Operation Artemis to the Democratic Republic of Congo with a strong Swedish contribution marked a turning point that led to a significant shift in attitude towards ESDP and the military dimension of the EU. Three aspects contributed to the strong Swedish support for Operation Artemis: the UN as Sweden's traditional foreign and security focus was asking the EU for help; it was autonomous of NATO and the Berlin plus arrangements, and Africa presented traditionally a Swedish priority (Lee-Ohlsson 2009, p. 130).

ment also shows in the Swedish participation in EU crisis management activities. Not only has Sweden contributed to all EU operations and missions, both civilian and military, the country also provided relatively high number contributions.[498] Sweden began "to realize that they (…) have at hand a greater array of instruments as a result of being part of a larger actor" (Strömvik 2006, p. 204).

In the context of Europeanization on the Swedish national level, commitments made by the Swedish government to ESDP have been utilized to introduce controversial defence reforms. The reform of the SAF was closely linked to the development of EU crisis management capabilities, to a large part took place as a result of ESDP developments and clearly shows EU impact (Lee-Ohlsson 2009, p. 134).[499] A clear sign of how the Swedish attitude toward a comprehensive crisis management approach changed became apparent in 2004, when Sweden declared that it would not only participate but actually act as a framework nation leading one of the EUBGs. The proactive stance to build up a battle group underlines the Swedish "willingness to contribute to an active and comprehensive EU external security policy" (Rieker 2006c, p. 311). In doing so, the reform of the SAF was directed toward achieving operational units in the context of EU crisis management capabilities. Sweden fully adopted the EUBG concept and the *Nordic Battle Group* (NBG) has been on standby since spring of 2008 (Government Offices of Sweden 2007).[500] As stated by interviewees, leading a EUBG also presented an opportunity for Sweden to both contribute to the development of a military capacity for international crisis management operations and to improve cooperation and integration in the military area. The EUBG was seen as a window of opportunity to push for crucial defence reforms using the EUBG concept and the ESS to realize politically difficult domestic reforms under cover of the EU prerogative. It was a tool, positive vehicle and catalyst to transform the SAF. An interviewee also assessed the creation of SWEDINT as a major step of the SAF to improve their capabilities to coordinate and work in an international alliance. The then minister of defence explicitly stressed that "[s]imilar reforms

498 See SIPRI Multilateral Peace Operation Database. For a detailed analysis of the willingness of EU member states to contribute troops to EU military operations see Soder (2010).

499 This argument is supported by the development of the Swedish defence industry, which no longer upholds the technological independence necessary for the credibility of neutrality but opened to European cooperation and competitiveness (Rieker 2006b, p. 78).

500 Sweden contributes the share of 2300 personnel out of 2700 personnel to the Nordic Battle Group (Government Offices of Sweden 2007). The other NBG participating states are Finland, Norway, Estonia and Ireland. Although Sweden preferred a national battle group, policymakers wanted to share risks and costs of putting up a battle group, while also easing the public debate and to build consensus. In the end, the NBG's multi-nationality became a political objective of the concept itself as a way to promote integration and to provide more member states with an opportunity to participate in ESDP (Norheim-Martinsen 2009, p. 14).

have been implemented in other European countries. (…) This is a reform that follows the development in Europe and which will give Sweden a modern defence for the future" (Björn von Sydow, as cited in Rieker 2006 b, pp. 79/80). Also, the establishment of the FBA resulted partly in response to the EU efforts to improve its crisis management capacity (Rieker 2006b, p. 85). While the participation in the NBG influenced the SAF transformation, the implementation of the EUBG concept also implied that Sweden took on an active role in discussion at EU level. This was not considered as a critical political issue as long as the battle group's focus has remained on peacekeeping tasks (Lee-Ohlsson 2009, pp. 131/132/135).[501]

Also, within the Swedish PGD as the crisis management area coherence is supported strongest in, coherence explicitly not only applies to the national level but also to the international level of the EU. In line with its domestic coherence focus, Sweden aims at the formulation of a more coherent EU policy framework in the development policy field (Patrick/Brown 2007a, p. 116). The PGD explicitly states that development cooperation has to be improved through multilateral coordination particularly with the EU. "Enhanced collaboration and coordination are also required between various actors if a coherent policy pursued by e.g. Sweden and the EU is to succeed in promoting development" (Government Offices of Sweden 2003, p. 18). In its PGD implementation, Sweden also deals with cooperation and relations to the EU, thereby aiming at strengthening the EU's work on PCD and increasing knowledge about coherence for development within EU member states, the Council, the European Parliament and the Commission. To contribute to EU coherence for development, the Swedish government's objective is to ensure that other EU member states as well as the EU Commission fulfil commitments made in EU Council conclusions and programmes to enhance PCD (Government Offices of Sweden 2008a, pp. 57/58). Just as Sweden considers greater coherence between different policy areas essential, it sees coherence between different actors as crucial to make the fight against poverty more effective.

Adjustments in Swedish international crisis management policy can be explained by the nature the EU activities in this field and also by the way it evolved. The EU's crisis management decision-making system is based on broad multilateralism. It is characterized by a high level of internationalism and activi-

501 With regard to a possible discrepancy of Swedish EUBG participation and non-alignment policy, Sweden refers to the Memorandum of Understanding, which states that "[t]he commitment of forces to the Battle Group will be without prejudice to each participant's sovereign right to make an autonomous decision on whether to deploy its forces to a specific crisis management operation" (Memorandum of Understanding 2005, p. 5).

ties demand to improve conflict situations by the application of values thought to be universally valid. Additionally, the EU is highly supportive of the UN system. For Sweden, these factors mean that in terms of the evolution of EU crisis management policy, adaptation has been fairly unproblematic and actually rarely provoked the demand to choose between a Swedish crisis management approach and a different policy on the EU level (Strömvik 2006, p. 204).

Organizationally, however, the Swedish government system placed some considerable strains on EU coordination. In particular, the often time-consuming consultative and consensus-oriented structures associated with the Swedish policy style, embodied in the political-administrative system, do not match the faster pace of EU decision-making. At the same time, "there has been a clash between the pragmatic and non-hierarchical Swedish administrative culture and the more legalistic EU culture" (Mazey 2001, p. 260). In this context, EU membership resulted in changes and organizational adjustments in the Swedish administrative system and in a comparably high degree of centralization.[502] From the Swedish perspective as a relatively small member state with small national ministries, the EU crisis management policy-making process presents a time-consuming and demanding area that takes up substantial parts of the Government Offices. A lot of resources are needed just to constantly respond to the EU-agenda and to keep up with EU-processes. Respective responses and the formulation of a Swedish position vis-à-vis other member states thus have been prioritized over the formulation of own crisis management initiatives or an in-depth analysis of the problem (Strömvik 2006, p. 208).

Originally, EU coordination in Sweden was handled by the department for trade in the MFA. Yet, this rather traditional MFA centred international affairs government coordinating mechanism proved inappropriate and since 1996 several reforms increasingly abolished the MFA's position as EU coordinator and reassigned it to the Prime Minister's Office (PMO). The Swedish government recognized "that effective central coordination of EU policy was important in maximizing Swedish influence within the EU decision-making arena" (Mazey 2001, p. 260). The new political and administrative coordinating structures aimed at horizontally streamlining and strengthening central coordination of EU policy (Mazey 2001, p. 260). The organizational reform toward "prime ministe-

502 A vital question concerning the organizational set-up of EU coordination in Sweden was whether EU matters should be treated as domestic or foreign policy. The debate questioned the legitimacy of the foreign ministry to handle EU matters, since it was not considered competent to be responsible for the coordination in policies of sectoral ministries other than its own and furthermore not being aware what was going on in other Government Offices. The Swedish organizational set-up of EU coordination and reforms were inspired by the UK system (Johansson/Raunio 2010, pp. 657/659).

rialisation" (Ekengren 2004, p. 216) was in particular pushed by the Swedish target and priority of running a good EU-presidency in 2001, where the EU department in the PMO had a central coordination role.[503] In 2004, responsibility for the coordination of EU issues lastly moved from the MFA to the PMO which also got engaged in all EU issues through a state minister for EU affairs. Yet, the MFA EU department was retained. Under the guidance of the PMO, the EU department in the MFA is concerned with the day-to-day EU coordination and supports the PMO in planning European Council meetings and with the coordination of EU issues within the Government Offices and within the MFA itself (Government Offices of Sweden 2005). With respect to the development of the EU's foreign and security dimension, due to the CFSP's broad scope, nearly all MFA departments need to be involved which led to increased coordination within the MFA (Rieker 2006b, pp. 74/75). The MFA thus maintains a central role with regard to routine coordination of EU policy and in the Council of Ministers. In addition, it also holds exclusive authority in formulating the Swedish position in the CFSP area (Mazey 2001, p. 263). The MFA's *European Security Unit* is in charge for drafting and preparations of CFSP instructions and has a lot of contact with the respective unit in the MoD. The MoD's Unit for Security Policy and International Affairs coordinates CFSP issues in cooperation with the MFA's Unit for European and Security Policy (Eriksson 2006, p. 91).[504]

Despite Swedish administrative tradition that the PMO should be rather small at the staffing level and not develop into an individual ministry, Sweden's EU policy today is led and horizontally and vertically coordinated from the PMO. The restructuring took place with the objective of an overall improvement of EU-coordination, but particularly to strengthen political steering. Other explanations refer to the increasing inter-ministerial and technical nature of EU-issues questioning the MFA's steering competence. The main argument behind the restructuring was that the individual Government Offices already handled EU matters as part of their daily work. The set-up of EU coordination with the overall responsibilities in the PMO was also supposed to establish closer links to the

503 Centralized EU-coordination was thus accepted in the cause of the clearly defined national interests of running a well-performed presidency (Johansson/Raunio 2010, pp. 657/658). For a detailed analysis of the Swedish EU-coordination during the 2001 EU presidency see Ekengren (2004).

504 In general terms, however, Sweden left large parts of decisions about how EU coordination should proceed to individual ministries. This resulted in looser and more informal networks in line with general coordination in Sweden being carried out rather informally (Jacobsson/Lægreid/Pedersen 2004, p. 113). For a detailed, chronologic elaboration and analysis of the restructuring of the Swedish organization for EU affairs see Johansson (2003), as well as Johansson and Raunio (2010). Eriksson (2006, pp. 88-108) provides a detailed description of the decision-making and implementation of Europeanized defence matters in Sweden.

Permanent Representation in Brussels.[505] The reorganization of the MFA and the PMO also meant a specialization in the area of EU decision-making that turned out to function as a legitimating force toward the other Government Offices (Ekengren 2004, p. 219). In addition, Sweden handled the demands deriving from EU-membership by an increased use of informal linkages. This relative informality of the Swedish coordination of EU-affairs matches Swedish administrative organization relying on more *ad hoc* and informal networks (Larue 2006, p. 129). From the Swedish point of view, informal mechanisms and measures of influence applied in the early decision-making stages are the way of having most impact on EU legislation (Larsson/Bäck 2008, p. 251).[506]

Apart from the horizontal coordination between different ministries, Sweden's EU membership also influenced vertical coordination between Government Offices and central administrative agencies. The country's EU accession has placed high demands on coordinating the Swedish administrative dualism and has been amplified in the Swedish case by the concentration of technical expertise in the much larger administrative agencies. Since the Government Offices are rather small, they have to rely on the expertise and support from the specialized agencies. This has become problematic as the decentralization of relevant expertise and capacities being placed in the government agencies make a quick reaction to EU demands challenging (Mazey 2001, p. 261). In order to counter these problems, Swedish government representatives on the EU level also include agency personnel or representatives from the Government Offices who are accompanied by agency civil servants to secure permanent expertise especially needed in contacts with the Commission (Larsson/Bäck 2008, p. 251). Due to the demand on coherence and complementarity on EU issues the ties between Government Offices and administrative agencies have generally become closer because of EU member-

505 The Permanent Swedish representation in Brussels is the largest Swedish representation abroad and handles daily contact and coordination of EU issues (Eriksson 2006, p. 90).

506 An important mechanism for broad coordination is the EU-drafting meeting which takes place almost every Monday between the state secretaries of all ministries to discuss and to share information on EU issues (Larue 2006, p. 129). An informal gathering on Friday (the so called Friday group) consists of the head of the Permanent Representation in Brussels who travels to Stockholm for Friday lunch meetings held at the Prime Minister's Office. Though not being a forum for formal decision-making, the meetings came forward as the main forum for informal information sharing and interministerial consultation and coordination of EU affairs (Johansson/Raunio 2010, p. 658). Another institutional adjustment is a coordination group for EU relations to determine inter-ministerial disputes over EU policies and to coordinate EU policy positions which is chaired by the state secretary on European affairs in the MFA and was composed of the state secretaries of the cabinet office, the Ministry of Finance and the Ministry of Justice (Larsson/Bäck 2008, p. 251).

ship.[507] Even more so, being closely involved in the EU policy-making process, not just the Government Offices but also the agencies have become Europeanized (Mazey 2001, p. 261). However, agencies lack political authority within EU institutions and work with and through the Government Office. As indicated by an interviewee, the Swedish agency model as a key characteristic of the country's politico-administrative system can be considered rather badly suited to the demands of the EU decision-making system. Furthermore, EU activities seem to have advanced vertical internal coordination within the same policy area, rather than advancing horizontal cross-sectoral coordination.

4.3.4 Synopsis: Sweden and Comprehensive European Crisis Management

4.3.4.1 Crisis Management Approach and Conceptual Dimension

As the above analysis showed, issues of policy coherence and coordination of crisis management are highlighted in various major Swedish documents and allow the reasonable account of the CA being a large strategic concern and aim in Sweden. Sweden is not reluctant to introduce notions of comprehensive crisis management in documentation pointing to the importance placed on coherence. The analysis of the Swedish crisis management and CA conceptions demonstrate that the Swedish discourse on the issue developed from an emphasis on traditional territorial defence conceptions to a focus on broader crisis management activities and a recognition of the contextual complexity. Yet, in strategic terms, Sweden does not have one particular cross-governmental guiding document or an explicit policy notion to implement the CA. However, while not having a strategy as such, a wide definition of security and multiple crisis management instruments are highlighted in Swedish crisis management documentation. Specific initiatives such as Sweden's Afghanistan Strategy also produce an important framework to support conceptual CA implementation. CA ideas are incorporated in numerous key government documents utilizing different starting points and several government initiatives set out policy and practice that seek to bring development, diplomacy, security and defence policy into line and lay CA foundations.

In doing so, Sweden's CA conceptualization to a great extent reflects the country's highly normative crisis management framework and is respectively coined. It is based on the internationalist ambition to export Swedish crisis management norms such as respect for human rights and mirrors the Swedish prefer-

507 Johansson (2003, p. 376) in this regard points to the problematic constitutional implications closer linkages may also have given the formal independent status of the government agencies.

ence for conflict prevention rather than dealing with the symptoms of conflict. As pointed out during a background interview, it does not matter whether initiatives are formulated by the MFA, the MoD or under the guidance of commissions, as all documents broadly contain numerous subjects and issues which neither of the document drafting ministries is directly responsible for. In doing so, the Swedish approach to crisis management conceptually not only comprehensively includes the whole spectrum of crisis management issues, but also the whole of government with only few conceptual departmental disconnections. Another coining characteristic of the Swedish CA model on conceptual side is a 'deliberative manifestation' which largely builds on conviction. Conceptual formulations of crisis management coherence appear more like a vision than an obligatory plan.

However, this results in respective government documents not being clear about how to practically implement the CA. It is difficult for Government Offices and agencies to pursue coherence if they do not have clear points of reference for measuring whether they are acting in accordance to a larger CA strategy. Indeed, while also pointing out the overall presence of the CA concept in Swedish documents, interviewees mentioned a continuing confusion on what the CA actually means for their role in international crisis management and how it should be implemented organizationally. While the reluctance to develop such specific strategies can be interpreted symptomatic of Swedish foreign policy in general,[508] the lack of a precise CA strategy results in the problem that Government Offices do not assume their responsibility and formal guidance authority with regard to CA implementation. According to an interviewee, the absence of a guiding policy is sensed. While the conceptual development of the CA in Sweden creates relevant 'boxes' to increase crisis management coherence, it is considered too unspecific on the consequences in organizational terms.

Yet, the Swedish discourse on the CA resulted in some changes in its security identity and in a movement away from traditional security formulations. In this context, the narrow definition of military territorial defence has gradually shifted toward an emphasis on the military's role as a part in the wider Swedish crisis management efforts. Civil-military relations gained renewed interest and the function and mission of the SAF were transformed and expanded from a territorially based organization into an instrument in the conduct of a new kind of crisis man-

508 As Bailes points out with regard to the role of strategies in Scandinavian countries, a "Nordic document plays the purely instrumental and tactical role of a 'strategy' designed to please others and/or win entry to an institution (…), none of these examples can easily be read either as the direct expression of a true, 'deep' national strategy" (Bailes 2009, p. 27). This argument by Bailes reflects the lack in the respective Swedish documents on how to organizationally implement the CA and what role different actors should play.

agement. MoD documents adopt an integrated view on security and Swedish security parameters moved away from a focus on military non-alignment and Total Defence. On the conceptual side of CA development in Sweden, the Total Defence tradition can be considered to have met demanding criteria of comprehensive crisis management by facilitating inter-ministerial coordination and thus Swedish adaptation to the new security context and CA implementation. As described by an interviewee, the Nordic perspective on international crisis management being softer on the military side and focusing on civilian crisis management is also mirrored in Swedish conceptual CA implementation.

While Sweden does not have a clearly identifiable CA concept, various government policy documents identify the CA as the ideal way to deal with contemporary conflict. Sweden frames crisis management initiatives in the CA context in particular by calling for greater coherence between its development policy and other crisis management fields. The dominating overarching aim of coherence and increased coordination of Swedish crisis management policy is to advance the effects of its development policy and to ensure that decisions taken in Sweden take development effects into account. As several interviewees were keen to point out, from Swedish perspective, international development policy is viewed as the best conflict prevention and security policy. The development perspective dominates the Swedish CA and cross-governmental crisis management efforts and comprehensive crisis management is first of all legitimized by referring to how it strengthens Sweden's development efforts. Swedish efforts toward crisis management coherence mainly focus on achieving the country's key goals of equitable and sustainable development. Framing the policy coherence debate, the PGD is the point of departure of Swedish CA implementation. With its focus on international development, Sweden has adopted an explicit but rather narrowing CA with regards to content and focus.

In spite of the policy's overall positive evaluation, also the PGD fails to address questions on political commitment and leadership, policy objectives and adequate institutional structures to promote inter-ministerial coordination. A conceptual deficit of the policy in this regard is that strategies on development that result from the PGD resemble the strategies of individual actors, "accompanied by little more than a description of what other Swedish actors are doing appended as an afterthought" (Patrick/Brown 2007a, p. 121). Critics also identify a lack of cross-governmental analysis and an overly focus on synergies while not paying enough attention to policy inconsistencies.[509] Nonetheless, the PGD resulted in increased inter-ministerial understanding and exchange among the

509 Also see Oden (2009, pp. 9/10).

MFA, the MoD and also the Ministry of Justice on development challenges by broadly addressing relevant issues.

While the analysis of Swedish efforts to improve crisis management coherence and the above summary showed that CA ideas conceptually materialized in Sweden, it also revealed some limitations. With the Swedish CA focus leaning very much toward development concerns, it can be observed that Swedish conceptual initiatives "seek to encompass the security, governance, economic and humanitarian aspects of the engagement, but show weaknesses when it comes to coordination with military" (Friis 2010, p. 42). While Sweden can be regarded as a role model in implementing a CA to global development it lacks similar approaches in other policy areas relevant to international crisis management. The Swedish PGD succeeded to mainstream development concerns across the Swedish government.[510] Yet, other aspects and coherence requirements of crisis management fall back. The transformation of the integrated PGD and coherence for development toward a CA that includes all crisis management dimensions on an equal level remains a conceptual challenge.

4.3.4.2 Organizational Dimension

Formal organizational structures and coordination mechanisms to achieve crisis management coherence do hardly exist in Sweden. Civilian crisis management largely is the competence of the MFA and the MoD heavily emphasizes the importance of civil-military relations and comprehensive crisis management. This showed in the creation of the FBA as a civilian-military coordination centre, apart from which strictly speaking no joint and formalized mechanisms with the specific aim to achieve more coherence in Swedish international crisis management between relevant organizational entities have been created. Government Offices have a minimum of formal arrangements to promote inter-ministerial coordination and regulated procedures are rarely used. However, cross-ministerial coordination has been made possible by a number of factors. In this regard, the Swedish approach of CA implementation reflects the network theme and respective ideas of coordination. Accordingly, policy coherence in Sweden very much depends on individual initiatives, informal mechanisms and the existing consensual culture of cooperation. Attempts to crack the departmentalization of Government Offices by setting up committees or other issue specific bodies are *ad hoc* and short-term (Larsson/Bäck 2008, pp. 185/186). Instead of establishing organizational structures in its

510 As assessed by Patrick and Brown (2007a, p. 127).

core executive in order to generate coherent government policies, Sweden focuses on procedures to structure the ways established in central government to increase policy coherence. The examination that "[t]he pursuit of 'joined-up-government' is not a major public goal in Sweden" (Ling 2002, p. 621) matches statements made by interviewees in this context. Efforts to improve crisis management coherence take place in an unobvious and indirect way with almost no visibility.

As analyzed above, the Swedish administrative culture can be described as cooperative, consensus seeking, problem-oriented and pragmatic and is characterized by a high degree of informality, bridging organizational boundaries and hierarchies. Studies on public organization in Sweden have suggested, that the smallness of the Swedish state and its administration allows this personalization of relations. Also in the CA context, the small size of the relevant crisis management Government Offices encourages inter-ministerial cooperation (Patrick/Brown 2007a, p. 119). These cultural characteristics have also been excessively used in an almost stereotype way with regard to structural attempts to implement the CA which are mainly aimed at through these features. There proves to be an important core to this cliché impacting on CA implementation in Sweden. In such oriented systems, divergent normative and political criteria lead actors to seek compromises and combined solutions that all sides can support. As assessed by interviewees, due to its consensual political culture and constant communication among key actors and institutions, inter-ministerial coordination does not pose many struggles in Sweden. The Swedish CA is essentially about inter-ministerial meetings to keep actors engaged and mutually informed and largely occurs in terms of positive coordination. On the downside, however, this focus on positive coordination disregards existing conflicting issues and policy inconsistencies. The emphasis on consensus and pragmatism in the domestic context not just seems to spill over into the making of foreign policy activities in general, interviewees also emphasized that it in particular advances difficult debates along dividing lines of civilian-civilian and civilian-military aspects of crisis management from the beginning.[511] A noteworthy aspect is the highly positive and open atmosphere toward CA implementation and cross-governmental coordination in Sweden which became apparent during background talks. This positive stance on the CA seems to be transferred to actual cross-governmental working and *vice versa*. However, the emphasis on governmental consensus can also result in agreements on the least common denominator und leads to time-consuming processes. This entails the danger that coordination and negotiation lead to stalemates at some point.

511 Patrick and Brown (2007) come to a similar conclusion in their study of donor government efforts towards policy coherence in fragile states.

Coherence in Sweden thus is aimed at by favouring informal processes rather than formal structure. Informal coordination in Sweden largely evolves through social mechanisms and *ad hoc* responses to interdependence outside formal structures on the personal level. Yet, this approach still includes formal legal requirements in high-level policy and it involves different relevant governmental actors in joint assessment and analysis processes via the Joint Preparation Process which leads to structured dialogue between different ministries and ultimately to a common understanding. With its Joint Preparation Process, Sweden's approach to crisis management coherence entails a requirement by law which proves a useful way to exchange information As a legal and high-level policy basis for the CA, the Joint Preparation Process serves coherence very well because it formalizes processes that need to be followed to ensure coherence of policy decisions. Rather than having an explicitly declared CA concept and a standing cross-governmental unit, it is this strategy-drafting and planning process that aims at creating coherence in Sweden. It ensures coordination and agreement of various departments at the working level, essentially helps to achieve a better common understanding and brings Government Offices together to a clear and common approach.

In addition, the feature of agencification attaches much importance to professional and expert considerations to Swedish crisis management. However, from coherence perspective, the Swedish administrative dualism with its functional differentiation and vertical specialization leads to some particular difficulties. The additional vertical division between Government Offices and agencies responsible for international crisis management generates high demands on the coordination and planning of the core executive. In this regard, the organization of Swedish crisis management combines two organizational principles. Firstly, a specialized bureaucracy at the bottom and secondly, a collective decision-making body on top. Demanding increased direction by the centre of government in terms of vertical coordination, such a system heavily relies on openness, mutual responses and feedback between actors (Larsson 2002, pp. 206/207). It also provides some fundamental barriers for structural changes to achieve greater crisis management coherence such as the limit to further integrate agencies due to their specialized tasks. An assessment made by public administration scholars concludes that in Sweden much importance is attached to "the preparation of decisions at the expense of their execution and feedback" (Arter 2008, p. 162). This issue can also be examined with regard to the Swedish CA implementation. The loss of political steering as a result of agencification has resulted in a lack of structured formal coordination and response that is felt between the Government Offices and agencies, especially between the MFA and Sida (De Coning et al. 2009, p. 16). In line with its pragmatic policy approach Sweden "seems to prioritize efficiency and usability rather than providing a more visionary leadership"

(Sundberg/Nilsson 2009, p. 1). While rhetoric on coherence and comprehensive crisis management is strong, as the conceptual analysis showed, its concretization and actual implementation is comparably weak. Relevant Government Offices are too focused on legislation, with the actual implementation falling short of being coordinated. The preparation of measures that aim at implementing the CA, such as the PGD, illustrates this point and while the MFA has a coordinating role, a lack of hierarchy is nonetheless noticeable.

As the CA in Sweden highly depends on informal networks in the crisis management area, flexibility is very important to the Swedish approach to support the open-ended quality of the networks. On the contrary, it is exactly the Swedish model's flexibility and vagueness on the CA that partly seems to obstruct its success. While Swedish informal inter-ministerial coordination is an effective way to reduce the structural formal distance between the Government Offices and hierarchies, such informal coordination and reliance on personalities bears the risk of suffering from personality clashes as much as they can benefit from good relations among the officials involved. Too much *ad hoc* and informal network characteristics thus can also become a weakness. Relying heavily on informal contacts, the Swedish CA model runs the risk of being too dependent on goodwill of the current officials and the personal chemistry of the personnel. An interviewee pointed out that another downside of this personalization is created by the regular rotation of staff. These risks, however, seem to be reduced by the amount of communications channels on different levels that allow regular exchange of information and preparation of joint activities. As evaluated by respondents, inter-ministerial coordination works smoothly.

As a result of a lack of a standing inter-ministerial organizational body, the Swedish approach to comprehensive crisis management is institutionalized in a conceptually driven way. However, exactly this approach allows for Sweden's consensus oriented political culture, the possibility and flexibility of *ad hoc* communication between small Government Offices which occurs on a regular basis, as well as a mutual understanding of each other's roles in crisis management. This results in organizational circumstances in which the creation of new standing inter-ministerial structures to coordinate crisis management and to advance the organizational implementation of the CA appears unnecessary.[512] Despite some obstacles, informal coordination mechanisms and personal networks

512 Patrick and Brown come to a different conclusion analyzing coherence of the Swedish PGD and call for "[e]stablishing a body outside of any one ministry might be most effective in terms of bringing about genuine country strategies based on the various comparative advantages of each department, but it would take leadership from the highest levels to articulate how engagement in select countries is vital for global development and security and stability" (Patrick/Brown 2007a, pp. 121/122).

compensate for what can be viewed as a lack of formal structure. The Swedish approach to implement the CA in organizational terms is characterized by wide forms of informal network type organization that is able to self-determined managing itself without intervention by hierarchical leadership or any steering role. Swedish institutional structures do not require a fundamental structural reorientation for the implementation of new ideas. Nonetheless, the faith in informal mechanisms to promote the CA in the belief that the consensus built up by consultation does not mean that no strong policy statement is required, as demanded by interviewees.

The informal and *ad hoc* character of the Swedish CA can be considered one of its strengths because it avoids unintended consequences that most organizational changes and the creation of new coordinating bodies imply. Dynamics of change in terms of signaling, framing, iteration and exchange as identified by scholars of organization theory seem to lose relevance in the context of the Swedish case. Furthermore, this approach appears to be optimal for the highly complex and dynamic international crisis management environment. Nonetheless, as the downgrading of formalism and the upgrading of informal contacts and coordination is a distinctive characteristic of Swedish public administration, designing CA implementation in Sweden promises to be more successful by stressing the objectives and results of coherent crisis management activities and not the mechanisms to achieve coherence themselves. Methods which aim at ensuring coherence and ultimately at implementing the CA in Sweden result in an ever increased need for informal coordination mechanisms, which could also give the government the opportunity to supervise CA implementation and crisis management coherence more closely.

4.3.4.3 European Dimension and EU-Interface

Since its EU accession, Sweden has adopted a proactive stance toward the EU that also aimed at contributing to develop and change the EU's increasing crisis management capacities in line with the Swedish approach. The role of the EU in Swedish crisis management and the positive effects of increasing EU engagement are heavily emphasized in Sweden. The Swedish contribution to EU crisis management is very high on both the conceptual side as well as on the structural side. Sweden has taken very active part in the development of both the military and civilian dimension of EU crisis management capacities and the country particularly focused on the balance between both. It sees itself at the core of the EU and the EU largely infuses Swedish policy as a means and an end. Providing a strong degree of Europeanization, above analysis points to an interwoven process, of both the EU influencing Swedish crisis management as well as Sweden actively shaping the EU's crisis management development in a "bi-directional process of Europeanization" (Lee-Ohlsson 2009, p. 123). Showing many factors in both the downloading and uploading direction of Europeanization, changes and efforts toward policy coherence in Swedish crisis management are influenced by the country's EU membership. A growing salience of the EU crisis management agenda and a clear adherence to common objectives took place. In addition to the traditional Swedish focus on the UN, the EU became the focal point with EU efforts increasingly substituting national policy. Sweden devised its security traditions in line with broader European crisis management priorities. This particularly applies the country's non-alignment policy. Despite initial concerns, Sweden has developed into a proactive EU crisis management participant, in both the civilian as well as the military dimension. Moreover, there seems to have been a perceived need to demonstrate Swedish willingness to engage in this area (Eriksson 2006, p. 140). The EU became an integrated part of the Swedish crisis management policy that on the other hand allows Sweden to become more actively involved in international decision-making processes.

In parallel with the changes toward crisis management coherence on the national level, Sweden also contributed to the development of a comprehensive EU crisis management capacity. After a skeptical and reluctant start, Swedish policy shifted toward becoming one of the most active ESDP participants. While the focus in this regard has clearly been placed on the civilian crisis management dimension, which Sweden demanded to be on the same level as the developing military dimension of ESDP, the country also contributed to the process of building up EU military capabilities for crisis management. In doing so, the Swedish government assured that a CA and a broad range of both civilian and military crisis management instruments were included in EU crisis management and became a strong CA-advocate. Respondents underlined that Sweden today strongly argues in favour of comprehensive EU crisis management including the whole arsenal ranging from development, diplomatic and humanitarian to military instruments in particular with regard to the Lisbon Treaty implementation.

With respect to the EU's role in Swedish crisis management, the country's perspective developed from a high skepticism against the military dimension of EU crisis management, to a reluctant recognition combined with strong support for the civilian dimension toward becoming proactive and pro-integrative on the military side of EU crisis management. While shortly after EU accession, the EU was mainly seen as a forum for dealing with the whole range of wide security challenges to improve Swedish security, this changed and developments in Sweden's national crisis management area were explicitly related to the EU dimension. The most obvious case of EU impact on Swedish crisis management took place with regard to defence reforms, where the EU was used a cover to push reforms (Lee-Ohlsson 2009, p. 134). The EU became a source of transformation itself, particularly, with respect the SAF reforms. The NBG was a cornerstone of the reform and impacted on major changes. While civilian crisis management aspects became a way of shaping and influencing EU crisis management in a manner corresponding to traditional Swedish foreign and security policy thinking, on the contrary, "the outspoken Swedish readiness towards international military participation, (…) reinforce[d] the image of Sweden having an appropriate international role. It compensates for the military non-aligned position" (Möller/Bjereld 2010, p. 378) and pushed Sweden to demonstrate credibility.

> "Swedish membership in the EU and the Swedish Ministry of Foreign Affairs' promotion of comprehensive security within the EU (…) have contributed to an increased awareness of the need to adapt to the new security context at home and especially within the ministry of defence" (Rieker 2006b, pp. 83/84).

Thus, due to the country's desire to project the Swedish way upon the international scene, a logic of domestic self-interest behind the Swedish EU crisis management activities becomes obvious. The instrumentality of the EU for Sweden also becomes apparent in the country's strategic priority to strengthen the EU. The Swedish active engagement in the EU's international crisis management efforts can also be viewed as a way of achieving traditional Swedish crisis management goals. In fact, Sweden succeeded repeatedly in embedding its own values and objectives in EU positions and the country pushed issues that suit the Swedish crisis management policy approach and objectives, such as conflict prevention and development matters on the EU level (Lee-Ohlsson 2009, p. 137).

Sweden's non-alignment policy impacted on its policy to EU crisis management in the first years and was brought up following the creation of the ESDP and again in connection with discussions of mutual defence guarantees in the Lisbon Treaty. By shifting its definition of military non-alignment and making it equivalent to not signing agreements on mutual defence guarantees, leaves the concept meaningless with regard to the EU.[513] However, the Swedish background as a non-aligned country, can serve as an explanation that efforts to influence EU crisis management have been more focused on activities and less on institutions on the EU level (Soder 2010, p. 14), which mirrors Swedish attempts to increase crisis management coherence on the national level. With Swedish

513 Strömvik (2006, pp. 204/205) stated in 2006 already that "the issue of non-alignment is simply no longer relevant for these states' [Sweden and Finland, S.P.] attitudes towards the EU's security and defence policy."

crisis management activities increasingly being linked to the EU, also the Nordic dimension is blurring with a European profile. But Nordic ideas have also become a broader European concern, "with the EU becoming to some extent 'Nordicized' in this respect" (Browning 2007, p. 40) while at the same time the Nordic model, has become Europeanized and acknowledged as a part of the EU's crisis management profile (Browning 2007, p. 44).[514]

As such, Swedish efforts toward European crisis management as well as its own policy coherence very much reflect Swedish overall crisis management preferences, especially on the conceptual and normative side. Both, the domestic political interest on the role of the EU in international crisis management as well as a genuine belief that civilian instruments are necessary for the EU to become an effective crisis management actor were rationales behind Swedish policy. As the analysis showed, the EU's soft and normative security power image is similar to the Swedish security political nature and crisis management approach as well as the internationalism Sweden became renowned for. According to an interviewee, due to the Swedish internationalism and concentration to work via the UN, it was natural for Sweden already before the country's EU accession to work together in international partnerships which made it easier for the country to see the EU in a bigger perspective, as the main trigger for Swedish goals and interests on the international level and a natural platform for Swedish crisis management policy.

In organizational terms it became clear that the development of EU international crisis management had a decisive impact on Swedish foreign and security policy administrative coordination. Sweden's administrative structures have been affected by organizational adjustments due the EU policy processes regarding communication, steering as well as vertical and horizontal coordination. There has been a clear tendency in Sweden to the centralization of EU policy making and coordination as they have been transferred from the MFA to the PMO. The weak spot of coordination in the Swedish model does not lie horizontally between the different ministries but vertically between the Government Offices and respective agency. At the same time, EU membership also enhanced interdependency between Government Offices and agencies and created new mechanisms of coordination which brought both closer together. The increased activity with EU issues, however, resulted in advanced vertical internal coordination within the same policy, rather than advancing horizontal cross-sectoral coordination. Furthermore, Sweden handled the demands deriving from EU-membership by an increased use of informal linkages. This relative informality of the Swe-

514 This also reflects Lee-Ohlsson's (2009, p. 123) bi-directional process of Europeanization as elaborated above.

dish coordination of EU-affairs matches Swedish administrative organization relying on more *ad hoc* and informal networks.

Sweden made a lot of efforts to assure the comprehensive character of EU crisis management. Yet, it is precisely this characteristic that allowed for such an active Swedish participation and that shows that the CA has indeed become an important component of Swedish crisis management. Coherence striven for at the national level is reflected in Swedish position at EU level. Furthermore, the Europeanization of Swedish crisis management also increased coordination between the Swedish actors and led to closer relations between participating bodies. Swedish efforts toward crisis management coherence on national as well as on the European level can be described in terms of active balancing and innovative adaptation. With influences going both ways it is possible to argue that more than by changes in the overall Swedish security environment the country's CA to international crisis management is coined by crisis management developments on the EU level in a process of parallel influence toward European crisis management.

5 Conclusion

The purpose of this dissertation was to investigate how the EU and its three member states UK, Germany and Sweden attempt to implement a Comprehensive Approach to their crisis management systems and how far they have come in doing so. The analysis has focused on four predominant dimensions and conditions: the particular crisis management approach, the conceptual and organizational CA application and the European dimension in this. The impact of the CA concept in approaches to international crisis management is highly evident both at the EU and the country case level. Respective national and EU-efforts expose numerous discernible characteristics in the vision for crisis management policy coherence and there are certain characteristics all three country case studies and the EU have in common, albeit to different degrees and in different variations. Also, efforts to implement the CA seem to be meeting the same obstacles. However, there are major differences in relation to both the conceptual and organizational dimensions of the particular attempts to achieve crisis management coherence and to implement the CA. A range of models for making responses to crisis of national governments and EU institutions more coherent emerged.

The following subchapters review the results and present conclusions emerging from the preceding examination. To facilitate an overall synthesis, findings are summarized according to the different dimensions of CA implementation that were applied throughout the empirical analysis. First, the different ways in which the EU and the analyzed member states aim at implementing the CA and achieve crisis management coherence on the conceptual and organizational level are summed up and major differences are highlighted. Subsequently, conclusions are drawn about the linkages between the EU and the national crisis management level. Findings from the previous chapters are brought in relation to identify both progress and existing deficits. In doing so, an overview of the challenges, problems and dilemmas to achieving a CA that relates to the pursuit of the concept both in the national and the international context is provided. While a comparison of the different approaches to achieve crisis management coherence is not intended in this thesis,[515] a

515 On the difficulty to meaningful compare cases see chapter 1.4.4 on challenges and limitations of this dissertation.

contrasting summary of findings is nonetheless valuable to some extent to clarify common strands and different approaches chosen by the analyzed actors and the EU. With the explanatory account in place offered in the previous chapters, it is possible to tentatively draw some practical lessons-learned and to present some recommendations for future CA efforts. These recommendations are specifically formulated against the theoretical background taken in this dissertation.[516] After pointing out limits and constraining factors of a Whole-of-Europe Approach, the chapter then makes suggestions for further research. The dissertation concludes with a general reflection and concluding personal remarks on the research subject and approach used.[517]

5.1 Toward a Whole-of-Europe Approach – Modelling Comprehensive European Crisis Management

> "Organizational analysis is dull until the crucial and central role of organizations is understood. If organizations are understood, then individuals have a tool by which they can deal with the reality that they face."[518]

5.1.1 Crisis Management Approaches and Conceptual Dimension

Actors at both the EU and the national country case study level identify the CA as a highly important concept and recognize the limited effect of individual crisis management efforts and crisis management instruments when implemented in isolation. The CA concept and ideas to increase crisis management coherence are inscribed in various official documents, strategies and statements. Analyzed documents demonstrate a broad CA commitment which acknowledges that a CA is needed to effectively address the diverse crisis management challenges but which also see the CA as one of the challenge itself. All institutional key actors are explicitly interested in improving crisis management coherence leaving the

516 The following recommendations are culled from above case studies, from relevant documents as well as interviews undertaken for this dissertation. More than in terms of a comprehensive directory they should be viewed as food for thought in the CA debate. Providing clear empirical answers is rather difficult given the complexity of the subject but following recommendations indicate a direction for development of CA implementation.

517 As each case study ends with a detailed conclusion on the case's particular findings and specific recommendations, these are not taken up in detail in the following chapters present the overall conclusion of the research subject.

518 Observation by Hall (1991, p. 27).

impression that the CA has been continually on the agendas of policy makers since the end of the 1990s. However, despite this broad agreement, the conceptual outcome and detailing as well as ideas about the way to proceed vary greatly. Furthermore, lately the CA concept has been increasingly questioned and contested as relevant actors often disagree on the definition and controversial utility of the concept, and, moreover, find the term a distraction from concrete challenges of crisis response.

The EU's crisis management growth promoted a strong conceptual CA commitment and crisis management coherence represents a major part of its larger conceptual development in this area. Yet, analyzed documents which outline objectives for the EU's comprehensive crisis management activities remain vague. As CA issues are spread over many policy documents by different actors, the relevance of the concept itself looks rather rudderless. EU crisis management policies developed conceptually disconnected from each other, following their own logic and with minor attempts to place them in a single integrated strategy. This situation has been challenged with the Lisbon Treaty. Policy coherence is a key objective of the treaty which offers scope for a more coherent approach of the EU's communitarian and intergovernmental crisis management tools. Triggers of the CA in the EU were increased demands of its crisis management engagement and specific failures in missions. Barriers, on the contrary, derived from incoherent efforts on the part of the EU's diverse crisis management institutions. As a consequence, the CA debate in the EU has developed from emphasizing a culture of coordination to a coordination-oriented approach which largely focuses on formal structural provisions to improve inter-institutional coherence. In doing so, the EU consistently refers to its own special capacity to activate a broad range of crisis management instruments contributing to the EU's credibility as a crisis management actor.

On the country case study level, the UK developed a number of key policies in order to deliver a CA in crisis management. The CA plays a big role in the country's attempts to adapt to the changing international environment and it is at the forefront of showing willingness and flexibility to experiment with new ideas in its CA efforts. CA developments in this regard have to be viewed as a part of an overall move of the UK to increase cross-departmental collaboration of its government as a whole. The CA was very much brought to the fore by the country's operations in Afghanistan and Iraq and efforts to implement the CA emerged predominantly in response to military crisis management operations. Thus, the concept's evolution was largely MoD-driven rather than resulting from an actual cross-departmental deliberative process. Attempts on a conceptual strategic level for some time were largely and one-sidedly informed by the military. Consequently, the underlying motivation focuses on crisis management coherence for national security reasons, but is also framed by the country's economic and great power decline. However, the UK's conceptualization is developed to focus on inter-departmental consultation, practical measures and operational issues concerned with doctrinal CA implementation. In doing so, the UK broadly interprets the CA as a concept concerning all levels and dimensions of crisis management as well as all actors. At the same time, the UK's CA underlines individual departmental strengths and emphasizes the unique competencies and resources of each department. The establishment of the NSS can be considered an important step to implement the CA in terms of this comprehensive and cross-departmental manner and to renew the process of setting out the CA in the UK. Pushed by heavy budgetary pressures and a perceived strategic deficit of the UK's foreign and security policy, the NSS was established as a comprehensive strategic framework explicitly committed to the CA in its background planning process and content.

Compared to the UK, which focused its efforts very much on strategically developing the CA, Germany and Sweden pursued less radical and strategic approaches, which rather appear as a loose connection of initiatives and ideas than a coordinated unified plan toward a particular end. The highly sensible political environment created by the ongoing German aversion of military force and a policy of restraint complicate conceptual CA application in Germany. Efforts to enhance crisis management coherence do so from a civilian crisis management perspective and sphere. The German CA model has materialized as an instrument for the country to express its positioning in relation to its military commitment as close as possible to its particular security policy background. While the German government aimed at improving crisis management coherence with several documents which show the importance Germany places on the subject, this intention did not result in a coherent conceptualization. Conceptual agreement on the CA is largely missing and the logic of strategic CA thinking is far from being recognized and appreciated in Germany. Respective documents and concepts do not take any form of hierarchy or a regular series, but are developed and formulated as required. This can largely be accredited to the conflict between different CA models developed in Germany, which were not able to find a remedy in this regard. Quite contrary, the co-existence of the Action Plan and its Civilian Crisis Prevention CA conception and the White Paper's concept of *Vernetzte Sicherheit* largely exacerbate tensions that 'naturally' come with the CA idea into a level of deep conflict. Moreover, German conceptual discussions and terminological debate in the CA context did not result in an advancement of the concept's objectives but rather in a blockade coined by dogma. There is no strong protagonist or lobby for the CA and related issues but instead and increasing lack of interest. As the CA in Germany is merely adopted in terms of a reform symbol rather than for problem-oriented reasons, the institutional and structural implications of the objective to improve crisis management coherence are ignored by political decision-makers.

Conceptually, Sweden does not have any particular cross-governmental guiding document or an explicit policy and cross-ministerial definition to implement the CA. Nonetheless, the Swedish discourse on the CA resulted in some modifications in its security identity and in a movement away from traditional security formulations. The analysis of the Swedish crisis management and CA conceptions demonstrate that Swedish crisis management discourse developed from an emphasis on traditional territorial defence conceptions to a focus on broader crisis management activities. In this regard, Sweden treats global development as the centerpiece of its international crisis management and it is development matters which promote cross-governmental coordination. Sweden's CA model does not focus on crisis management in terms of international peace and security interventions as such, but covers international development cooperation and policy and is largely coined by the country's PGD. With its focus on international development, Sweden has adopted a rather narrow CA with regard to content and focus. Despite of the PGD's overall positive evaluation, it fails to address questions on political commitment and leadership, policy objectives and adequate institutional structures to promote coordination. Another coining characteristic of the Swedish CA model is the 'deliberative manifestation' of the concept which is based on conviction. Conceptual formulations of crisis management coherence appear more like a vision than an obligatory plan. However, this results in respective government documents not being clear about how to practically implement the CA.

This brief summary demonstrates, that multiple agendas have been advocated under the CA heading. The topic is approached from different angles which reflect different political cultures and crisis management understandings. Disparities become obvious in terms of conceptual origins, priority-setting and incentives to apply the CA. In this regard, the CA has for the most part been advanced on the particular national basis or, in the EU case, in the EU's particular context as an IO. The individual conceptual CA application depends on the particular crisis management culture, basic crisis management assumptions and values. A distinction must be made, for instance, between interpreting CA as a mindset like Sweden or or in institutional terms with a focus on structure and precesses like the EU and the UK do. In addition, each actor and also their respective crisis management institutions themselves represent different CA models based on different motivations and objectives. This comes to no surprise as also the rationale for engaging in crisis management differs between actors. At the same time, the attempts to create a more coherent crisis management policy reflects a lack of clarity between individual institutions on how CA implementation should be modeled and moreover, what the CA is supposed to accomplish. Case studies also revealed that conceptual CA efforts are mostly dominated by one of the relevant institutions which entails the risk that these are likely to emphasize on the crisis management instruments in which they see a comparative advantage. Analysis also showed that actors mobilize the concept for their own purposes. The extent every actor tried to pull the CA in his own direction to increase legitimacy or visibility further blurred its nature. The conceptual analysis of CA implementation thus exposes that actors have largely developed and conceptualized a CA so as to be compatible with their own crisis management background and needs. They have different objectives, use different language and instruments and have specific institutional policy imperatives and approaches. The conceptual implementation of the CA reflects competing, value-laden tensions between the cultural influences of constituencies holding strong identities and role-conceptions and a strong influence of these identity ambitions on the CA. These different subcategories behind the CA constitute themes on their own right. This has resulted in a cluster of competing CA strategies.

Approaching the CA from different perspectives, the variety of models and aims of actors and the lack of a strategic formula lead to a troublesome duality of strategic independence and emphasized interdependence of individual actors. Reforms toward comprehensive crisis management in Europe not only differ with regard to content but also with respect to their formulation and development. Thus, actors define the CA divergently and are at different stages with respect to overall conceptual development and clarification. It should also be noted that there is considerable variation in the use of CA terminology. Key CA terms are not determined themselves and not shared at a strategic level between actors. In this context, different approaches have not just become identified with different labels,[519] but also ownership and understanding of the particular concept varies revealing misunderstandings considering the CA's formation. Furthermore, concepts and relevant official documents are not systemized. The conceptual confusion of the CA and lack of clarity of relevant terms and terminology are directly linked with questions of responsibility and felt competence.

The different CA conceptualizations played an important role and created inter-organizational tensions. One of the most evident fault lines concerns the triggering background of the respective agenda toward crisis management coherence. The analysis showed that different vehicles have driven CA development.

519 Efforts to implement the CA are known under the label of Comprehensive Approach in the EU, Joined-Up government, Comprehensive Approach or Integrated Approach in the UK, the Vernetzte Sicherheit framework in Germany and no particular label in Sweden.

In this regard, development bodies in the EU and development ministries, for example, resisted the CA because they perceived the concept as an instrumentalization of development for security purposes. When emanating from the military side, civilian crisis management actors preferred their own CA interpretations. Generally, concepts imported from other institutions are received as imposed and come with associated prejudices, assumptions and visions. Moreover, they are perceived to create a hierarchy of ownership which decides to view the CA either as a purely civilian function or from a military perspective. While the analysis revealed civil-military antagonism *vis-à-vis* the CA, disagreement also occurred between civilian actors, in particular at the interface of humanitarian and political dimensions of crisis management. At the same time, the awareness of these fronts resulted in profound insecurities regarding an appropriate behavior to support the CA. This anxiety becomes most obvious on the military side which is concerned that pushing too much toward the CA could be interpreted as a militarization of the concept and its instrumentalization for military goals. A 'wait-and-see attitude' on the contrary, is feared to be perceived as indifference.

Policies and objectives for improving crisis management coherence have been defined in rather general terms without providing a clear definition of the CA and demands of its implementation more specifically. The concept of coherence, while being extensively used, has proved to be flattened in its design and implementation. While the CA is rhetorically all over the place, the analysis showed that there is not much alignment between the actors. While actors agree on a more general level, they take on different views on various characteristics of the CA which contributed to the difficulty of a coherent development of the CA agenda. Conceptual orientations at the policy level have generally not been translated, were not univocal and shared. The studies for this thesis suggest a number of reasons for this. The first is that neither the EU nor the analyzed countries have what could be characterized as 'a policy' toward implementing the CA and achieving coherence of its crisis management efforts.[520] On a conceptual level, only the UK follows a more specific approach that could be used for mandating the CA or as a framework or setting specific guidelines for implementing the CA organizationally. Thus, the EU as well as national core executives have tried to operationalise CA efforts without strategic guidance. This situation left relevant actors without a reference to guide their action and acted as a barrier to address complex CA issues. The deficiency of a clearly guiding policy and an overall concept leads to a situation in which everyone can project different aspects into the concept. The lack of clarification thus resulted in heavy discussion and disa-

520 This also becomes apparent in the wide variety of official sources the research for this dissertation's case studies on conceptual level was deducted from.

greement on what a CA means for the different actors. As particularly the country case analysis showed, debates how the CA relates to crisis management roles entails the risk to maintain existing approaches. In this regard, the empirical analysis also revealed that CA implementation is to a great extent about the attitudes and ways of approaching other actors. Traditional guiding principles of actors continue to exist and to potentially undermine a CA. In addition, too little awareness of strategic government priorities also puts governments in a rather reactive position (Bourgon 2011, p. 1).

Thus, a more precise strategic conceptualization of the CA could provide more guidance. Realizing a CA and working together toward this end assumes some degree of agreement and a shared conceptual platform of values, principles and priorities. The analysis showed that, conceptually, it is important to ensure clarification and a common understanding of concepts among actors in key reference documents. This also entails the development and realization of a more structured and systematic approach. A common strategy of joint efforts and collaborative inter-organizational working should function like a handrail for actors by providing brief principles and clarifying relationships and providing a ranking order between existing documents. Also, the naming or labeling of efforts to implement the CA is an important conceptual issue with implications for the further development of the process. Strategizing the CA is a necessity for strategic guidance as a basis for coordination. Moreover, strategic clarification is the precondition to create the appropriate structures. As mentioned above, institutions prefer their own concepts and names and imposing a concept also brings associated assumptions and hierarchies. An overarching CA strategy is thus also necessary to avoid a situation in which each actor pulls it into its own direction, interpreting the CA according to his own interests and avoiding utilization to increase own legitimacy and prominence at the expense of other actors. An overarching comprehensive strategy such as the British NSS might solve the repeated problem of lacking inter-institutional strategic agreement and a common CA rationale and awareness.[521]

On the contrary, while the empirical analysis pointed to the need of more strategic CA guidance and the CA has been criticized for its strategic vagueness, this might be right on target given the crisis management and organizational complexity and dynamics as well as the deep disagreement between some actors on the issue. It is this observation which points to the need of planning the CA idea as broad and open as possible. To avoid a too specific, detailed and narrow understanding, despite the need of strategic guidance, the CA should not be

521 See also Stepputat (2009b, p. 55) who also argues that CA actors need to agree on strategic objectives and the definition of a strategy to achieve them.

viewed as a static, inflexible model, but rather as a set of principles to guide crisis management actors. To this end, the CA's character as a mindset or philosophy of the same imperative concerning the need to better coordinate political, economic, civilian and military contributions to crisis management has to be underlined. Considering the role of overarching high-level strategies, the negotiation of one doctrinal set of guidance that meets every actor's needs appears unrealistic. It is thus questionable, if a CA language can be more precise. While an overall strategic framework provides orientation and also shows support at the highest strategic level, the additional value on a practical level is doubtful. Reflecting a middle ground between strategic guidance and conceptual flexibility, an overarching inter-institutional CA strategy would have to be complemented by concepts oriented on practical implementation concerning the specific conflict management task at hand. Multiple minor strategies are more likely to lead to actual conceptual coherence than a centrally sealed master plan. This conceptualization underlines the significance of independent action, which actor's try to coordinate but not to integrate.[522]

In line with this discussion, the analysis indicated that the process to conceptually implement the CA has to include all potential actors from the start in order to ensure organizational and conceptual ownership. The CA tends to work best as a partnership of equal and a strategic CA development in terms inter-institutional consensus. Inclusiveness is a key principle of the conceptual CA formulation as the UK's NSS development showed.[523] On the contrary, it can be argued that only political leadership can provide a solid foundation, can identify a clear focus and a coherence agenda for the cross-cutting factors involved. Analysis in this dissertation indeed points toward the necessity of political leadership as an important superior factor essential for the successful implementation of the CA. In its absence, individual institutions tend to pursue their own objectives under the CA and coherence is achieved either in terms of the lowest common denominator or the strongest voice. With regard to the role of inter-institutionally agreed documents, guidance on common document elements and necessary mutual consultation would facilitate comprehensive crisis management. Due to the lack of a coherent CA framework, actors depend on a kind of implicit consensus which can only be ensured by political authority (Major/Schöndorf 2011, p. 6). Political leadership is essential for the conceptual and strategic developments in order to buttress these with the necessary resources and to actually demand structural changes even against opposition. Of the cases

522 As argued by Teuten and Korski (2010, p. 150) in view of the UK, "departments do not need a new hybrid language that replaces their current tongues."

523 This approach is also advocated by Patrick and Brown (2007a, pp. 129/130).

analyzed, only the UK has demonstrated high-level political support behind CA initiatives. Underlining the need of strong political will and commitment, the political backing of the CA in the UK has made essential reform processes possible. The necessity of political leadership and direction also applies for the EU on the international level. On the national level, the respective center of government should behave proactively promoting and ensuring interlinking of strategic planning.[524] No general rules should determine which region, country or issue should be subject to comprehensive strategy planning. To avoid unnecessary coordination and transaction costs "the centre of government needs to determine which strategies are to be produced on the basis of a government wide assessment of threats and opportunities" (Teuten/Korski 2010, p. 27).

An important aspect in this regard is the development of an 'intellectual foundation' for CA concepts which seems necessary before a 'strategic synchronization' in terms of conceptually moving into one CA direction can take place. Frank debate over the goals of crisis management coherence should not be avoided in this regard. However, so far, the CA debates on the organizational dimension of its implementation reveal the tendency to develop back to procedural discussions to avoid controversy rather than tackling more substantive problems. The risk that any CA conceptualization entails in this regard is that the concept has to remain tailored to the specific crisis management circumstances. Otherwise, the CA is lifted into merely conceptual spheres without being manageable in reality. Furthermore, presented as good in itself so far has covered "the fact that the values driving it are not shared universally" (Macrae/Leader 2000, p. 5). Interviewees pointed out that it is easy to support and to be in agreement on the CA and its overall goals on a conceptual level as it is a reasonable approach. While this contains the risk of overusing the CA as a buzzword, as soon as details on the CA's means and ends are officially stated, different policy approaches and disagreement become apparent.

As indicated in several interviews, another major issue of the CA debate is that 'everybody just talks'. Talks in CA terms, however, should not be about the CA itself but about comprehensive efforts. At present, a kind of 'conceptual double consciousness' can be observed, with crisis management actors thinking in terms of a multi-dimensional reality. On the one hand, the CA is handled along traditional lines of crisis management, such as existing principles and objectives, status and the position of different actors in the crisis management system. On the other hand, actors heavily emphasize collective crisis management responsibilities and the importance of taking into account other actors to increase

524 Referring to WoG approaches, the OECD's study (OECD 2006, p. 22) affirms that "effective political leadership is key for the implementation of all-encompassing approaches."

crisis management coherence and adapt to CA demands. According to respondents, a CA mentality on conceptual level would also demand that actors accept weaker positions and to set own interests aside for the 'greater good'. Efforts to implement the CA in crisis management both on national as well as on international level illustrate this paradox. The crisis management machinery is engaged to defend own interests while at the same time being highly conscious of common crisis management purposes and objectives. As it turned out, conceptual CA dominance does not necessarily add up to comprehensive crisis management *per se* and a major division between rhetoric demanding a CA and policy realities of implementing coherence at international and national level exists. In this regard, the overall value of existing CA concepts can be questioned mainly due to the fact that it mostly expresses what should be obtained but rarely describes how.

5.1.2 Organizational Dimension

Looking across the cases with regard to the organizational dimension of CA implementation, it is first of all worth mentioning that the empirical analysis underlined initial theoretical assumptions, namely, that institutional setups matter in making CA implementation and crisis management policy coherence a reality. It also revealed that variations of organizing the CA in existing crisis management structures have emerged. In this context, also the signaling, framing, iteration and exchange as the facilitating factors of organizational reforms proved relevant factors with regard to CA implementation and have been discernable to different degrees in the individual cases. Though the EU and the three member states analyzed all can be regarded as key supporters of the CA agenda, the analysis has demonstrated major differences between individual efforts to organizationally adjust to the concept. National governments and also the EU experiment with different organizational models in an attempt to improve their crisis management coherence and answer the questions how to implement the CA in different ways. The widely shared CA premise led to quite different organizational results and despite the apparently consensual goal of crisis management policy coherence, means and results of CA endeavors varied for each case. Yet, in remodeling the organizational CA setting and positioning structures, all analyzed cases presently have better conditions than they had at the beginning. While the form and utilization and also the visibility or invisibility of cross-governmental institutions and coordination mechanisms constitute a major discrepancy between the actors analyzed, generally, in all cases some sort of support structures to facilitate crisis management coherence were created. At the same time, several interviewees pointed out that while the CA concept is omnipresent on the conceptual side, there still is a huge confusion on what it actually means in practical organizational terms. The great concern of practitioners is that they want to know more exactly how organizational design and redesign affect a more or less successful implementation of the CA and how it affects coordination choices in their respective ministries and institutions. Though the analysis does not allow a direct derivation of how CA coordination mechanisms have to be organized, some generic lessons can be learned as regards future efforts to implement the CA. These provide a menu of tools that can be used to improve coordination.

Different options exist for enhancing the implementation of the CA crisis management coherence such as the establishments of legal requirements to interact, institutional arrangements that foster coordination and initiatives that raise the awareness about the benefits of coordination. Efforts to adjust existing crisis management structures to the demands of comprehensive crisis management tended to consist of the creation of additional coordinating institutions or mechanisms and institutions responsible for complex crisis management efforts. The spectrum of these range from enlarged possibilities of horizontal coordination between crisis management actors centrally in the core executive in the UK, to a focus on increased vertical coordination between Government Offices and agencies in Sweden. Sometimes the CA is about inter-ministerial meetings to keep relevant actors mutually informed. This is the coordination approach dominant in Sweden with its dual administrative system and an emphasis on consensus and joint ownership, collegial steering and negotiation bodies in which informal forms of coordination are dominant and hierarchically structured instruments for coordination are rare. Also in Germany, CA efforts include the establishment of 'softer' forums for interaction through training and seminars. This has given birth to institutions such as the BAKS or the ZIF. Steps taken in Germany on structural level represent minor and cautious changes which are impeded by a set of problems on the strategic level. CA implementation in Germany is also set-up by the Inter-Ministerial Steering Group for Civilian Crisis Prevention as a specific mechanism of inter-ministerial interaction to determine core areas of cooperation. Besides the steps to implement the CA taken on the member state level, in parallel numerous initiatives were taken by the EU to strengthen coherence and coordination. The nature and structure of EU crisis management provided the Union with a fundamentally different starting point for CA implementation compared to the efforts of the analyzed country cases. The provisions of the Lisbon Treaty, such as the creation of the EEAS and the HR/VP provide for a framework to strengthen the EU's capacity to address the upcoming challenges through an improved coherence, better institutional coordination and enhanced strategic decision-making.[525]

525 The specific measures of CA organization are evaluated in the conclusions of the respective case study and not taken up again in this overall conclusion.

In doing so, CA implementation took place in a way that corresponds with the particular background of organizational form and is influenced by respective features of administration. CA approaches and policies set up to manage the performance of crisis management coordination are highly institution and context-based. Thus, approaches adopted differ in accordance with the particular administrative-political framework and crisis management context in which the respective institution or government operates. Except for efforts on EU-level, none of the outlined possibilities to enhance coordination and coherence modify or reconstruct the organizational set-up among the entities involved. Instead, they present approaches toward a coherent and joint effort to comprehensive crisis management that predominantly focus on the relationship between existing crisis management bodies. In this regard, institutional innovation strictly followed traditional organizational and ministerial logics. Domestic administrations responded to the requirements of CA implementation through their own structural repertoires and operating procedures. This might be a reason that despite these measures, inter-institutional rivalries and different institutional weights continue to play an important role when it comes to efforts to achieve crisis management coherence. The most fundamental reworking of crisis management governing machinery took place in the UK with profound structural innovations and institutional transformations to improve inter-departmental crisis management coordination. These included the establishment of joint funding mechanisms as central financial coordination mechanism across departments and the creation of cross-departmental and tri-departmental coordination bodies. In this respect, the UK's SU can serve as an exemplary model of organizational CA implementation and sheds some light on potential benefits and drawbacks of the creation of new, permanent organizational entities to improve coherence.

From a coordination and coherence viewpoint, it makes sense to merge organizations and to bring officials into one organization (Drent 2011, p. 13). The creation of such a unit seeks to enhance the resolution of differing views by internalizing discussion. Indeed, the UK experience from this effort suggests that standing inter-institutional bodies increase CA sensitivities, institutionalize and facilitate systematically joint analysis and more systematic coordination of crisis management matters and improve inter-organizational awareness and institutional learning. Experience indicated that standing dedicated cross-organizational bodies can generate a focal point for crisis management coordination. Merging officials from different institutions additionally facilitates 'reach-back' to their home institutions. On the contrary, experiences prove the difficulty of coordinating other relevant actors from one single institution, even if it draws officials from these other actors. Establishing organizational structures, in particular when drawing on existing distinct ones, seems to provide only minor incentives to work out coherent responses. As the UK case showed, it seems important, that inter-institutional bodies are created under the highest possible authority to upstream interaction that in return also facilitates cooperation and coherence on lower levels. On the contrary, understaffed und underfinanced, new organizational entities lack political weight and remain too weak to generate policy coherences outside their own entity, as the German case study demonstrated where the Inter-Ministerial Steering Group as a coordinating institution has not been empowered with its own support structures and financial resources needed to become operational. Without distinct competencies to make decisions, a critical amount of financial and human resources and demonstrable political backing, newly created coordination bodies cannot function effectively and rather duplicate than synergize actors involved.

Interestingly, similar observations could be made as a consequence of the creation of the EEAS on EU level. Yet, the creation of the EEAS firstly showed that the establishment or imposition of a new institution with own competencies and resources, which are partly newly created or drawn from existing bodies, can create rivalry and envy and increase coordination problems. The above analysis both on the EU as well as member state level made clear that the creation of a new institution in order to increase crisis management coherence entails a high risk of competition with existing bodies. The establishment of a new organizational body from scratch leaves it with a lack of an accepted power-base and clear authority position compared to other actors. While the creation of such bodies has highly visible effects on coordination, their 'visibility' may also be viewed as undesirable and threatening by existing actors which stand to lose their status or position. Furthermore, the analysis revealed that while such bodies set a clear framework for interaction, there is a risk that permanent coordination bodies become just as heavily institutionalized themselves and as such simply present another institution that needs to be coordinated.[526] While the construction of new institutional levels reflects efforts to enhance crisis management coherence through structural responsiveness, it also increases already complex inter-organizational relations. To an extent, the creation of whole new coordination machinery adds to fragmentation and complexity and can actually complicate crisis management organization further. Moreover, it entails the danger of creating machinery and structures with overlapping mandates raising accountability problems. Another drawback is the organizational cost of creating new organizational structures. In addition, the empirical examination of the SU showed that the institutional creation of new structures demands time to achieve effects such

526 As noted in Bourgon (2011, p. 94), there is a risk that "[r]e-organization simply creates new boundaries that will need to be crossed to address the next set of issues."

as the reduction of organizational tensions and the establishment of coordination routines. Consequently, various potential conflicts between the CA agenda and the establishment of new, permanent organizational entities can be observed.[527]

The analysis revealed that yet another 'institutional add-on' does not in itself generate incentives for improved interaction. A key ongoing challenge is the further weakening of strong cultural barriers between actors and the need for a capacity to understand the whole picture persists. This points to the need of informal coordination mechanisms in terms of a 'culture of coordination' to take root. Changing the culture of institutions is a popular theme of analyzed cases and contact between officials by joined training and respective competencies for staff as well as appraisal systems have increased. Also, career incentives are important mechanisms to this end. Indeed, the importance of joint education and training to overcome cultural differences has been pointed out in most interviews. Respondents also underlined the importance of creating platforms for subject- and goal-oriented dialogue and exchange. In this way, the implementation of the CA also includes the establishment of an appropriate human resources policy which is increasingly acknowledged. The empirical investigation showed that the effective implementation of the CA should involve to expect staff to move across institutions in their careers. In this context, the benefits of staff exchange, cross-cutting activities and inter-organizational experiences should be highlighted and more visibly rewarded. Incentives should be given to avoid the aversion of staff toward such exchanges. In this regard, more formal and well-communicated programs could persuade officials to gain such experience across the wider crisis management sector. Career progression, skills and competencies of professionals should be judged and reviewed on their performance to guarantee that appropriate weight is given to understanding inter-organizational working (Teuten/Korski 2010, p. 101).[528] This would require a significant shift of acknowledgement across relevant institutions while at the same time such reforms are difficult to quantify and it takes some time until benefits become ob-

527 See also Patrick and Brown (2007a, p. 133) for pros and cons of standing cross-organizational bodies.

528 As suggested in the UK's House of Common's report (House of Commons 2010, Ev 148) on the CA "[M]ore 'comprehensiveness' will need to be built into the way the British government recruits, trains, gives incentives and promotes civil servants – so a new generation of officials can be brought up to work differently. But rather than developing diplomats, soldiers, or development workers – or trying to teach each cadre of about the other – it may be necessary in the long-term to go further and develop officials who are equally at home in all government departments and who are encouraged to work together and in each other's offices in order to progress through the ranks." See also arguments in Ling (2002, p. 629) Patrick and Brown (2007b, p. 3) as well as Penttinen (2010, p. 48).

servable. However, the case studies also revealed restraints of such efforts. The benefits of staff exchange, for instance, largely depend on their integration into the new working environment and how both institutions utilize the officers' insights on the other actor respectively. It is important to clarify the function and responsibility of liaison officers as well as their integration into day-to-day work.

Of the analyzed cases, only the UK has a permanent pooled funding system for crisis management. Such funds reflect the fact that today's complex crisis management efforts involve operating expenses that are neither entirely military nor merely focused on development, but rather politically driven (Stepputat 2009a, p. 9). In addition, coordination mechanisms which pool resources play a crucial role in the CA context by supporting its fundamental objective of making efficient and coherent use of available resources. The set-up of inter-organizational funds is considered to promote policy coherence, encourage actors to work more closely together and increase joint activities of relevant institutions. Following a consensual allocation process, pooled funds can be considered a rather successful instrument to facilitate shared understanding, inter-institutional awareness and coordination.[529] On the contrary, also structures of joint-funding mechanisms need to be understood within their own limitations and are not able to break down different views on crucial crisis management objectives and compensating deep disagreements. As the CPPs and also the *Nachtwei-Millionen* in Germany revealed, joined funding entails the risk of being used merely in terms of an additional funding possibility alongside the regular budget. Thus, a pooled budget is not necessarily perceived and used as resources for integrated activities and can discourage common efforts. From this perspective, a joint budget creates insular projects and leads to competition. Also the amount of money available in pooled funds plays a role as this has to be high enough to actually serve as an incentive for coordination. From this perspective, despite their potential, pooled funds so far remained rather modest in scope. The increasing lack of financial resources nonetheless will in the long term probably be the CA's greatest ally. Generally, interviewees were keen to point out that the appreciation and effectiveness of CA reforms largely rely on resources.

Various steps toward more comprehensive crisis management have been taken and it is justified to state that structural measures have been going in the

529 As evaluated by Patrick and Brown (2007a, p. 134) with regard to the UK's CCPs, these show "that such instruments can bring relevant agencies to the table and encourage buy-in, lead to compromises on objectives, reduce time lags for addressing urgent needs, and facilitate the conducting of joint assessments and the formulation of genuinely integrated country strategies. Over time, they may play a socializing role among agencies, in helping participants understand the perspectives of their counterparts."

right direction. In a differentiated way, these have made useful contributions toward more coherence. However, while organizational reform efforts toward a CA in crisis management had some positive effects in particular in raising inter-institutional awareness, they did not solve some of the fundamental problems and more essentially, are not aligned with the complex problems they are expected to address anymore. In some regards, both on EU and on national level, crisis management structures appear to be remarkably resistant to changes (Major/Schöndorf 2011, p. 7). In an attempt to explain why crisis management coherence appears to remain limited despite these efforts, several related and mutually reinforcing factors can be put forward. An important aspect is the deep-running institutional fragmentation of crisis management systems with each actor generating autonomous subsystems. Furthermore, legal restraints and the ideological conviction among crisis management actors that their capabilities should only be used to narrowly serve their own goals and interests oppose efforts to increase coherence. According to background talks, the CA is dealt with individually in the various institutions, ministries, committees, cross-cutting units and working groups. Various crisis management elements are not managed as a coherent system but are addressed as discrete components. Thus, the protection of vested rights and particular interests ensured through constitutional provisions dominates new coordination mechanisms. Structurally, coordination also implies a subsumption in the larger context. This may mean that freedom of action, competencies or financial resources can be limited. Often newly created instruments to ensure coordination lack power themselves and are also insufficiently linked to higher levels of power and access to financial resources to be effective.

In this context, an observation by Christensen and Lægreid (2006, pp. 19/20) is particularly applicable to the CA context where "[t]he desire to coordinate is greater than the wish to be coordinated. In other words we are confronted with a coordination paradox, which assumes that all agree on the need for coordination, so long as it does not apply to their own organization." The empirical analysis suggests that the CA also is a matter of guidance and that CA buy-in and absorption created at the highest level can be helpful for coordination. Conceptual and structural deficiencies found in this analysis point to a superior factor central to the success of CA implementation, namely political leadership to structurally implement the CA which has been a crucial deficit in most cases analyzed.[530] Leadership is also a factor which has been demanded by respondents. In cases of limited strategic and political direction, CA arrangements en-

530 See also a report by the OECD (2006, p. 20). Only efforts to produce crisis management policy coherence in the UK entailed a comparably high degree of policy direction and high-level commitment due to the prime minister's strong role on high-level issues.

tailed a high degree of *ad hoc* and informal arrangements. While such an approach can be valuable and positive, as the Swedish model revealed, often high-level remit seems necessary to complement efforts.[531] Leadership is necessary to develop strategies, to substantially underpin these structurally and financially and to defend reforms against opposition. In particular, as the CA does not provide one coherent framework, actors depend on a kind of implicit consensus which can only be ensured by political authority. Institutional CA reorganization seems to necessitate promotion in a top-down direction in which high-level political backing, clear political guidelines and input are essential to create momentum for coherence and cross-cutting coordination. The higher the position of the actors involved, the more impact an institution can have in raising and enhancing views or issues and "to direct rather than just facilitate the coordination of the contributions of the different entities" (Stepputat 2009a, p. 8).[532]

At the same time, this political direction has to ensure that whatever policy is developed needs to engage all relevant actors on an equal level from the outset to ensure acceptance. In addition to the leadership demand, interviewees underlined the importance of flexibility emphasizing that the CA is not just a matter of strong leadership because that way the concept would lose its strength on the lower working levels. It has to be ensured that everyone feels responsible. Rather than strictly command and control based, CA leadership has to balance the independence and interdependencies of different actors. In this regard, unavoidably the question of who should take the lead arises. The institutional background and hierarchical status of a potential driver of efforts toward policy coherence seems crucial. While there cannot be one answer to this question, it needs to be a particularly strong actor, either in terms of resources or legitimacy (Rintakoski/Autti 2008, pp. 31/32). An additional aspect is that what is not accepted and operationalised cannot be coordinated. Political acceptance is an essential requirement for any actor entitled with coordination. In this regard, Friis and Jarmyr (2008, p. 16) point out that

> "in the end, the answer to who gets to lead a comprehensive approach should depend on whether or not this actor possesses the necessary qualities for coordinating others. This includes the power to persuade other, potentially reluctant, actors (…).

531 Indeed, the Swedish country case suggests in case of a strong consensual political culture, such heavily directing political guidance and leadership may be unnecessary.

532 As also advocated by Patrick and Brown (2007a, pp. 132/133), Weiss, Spanger and Van Meurs (2009, p. 48), Johannsen (2011) and Wittkowsky et al. (2011, p. 3). See also a Chatham House report (Evans/Steven 2010, p. 16) which argues that the UK government "should appoint more ministers for cross-cutting issues. By applying this approach to foreign policy issues, the government can break down barriers between sectors, departments and countries, helping create the momentum needed to engage (…), or driving a new approach to a key policy issue across all members."

> Moreover, a leader will have to be able to serve as broker between several actors with conflicting interests. Actors might have different incentives for seeking coordination. So, to act as leader in a comprehensive approach and to reconcile these interests, it is necessary to have insight into the views and values of different actors."

An important question that has to be reflected on in this conclusion on the organizational CA dimension concerns dominant organization forms and modes of coordination. In the UK, clear hierarchical components in terms of political leadership and formal structures dominated the country's style of CA implementation. Mechanisms to improve cross-departmental coordination have been developed as overarching authority mechanisms to coordinate different parts of government. The country emphasizes strong structural incentives in order to enhance crisis management coherence. Also networks are being pressed for under the top-down style and hierarchical direction of the CA application. While these reforms largely aimed at achieving positive coordination, the UK did so in the most formal way with administrative and regulated measures. Though increasingly attracting the necessary attention, informal coordination mechanism were long regarded as being of minor importance. As an answer to unfavorable institutional preconditions, the German government chose a careful approach to build governmental substructures to organize crisis management in a coherent way. Organizational CA reforms in Germany have been mainly triggered by sectoral developments without hierarchical characteristics, political leadership and guiding decision-making. Coordination occurs irregularly, only for the time acutely needed and without an organizing or directing centre. However, while mechanisms of negative coordination dominate between ministries, platforms and networks for positive coordination have been created and informal coordination mechanisms have been considered highly valuable. In Sweden, hierarchical and market types of organization were extremely weak due to the strong position of informal coordination mechanisms and the agency system. Sweden in particular uses informal forms of network-type coordination, stimulated by the importance of consensual decision-making. Though only loosely linked, coordination is built up and maintained voluntarily and rests upon shared values, trust, and solidarity. Formal mechanisms to coordinate crisis management were established in an implicit and almost hidden way. In the EU, the Lisbon Treaty reforms follow a coordination oriented approach which largely focuses on formal structural provisions to improve inter-institutional coherence. In this context, the EU's organizational crisis management form has been reorganized merging organizational entities and by moving them both horizontally and vertically to improve coordination. In doing so, the EU showed an activism in the organizational dimension similar to the UK, with a focus on formal coordination mechanisms. However, in contrast to UK efforts, these efforts lack decision-making authority and political leadership support, both from other EU institutions as well as from member states contributing to the uncertainties surrounding competencies of newly created bodies and posts.

This summary of the above analysis shows that when it comes to CA implementation and talk about the need for more coherence and improved coordination, for some actors this signifies moving toward this objective with more hierarchical arrangements, the functional centralization and reorganization in formal and visible structures. For others, CA objectives and improved coordination can be achieved by better information sharing, personal exchanges and by working toward an informally agreed set of common principles which mostly takes place on an informal level. These are quite different visions of how to implement the CA and achieve greater crisis management coherence and coordination. The former involves the reorganization of crisis management structures, the institutionalization of new coordination entities with new hierarchies, whereas the latter avoids hierarchies and seeks to make crisis management work more efficiently as a network. Market type organization forms can be neglected in this context as the competition based approaches to interaction of market type organization forms oppose the CA idea from the beginning. Likewise, the overall objective of CA implementation cannot be to merely build hierarchies.

The analysis indicates a shared tendency in the cases analyzed to increasingly utilize network-based coordination mechanisms in a formalized manner by creating platforms fostering coordination, inter-institutional awareness and joint working as well as collegial bodies and instruments for information sharing. Indeed, according to the elaboration of the three ideal type organizational forms and their modes of coordination, network-type organization presents itself as the appropriate model for effective CA implementation from a structural perspective. The complex crisis management environment largely fits conditions that have been identified as being conducive to a shift from hierarchical coordination structures to networks. Given the demands of the contemporary crisis management scenery with its multitude of actors, tasks and timelines, the implementation of the CA in form of network organization has several advantages in dealing with above-outlined problems to achieve crisis management coherence compared to the other two organization forms.[533] The crisis management architecture consists of diverse organizations with a varying level of interaction and interdependencies as well as relationships that exist over time but in a more flexible way than in a hierarchy. Network-based, flat and flexible structures and mechanisms

533 For considerations on the different organization forms see Ness and Brechin (1988) and Bourgon (2011, p. 94). Also Herrhausen (2009) argues towards the network as the most suitable model for UN peacebuilding.

seem appropriate to deal with the high degree of complexity inherent to CA implementation. Moreover, the loose structure of a network and respective forms of coordination allows adaptation without disruption of the whole existing system. The starting point for the following deliberations is to recognize that the CA can organizationally neither function purely like a market in which individual actors pursue their individual goals with little sense of sharing common objectives, nor is it purely a hierarchy or a system of top-down structural innovation.

There are several factors promoting network type organizations in the CA context which are critical components of comprehensive crisis management. First of all, the network organizational form recognizes the existence of different working cultures. As pointed out by respondents, overcoming these by increasing formal hierarchical coordination in newly created structures is an illusion and, moreover, is not what should be aimed for in the CA context. Particularly in the crisis management field from the perspective of civilian and military crisis management actors, there are good reasons for this functional division and different specific professional views and objectives. Each actor should keep its identity and these should not become irrelevant. It should be acknowledged that some incoherence responds to legitimate interests. According to interviewees, different security cultures on the one hand make CA difficult but different points of view are also needed for a CA and single institutions and actors should not lose their sense of identity. Depending on the background and context, clear characteristics between military, political and humanitarian objectives need to remain. However, to the extent that shared backgrounds form the basis of trust within networks, different working cultures can be a limiting factor, but through regular interaction such differences can be softened by ongoing activities increasing awareness and trust. Networks allow for building up consensual knowledge. Moreover, coordination only takes place under certain circumstances and not for the sake of it. This flexibility is a major benefit given by a network due to its open-ended quality.

At the same time, as experience in the analyzed cases demonstrated, the challenge to achieve coherence in European crisis management is not simply to strengthen informal coordination and network type organization forms. While there are clear benefits to network types of organization in the CA context, the crisis management architecture and actors have been unable of effective self-organizing in the network sense. To assume that better exchange of information solves the problems of interorganizational coordination and better communication alone leads to a more coordinated and collective self-organizing network of crisis management actors seems illusory. Methods have to be adjusted to the distinctive characteristics and requirements of European crisis management. While greater coordination in the form of network type coordination is needed, it alone cannot sufficiently address the deep running differences and fragmentation as well as strategic CA gaps in the area of crisis management. In fact, one of the principal scholarly criticisms of theories of network organization is that they have "neglected [the, S.P.] role of power as an instigator of cooperation and have placed too high an expectation on consensus" (Paris 2009, p. 62). This suggests that combining the types of organization for and coordination mechanisms in a systemic way is necessary for CA progress. In fact, a hierarchical authority structure is difficult to avoid in governance matters and the framework for crisis management on the national and EU level is akin to traditional top-down approaches for achieving coordination. Governmental crisis management is hierarchical in its set-up. Furthermore, hierarchies have the benefit of dealing with situations of complexity, novelty and uncertainty by shifting these discussions into more familiar terrain. More hierarchical arrangements also satisfy political pressures to demonstrate commitment to coordination by formalizing it. While a hierarchical focus and too much centralization hold the risk of reducing policy innovation and flexibility, achieving coordination through networks cannot be taken for granted, even in networks whose members share a high level of trust and common goals.

As organization theory acknowledges the existence of mixed organization forms, coordination problems can be avoided by establishing a hierarchical element in the network. The CA does not necessitate the integration of all actors into one hierarchical structure. Horizontal coordination in networks is already taking place as the analysis showed, and this is the model on which the CA can be developed further in the organizational dimension in which some elements of hierarchy or central leadership direction are necessary to deal with the deep-running problems of incoherence and inter-organizational conflict over goals and strategies as well as turf wars that have undermined implementation so far. As mentioned previously in the conceptual leadership discussion, adding elements of hierarchy is necessary in order to address substantive disagreements. Without undermining the flexibility of networks these need to mediate rather than command. The hierarchical element enhances network efficiency and prevents it from breaking apart. It is necessary to institute elements of top-down direction, such as a lead organization to devise network-wide strategies and monitor the performance of network members. In this context, a systematic combination of both positive and negative coordination may be the optimal approach. Using both organization forms can ensure that CA implementation is as significant, inclusive and flexible as possible and that it matches cultures, mandates, structures and situational factors of the various actors.[534] Consequently, the hierar-

534 This approach to government transformation is also underlined by Bourgon (2011, pp. 103-105) and is advocated by Paris (2009) with regard to the organization of the international statebuilding organization.

chical component should establish the rules under which network coordination forms operate. The CA has to be implemented in a semi-hierarchical order that provides the framework in which coordination, consultation and negotiation takes place that resolves disputes through a deliberative forum (Thompson 2003, pp. 179/180). A balanced approach to improving coordination in comprehensive crisis management would thus entail network forms of coordination by greatly increasing information-sharing and consultation, while at the same time, modestly strengthening the hierarchical features CA implementation. Both, network and hierarchical approaches should run in parallel to each other, each highlighting certain aspects of the complex whole of CA implementation.

Concrete steps to support the existence of both hierarchical and network types of organization to implement the CA could include visible high-level recognition of the importance of the CA, guiding policy statements and documents from the highest authority which clearly articulates expectations of inter-organizational coordination and recognizes this with funds created to cover coordination costs and also reflects the CA in the human resources policy as outlined above. The role of leadership is to create an environment conducive to the CA. At the same time, network patterns should be institutionalized that go beyond mere information sharing and support dialogue so that actors consider and adapt CA approaches in their policies and programmes. Yet, it is hard to envisage that the issues which prevent a successful CA implementation will be resolved merely by such reforms. As Page (2008, p. 149) observed in the context of Joined-Up Government reforms in the UK, "any silos underpinned by political conflicts – of the turf, budget, or bureaucratic politics types – are unlikely to addressed significantly by measures which address the civil service alone and fall short of such a political accommodation." Indeed, the recurrence of a consistent set of critical but unsolved issues suggests that the process of organizational CA implementation is a lethargic and halting one. Crisis management actors should strengthen the strategic as well as the organizational dimension of the CA with the objective to fill the gap between the high conceptual policy commitments to crisis management coherence and the actual realization of recognized goals.[535] In this regard, the successful implementation of the CA and the effective transformation of crisis management policy toward increased coherence is not just about establishing the 'right' concepts and structures but about cultivating the right background for coordination. Building appropriate mechanisms among relevant insti-

535 One interviewee compared the situation exemplarily with the situation of an alcoholic who, even if he acknowledges to himself being one, still has to stop drinking which is the hardest thing.

tutions is important, but not sufficient. A greater sense of a CA community should be promoted.

These elaborations show that the challenge of CA implementation is not simply to increase coherence by strengthening coordination, but relies on the capability to develop organization methods that are calibrated to the distinctive characteristics and requirements of comprehensive crisis management. The most effective approach is likely to be a combination of both hierarchical and network types of coordination to implement the CA. Despite the attention paid to the CA and crisis management coordination, the major problem toward policy coherence is the lack of structures, practices and policy that facilitates the co-existence of both, hierarchy and networks. So far, none of the measures to increase crisis management coherence as analyzed in this dissertation seems to strike the right "balance between an overly rigid hierarchy and an under-organized network" (Paris 2009, p. 71). It is the mix of organization mechanisms that matters in implementing the CA organizationally. At the same time, the variety of actor-specific structures, practices and political and administrative cultures means that there is no single best model.

5.1.3 European Dimension and EU Interface

The empirical analysis of this dimension involved two steps. Firstly, the EU's CA efforts have been analyzed. Assessments from this scrutiny have been incorporated in the previous chapters of this conclusion. Secondly, the EU's impact on national crisis management and CA attempts of three EU member states and vice versa have been looked at. It is the wrapping up of the latter under the concept of Europeanization with which this chapter is concerned.

As the analysis showed, the UK, Germany and Sweden all approach the EU to a different extent and for different reasons in the context of international crisis management. In the UK, Europeanization of crisis management takes place in terms of potential operational benefits which would be difficult for the UK to obtain independently. The EU's role is limited to the extent that the success of crisis management initiatives rests on extending or withholding resources and to influence the behavior of other states. For Germany, the EU provides an important source of legitimacy to crisis management efforts initiated by the country. The EU's backing becomes particularly important in situations which concern the German principles of international engagement and is critical for sustaining domestic support. For Sweden, the EU offers an institutional platform to formalize, broaden and also to enforce the country's norms and principles. As the analysis of the characteristics of the EU's CA showed, the German and Swedish approaches to crisis management coherence come very close to the characteristics of the Union's. EU crisis management, from a member state perspective, is characterized by different expectations and assumptions about its crisis management responsibilities and priorities. The empirical analysis of CA implementation has shown that the limits toward a Whole-of-Europe approach clearly lie in the persistency of national characteristics of international crisis management.

In addition to overall differences as regards EU crisis management commitment, another barrier to the development of and progress of the CA lies in divergent modes of implementation and progress being made in conceptual thinking on the part of the member states. Whereas the UK has a relatively advanced CA thinking and implementation and kept the CA issue as a whole on the priority list with the necessary commitment, Germany and Sweden have traditionally emphasized civilian crisis management and aspects of development cooperation. While the international framework is taken into account, national CA conceptions predominantly focused on domestic cross-governmental coherence. How they relate to the international, in particular the EU set-up, differed widely. Particularly, CA implementation in the UK evolved with a very limited projection toward the EU and a clear focus on NATO and *ad hoc* coalitions. Due to the strong German preference for multinational frameworks in crisis management, the German government absorbed the EU agenda of policy coherence. Germany leans heavily on the EU in devising its own crisis management objectives also in terms of identifying and developing policy for a CA. On the contrary, though Germany faced severe pressures to engage more actively in EU military crisis management, it maintained many of its crisis management characteristics. The limitations to Europeanization of German crisis management become evident when it comes to the operationalization of nevessary institutional structures. Also in Sweden, a coherent EU crisis management agenda and effort constitute an important driver. Sweden has been active developing a CA to crisis management, but insisted on strengthening the civilian side of the EU's crisis management. Particularly the Swedish country case study showed that the Europeanization of national crisis management structures at the same time implies a need for increased coordination among national actors taking part and can thus lead to closer relationships.

CA responses in all analyzed country cases largely rely on national patterns. As already observed in the analysis of overall CA approaches, crisis management positions and preferences are different in some key defining aspects. These are mirrored in the European dimension of CA implementation, such as how to strike the balance between military and civilian crisis management means in order to enhance and contribute to EU crisis management coherence. These shifting views are based on factors that originate from power and influence considerations as well as from different historical points of reference and values which have made them develop different preferences with regard to both the EU and the CA. From this perspective, it can be argued that the implementation of the CA at a domestic level can actually be counterproductive in seeking to work out a European CA as national approaches are based on national objectives and concerns as a primary state-centric consideration. Attempts to create a more coherent EU crisis management policy reflect the lacking clarity between and in member states on what the CA implementation should be modeled after and moreover, what EU crisis management is really supposed to accomplish. Unclear terminology and lack of conceptual clarity both on national and international level hampers coordination and communication. Member states' diverging positions to EU crisis management and CA efforts are rooted in conflicting national institutional crisis management logics. None of these logics has become dominant at the EU level. This suggests that as long as several logics continue to co-exist, disagreement on the future development of European comprehensive crisis management will persist. It is the member states' different striving at EU level which impedes the development of a more coherent framework.

Particularly relevant in this regard is that while the EU's CA model can be described to be directed by a norm-driven crisis management approach, the member states almost suspiciously evaluate these normative CA features to match respective national interests and to be in accordance with national traditions. The nature of crisis management of the EU makes it possible for domestic crisis management policies to show signs of both taking the EU into account and rejecting it at the same time. In addition to the availability of respective structures and institutions and adequate capabilities to put them into place on EU level, the key prerequisite for the implementation of the CA under a Whole-of-Europe approach is clarity among member states how they attempt to respond, both with which measures and in which multinational framework. The analysis also indicates that EU crisis management and CA efforts are used in different ways to positions themselves on the international stage. It can be an instrument for strengthening or adding another 'muscle' to the existing national crisis management approach and to promote national interests and objectives, as it is the case in the UK and partly also in Sweden. For Germany and Sweden, the EU is approached due to the importance of strengthening the EU's crisis management identity and coherence as a whole and for reason of legitimization. The analysis showed that the concept of the CA and objectives behind it allowed member states to project respective national approaches to both crisis management itself and to its coherence onto EU level aiming at modeling respective EU efforts. In this regard, the bottom line is that EU CA implementation efforts reflect a number of compromises by member state governments and is the lowest common denominator upon which could be agreed on.

In addition, the possibility of the EU to develop crisis management coherence conflicts with its member states' insistence on maintaining as much competences as possible (Weiss/Spanger/Van Meurs 2009, p. 35). As the analysis on the EU level showed, the Lisbon Treaty reforms generally strengthen the intergovernmental dimension of comprehensive crisis management and reinforce the role of the member states. Although the structural and functional borders between the competencies of the member states, the Commission and the Council have become blurred with the creation of new coordinating provisions, member states continue to hold their crisis management domains. Thus, also after Lisbon, the EU's international crisis management policy continues to be dominated by intergovernmental procedures. This indicates that the EU member states continue to largely regard their interaction in strategic national and declaratory manners. The division of EU crisis management, with civilian capacities partially and military capabilities fully in the responsibility of the member states, is a major factor that challenges envisioning a Whole-of-Europe approach to crisis management. As long as EU member states oppose giving up competencies in the area of crisis management policy and are not willing to abandon sovereignty in important areas, crisis management coherence on EU level remains an unfulfilled promise.

However, at the same time, the Lisbon Treaty framework weakens crisis management interaction between the EU and its member states by reducing member state access to relevant structures and decreasing opportunities for visible action, agenda-setting and for lobbying their issues due to the abolition of the rotating presidency. This is likely to result in less committed member states and more reluctance toward EU crisis management engagements. Indeed, four years after the entry into force of the Lisbon Treaty, member states seem to have lost interest in EU crisis management instruments and in addition, increasingly question their usefulness.[536] Member state interest in this regard is not likely to grow which will ultimately result in a loss of significance of the EU as a crisis management actor. The purpose of the EU's Lisbon Treaty was to enhance the unity of the member states' foreign policies and the coherence of EU external action. As manifested recently by the lack of unity and delayed action in the wake of the Arab revolutions, crisis management stalemate exists since Lisbon despite the aim of an increased capacity for the EU to act.

From the explorations in this dissertation it may be concluded that there is no such thing as a Whole-of-Europe approach and also that there is no Whole-of-Europe model in crisis management in sight. It becomes clear from the above that the EU itself does not provide a model of comprehensive crisis management for member states to follow nor that it does offer them guidance for handling CA implementation. Quite the contrary, despite a great deal of emphasis on a com-

536 See, for instance, the assessment by Major and Bail (2011, p. 35) in the dimension of civilian crisis management.

mon European approach and common values, crisis management backgrounds and actual mechanisms vary from country to country. The three countries in question have differed in their approaches to the CA as a concept and its organizational implementation. These differences stem from domestic organizational and political cultures and international affiliations. Moreover, individual member states are protective of their ways of crisis management and respective values and objectives. Just as the EU has not managed to iron out the wide variations in crisis management approaches among its own institutions, wide variations continue to exist among individual member states. Consequently, every actor adjusts its crisis management policy more or less on its own conditions. The survey of the different approaches underscores that its unevenness is possibly the Europe's greatest challenge with regard to comprehensive crisis management.

EU crisis management shortcoming and coherence difficulties in successfully implementing the CA and the incoherent picture of EU conflict management are often explained with the specific characteristics of the EU, its lacking organizing ability and institutional barriers. However, after the analysis undertaken in this thesis, it seems that it is not the structural and institutional particularities of the Union that explain incoherence, but rather the intricacy of merging respective crisis management policies and linking respective units as such. Not just the EU but also none of the member states managed to easily adopt the CA and increase crisis management coherence. Furthermore, it is notable, that while individual EU member states worked to increase crisis management coherence by bringing separated ministries closer together, the structures developing on EU level remain deliberately disconnected, putting severe limitations to EU CA efforts from the outset in the name of member states protecting their influence. This observation allows the conclusion that it is not the EU institutional level inhibiting EU crisis management coherence, but that the real dilemma has to be located at the level of member state will to enable EU crisis management institutions to function in a coherent way and to actually work toward a Whole-of-Europe Approach. Despite an extraordinary collection of political and institutional crisis management structures that span Europe, the delivery of the CA relies on individual nation states. EU member state governments command more power than any other set of actors to exercise a positive influence on EU crisis management coordination. Constructing a CA is essentially a bottom-up process that must be transported on to the European agenda.

The Europeanization of the CA thus ultimately depends on the commitment of governments. Moreover, the increasing demands by EU crisis management places on the responsibility on EU member states to deliver this commitment to EU coordination. This being said, European crisis management coherence is only possible if the member states agree on a common strategy to overcome specific conceptual obstacles and at the same time aim at national policies which are conform to EU positions. Linking the work of national ministries as well as national concepts must go hand in hand with EU level development and implementation. To incorporate a suggestion by Ekengren and Sundelius (2004, p. 111) coordination in crisis management should be seen "in terms of a notion of *complementary* sovereignty between the national and the EU levels, rather than *shared* or *divided* sovereignty". To this end, it is also highly important that EU institutions clarify their ability to bring added value. Achieving multidimensional European coherence in this regard also depends on strategic guidelines and priorities addressed by the whole crisis management apparatus and concerted efforts by EU member states to ensure coherence in the formulation and implementation of the CA (De Vasconcelos 2010, pp. 12/13).

5.2 Dilemmas of European Crisis Management Coherence

The analysis has pointed to various unresolved, and perhaps unsolvable, problems and dilemmas which became evident in the practice of CA implementation. Against this background, it seems of major importance to also acknowledge these dilemmas and possible limits of the CA at the national and European level.

Increasingly, perceived deadlock and disappointing experiences combined with unrealistic expectations result in a frustration about the CA and growing questioning of the aspiration and advantages of crisis management coherence and coordination. This development is largely based on the gap between CA rhetoric and available concepts and their actual organizational implementation revealed in the above examination. Indeed, in the face of complex crisis management challenges, the CA is based on the tempting thought that the variety of relevant actors and instruments merely has to be closely coordinated to solve any problem. As articulated by Mulgan (2008, p. 187)

> "[s]ome would suggest that the quest for coordination is the twentieth-century equivalent of the medieval search for the philosopher's stone in the sense that if only we can find the right formula for coordination we can reconcile the irreconcilable. There is of course not such formula."

Crisis management coherence and the improvement of coordination cannot serve as a catch-all for deeper running problems. While being a condition for success in crisis management, coordination should not be an end in itself. "Comprehensive approaches are not a universal remedy" (Major et al. 2011, p. 136).[537] Indeed, it has been noted in various background talks that too much of a focus on

537 See also Paris (2009) addressing the coordination problem as a convenient catch-all.

CA implementation and concern for formal coordination structures, can have the effect that actors disregard their original task and important work is left undone because too much time and energy is spent on coordination. While calls for more crisis management coherence are correct, the demands to aim for a CA to the maximum has the effect that actors overlook operational contexts, deeper running disagreements and redefine interfaces and interorganizational relations in merely procedural-technocratic terms, namely as coordination problems. Furthermore, time and resources dedicated to coordination at a certain point make it necessary "to consider the point at which the cost benefit ration becomes negative. Coordination activities would risk being disproportional to the overall effort" (Johannsen 2011, p. 348). Also, the more coordination takes place on one level, the higher the risk of ignoring actors outside the coordination circle. The CA is not a mechanical process that can be created through organizational tinkering and there are limits to the extent of coordination actors can and should be occupied with. Closely related to this issue, another aspect worth being questioned is the question to what degree increased coherence and coordination come with the danger of confused lines of accountability and a weakening of important specialization and expertise of actors. It is unavoidable that different actors have widely different approaches to crisis management and particularly military and civilian personnel operate under different terms and conditions. In this regard, crisis management actors must also learn to live with partially conflicting goals and values. A starting point could be to recognize that diversities in mandates, strategies, approaches and practices do exist for good reasons.

Apart from the risk of overlooking deeper disagreements between actors and diverting attention away from areas of importance, another problematic issue is the excessive focus on matters of coordination which leads to expectations which are raised too high regarding organizational fixes of problems at hand. Considering deep-running contradictions between talk and actions owing to incompatible 'material' and normative demands placed on the CA, Christensen et al. (2007, p. 152) observe that organizational reform and change mostly takes place in face of this gap between performance and expectations. In these terms, the dynamics of reform can also result in a "problem of over-selling" (Christensen et al. 2007, p. 152). As reforms are often met with skepticism and opposition by relevant actors and organizations, in order to enable support, reform advocates often promise more than in fact possible to achieve. In consequence, expectations grow to be unrealistically high, with consequent frustration among actors when results are not reached.

> "In other words, reforms can increase expectations more than they can increase performance, and the gap between expectations and performance can be even larger after the reforms have been initiated, even though performance may have improved" (Christensen et al. 2007, p. 152).

An unfortunate effect of this is that implementing respective reforms becomes as much about correct interpretations as about working out the original issues the reform was meant to consider. This theoretic scenario can be detected in efforts to implement the CA. Analysis revealed that CA reforms, in their idealistic exuberance, always demanded more than the reality of the crisis management policy-making process yields. The CA has been pushed heavily rhetorically and carried around as the new mantra. The hold of the normative ideal of the concept and overall crisis management coherence by almost all actors resulted in a 'selling' of these which has become overblown and oversold. As the empirical analysis showed, the CA is very hard to implement and the perfect and all-fitting CA plan or coordination mechanism does not exist. The concept suffers from exaggerated expectations which can not be achieved and in the end prove counterproductive by reinforcing mechanisms to oppose reforms. Considering the limits of coherence in terms of the coordination-expectations gap and to avoid risks of overstretches and unrealistic expectations, it is essential for CA ambitions to have a realistic mandate and pragmatic approach. Such a mandate builds more on observed lessons learned of past CA efforts than on faith and reputation. To this end, a basic and open debate between the various actors is needed to overcome hardened standpoints, accusations and recriminations and to find a sense of direction for an adequate and pragmatic approach. Such debates, however, seem particularly hard in the CA context.

Another dilemma of CA implementation and largely unpredictable variable is the role of political interest and interrelated questions on the availability of resources.[538] "As is so often the case, coordination (or any other virtue) may be achievable without special mechanisms if there is the will to coordinate, but no mechanism is sufficient if there is an absence of will" (Peters 1998b, p. 52). This quote proves very true concerning the research subject and often, the CA largely is a political problem. All conceptual and structural efforts cannot compensate for a lack of political will and commitment when it comes to CA implementation.[539] Indeed, disagreement and fragmentation in crisis management often seems more political than practical and also the answers to questions of coordination are essentially political. The degree of political importance ascribed to a conflict or a particular country also determines the level of CA efforts and the use of crisis management instruments in CA terms. As the then Minister for the British Armed Forces Bill Rammell stated when giving evidence to the UK House of Commons Defence Committee (House of Commons 2010, Ev 68),

538 See also studies in OECD (2006) which indicate this.

539 See also remarks on the role of political leadership and guidance in CA implementation.

> "[u]ltimately, this is about political will. It is about the relevant secretaries of state coming together and pushing and persuading their Whitehall departments to break down the barriers, to cut through the bureaucracy, to challenge the cultures and say, 'you have got to work at it in this way'. No amount of structural re-orientation is a substitute for that."

This quote points to another CA dilemma. The extent to which actors and institutions are willing to implement the CA and to promote cross-departmental coherence heavily depends on personalities, personal chemistry, sensitivities and understanding reached between relevant staff. A recurring theme of interviews has been that the task of coordination is as personal and idiosyncratic as it is technical and demanding. Successful CA application essentially is about people, individual competence and willingness to act according to the CA paradigm, how they work together, the beliefs they hold and the way their behaviors are influenced. As stated by a respondent, one can force meetings but cannot change the minds of people. Reformers should be aware of this 'certain amount of anarchy' that makes the system less efficient. The personal understanding between individuals in the CA context is particularly exacerbated by lack of confidence between crisis management institutions in terms of mistrust and suspicion. While some structural changes can foster and complement this interpersonal dimension, however, possibilities to organizationally balance the human factor in CA implementation is limited and can hardly be controlled and entirely overcome. 'CA common sense' in social interaction is in fact not common. In CA reforms, the resistance to change by human beings often seems to be underestimated. However, the willingness and motivation for officials to take up the opportunities created to improve coordination and increase inter-organizational awareness is even more important than the institutions creating those opportunities. While not controllable, this personality-dependent nature of CA implementation points toward the increased need for network structures as outlined above.

Other contextual factors are regularly and seriously underestimated. In this regard, for instance, the time required to reach a CA on any level, nationally or internationally, should not be underrated. It should be kept in mind that the CA needs sufficient time to change working routines or organizational cultures and to build up a common understanding among institutions. As stated by respondents, practice needs a lot of time organizing the new structures. Its implementation thus needs to be seen as an ongoing process. This also requires sustained effort to explore and experiment to find out what works in practice. It should be kept in mind that the application of the CA still is "a work in progress, and has to a great extent been a case of trial and error" (Nilsson et al. 2008, p. 58).

Related to this, the major limit of CA as experienced throughout the research for this thesis is the profound exhaustion on Comprehensive Approach matters, in particular among practitioners. Frustration and exhaustion seems to be particularly fed by discussions about CA role models which as mentioned in background talks, are interesting merely within the academic ivory tower while practitioners are confronted with the reality at hand. This tiredness is largely counterproductive to accommodating the CA. Negative attitudes largely limit beliefs about the possibilities of the CA in practice, which are hardly ever addressed.[540] Observing this development at the time of writing, it seems doubtful that the above-outlined challenges and problems will be adequately addressed and that the political appetite exists toward a Whole-of Europe approach.

5.3 Further Research

The previous analysis and its findings make it possible to highlight some areas of relevant and interesting future research. There are a number of challenges and topics that this dissertation alluded to but which, in the interest of thematic precision, could not be discussed further. Yet, some of these subject matters deserve further exploration.

Research experiences from writing on this subject point to the necessity to further narrow the topic down by focusing on specific crisis management challenges characteristic for the CA, for instance a thematic study on SSR or DDR, or by concentrating analysis on specific conflicts on a case-by-case-basis. With regard to the analysis of crisis management activities in specific regions, a study of the EU's long-term crisis management engagement in Bosnia and Herzegovina promises particular insights. Here, the EU has utilized a large set of its crisis management instruments comprising both military and civilian components as well as Commission activities. Analyzing the outcomes and impact in concrete crisis management activities also allows judging the utilization and effectiveness of mechanisms to ensure crisis management coherence, such as pooled funding or cross-governmental coordination arrangements. Another interesting thematic narrowing could be to look at the coherence of specific crisis management mechanisms such as early warning or evaluation measures. As these aspects were left out of the picture, analyzing the CA in more concrete operational crisis management activities

540 See also article by Penttinen (2010) on the role of positivity in individual dedication towards a CA and the need to place attention on the relevance of positive emotions during CA trainings. Penttinen (2010, p. 46) states that "[t]he possibility for the comprehensive approach to be operationalised is finally about individual competence and willingness to act according to the new security paradigm." She recommends to enable space in which the CA successes can be shared "in line with positive psychology, turning the positive exception into a norm" (Penttinen 2010, p. 56).

would contribute to an analysis of the arguments that are made in this thesis and present an avenue for further research.[541]

Another potential future research trajectory concerns overall changes and adjustment of the functional ministerial structure in nation states due to internationalization and the overall speedup of events such as the economic and financial crisis and the Arab Spring revolutions. These raise the overall question of how and to what degree crisis management actors on national level can maintain autonomy in decision-making in an age where every actor and ministry develops own foreign policy perspectives. The CA questions the construction of such sectors as the concept is in fact about opening up sectors and departmentalization. In particular, national ministries of foreign affairs should be prepared to support the internationalization of formerly mainly domestically oriented ministries.[542] In addition, with the establishment of special organizational bodies, questions of the division of labor and cut-off points emerge. Furthermore, internationalization results in direct links between sectoral ministries and respective international institutions which questions bilateral diplomacy via traditional embassy representation.[543] Thus, also researchers must increasingly be able to go beyond the framework of sectors or expertise on specific sectors, regions and organizations. Academics need to increasingly be able to cross various field of research, depending on the issue studied.

CA thinking needs to build on 21st century technology. Thus, another aspect deserving more attention is research toward better understanding of the

---

541 Some studies in this direction have already been undertaken to analyze the CA on the ground. See, for instance, Mustonen (2008) on the coordination of civilian and military ESDP instruments in Bosnia and Herzegovina, Johannsen's (2011) study analyzing the EU's CA in the Democratic Republic of Congo, Wendling's (2010) as well as Ehrhart and Petretto's (2012) studies which analyze Operation Atalanta as part of the EU's CA as well as Gauster's (2012) analysis on CA effectiveness on the ground in Afghanistan.

542 In a recent article, Faust and Messner (2012) question the role of the BMZ and the AA in the German ministerial structure due to the broadening of traditional questions of global development and unclear differentiations between development policy and other externally oriented policies. Faust und Messner (2010, pp. 170-174) develop four scenarios of organization between the BMZ and the AA which range from an integration of the BMZ in the AA to the establishment of a Ministry for global development covering all crucial areas of global development. They conclude that whatever form ministerial reorganization takes on in Germany, reforms are essentially needed as the BMZ in its present form will be increasing less able to effectively implement such a broad development policy portfolio. Ulrich (2011. p. 4) in this regard predicts an increase of power for operative departments and ministries such as the German Chancellery, the Ministry of Finance and the Ministry of Defence which can move financial and human resources and capabilities. Paradoxically the department suffering most from internationalization is likely to be the ministry for foreign affairs.

543 See, for instance, the paper by Bratberg (2007) which deals with the status of bilateral diplomacy and foreign ministries' foreign services in the EU.

importance of comprehensive communication as part of efforts toward crisis management coherence. The set-up of an appropriate technological infrastructure and network technologies connects policies and results in a greater level of inter-policy coordination. However, the task of establishing new infrastructures in crisis management is highly sensitive as these often go across several policy arenas with different jurisdictions (Knoepfel et al. 2011, p. 295).[544] An opposing imperative for cyber security means that the ability to communicate with each other within the government is obstructed, for instance by insufficient levels of staff security clearance impeding information access. However, despite existing cyber threats, it is important to find ways to communicate, in particular in civil-military environments. Inter-organizational information technology and information management systems represent a highly interesting area for future inter-disciplinary research and joining-up government by the creation of an 'information architecture' which is central to the question of CA implementation on the domestic, international and field level.

Technological designs of the CA should not be neglected. Technical processes and innovations impact on organizational models. Enabling information sharing and data access while maintaining confidential information through filters amongst actors by creating collaborative computer tools is the key to concretizing the CA on this level. Coherence demands actors "to have systems for managing and sharing information and to communicate electronically in a timely manner" (Teuten/Korski 2010, p. 156). The possibilities and capacity to communicate and share data poses a key challenge to the successful implementation of the CA. Operational field experience and exercises carried out underscore the need to improve alignments and links of governmental information technology and information management systems to ensure connectivity and advance communications in the area of crisis management (House of Commons 2010, p. 41).[545] Despite being connect-

544 For an introduction of the role of the internet and virtual communities in public administrations see Bourgon (2011, pp. 28-30).

545 For instance, the UK government "uses at least five different computer and network systems and can therefore only transfer classified documents and briefs inter-departmentally with the greatest difficulty" (House of Commons 2010, Ev 147). Looking at the IT dimension, the UK's FCO, MoD and DFID work towards "an action plan for how they intend to remedy the deficiencies in communication, information systems and data sharing between their Departments" (House of Commons 2010, p. 41). In addition, the UK established cross-departmental Information Steering Groups for Iraq and Afghanistan (Teuten/Korski 2010, p. 109). For a detailed analysis of the development of knowledge management capabilities, internal communication systems and technological initiatives in the UK see also Teuten and Korski (2010, pp. 156-159). On EU level, the EEAS, for instance, is included into the different IT management systems used by the Commission to manage its work to coordinate and participate in diverse activities. However, the launch of the EEAS was also accompanied by various IT problems resulting in technical delays (Helwig,/Ivan/Kostanyan 2013, p. 37). See also the OECD (2006, p. 11) briefly underlining the importance of joined-up information management systems.

ed through policy action, actors are often isolated in this regard. Thus, the necessity for institutional innovation and new management forms to ensure digital coordination will largely increase in the future (Knoepfel et al. 2011, p. 295). Interaction and interdependency of actors has to be reflected in the networking of technological systems making it necessary to respectively enforce academic thinking on the integrative power such technologies can provide.

In addition to these practical research approaches, there are several specific theoretical challenges for the research community. The multi-theoretical framework chosen in this dissertation has demonstrated some great potential and a fruitful exercise. In particular organization theory entails a range of subdisciplines, such as network theories, that could offer additional useful theoretical insights into dynamics of organizational CA implementation. Reproducing what has been dealt with earlier this chapter, network studies could provide useful insights especially with regard to questions of information technology as "there is a fascination [of network studies, S.P.] with the effect of the world wide web (www) on a whole range of political, social, and cultural relations" (Thompson 2003, p. 189).[546]

Another perspective worth further investigation is a research approach based on organizational psychology. This became apparent during interviews conducted for this dissertation and the observation has already been picked up in the previous chapter on the 'human factor' as one of the CA dilemmas in terms that individuals in organizations are highly influenced by social practices. Reform and reorganization is experienced differently and often takes place in emotional ways and in terms of perceived loyalties of officials. On a personal level, organizational reforms impact career structures and paths as well as collegiality and essentially change the familiar working environment. It is often difficult for affected staff to make judgments on the ultimate effect reforms have for them. Thus, uncertainties resulting from change are part of the problem which results in destabilization of staff morale and a detraction from coherence. For instance, the current view in Brussels seems to be that given the amount of discussion and frustration which accompanied the establishment of the EEAS, it will take rather long for an 'esprit de corps' to materialize in the new institution. In this regard, insights from Bleicher's (1991, pp. 70-83) research on *Mißtrauens-* and *Vertrauensorganisation* might provide a helpful analytical starting point and basis of orientation. It might be worth investigating this approach further and this dissertation is to be understood as a stimulus for further inquiry of this kind.

546 See Thompson (2003, pp. 189-200) on networks and information and communication technologies.

5.4 Final Reflections

This dissertation has sought to analyze the implementation of the Comprehensive Approach in European crisis management and to examine what steps have been taken on EU and on national member state level to improve crisis management coherence. In doing so, this research has sought to contribute to the ongoing debate about CA issues through a systematic discussion. It is certainly not exhaustive as the CA as a research subject reflects the nature and complexity of today's international conflicts. What it has undertaken, however, is to comprehensively highlight areas of challenge that cannot be ignored by those tasked with advancing coherence and coordination in international crisis management. Among the contributions made by the approach used in this thesis is the aim to cover European crisis management complexity and to provide a new angle that enables this study to overcome traditional narrow approaches of CA research. While being a difficult undertaking, the analysis largely benefited from this comprehensive grasp of the research subject.

A number of reforms on conceptual and on organizational levels to achieve comprehensive crisis management have been taken and practical steps indicate the importance of the CA. But this dissertation has also argued that they have not amounted to a sufficiently 'Comprehensive Approach to the Comprehensive Approach'. A statement of the 2007 report by Patrick and Brown (2007a, p. 128) on WoG approaches to fragile states in diverse OECD countries which notes that "the whole is not yet as great as the sum of its parts" still is very much state-of-the-art. Rhetorical steps have outpaced practical ones and despite various innovations, actors struggle in their CA on various levels, starting with the definition of the purposes of policy coherence, strategy formulation and guidance, the establishment of appropriate structures to facilitate and when necessary increase coordination tailored to the condition of crisis management policy and to finally ensure the alignment of their national efforts with the European level.

However, managing conflict has long been a dominant issue in international policies and is likely to remain at the centre of concerns. Actors are thus obliged to ensure that they are prepared to handle future challenges by improving strategy and structure. To achieve the scope and ambition necessary to successfully implement the CA, a paradigm shift in the understanding of crisis management is necessary. In this regard, the CA should not be perceived as a wholly new method of organizing international crisis management. Rather, it is a means of modifying and modernizing it which remains predominantly organized in traditional structures. Thus, the aim of promoters of the CA should not so much be to change organizational arrangements as to ensure that arrangements are flexible enough to accommodate the policy needed to confront complex crisis management issues. In this regard, the analysis undertaken above has shown that it is worth drawing experiences of diverse crisis management actors when further developing the CA and linking it to the multilateral level.

The aim of the CA is highly ambitious, indeed much more ambitious than a mere change of crisis management strategies and institutional arrangements, in particular on the European level. At a time when the EU is under tremendous stress, it is essential to keep the historical perspective in mind that, above all else, the EU itself has been one of the greatest crisis management mechanisms ever devised. The Nobel committee's decision to award the 2012 peace prize to the EU, which resulted in mixed responses in its member states, should be seen as a morale boost for the Union to resolve the Euro-crisis but should also give some self-confidence to the Union to further develop and re-increase its role in relation to crisis management. Responses of EU leaders point in this very direction.[547]

The Comprehensive Approach and the endeavor for more coherence and coordination in international crisis management may in some respect resemble what Churchill said about democracy: the worst alternative except for all the others. After the in-depth analysis undertaken in this thesis, it can be concluded that a major challenge CA implementation struggles with is posed by practicioners and scholars dealing with the subject. In this regard, difficulties of CA implementation must not be overestimated, the potential not oversold. In view of organizing comprehensive crisis management in a meaningful way, it should be done with a "rational organizational perspective" (Christensen et al. 2007, p. 29), where institutions and actors are instruments to achieve goals and objectives. It is important that the organizational realities of crisis management in Europe are taken into account with its architecture comprised of a collection of more or less independent actors both on national governmental and on EU level. CA efforts and initiatives so far have rather given an idea of a desired crisis management coherence end-state which did not take the status quo into account and which, moreover, is difficult to achieve in face of present strategic, structural and political constraints. Strengthening this rationality in the thinking of crisis management actors would better enable CA implementation both on EU and on national levels. More than a Comprehensive Approach, a *Whole-of-Reality Approach* might be what is needed to make European crisis management more effective.

547 See Joint statement (European Council/The President 2012) by the President of the European Council Joint statement Van Rompuy and Commission President José Manuel Barroso on the award of the 2012 Nobel Peace prize to the EU as well as the statement by EU HR/VP Ashton (Ashton 2012).

6 References

If not indicated otherwise all internet addresses have been reapproved valid on June 18th, 2013. This applies for both primary and secondary sources which can be retrieved online. Some relevant official documents were not available in English and were thus analyzed in the language of their original announcement. This, however, led to a referencing system which, for instance, distinguishes between the Federal Foreign Office and Auswärtiges Amt as publishers, though it is the same institution issuing the documents.

6.1 Primary Sources

6.1.1 Official Documents, Speeches and Press Releases

Ashton, Catherine (2010): *Joint Debate on Foreign and Security Policy*. Speech/10/82, 10 March 2010. Strasbourg: European Parliament. Retrieved from http://europa.eu/rapid/pressReleasesAction.do?reference=SPEECH/10/82&type=HTML

Ashton, Catherine (2012): *Statement by EU High Representative on the Award of the 2012 Nobel Peace Prize to the European Union.* A 454/12. Brussels: European Union. Retrieved from http://www.consilium.europa.eu/uedocs/cms_Data/docs/press data/EN/foraff/132806. pdf

Auswärtiges Amt (2012a): *Auswärtiges Amt richtet Ressortübergreifende Task Force Syrien ein.* Berlin: Auswärtiges Amt. Retrieved from http://www.auswaertiges-amt.de/cae/servlet/contentblob/623232/publicationFile/170729/120803-TF_SYR_DL.pdf

Auswärtiges Amt (2012b): *Fragile Staaten – Eine Herausforderung für die Sicherheit.* Berlin: Auswärtiges Amt. Retrieved from http://www.auswaertiges-amt.de/DE/Aussenpolitik/ AktuelleArtikel/120919_Fragile_Staaten_node.html

Auswärtiges Amt/Bundesministerium der Verteidigung/Bundesministerium für wirtschaftliche Zusammenarbeit und Entwicklung (2012): *Für eine kohärente Politik der Bundesregierung gegenüber fragilen Staaten. Ressortübergreifende Leitlinien.* Berlin: Auswärtiges Amt/Bundesministerium der Verteidigung/Bundesministerium für wirtschaftliche Zusammenarbeit und Entwicklung. Retrieved from http://www.aus waertiges-amt.de/cae/servlet/contentblob/626452/publicationFile/171874/ 1209 19_Leitlinien_Fragile_Staaten.pdf

Blair, Tony (1995): The Rights We Enjoy, the Duties We Owe. Speech Given at the Spectator Lecture. Reprinted in Id. (1996): *New Britain. My Vision of a Young Country*. London: Fourth Estate.

Bundesakademie für Sicherheitspolitik (2008): *Our College – Task*. Berlin: Bundesakademie für Sicherheitspolitik. Retrieved from http://www.baks.bund.de/EN/Our_College/Task/ task_ node.html

Bundesministerium für Wirtschaftliche Zusammenarbeit und Entwicklung (2005): *Übersektorales Konzept zur Krisenprävention, Konfliktbearbeitung und Friedensförderung in der deutschen Entwicklungszusammenarbeit. Eine Strategie zur Friedensentwicklung (Strategy for Peacebuilding)*. Berlin/Bonn: Bundesministerium für Wirtschaftliche Zusammenarbeit und Entwicklung. Retrieved from http://www.bmz.de/de/zentrales_downloadarchiv/themen_und_schwerpunkte/frieden/krisenpraevention.pdf

Cabinet Office (n.d.): *National Security Council*. London: Cabinet Office. Retrieved from http://www.cabinetoffice.gov.uk/content/national-security-council

Cabinet Office (1999): *Modernising Government*. Norwich: The Stationery Office. Retrieved from http://www.archive.official-documents.co.uk/document/cm43/4310/4310-00.htm

Cabinet Office (2000): *Wiring It Up. Whitehall's Management of Cross-Cutting Policies and Services. A Performance and Innovation Unit Report*. Norwich: The Stationery Office. Retrieved from http://webarchive.nationalarchives.gov.uk/+/http://www.cabinetoffice. gov.uk/upload/assets/www.cabinetoffice.gov.uk/strategy/coiwire.pdf

Cabinet Office (2005): *Investing in Prevention: An International Strategy to Manage Risks of Instability and Improve Crisis Response*. London: Prime Ministers Strategy Unit. Retrieved from http://webarchive.nationalarchives.gov.uk/+/http://www.cabinetoffice.gov.uk/media/ cabinetoffice/strategy/assets/investing.pdf

Cabinet Office (2008): *The National Security Strategy of the United Kingdom. Security in an Interdependent World*. Norwich: The Stationery Office. Retrieved from http://www.defesa. gov.br/projetosweb/livrobranco/arquivos/pdf/UK04.pdf

Cabinet Office (2009): *The National Security Strategy of the United Kingdom: Update 2009. Security for the Next Generation*. Norwich: The Stationery Office. Retrieved from http://www.official-documents.gov.uk/document/cm75/7590/7590.pdf

Cabinet Office (2010): *A strong Britain in an Age of Uncertainty. The National Security Strategy*. Norwich: The Stationery Office. Retrieved from http://www.direct.gov.uk/prod_consum_dg/groups/dg_digitalassets/@dg/@en/documents/digitalasset/dg_191639.pdf?CID=PDF&PLA=furl&CRE=nationalsecuritystrategy

Cameron, David (2010a): The Conservative Party. UK General Election: Defence Reviews. *Rusi Journal*, 155 (2), pp. 9/10.

Cameron, David (2010b): *Written Ministerial Statement: National Security – The Context*. London: Number 10 – The Official Site of the British Prime Minister's Office. Retrieved from http://www.number10.gov.uk/news/written-ministerial-statement-national-security-the-context/

CDU/CSU/FDP (2009): *Growth, Education, Unity. The Coalition Agreement between the CDU, CSU and FDP*. Berlin: CDU/CSU/FDP. Retrieved from http://www.cdu.de/sites/default/files/media/dokumente/091215-koalitionsvertrag-2009-2013-englisch_0.pdf

Council of the European Union (1999): *Helsinki European Council. 10 and 11 December 1999. Presidency Conclusions*. Brussels: European Parliament. Retrieved from http://www.europarl.europa.eu/summits/hel1_en.htm#b

Council of the European Union (2001): *Draft Council Conclusions on EU-UN Cooperation in Conflict Prevention and Conflict Management*. Brussels: Council of the European Union. Retrieved from http://www.consilium.europa.eu/uedocs/cms_data/docs/pressdata/en/ misc/09528-r2.en1.html

Council of the European Union (2002): *Civil-Military Coordination. Forwarding Note from Secretariat/Commission*. Doc. 12307/02. Brussels: Council of the European Union. Retrieved from http://register.consilium.europa.eu/pdf/en/02/st12/st12307.en02.pdf

Council of the European Union (2003a): *A Secure Europe in a Better World. European Security Strategy*. Brussels: Council of the European Union. Retrieved from http://www.consilium.europa.eu/uedocs/cmsUpload/78367.pdf

Council of the European Union (2003b): *Civil Military Co-ordination (CMCO)*. Doc. 14457/03, Brussels: Council of the European Union. Retrieved from http://register.consilium.eu.int/pdf/en/03/st14/st14457.en03.pdf

Council of the European Union (2003c): *Suggestions for Procedures for Coherent, Comprehensive EU Crisis Management*. Doc. 11127/03. Brussels: Council of the European Union. Retrieved from http://www.google.de/url?sa=t&rct=j&q=&esrc=s&source= web&cd=1&ved=0CDUQFjAA&url=http%3A%2F%2Fesdc.mil-edu.be%2Findex.php%2Fcomponent%2Fdocman%2Fdoc_download%2F265-suggestions-for-procedures-for-coherent-comprehensive-eu-crisis-management&ei=Jh3AUYqAJ8WntAaa3oHgCQ &usg=AFQjCNE-5IPthZmh6HjQ69HWo7Klcb1Nxw&sig2=3Dy2pN6dEsOVAVD6o ZTPqg&bvm=bv.47883778,d.Yms

Council of the European Union (2004a): *Civilian Headline Goal 2008*. Doc. 15863/04. Brussels: Council of the European Union. Retrieved from http://register.consilium.eu.int/ pdf/en/04/st15/st15863.en04.pdf

Council of the European Union (2004b): *Headline Goal 2010*. Brussels: Council of the European Union. Retrieved from http://ue.eu.int/uedocs/cmsUpload/2010%20 Headline%20Goal.pdf

Council of the European Union (2005): *Draft EU Concept for Comprehensive Planning*. Doc. 13983/05. Brussels: Council of the European Union. Retrieved from http://register. consilium.europa.eu/pdf/en/05/st13/st13983.en05.pdf

Council of the European Union (2006): *Civil-Military Co-ordination (CMCO): Possible Solutions for the Management of EU Crisis Management Operations – Improving Information Sharing in Support of EU Crisis Management Operations*. Doc. 13218/5/06. Brussels: Council of the European Union. Retrieved from http://www.europarl.europa.eu/meetdocs/2009_2014/documents/sede/dv/sede260410cmcoinformationsharing_/sede260410cmcoinformationsharing_en.pdf

Council of the European Union (2008): *Report on the Implementation of the European Security Strategy. Providing Security in a Changing World*. Brussels: Council of the European Union. Retrieved from http://www.consilium.europa.eu/ueDocs/cms_Data/docs/pressdata/EN/reports/104630.pdf

Council of the European Union (2010): *Council Decision Establishing the Organisation and Functioning the European Action Service*. Doc 11665/1/10. Brussels: Council of the European Union. Retrieved from http://register.consilium.europa.eu/pdf/en/10/st11 /st11665-re01.en10.pdf

Department for International Development (1997): *Eliminating World Poverty: A Challenge for the 21st Century. White Paper on International Development*. Norwich: The Stationery Office. Retrieved from http://webarchive.nationalarchives.gov.uk/+/http://www.dfid.gov. uk/PolicieAndPriorities/files/whitepaper1997.pdf

Department for International Development (2000): *Eliminating World Poverty: Making Globalization Work for the Poor. White Paper on International Development*. Norwich: The Stationery Office. Retrieved from http://webarchive.nationalarchives.gov.uk/+/http:// www.dfid.gov.uk/policieandpriorities/files/whitepaper2000.pdf

Department for International Development (2005): *Fighting Poverty to Build a Safer World. A Strategy for Security and Development*. London: Department for International Development. Retrieved from http://webarchive.nationalarchives.gov.uk/+/http:/www.dfid. gov.uk/pubs/files/securityforall.pdf

Department for International Development (2009): *Eliminating World Poverty: Building our Common Future*. White Paper on International Development. Norwich: The Stationery Office. Retrieved from http://www.official-documents.gov.uk/document/cm76/7656/ 7656.pdf

Department for International Development (2010): *DFID Briefing. A DFID Practice Paper. Working Effectively in Conflict-Affected and Fragile Situations*. London: Department for International Development. Retrieved from http://www.dfid.gov.uk/Documents/ publications1/governance/summary-note-briefing-papers.pdf

Deutscher Bundestag (2011): *Antrag der Abgeordneten Edelgard Bulmahn, Lothar Binding (Heidelberg), Klaus Brandner, Dr. h.c. Gernot Erler, Petra Ernstberger, Dagmar Freitag, Iris Gleicke, Günter Gloser, Dr. Barbara Hendricks, Hans-Ulrich Klose, Dr. Bärbel Kofler, Ute Kumpf, Burkhard Lischka, Dr. Rolf Mützenich, Thomas Oppermann, JohannesPflug, Dr. Sascha Raabe, Karin Roth (Esslingen), Frank Schwabe, Franz Thönnes, Wolfgang Tiefensee, Heidemarie Wieczorek-Zeul, Uta Zapf, Manfred Zöllmer, Dr. Frank-Walter Steinmeier und der Fraktion der SPD: Deutschland braucht dringend eine kohärente Strategie für die zivile Krisenprävention*. 17. Legislaturperiode. Berlin: Deutscher Bundestag.

DFID/FCO/MoD (2003): *The Global Conflict Prevention Pool. A Joint UK Government Approach to Reducing Conflict*. London: Foreign and Commonwealth Office. Retrieved from http://webarchive.nationalarchives.gov.uk/+/http:/www.dfid.gov.uk/pubs/files/global-conflict-prevention-pool.pdf

DFID/FCO/MoD (2011): *Building Stability Overseas Strategy*. London: Department for International Development/Foreign and Commonwealth Office/UK Ministry of Defence. Retrieved from http://www.dfid.gov.uk/Documents/publications1/Building-stability-overseas-strategy.pdf.

Die Bundesregierung (2000): *Gesamtkonzept der Bundesregierung: Zivile Krisenprävention, Konfliktlösung und Friedenskonsolidierung*. Berlin: Die Bundesregierung. Retrieved from http://www.cpti.ws/conf/02/ger/zusatz/gesamtkonzept.pdf

Die Bundesregierung (2004): *Action Plan. Civilian Crisis Prevention, Conflict Resolution, and Post-Conflict Peacebuilding*. Berlin: Die Bundesregierung. Retrieved from http://www.auswaertiges-amt.de/cae/servlet/contentblob/384232/publicationFile/4346/Aktionsplan-En.pdf

Die Bundesregierung (2006): *Sicherheit und Stabilität durch Krisenprävention gemeinsam stärken. 1. Bericht der Bundesregierung über die Umsetzung des Aktionsplans „Zivile Krisenprävention, Konfliktlösung und Friedenskonsolidierung"*. Berlin: Die Bundesregierung. Retrieved from http://www.auswaertiges-amt.de/cae/servlet/contentblob /384196/publicationFile/4343/Aktionsplan-Bericht1-de.pdf

EU Glossary (n.d.): *Petersberg Tasks. Summary of EU Legislation/Glossary*. Brussels: European Union. Retrieved from http://europa.eu/legislation_summaries/glossary/petersberg_tasks_en.htm

EU Glossary (n.d.): *Qualified Majority. Summary of EU Legislation/Glossary*. Brussels: European Union. Retrieved from http://europa.eu/legislation_summaries/glossary/qualified_majority_en.htm

European Commission (2005): *Communication from the Commission to the Council, the European Parliament and the European Economic and Social Committee. Policy Coherence for Development. Accelerating Progress Towards Attaining the Millennium Development Goals*. COM(2005) 134 final. Brussels: European Commission. Retrieved from http://eur-lex.europa.eu/LexUriServ/LexUriServ.do?uri=COM:2005:0134:FIN :EN:PDF

European Commission (2007): *Communication from the Commission to the Council and the European Parliament. EU Code of Conduct on Division of Labour in Development*. COM(2007) 72 final. Brussels: European Commission. Retrieved from http://eur-lex.europa.eu/LexUriServ/LexUriServ.do?uri=COM:2007:0072:FIN:EN:PDF

European Commission (2009): *Communication from the Commission to the Council, the European Parliament, the European Economic and Social Committee and the Committee of the Regions. Policy Coherence for Development – Establishing the Policy Framework for a Whole-of-the-Union Approach*. COM(2009) 458 final. Brussels: European Commission. Retrieved from http://ec.europa.eu/development/icenter/repository/COM_ 2009_458_part1_en.pdf

European Commission Conflict Prevention and Crisis Management Unit (2003): *Civilian Instruments for EU Crisis Management*. Brussels: European Commission. Retrieved from http://eeas.europa.eu/cfsp/crisis_management/docs/cm03.pdf

European Council/The President (2012): *Joint Statement by Herman Van Rompuy, President of the European Council, and José Manuel Barroso, President of the European Commission, on the Award of the 2012 Nobel Peace Prize to the European Union*. EUCO 186/12. Brussels: European Council/The President. Retrieved from http://www.consilium. europa.eu/uedocs/cms_Data/docs/pressdata/en/ec/132807.pdf

European External Action Service (n.d.): *Crisis Response*. Brussels: European External Action Service. Retrieved from http://www.consilium.europa.eu/eeas/foreign-policy/crisis-response

European External Action Service (2011): *Report by the High Representative to the European Parliament, the Council and the Commission*. Brussels: European External Action Service Retrieved from http://www.eeas.europa.eu/images/top_stories/2011_eeas_report_cor_+ _formatting.pdf

European Parliament (2009): *Report on the European Security Strategy and ESDP (2008/2202(INI))*. Committee on Foreign Affairs. Rapporteur Karl von Wogau. Brussels: European Parliament. Retrieved from http://www.europarl.europa.eu/sides/getDoc.do?pubRef=-//EP//NONSGML+REPORT+A6-2009-0032+0+DOC+PDF+V0//EN

European Union (n.d.): *The Union's Decision-Making Procedures. Extension of Qualified Majority Voting*. Brussels: European Union. Retrieved from http://europa.eu/scadplus/constitution/ majority_en.htm

Federal Foreign Office (2012a): *National Structures*. Berlin: Federal Foreign Office. Retrieved from http://www.auswaertiges-amt.de/EN/Aussenpolitik/Friedenspolitik/Krisenpraevention/NationaleStrukturen_node.html

Federal Foreign Office (2012b): *Interministerial Agreement between the Federal Foreign Office and the Federal Ministry for Economic Cooperation and Development Enters Into Force*. Press Release. Date of Issue: 24.05.2012. Berlin: Federal Foreign Office. Retrieved from http://www.auswaertiges-amt.de/EN/Infoservice/Presse/Meldungen/2012/120524 _AA_BMZ_Ressortvereinb.html

Federal Ministry of Defence (2003): *Defence Policy Guidelines 2003*. Berlin: Federal Ministry of Defence. Retrieved from http://www.bmvg.de/portal/a/bmvg/!ut/p/c4/NYyxDoMwDET_yCYSrUS3Vl06sJSB0i2AFSxIHAVDl358w9A76Q33pMM35ga7s7PKEuyCL-wGvvQf6P3uYOVhojQR6xplYeUZbHDUixKMMm-eghK2x8tIMEggPah55kyXrEqCKEmXw2wpZQM8YleY-60wxT_mey7bpnxW5lTXjwaj99cfzhxv5A!!/

Federal Ministry of Defence (2006): *White Paper 2006 on German Security Policy and the Future of the Bundeswehr*. Berlin: Federal Ministry of Defence. Retrieved from http://www.bmvg.de/portal/a/bmvg/!ut/p/c4/Dca7DYAwDAXAWVgg7unYAuicYCVP-Qrnsz7omqObfoUnPHfUwolOuhx2u4zN0xuFC_IGQddWEzqi4eLF1i7mqXFkKf-WQNUOF6jFY_sAY_7e5g!!/

Federal Ministry of Defence (2011): *Defence Policy Guidelines 2011*. Berlin: Federal Ministry of Defence. Retrieved from http://www.bmvg.de/portal/a/bmvg/!ut/p/c4/LYsxEoAgDATf4gdIb-cv1MYBzcQbMDgQ8ftSONtssUsrddQ3iDdk9YlmWnaM4XXhaulq9pPLybB65wRDdF6FQzZ2R47PxdqtcTHGAXlU_q72byv9tgQFK91xGj6tRgx1/

Foreign and Commonwealth Office (2003): *UK's International Priorities. A Strategy for the FCO*. Norwich: The Stationery Office. Retrieved from http://collections.europarchive.org/tna/20080205132101/http://www.fco.gov.uk/servlet/Front?pagename=OpenMarket/Xcelerate/ShowPage&c=Page&cid=1142705307742

Foreign and Commonwealth Office (2006): *Active Diplomacy for a Changing World: The UK's International Priorities*. London: Foreign and Commonwealth Office. Retrieved from http://www.official-documents.gov.uk/document/cm67/6762/6762.pdf

Foreign and Commonwealth Office (2008): *Better World, Better Britain*. Norwich: The Stationery Office. Retrieved from http://www.official-documents.gov.uk/document/cm73/7398/7398.pdf

Foreign and Commonwealth Office (2011a): *HMG Conflict Funding*. London: Foreign and Commonwealth Office. Retrieved from http://www.fco.gov.uk/en/about-us/what-we-do/spend-our-budget/funding-pro grammes1/conflict-funding/

Foreign and Commonwealth Office (2011b): *Conflict Pool*. London: Foreign and Commonwealth Office. Retrieved from http://www.fco.gov.uk/en/about-us/what-we-do/spend-our-budget/funding-programmes1/conflict-funding/conflict-pool/

Government Offices of Sweden (1995): *Sverige i Europa och Världen*. DS 1995:28. Stockholm: Ministry of Defence/Defence Commission. Retrieved from http://www.regeringen.se/content/1/c4/13/63/a8c2a453.pdf

Government Offices of Sweden (1999a): *Preventing Violent Conflict – A Swedish Action Plan*. DS 1999:24. Stockholm: Ministry for Foreign Affairs. Retrieved from http://www.sweden. gov.se/content/1/c6/09/37/14/aad1f9e6.pdf

Government Offices of Sweden (1999b): *A Changing World – a Reformed Defence. Executive Summary of the Defence Commission's Report*. DS 1999:2. Stockholm: Ministry of Defence/Defence Commission. Retrieved from http://www.sweden.gov.se/content/1/c6/ 02/56/73/e028eb9c.pdf

Government Offices of Sweden (1999c): *The New Defence*. Stockholm: Ministry of Defence. Retrieved from http://www.regeringen.se/content/1/c6/02/56/39/526a92a6.pdf

Government Offices of Sweden (2003): *Shared Responsibility. Sweden's Policy for Global Development*. Government Bill 2002/03:122. Stockholm: Ministry for Foreign Affairs. Retrieved from http://www.sweden.gov.se/content/1/c6/02/45/20/c4527821.pdf

Government Offices of Sweden (2004): *Our Future Defence – The Focus of Swedish Defence Policy 2005–2007*. Government Bill 2004 /05:5. Stockholm: Swedish Ministry of Defence. Retrieved from http://www.sweden.gov.se/content/1/c6/03/21/19/224a4b3c.pdf

Government Offices of Sweden (2005): *Departments and Their Functions (Last Updated November 2011)*. Stockholm: Ministry for Foreign Affairs. Retrieved from http://www.sweden.gov.se/sb/d/3367/a/37735

Government Offices of Sweden (2006): *Sweden's Policy for Global Development*. Government Communication 2005/06:204. Stockholm: Ministry for Foreign Affairs. Retrieved from http://www.sweden.gov.se/content/1/c6/07/01/68/3e990ee4.pdf

Government Offices of Sweden (2007): *Questions and Answers Concerning the Nordic Battlegroup and EU's Battlegroups (Last Updated May 2007)*. Stockholm: Ministry of Defence. Retrieved from http://www.sweden.gov.se/sb/d/9136

Government Offices of Sweden (2008a): *Global Challenges – Our Responsibility*. Government Communication on Sweden's Policy for Global Development. Government Communication 2007/08:89. Stockholm: Ministry for Foreign Affairs. Retrieved from http://www.sweden.gov.se/sb/d/9807/a/113283

Government Offices of Sweden (2008b): *National Strategy for Swedish Participation in International Peace Support and Security Building Operations*. Government Communication 2007/08:51. Stockholm: Ministry for Foreign Affairs. Retrieved from http://www.sweden.gov.se/content/1/c6/10/80/95/73e64223.pdf

Government Offices of Sweden (2009): *Press Release. A Functional Defence – With a Substantially Strengthened Defence Capability*. Stockholm: Ministry of Defence Press Release. Retrieved from http://www.sweden.gov.se/sb/d/10448/a/123010

Government Offices of Sweden (2010a): *Meeting Global Challenges*. Government Communication on Policy Coherence for Development. Government Communication 2009/10:129. Stockholm: Ministry for Foreign Affairs. Retrieved from http://www.sweden. gov.se/content/1/c6/15/27/91/ea0dc207.pdf

Government Offices of Sweden (2010b): *A Functional Defence, Summary of the Bill Dated March 19 (Last Updated January 2010)*. Stockholm: Swedish Ministry of Defence. Retrieved from http://www.sweden.gov.se/sb/d/574/a/137705

Government Offices of Sweden (2010c): *Strategy for Swedish Support for International Involvement in Afghanistan*. Stockholm: Government Offices of Sweden. Retrieved from http://www.swedeninafghanistan.se/LinkClick.aspx?fileticket=r3uSLwkz2N4 %3d&tabid=189

Government Offices of Sweden (2011a): *Carl Bildt, Minister for Foreign Affairs. Statement of Government Policy in the Parliamentary Debate on Foreign Affairs*. Stockholm: Ministry for Foreign Affairs. Retrieved from http://www.regeringen.se/ content/1/c6/16/11/48/dc78c 337.pdf

Government Offices of Sweden (2011b): *Frequently Asked Questions on the Aid Budget 2012*. Stockholm: Ministry for Foreign Affairs. Retrieved from http://www.sweden. gov.se/sb/ d/15180

Government Offices of Sweden (2011c): *Facts & Figures. Swedish Government Offices Yearbook 2010*. Stockholm: Office for Administrative Affairs. Retrieved from http://www.sweden.gov.se/content/1/c6/17/17/06/39c20e86.pdf

Government Offices of Sweden (2011d): *Peace and Security for Development. Policy for Security and Development in Swedish Development Cooperation 2010-2014*. Stockholm: Ministry for Foreign Affairs. Retrieved from http://www.sweden.gov.se/ content/1/c6/ 16/45/85/88f00571.pdf

Her Majesty's Government (2007): *PSA Delivery Agreement 30: Reduce the Impact of Conflict through Enhanced UK and International Efforts*. London: HM Treasury. Retrieved from http://www.hm-treasury.gov.uk/d/pbr_csr07_psa30.pdf

House of Commons (2010a): *The Comprehensive Approach. The Point of War is Not Just to Win but to Make a Better Peace*. Seventh Report of Session 2009-2010. London: House of Commons – Defence Committee. Retrieved from http://www.publications. parliament.uk /pa/cm200910/cmselect/cmdfence/224/224.pdf

House of Commons (2010b): *DFID's Performance in 2008-09 and the 2009 White Paper*. Fourth Report of Session 2009-10. London: House of Commons – International Development Committee. Retrieved from http://www.publications.parliament.uk/pa/ cm200910/cmselect/cmintdev/48/48i.pdf

House of Lords (2012): *European Union Sub-Committee for External Affairs. European External Action Service. Evidence – Written*. London: House of Lords. Retrieved from http://www.parliament.uk/documents/lords-committees/eu-sub-com-c/EEAS/ WrittenevidencevolumeEEAS.pdf

Memorandum of Understanding (2005): *Memorandum of Understanding between the Ministry of Defence of the Republic of Estonia and the Ministry of Defence of the Republic of Finland and the Ministry of Defence of the Kingdom of Norway and the Government of the Kingdom of Sweden Concerning the Principles for the Establishment and Operation of a Multinational Battle Group to be Made Available to the European Union*. Brussels. Retrieved from http://www.regeringen.se/content/1/c6/ 04/49/80/60960599.pdf

NATO (2002): *NATO Military Policy on Civil-Military Co-operation*. MC 411/1. Brussels: NATO International Military Staff. Retrieved from http://www.nato.int/ims/ docu/mc411-1-e.htm

NATO (2006): *Riga Summit Declaration.* NATO Press Release (2006)150. Riga: North Atlantic Council. Retrieved from http://www.nato.int/docu/pr/2006/p06-150e.htm

NATO (2008): *Bucharest Summit Declaration.* Bucharest: North Atlantic Council. Retrieved from http://www.nato.int/cps/en/natolive/official_texts_8443.htm

NATO (2009): *First NATO Press Conference by Secretary Anders Fogh Rasmussen.* Brussels: North Atlantic Treaty Organization. Retrieved from http://www.nato.int/cps/en/ natolive/opinions_56776.htm

NATO (2010): *Active Engagement, Modern Defence. NATO's New Strategic Concept.* Lisbon: North Atlantic Treaty Organization. Retrieved from http://www.nato.int/lisbon2010/strategic-concept-2010-eng.pdf

NATO (2012): *A "Comprehensive Approach" to Crisis Management.* NATO A-Z. Brussels: North Atlantic Treaty Organization. Retrieved from http://www.nato.int/cps/en/natolive /topics_51633.htm

OECD (1997): *Managing Across Levels of Government.* Paris: OECD Publications Service. Retrieved from http://www.oecd.org/dataoecd/10/14/1902308.pdf

OECD (2002): *Glossary of Key Terms in Evaluation and Results Based Management.* Paris: OECD Publications Service. Retrieved from http://www.oecd.org/dataoecd/29/21/ 2754804.pdf

OECD (2003): *Policy Coherence: Vital for Global Development.* OECD Policy Brief. Paris: OECD Public Affairs Division. Retrieved from http://www.oecd.org/dataoecd/11/35/ 20202515.pdf

OECD (2006): *Whole of Government Approaches to Fragile States.* DAC Guidelines and Reference Series. A DAC Reference Document. Paris: OECD Publications Service. Retrieved from www.oecd.org/dataoecd/15/24/37826256.pdf

OECD (2008): *Is It ODA?* OECD Factsheet. Paris: OECD Publications Service. Retrieved from http://www.oecd.org/dac/stats/34086975.pdf

OECD (2010): *Germany. Development Assistance Committee (DAC), Peer Review 2010.* Paris: OECD Publications Service. Retrieved from http://www.oecd.org/development/peerreviewsofdacmembers/46439355.pdf

OECD (2011): *DAC Mid-term Review of Sweden.* PEER/KJ(2011)22. Stockholm: OECD Development Co-operation Directorate. Retrieved from http://www.sweden.gov.se/content /1/c6/17/30/39/415ec365.pdf

Office Journal of the European Union (2010): *Consolidated Versions of the Treaty on European Union and the Treaty on the Functioning of the European Union.* Doc. (2010/C 83/01). Luxembourg: Office for Official Publications of the European Communities. Retrieved from http://eur-lex.europa.eu/LexUriServ/LexUriServ.do?uri=OJ:C:2010: 083:FULL:EN:PDF

Prodi, Romano (2003): *The European Project in the World: Between Values and Politics.* Alessano: Fondazione Don Tonino Bello. Retrieved from http://europa.eu/rapid/press-release_SPEECH-03-303_en.htm

Regeringen (2009): *Work Programme for the Swedish Presidency of the EU. 1 July – 31 December 2009.* Stockholm: Regeringen. Retrieved from http://www.europarl.europa.eu/meetdocs/2009_2014/documents/envi/dv/sv_workprogramme_/sv_workprogramme_en.pdf

Solana, Javier (2000): *Towards a New International Morality: The Humanitarian Interventions*. Madrid: University of Alcalá de Henares. Retrieved from http://www.consi lium.europa.eu/uedocs/cms_data/docs/pressdata/en/discours/000707%20madrid%20en .htm

Solana, Javier (2005): *Remarks by EUHR Solana at Ministers of Defence Meeting*. Brussels. Retrieved from http://www.eu-un.europa.eu/articles/fr/article_5343_fr.htm

The Federal Government (2006): *Sicherheit und Stabilität durch Krisenprävention Gemeinsam Stärken. 1. Bericht der Bundesregierung über die Umsetzung des Aktionsplans Zivile Krisenprävention, Konfliktlösung und Friedenskonsolidierung*. Berlin: Die Bundesregierung. Retrieved from http://www.auswaertiges-amt.de/cae/servlet/ contentblob/ 384196/publicationFile/4343/Aktionsplan-Bericht1-de.pdf

The Federal Government (2008): *Crisis Prevention as a Shared Task. 2nd Federal Government Report on the Implementation of the Action Plan 'Civilian Crisis Prevention, Conflict Resolution, and Post-Conflict Peacebuilding'*. Berlin: Die Bundesregierung. Retrieved http://www.auswaertiges-amt.de/cae/servlet/contentblob/483640/ publication File/4341/Aktionsplan-Bericht2-en.pdf

The Federal Government (2010): *3rd Federal Government Report on the Implementation of the Action Plan 'Civilian Crisis Prevention, Conflict Resolution, and Post-Conflict Peacebuilding'*. Berlin: Die Bundesregierung. Retrieved from https://www. civcap.info/ fileadmin/user_upload/Germany/3rd_report.pdf

UK Ministry of Defence (1998): *Strategic Defence Review. Modern Forces in a Modern World*. Norwich: The Stationery Office. Retrieved from http://merln.ndu.edu/white papers/UnitedKingdom1998.pdf

UK Ministry of Defence (2002): *Strategic Defence Review: A New Chapter*. Norwich: The Stationery Office. Retrieved from http://merln.ndu.edu/whitepapers/UK_SDR _2002.pdf

UK Ministry of Defence (2003): *Delivering Security in a Changing World. Defence White Paper*. Norwich: The Stationery Office. Retrieved from http://merln.ndu.edu/ whitepapers/_UnitedKingdom-2003.pdf

UK Ministry of Defence (2004): *Delivering Security in a Changing World: Future Capabilities*. Defence White Paper. Norwich: The Stationery Office. Retrieved from http://merln.ndu.edu/whitepapers/UnitedKingdom-2004.pdf

UK Ministry of Defence (2006): *The Comprehensive Approach*. Joint Discussion Note 4/05. Swindon: The Development, Concepts and Doctrine Centre. Retrieved from http://www.google.com/url?sa=t&rct=j&q=&esrc=s&source=web&cd=1&ved=0CC QQFjAA&url=http%3A%2F%2Fwww.arrc.nato.int%2Fsystems%2Ffile_download. ashx%3Fpg%3D3313%26ver%3D1&ei=89VXUIORFvOv0AHyroCQDA&usg=AF QjCNEF5Hllu9tO_UUFuwzRhg7aludNtg&sig2=n-GiRzkzzqRSaKWAURLnIw

UK Ministry of Defence (2008): *British Defence Doctrine*. Joint Doctrine Publication 01-1. Swindon: The Development, Concepts and Doctrine Centre. Retrieved from http://tvernedra.ru/Voendoktrinavelikobr2011.pdf

UK Ministry of Defence (2010a): *Securing Britain in an Age of Uncertainty. Strategic Defence and Security Review*. Norwich: The Stationery Office. Retrieved from http://www.direct.gov.uk/prod_consum_dg/groups/dg_digitalassets/@dg/@en/docu ments/digitalasset/dg_191634.pdf?CID=PDF&PLA=furl&CRE=sdsr

United Kingdom Parliament (2011): *Written Ministerial Statements for 5 April 2011*. London: UK Parliament. Retrieved from http://www.publications.parliament.uk/pa/cm/cmtoday /cmwms/archive/110405.htm#d2e292

United Nations (1992): *An Agenda for Peace. Preventive Diplomacy, Peacemaking and Peace-keeping*. A/47/277 – S/24111. New York: United Nations. Retrieved from http://www.unrol.org/files/A_47_277.pdf

United Nations (2000): *Report of the Panel on United Nations Peace Operations*. A/55/305 – S/2000/809. New York: United Nations. Retrieved from http://www.un.org/ga/search/view _doc.asp?symbol=A/55/305

United Nations (2001): *Security Council Addresses Comprehensive Approach to Peace-building in Presidential Statement*. Press Release SC/7014. New York: United Nations. Retrieved from http://www.un.org/News/Press/docs/2001/sc7014.doc.htm

United Nations (2006a): *Note from the Secretary-General: Guidance on Integrated Missions*. New York: United Nations. Retrieved from http://www.regjeringen.no/upload/UD /Vedlegg/missions/sgnote.pdf

United Nations (2006b): *Delivering as One. Report of the High-level Panel on System-wide Coherence in the Areas of Development, Humanitarian Assistance, and the Environment*. A/61/583, 9, New York: United Nations, Retrieved from http://www.un.org/ga/president_/62/issues/resolutions/a-61-583.pdf

United Nations (2008): *Decision No. 2008/24 – Integration. Decisions of the Secretary-General – 25 June Meeting of the Policy Committee*. New York: United Nations. Retrieved from http://www.undg.org/docs/9898/Integration-decision-SG-25-jun-08.pdf

United Nations (2009): *A New Partnership Agenda. Charting a New Horizon for UN Peacekeeping*. New York: Department of Peacekeeping Operations and Department of Field Support. Retrieved from http://www.un.org/en/peacekeeping/documents/newhorizon.pdf

United Nations Development Programme (1994): *Human Development Report 1994*. New York/Oxford: Oxford University Press. Retrieved from http://hdr.undp.org/en/reports/ global/hdr1994/chapters/

Zentrum für Internationale Friedenseinsätze (2012): *Key Issues – Comprehensive Approach*. Berlin: Zentrum für International Friedenseinsätze. Retrieved from http://www.zif-berlin.org/en/focus-areas/comprehensive-approach.html

6.1.2 List of Interviewees

Ahlbrecht, Gerhard – Secretary General, European Organisation of Military Associations, Brussels

Birocchi, Frederico – DG EuropeAid Development and Co-operation, European Commission, Brussels

Bolin Pennegård, Ann Marie – Ambassador (Representative to CIVCOM), Permanent Representation of Sweden to the EU, Brussels

Braner, Christoph – Desk Officer, Common Security and Defence Policy, Federal Foreign Office, Germany

Bulmahn, Edelgard – Member of the German Parliament and the Parliament's Subcommittee on Civilian Crisis Prevention and Networked Security, Germany
Day, Catherine – Head of Press and Media, Information Office, House of Lords, UK
Detlefsen, Hans-Jörg – Captain, Defence Adviser, Permanent Representation of the Federal Republic of Germany to the EU, Brussels
Ehrenkrona, Olof – Ambassador, Senior Advisor to the Minister for Foreign Affairs, Ministry for Foreign Affairs, Sweden
Ekengren, Magnus – Associate Professor, Director of the Programme for European Security Research, Swedish National Defence College, Sweden
Ericson, Peter – Deputy Director General and Head of the Security Policy Department, Ministry for Foreign Affairs, Sweden
Finney, Mike – Commodore, Deputy UK Military Representative, Permanent Representation of the United Kingdom to the EU, Brussels
Fischer, Uwe Christian – Plans and Programs Division, Federal Ministry of the Interior, Germany
Fredriksson, Peter – Lieutnant Colonel, Permanent Representation of Sweden to the EU, Brussels
Galloway, Tom – Deputy Head of the Stabilisation Unit, Stabilisation Unit, UK
Geist, Wolfgang – Colonel, Director of Studies and Chief of Staff, Federal College for Security Studies, Germany
Hartzell, Carl – Adviser to the Permanent Chair of the EU Political and Security Committee, European External Action Service, Brussels
Heinrich, Rüdiger – Lieutnant Colonel, Strategy and Policy Branch, Federal Ministry of Defence, Germany
Heldt, Birger – Head of Research, Folke Bernadotte Academy, Sweden
Herbst, John – Director, Center for Complex Operations, Institute for National Strategic Studies, National Defence University, Washington DC
Hull, Cecilia – Analyst, Department of Peace Support Operations, Swedish Defence Research Agency, Sweden
Hunter, Charlotte – Civilian CSDP Desk Officer, EU Civilian Crisis Management and External Spend Team, Europe External Policy Department, Foreign and Commonwealth Office, UK
Johnston, Sandy – Asst. Head (EU, France and Benelux), NATO and Europe Policy, Ministry of Defence, UK
Kania, Beatrix – Deputy Head of Division, Crisis Prevention and Peacebuilding, Federal Foreign Office, Germany
Kettle-Williams, Adam – Civil-Military Strategy and Exercise Coordinator, Security Policy Department, Foreign and Commonwealth Office, UK
Klink, Eckart – Brigadier General, Chairman of the European Union Military Committee Working Group, Brussels
Krull, Ralph – MA to the Chairman, European Union Military Committee Working Group, Brussels
Miozzo, Agostino – Managing Director for Crisis Response and Operational Coordination, European Union External Action Service, Brussels

Moore, Michael – Director, Department for Military Capability and Operations, Ministry of Defence, Sweden
Muermans, Peter – Division 400 – Peace and Security, Federal Ministry for Economic Cooperation and Development, Germany
Nachtwei, Winfried – Former Member of Parliament and Chairman of the Advisory Board for Civilian Crisis Prevention, Germany
Nilsson, Claes – Analyst, Department of Peace Support Operations, Swedish Defence Research Agency, Sweden
Nilsson, Henrik – Deputy Director, Department for Security Policy, Ministry for Foreign Affairs, Sweden
Nordh, Christer – Colonel, Deputy Military Representative, Permanent Representation of Sweden to the EU, Brussels
Patrick, Stewart M. – Senior Fellow/Director, International Institutions and Global Governance Program, Council on Foreign Relation, Washington DC
Petropoulos, Stavros – Crisis Response and Operational Coordination Unit, European External Action Service, Brussels
Pung, Hans – Vice President, Rand Europe, UK
Ruiz Calavera, Genoveva – Head of Instrument for Stability Operations Crisis Response and Peace Building/Service for Foreign Policy Instruments, Brussels
Schäfer, Kai – DG EuropeAid Development and Cooperation, European Commission, Brussels
Spatz, Joachim – Member of the German Parliament and the Parliament's Subcommittee on Civilian Crisis Prevention and Networked Security, Germany
Stares, Paul B. – General John W. Vessey Senior Fellow for Conflict Prevention, Director of the Center for Preventive Action, Council on Foreign Relation, Washington DC
Stolt, Peter – Supreme Commander's Staff, Department for Future Capabilities, Swedish Armed Forces Headquarters, Sweden
Svensson, Charlotte – Department for Strategy and Security Policy, Ministry of Defence, Sweden
Van Bellinghen, Marc – Deputy Head of the Peacebuilding, Conflict Prevention and Mediation Unit, European External Action Service, Brussels
Vraila, Marina – Head of Section (Exercises, CMCO, Military Concepts), Crisis Management Planning Directorate, European Union External Action Service, Brussels
Wells II, Linton – Director, Center for Technology and National Security Policy, Institute for National Strategic Studies, National Defence University, Washington DC
Widman, Erik – Deputy Director, Security Policy Department/Europe Division, Ministry for Foreign Affairs, Sweden
Winnerstig, Mike – Dep. Research Director, Defence Analysis, Swedish Defence Research Agency, Sweden
Wittkowsky, Andreas – Project Manager Peace and Security/Coordinator, Working Group on Comprehensive Conflict Management, Analysis Division, Center for International Peace Operations, Germany

6.2 Secondary Sources

Advisory Council on International Affairs (2009): *Crisis Management Operations in Fragile States. The Need for a Coherent Approach.* The Haque: Advisory Council on International Affairs. Retrieved from http://www.aiv-advies.nl/ContentSuite/upload/aiv/doc/AIV_64 _ENG_webversie.pdf

Aggestam, Lisbeth (2000): Germany. In Manners, Ian/Whitman, Richard W. (Eds.): *The Foreign Policies of European Union Member States.* Manchester/New York: Manchester University Press, pp. 64-84.

Agius, Christine/Devine, Karen (2011): 'Neutrality: A really dead concept?' A Reprise. *Cooperation and Conflict*, Vol. 46, No. 3, pp. 265-284.

Aide à la Décision Économique (2011): *Thematic Evaluation of European Commission Support to Conflict Prevention and Peace-Building.* Final Report. Evaluation for the European Commission. Louvain-La-Neuve: Aide à la Décision Économique.

Aktipis, Michael/Oliver, Tim (2012): Europeanization and British Foreign Policy. In Wong, Reuben/Hill, Christopher (Eds.): *National and European Foreign Policy.* London/New York: Routledge, pp. 72-92.

Alamir, Fouzieh M. (2012): The Complex Security-Development Nexus – Practical Challenges for Development Cooperation and the Military. *S + F. Sicherheit und Frieden. Security and Peace*, Vol. 30, No. 2, pp. 69-75.

Allen, David/Oliver, Tim (2006): The Foreign and Commonwealth Office. In Bache, Ian/Jordan, Andrew: *The Europeanization of British Politics.* Basingstoke: Palgrave Macmillan, pp. 52-66.

Allison, Graham T. (1971): *Essence of Decision. Explaining the Cuban Missile Crisis.* New York: Harper Collins Publishers.

Allison, Graham T./Zelikow, Philip (1999): *Essence of Decision. Explaining the Cuban Missile Crisis.* 2nd Ed., New York: Longman.

Ankersen, Christopher (2008): Introduction. Interrogating Civil-Military Cooperation. In Id. (Ed.): *Civil-Military Cooperation in Post-Conflict Operations. Emerging Theory and Practice.* London/New York: Routledge, pp. 1-11.

Annual Review of Global Peace Operations (2012): *Briefing Paper.* New York: Center on International Cooperation/New York University. Retrieved from http://cic.es.its.nyu.edu/sites/default/files/gpo_2012.pdf

Arloth, Jana/Seidensticker, Frauke (2007): *The ESDP Crisis Management Operations of the European Union and Human Rights.* Berlin: Deutsches Institut für Menschenrechte. Retrieved from http://www.institut-fuer-menschenrechte.de/uploads/tx_commerce/study_the_esdp_crisis_management_operations_of_the_eu_and_human_rights.pdf

Arter, David (2008): *Scandinavian Politics Today.* 2nd Ed., Manchester: Manchester University Press.

Åselius, Gunnar (2005): Swedish Strategic Culture after 1945. *Cooperation and Conflict*, Vol. 40, No. 1, pp. 25-44.

Austin, Greg/Brusset, Emery/Chalmers, Malcolm/Pierce, Juliet (2004): *Evaluation of the Conflict Prevention Pools.* Synthesis Report. London: Department for International Development. Retrieved from http://www.oecd.org/countries/sierraleone/35094094.pdf

Bach, Tobias/Fleischer, Julia/Hustedt, Thurid (2010): *Organisation und Steuerung Zentralstaatlicher Behörden. Agenturen im Westeuropäischen Vergleich.* Berlin: Edition Sigma.

Bache, Ian (2008): *Europeanization and Multilevel Governance. Cohesion Policy in the European Union and Britain.* Lanham: Rowman & Littlefield Publishers.

Bache, Ian/George, Stephen/Bulmer, Simon (2011): *Politics in the European Union.* 3rd Ed., Oxford: Oxford University Press.

Bache, Ian/Jordan, Andrew (2006): Europeanization and Domestic Change. In Ead. (Eds.): *The Europeanization of British Politics.* Basingstoke: Palgrave Macmillan, pp. 17-33.

Bailes, Alyson J.K. (2006): Introduction. The European Defence Challenge for the Nordic Region. In Bailes, Alyson J.K./Herolf, Gunilla/Sundelius, Bengt (Eds.): *The Nordic Countries and the European Security and Defence Policy.* Oxford: Oxford University Press, pp. 1-26.

Bailes, Alyson J.K. (2008): Designing a Comprehensive Security Policy for Europe and European States. In Aldis, Anne/Drent, Margriet (Eds.): *Common Norms and Good Practices of Civil-Military Relations in the EU.* Harmonie Paper 21. Groningen: Centre of European Security Studies, pp. 141-161.

Bailes, Alyson J.K. (2009): *Does a Small State Need a Strategy?* Occasional Paper 2/2009. Reykjavík: Institute of International Affairs/University of Iceland. Retrieved from http://ams.hi.is/sites/ams.hi.is/files/Bailes_Final%20wh.pdf

Banim, Guy (2009): EU Responses to Fragile States. In Weiss, Stefani/Spanger, Hans-Joachim/Van Meurs, Wim (Eds.): *Diplomacy, Development and Defence: A Paradigm for Policy Coherence. A Comparative Analysis of International Strategies.* Gütersloh: Verlag Bertelsmann Stiftung, pp. 303-355.

Barnett, Michael N./Finnemore, Martha (1999): The Politics, Power, and Pathologies of International Organizations. *International Organization*, Vol. 53, No. 4, pp. 699-732.

Barnett, Michael N./Finnemore, Martha (2004): *Rules for the World. International Organisations in Global Politics.* Ithaca/London: Cornell University Press.

Bauer, Michael W./Knill, Christoph (Ed.) (2007): *Management Reforms in International Organisations.* Baden-Baden: Nomos.

Bengtsson, Rikard (2010): *The EU and the European Security Order. Interfacing Security Actors.* London/New York: Routledge.

Benner, Thorsten/Mergenthaler, Stephan/Rotmann, Philipp (2009): Internationale Bürokratien und Organisationslernen. Konturen einer Forschungsagenda. *Zeitschrift für Internationale Beziehungen*, Vol. 16, No. 2, pp. 203-236.

Bennett, Andrew (2004): Case Study Methods: Design, Use and Comparative Advantages. In Sprinz, Detlef F./Wolinsky-Nahmias, Yael (Ed.): *Models, Numbers, and Cases: Methods for Studying International Relations.* Ann Arbor: University of Michigan, pp. 19-55.

Bertea, Stefano (2005): Looking for Coherence within the European Community. *European Law Journal*, Vol. 11, No. 2, March 2005, pp. 154–172.

Björkdahl, Annika (2007): *Norm Advocacy: A Small State Strategy to Influence the EU.* Lund: Lund University/Department of Political Science. Retrieved from http://www.nupi. no/content/download/1342/37199/file/Paper-Bjorkdahl.pdf

Björkman, Torsten/Friman, Henrik (2011): Comprehensive Approach. The Swedish Case. In Neal, Derrick J./Wells, Linton (Eds.): *Capability Development in Support of Comprehensive Approaches. Transforming International Civil-Military Interactions*. Washington DC: National Defense University, pp. 229-243.

Bjurner, Anders (2003): Sweden. In Ojanen, Hanna (Ed.): *Neutrality and Non-Alignment in Europe Today*. FIIA Report 6/2003. Helsinki: The Finnish Institute of International Affairs, pp. 41-45. Retrieved from http://www.fiia.fi/en/publication/55/

Bleicher, Knut (1991): *Organisation. Strategien – Strukturen – Kulturen*. 2nd Ed., Wiesbaden: Gabler.

Blockmans, Steven (2011): Beyond Conferral: The Role of the European External Action Service in Decision-Shaping. In Larik, Joris/Moraru, Madalina (Eds.): *Ever-closer in Brussels – Ever Closer in the World? EU External Action after the Lisbon Treaty*. EUI Working Papers. Florence: European University Institute, pp. 5-17.

Blockmans, Steven/Wessel, Ramses A. (2009): *The European Union and Crisis Management: Will the Lisbon Treaty Make the EU More effective?* CLEER Working Papers 01/2009. The Hague: Center for the Law of EU External Relations. Retrieved from http://doc.utwente.nl/77157/1/Blockmans09eu.pdf

Bogdanor, Vernon (Ed.) (1987): *Blackwell Encyclopedia of Political Institutions*. Oxford: Blackwell Reference.

Bogdanor, Vernon (2008): Introduction. In Id. (Ed.): *Joined-Up Government*. 3rd Ed., Oxford: The British Academy/Oxford University Press, pp. 1-18.

Bogumil, Jörg/Jann, Werner/Nullmeier, Frank (2006): Politik und Verwaltung – Perspektiven der Politikwissenschaftlichen Verwaltungsforschung. In Ead. (Eds.): Politik und Verwaltung. *Politische Vierteljahresschrift (PVS)*. Sonderheft 37. Wiesbaden: Verlag für Sozialwissenschaften, pp. 9-26.

Booth, Ken (Ed.) (2005): *Critical Security Studies and World Politics*. Boulder/London: Lynne Rienner Publishers.

Borchert, Heiko (2006): Homeland Security and Transformation: Why It Is Essential to Bring Together Both Agendas. In Brimmer, Esther (Ed.): *Transforming Homeland Security: U.S. and European Approaches*. Washington D.C.: Center for Transatlantic Relations, pp. 3-22.

Borrás, Susana/Radaelli, Claudio M. (2011): The Politics of Governance Architectures: Creation, Change and Effects of the EU Lisbon Strategy. *Journal of European Public Policy*, Vol. 18, No. 4, pp. 463-484.

Börzel, Tanja A./Risse, Thomas (2000): When Europe Hits Home: Europeanization and Domestic Change. *European Integration online Papers (EIoP)*, Vol. 4, No. 15. Retrieved from http://eiop.or.at/eiop/texte/2000-015a.htm

Börzel, Tanja A./Risse, Thomas (2003): Conceptualising the Domestic Impact of Europe. In Featherstone, Kevin/Radaelli, Claudio M. (Eds.) (2003): *The Politics of Europeanization*. Oxford: Oxford University Press, pp. 57-80.

Bossong, Raphael/Benner, Thorsten (2010): The Case for a Public Administration Turn in the Study of the EU's Civilian Crisis Management. *Journal of European Public Policy*, Vol 17, No. 7, pp. 1074–1086.

Bourgon, Jocelyne (2011): *A New Synthesis of Public Administration. Serving the 21st Century*. Montreal/Kingston: McGill-Queen's University Press.

Bratberg, Øivind (2007): Bilateral Diplomacy in an Integrated Europe: the Co-Existence of Institutional Orders. In Egeberg, Morten (Ed.): *Institutional Dynamics and the Transformation of Executive Politics in Europe*. CONNEX Report Series Nr. 03. Mannheim: University of Mannheim, pp. 533-561.

Brauss, Heinrich (2006): A Comprehensive Approach to Crisis Management. The Civilian-Military Dimension of ESDP. In European Commission/External Relations Directorate-General: *Faster and More United? The Debate about Europe's Crisis Response Capacity*. Luxembourg: Office for Official Publications of the European Communities, pp. 48-53.

Bretherton, Charlotte/Vogler, John (1999): *The European Union as a Global Actor*. London/New York: Routledge.

Brimmer, Esther (Ed.) (2006): *Transforming Homeland Security: U.S. and European Approaches*. Washington D.C.: Center for Transatlantic Relations.

Browning, Christopher S. (2007): Branding Nordicity. Models, Identity and the Decline of Exceptionalism. *Cooperation and Conflict*, Vol. 42, No. 1, pp. 27-51.

Bulmer, Simon/Jeffery, Charlie (2010): Does Congruence Matter? Germany and Britain in the European Union. In Bulmer, Simon/Jeffery, Charlie/Padgett, Stephen (Eds.): *Rethinking Germany and Europe. Democracy and Diplomacy in a Semi-Sovereign State*. Basingstoke: Palgrave Macmillan, pp. 113-138.

Bulmer, Simon J./Radaelli, Claudio M. (2004): The Europeanisation of National Policy? *Queen's Papers on Europeanisation*, No. 1/2004. Retrieved from http://www.qub.ac.uk/schools/SchoolofPoliticsInternationalStudiesandPhilosophy/FileStore/EuropeanisationFiles/Filetoupload,38405,en.pdf

Burke, Edward (2012a): Deliberate 'Spoiler' or Misunderstood Pragmatist? Britain and the European External Action Service. In European Policy Institutes Network (Ed.): *Reviewing Member States' Commitment to the European External Action Service*. Working Paper No. 34. Brussels: Center for European Policy Studies, pp. 14-18. Retrieved from http://www.ceps.be/book/reviewing-member-states%E2%80%99-commitment-european-external-action-service

Burke, Edward (2012b): *Europe's External Action Service: Ten Steps Towards a Credible EU Foreign Policy*. Policy Brief. London: Center for European Reform. Retrieved from http://www.cer.org.uk/publications/archive/policy-brief/2012/europes-external-action-service-ten-steps-towards-credible-eu

Buzan, Barry/Wæver, Ole/De Wilde, Jaap (1998): *Security. A New Framework for Analysis*. Boulder/London: Lynne Rienner Publishers.

Calabrò, Andrea (2011): *Governance Structures and Mechanisms in Public Service Organizations. Theories, Evidence and Future Directions*. Berlin/Heidelberg: Physica-Verlag.

Carnegie Commission on Preventing Deadly Conflict (1997): *Preventing Deadly Conflict*. Final Report. Washington DC: Carnegie Corporation.

Chandler, J.A. (2000): Introduction. In Id. (Ed.): *Comparative Public Administration*. London/New York: Routledge, pp. 1-13.

Christensen, Tom (2003): Organization Theory and Public Administration. Introduction. In Peters, B. Guy/Pierre, Jon (Eds.): *Handbook of Public Administration*. London: SAGE Publications, pp. 109-115.

Christensen, Tom (2004): Modern State Reforms. In Heidar, Knut (Ed.): *Nordic Politics. Comparative Perspectives*. Oslo: Universitetsforlaget, pp. 24-39.

Christensen, Tom/Lægreid, Per (2006): *The Whole-of-Government Approach – Regulation, Performance, and Public-Sector Reform*. Working Paper 6. Bergen: Stein Rokkan Centre for Social Studies. Retrieved from https://bora.uib.no/bitstream/handle/1956/1893/N06-06%20Christensen-L%E6greid.pdf;jsessionid=3EA498BAB9BB5C90930046C8AD996043.bora-uib_worker?sequence=1

Christensen, Tom/Lægreid, Per/Roness, Paul G./Røvik, Kjell A. (2007): *Organization Theory and the Public Sector. Instrument, Culture and Myth*. London/New York: Routledge.

Christiansen, Thomas (2001): Intra-Institutional Politics and Inter-Institutional Relations in the EU: Towards Coherent Governance? *Journal of European Public Policy*, Vol. 8, No. 5, pp. 747-769.

Clarke, Michael (2010): *Preliminary RUSI Briefing: The National Security Strategy 2010*. London: Royal United Services Institute. Retrieved from http://www.rusi.org/analysis/ commentary/ref:C4CBC6D8637AB7/

Collier, Paul/Elliott, V.L./Hegre, Håvard/Hoeffler, Anke/Reynal-Querol, Marta/Sambanis, Nicholas (2003): *Breaking the Conflict Trap. Civil War and Development Policy*. Washington, DC: The World Bank/Oxford University Press.

Collier, Paul/Hoeffler, Anke/Söderbom, Måns (2006): *Post-Conflict Risks*. Oxford: Centre for the Study of African Economies/University of Oxford. Retrieved from http://www.csae.ox. ac.uk/workingpapers/pdfs/2006-12text.pdf

CONCORD – European NGO Confederation for Relief and Development (2011): *Spotlight on EU Policy Coherence for Development. A Lisbon Treaty Provision. A Human Rights Obligation*. Brussels. Retrieved from http://coherence.concordeurope.org/pdf/PCD_country_profiles_110711.pdf

Coppieters, Bruno/Emerson, Michael/Huysseune, Michael/Kovziridze, Tamara/Noutcheva, Gergana/Tocci, Nathalie/Vahl, Marius (Eds.) (2004): *Europeanization and Conflict Resolution. Case Studies from the European Periphery*. Gent: Academia Press.

Cornish, Paul/Dorman, Andrew (2012): Smart Muddling Through: Rethinking UK National Strategy Beyond Afghanistan. *International Affairs*, Vol. 88, No. 2, pp. 213-222.

Cowles, Maria G./Caporaso, James/Risse, Thomas (Eds.) (2001): *Transforming Europe. Europeanization and Domestic Change*. Ithaca: Cornell University Press.

Crocker, Chester A./Hampson, Fen O./Aall, Pamela (2011): Collective Conflict Management: A New Formula for Global Peace and Security Cooperation? *International Affairs*, Vol. 87, No. 1, pp. 39-58.

Daehnhardt, Patricia (2012): Germany in the European Union. In Wong, Reuben/Hill, Christopher (Eds.): *National and European Foreign Policy*. London/New York: Routledge, pp. 35-56.

Dahl, Ann-Sofie (2006): Sweden. Once a Moral Superpower, always a Moral Superpower? *International Journal*, Vol. 61, No. 4, pp. 895-908.

Debiel, Tobias (2004): Wie Weiter mit effektiver Krisenprävention? Der Aktionsplan der Bundesregierung im Vergleich zu den Britischen Conflict Prevention Pools. *Die Friedenswarte*, Vol. 79, No. 3-4, pp. 253-298.

De Coning, Cedric (2004): Coherence and Integration in the Planning, Implementation and Evaluation of Complex Peacebuilding Operations. *Conflict Trends*, pp. 41-48. Retrieved from http://www.accord.org.za/publications/conflict-trends/downloads/434-conflict-trends-20041.html

De Coning, Cedric (2008a): *The United Nations and the Comprehensive Approach.* DIIS Report 2008/14. Copenhagen: Danish Institute for International Studies. Retrieved from http://www.diis.dk/graphics/Publications/Reports%202008/Report-2008-14_The_United_ Nations_and_the_Comprehensive_Approach.pdf

De Coning, Cedric (2008b): The Coherence Dilemma in Peacebuilding and Post-Conflict Reconstructing Systems. *African Journal on Conflict Resolution*, Vol. 8, No. 3, pp. 85-110.

De Coning, Cedric (2008c): Civil-Military Coordination and Complex Peacebuilding Systems. In Ankersen, Christopher (Ed.): *Civil-Military Cooperation in Post-Conflict Operations. Emerging Theory and Praxis.* London/New York: Routledge, pp. 52-74.

De Coning, Cedric (2009): An Introduction to Integrated Crisis Management. In Henriksson, Kirsi (Ed.): *Crisis Management Centre Finland Yearbook 2009 on Civilian Crisis Management Studies.* CMC Finland Crisis Management Studies, Vol. 2, No. 6. Helsinki: Crisis Management Centre, pp. 110-132. Retrieved from http://www.cmc finland.fi/pelastus/cmc/images.nsf/files/DA2FE9ECA609975EC225767B0047C14C /$file/CMC_Yearbook_2009.pdf

De Coning, Cedric/Friis, Karsten (2008): Introduction. How to Conceptualize 'Comprehensive Approach'? In Friis, Karsten/Jarmyr, Pia (Ed.): *Comprehensive Approach. Challenges and Opportunities in Complex Crisis Management.* Oslo: Norwegian Institute of International Affairs, pp. 2-9. Retrieved from http://www.isn.ethz.ch/Digital-Library/Publications/Detail/?ots591=0c54e3b3-1e9c-be1e-2c24-a6a8c70602 33&lng=en&id=57375

De Coning, Cedric/Friis, Karsten (2011): Coherence and Coordination. The Limits of the Comprehensive Approach. *Journal of International Peacekeeping*, Vol. 15, No. 1-2, pp. 243-272.

De Coning, Cedric/Lurås, Helge/Nagelhus Schia, Niels/Ulriksen, Ståle (2009): *Norway's Whole-of-Government Approach and its Engagement with Afghanistan.* Security in Practice 8/2009. NUPI-Report. Oslo: Norwegian Institute of International Affairs. Retrieved from http://english.nupi.no/content/download/11635/113434/version/2/file/SIP-08-09-de+Coning+et+al.pdf

De Coning, Cedric/Lurås, Helge/Nagelhus Schia, Niels/Ulriksen, Ståle (2010): *Norway's Whole-of-Government Approach. Challenges for Engagement with Afghanistan and Other Fragile States.* Policy Brief 2/2010. Oslo: The Norwegian Institute of International Affairs. Retrieved from http://www.oecd.org/dac/evaluation/dcdndep/4710 7434.pdf

De Flers, Nicole A./Müller, Patrick (2012): Dimensions and Mechanisms of the Europeanization of Member State Foreign Policy: State of the Art and New Research Avenues. *European Integration*, Vol. 34, No. 1, pp. 19-35.

Dettke, Dieter (2009): *Germany says "No". The Iraq War and the Future of German Foreign and Security Policy.* Washington, D.C.: Woodrow Wilson Center.

De Vasconcelos, Álvaro (2010): Introduction. In Id. (Ed.): *A Strategy for EU Foreign Policy*. EU ISS Report No. 7. Paris: European Union Institute for Security Studies, pp. 12-24. Retrieved from http://www.iss.europa.eu/uploads/media/A_strategy_for_EU_foreign _policy.pdf

Devine, Karen (2011): Neutrality and the Development of the European Union's Common Security and Defence Policy: Compatible or Competing? *Cooperation and Conflict*, Vol. 46, No. 3, pp. 334-369.

Dimitrakopoulos, Dionyssis G./Passas, Argyris G. (2003): International Organizations and Domestic Administrative Reform. In Peters, B. Guy/Pierre, Jon (Eds.): *Handbook of Public Administration*. London: SAGE Publications, pp. 440-450.

Dingwerth, Klaus/Kerwer, Dieter/Nölke, Andreas (Eds.) (2009): *Die Organisierte Welt. Internationale Beziehungen und Organisationsforschung*. Baden-Baden: Nomos.

Dorman, Andrew M. (2011): Making 2+2=5: The 2010 Strategic Defence and Security Review. *Defense and Security Analysis*, Vol. 27, No. 1, pp. 77-87.

Douma, Pyt/Van de Goor, Luc/Van Klaas, Walraven (1998): Research Methodologies and Practice. A Comparative Perspective on Methods for Assessing the Outbreak of Conflict and the Implementation in Practice by International Organisations. In Cross, Peter (Ed.): *Contributing to Preventive Action*. CPN Yearbook 1997/98. Baden-Baden: Nomos, pp. 79-94.

Dover, Robert (2007): *Europeanization of British Defence Policy*. Aldershot: Ashgate.

Drent, Margriet (2011): The EU's Comprehensive Approach to Security: A Culture of Coordination? In Biscop, Sven (Ed.): *Studia Diplomatica – The Brussels Journal of International Relations*, Vol. LXIV, No. 2, Brussels: Egmont – The Royal Institute for International Relations, pp. 3-18.

Drent, Margriet/ Volten, Peter (2008): Civil Direction of the Military: Redefining the Balance in France, Germany, Romania and the United Kingdom. In Aldis, Anne/Drent, Margriet (Eds.): *Common Norms and Good Practices of Civil-Military Relations in the EU*. Harmonie Paper 21. Groningen: Centre of European Security Studies, pp. 15-38.

Drent, Margriet/Zandee, Dick (2010): *Breaking Pillars. Towards a Civil-Military Security Approach for the European Union*. The Hague: Netherlands Institute of International Relations Clingendael. Retrieved from http://www.clingendael.nl/sites/default/files/ 20100211_breaking_pillars.pdf

Duke, Simon (2002): *The EU and Crisis Management. Development and Prospects*. Maastricht: European Institute of Public Administration.

Dunn Cavelty, Myriam/Mauer, Victor (Eds.) (2010): *The Routledge Handbook of Security Studies*. London/New York: Routledge.

Dunne, Tim (2004): When the Shooting Starts: Atlanticism in British Security Strategy. *International Affairs*, Vol. 80, No. 5, pp. 893-909.

Dwan, Ranata (2001): Conflict Prevention and CFSP Coherence. In Missiroli, Antonio (Ed.): *Coherence for European Security Policy. Debates – Cases – Assessments*. ISS Occasional Paper 27. Paris: European Union Institute for Security Studies, pp. 17-22.

Dyson, Kenneth/Goetz, Klaus H. (2003): Living with Europe: Power, Constraint, and Contestation. In Ead. (Eds.): *Germany, Europe and the Politics of Constraint*. Oxford: Oxford University Press, pp. 3-35.

Dyson, Kenneth/Sepos, Angelos (2010): Differentiation as Design Principle and as Tool in the Political Management of European Integration. In Ead. (Eds.): *Which Europe? The Politics of Differentiated Integration.* Basingstoke: Palgrave Macmillan, pp. 3-23.

Edmunds, Timothy (2012): British Civil-Military Relations and the Problem of Risk. *International Affairs*, Vol. 88, No. 2, pp. 265-282.

Egeberg, Morten (2003): How Bureaucratic Structure Matters: An Organizational Perspective. In Peters, B. Guy/Pierre, Jon (Eds.): *Handbook of Public Administration.* London: SAGE Publications, pp. 116-126.

Egnell, Robert (2009a): *Complex Peace Operations and Civil-Military Relations. Winning the Peace*. London/New York: Routledge.

Egnell, Robert (2009b): Civil-Military Aspects of Effectiveness in Peace Support Operations. In Michael, Kobi/Kellen, David/Ben-Ari, Eyal (Eds.): *The Transformation of the World of War and Peace Support Operations*. Westport: Praeger Security International, pp. 122-138.

Ehrhart, Hans-Georg (2007): *Civil-Military Co-operation and Co-ordination in the EU and in Selected Member States*. Brussels: European Parliament. Retrieved from http://www.euro parl.europa.eu/meetdocs/2009_2014/documents/sede/dv/sede260410studycivilmilitarycoop_/sede260410studycivilmilitarycoop_en.pdf

Ehrhart, Hans-Georg (2011): Zivil-Militärisches Zusammenwirken und Vernetzte Sicherheit als Herausforderung Deutscher Sicherheitspolitik. Der Fall Afghanistan. *Zeitschrift für Außen- und Sicherheitspolitik*, Vol. 4, pp. 65-85.

Ehrhart, Hans-Georg/Petretto, Kerstin (2012): *The EU and Somalia: Counter-Piracy and the Question of a Comprehensive Approach.* Hamburg: Study for The Greens/ European Free Alliance. Retrieved from http://www.greens-efa.eu/fileadmin/dam/ Documents/Studies/ Ehrhart_Petretto_EUandSomalia_2012_fin.pdf

Eide, Espen Barth/Kaspersen, Anja Therese/Kent, Randolph/Von Hippel, Karen (2005): *Report on Integrated Missions. Practical Perspectives and Recommendations.* Independent Study for the Expanded UN ECHA Core Group. Oslo: Norwegian Institute of International Affairs. Retrieved from http://www.regjeringen.no/upload/UD/ Vedlegg/FN/report_on_ integrated_missons.pdf

Ekengren, Magnus (2004): National Foreign Policy Co-ordination: The Swedish EU Presidency. In Carlsnaes, Walter/Sjursen, Helene/White, Brian (2004): *Contemporary European Foreign Policy*. London/Thousand Oaks/New Delhi: Sage Publishers, pp. 211-226.

Ekengren, Magnus (2006): The Interface of External and Internal Security in the EU and in Nordic Policies. In Bailes, Alyson J.K./Herolf, Gunilla/Sundelius, Bengt (Eds.): *The Nordic Countries and the European Security and Defence Policy*. Oxford: Oxford University Press, pp. 267-287.

Ekengren, Magnus/Sundelius, Bengt (2004): National Foreign Policy Co-ordination. In Carlsnaes, Walter/Sjursen, Helene/White, Brian (2004): *Contemporary European Foreign Policy*. London/Thousand Oaks/New Delhi: Sage Publishers, pp. 110-122.

Eriksson, Arita (2006): *Europeanization and Governance in Defence Policy. The Example of Sweden*. Stockholm Studies in Politics 117. Stockholm: Stockholm University.

Eriksson, Pär/Foghelin, Jan (2002): European Crisis Management: A Swedish View. In Ehrhart, Hans-Georg (Ed.): *Die Europäische Sicherheits- und Verteidigungspolitik. Positionen, Perzeptionen, Probleme, Perspektiven.* Baden-Baden: Nomos, pp. 88-99.

Eronen, Oskari (2008): *PRT Models in Afghanistan. Approaches to Civil-Military Integration.* Kuopio: Crisis Management Centre Finland. Retrieved from http://www.cmcfinland.fi/pelastus/cmc/images.nsf/files/8442FEDCC134982CC225755A0059D3D0/$file/Studies_5_Eronen.pdf

European Peacebuilding Liaison Office (2012a): *Common Foreign and Security Policy Structures and Instruments after the Entry into Force of the Lisbon Treaty.* EPLO Briefing Paper 1/2012. Brussels: European Peacebuilding Liaison Office. Retrieved from http://www.eplo.org/assets/files/2.%20Activities/Working%20Groups/CSDP/EPLO_Briefing_Paper_1-2012_CFSP_After_Lisbon.pdf

European Peacebuilding Liaison Office (2012b): *The EEAS and Peacebuilding One Year on.* Brussels: European Peacebuilding Liaison Office. Retrieved from http://www.eplo.org/assets/files/2.%20Activities/Working%20Groups/EEAS/EPLO_Statement_EEASPeacebuildingOneYearOn.pdf

Evans, Alex/Steven, David (2010): *Organizing for Influence. UK Foreign Policy in an Age of Uncertainty.* A Chatham House Report. London: Chatham House. Retrieved from http://www.chathamhouse.org/sites/default/files/public/Research/Europe/r0610_stevens_evans.pdf

Faust, Jörg/Messner, Dirk (2012): Probleme Globaler Entwicklung und die Ministerielle Organisation der Entwicklungspolitik. *Zeitschrift für Außen- und Sicherheitspolitik*, Vol. 5, No. 2, pp. 165-175.

Featherstone, Kevin (2003): Introduction. In Featherstone, Kevin/Radaelli, Claudio M. (Eds.): *The Politics of Europeanization.* Oxford: Oxford University Press, pp. 3-26.

Felberbauer, Ernst M./Feichtinger, Walter E./Schmidl, Erwin A. (2011): Preface. In Ead. (Eds.): *International Crisis Management: Squaring the Circle.* Vienna/Geneva: Geneva Centre for Security Policy, pp. 5/6.

Fischer, Martina (2004): Der Aktionsplan Krisenprävention der Bundesregierung – Von der Bestandsaufnahme zur Selbstverpflichtung für eine Zivile Außenpolitik?, *Die Friedenswarte*, Vol. 79, No. 3-4, pp. 313-322.

Forsberg, Tuomas (2011): Normative Power Europe, Once Again: A Conceptual Analysis of an Ideal Type. *Journal of Common Market Studies*, Vol. 49, No. 6, pp. 1183-1204.

Forster, Anthony (2000): Britain. In Manners, Ian/Whitman, Richard W. (Eds.): *The Foreign Policies of European Union Member States.* Manchester/New York: Manchester University Press, pp. 45-63.

Franke, Volker (2012): Vernetzte Sicherheit. *S + F. Sicherheit und Frieden. Security and Peace*, Vol. 30, No. 2, p. III.

Frazier, Derrick V./Dixon, William J. (2009): Third-party Intermediaries and Negotiated Settlements, 1946-2000. In Bercovitch, Jacob/Gartner, Scott S. (Ed.): *International Conflict Mediation. New Approaches and Findings.* London/New York: Routledge, pp. 43-66.

Friis. Karsten (2010): Conclusion: Similar Challenges – But No Nordic Whole of Government Model. In Friis, Karsten/Rehman, Sanaa (Ed.): *Nordic Approaches to Whole-of-Government – In Afghanistan and Beyond.* Security in Practice 6/2010. NUPI-Report. Oslo: Norwegian Institute of International Affairs. Retrieved from http://www.cmcfinland.fi/pelastus/cmc/images.nsf/files/9BA84D6419379F1BC2257 75000468886/$file/Jari_M_in_SIP-06-10-Rehman-Friis.pdf

Friis, Karsten/Jarmyr, Pia (Eds.) (2008): *Comprehensive Approach. Challenges and Opportunities in Complex Crisis Management.* Oslo: Norwegian Institute of International Affairs. Retrieved from http://www.isn.ethz.ch/Digital-Library/Publications/ Detail/?ots 591=0c54e3b3-1e9c-be1e-2c24-a6a8c7060233&lng=en&id=57375

Fukuyama, Frank (1992): *The End of History and the Last Man.* New York: The Free Press.

Gärtner, Heinz/Hyde-Price, Adrian/Reiter, Erich (Ed.) (2001): *Europe's New Security Challenges.* Boulder, CO: Lynne Rienner.

Gauster, Markus (2012): Comprehensive Approach in Zivil-militärischen Einsätzen – (k)ein Mehrwert in Afghanistan? *S + F. Sicherheit und Frieden. Security and Peace,* Vol. 30, No. 2, pp. 75-81.

Gebhard, Carmen (2008): *Civil-Military Coordination and Cooperation in the Context of the EU's Crisis Management. CMCO versus CIMIC. Conceptual and Terminological Clarifications.* Vienna: National Defence Academy/Institute for Support and Conflict Management.

Gebhard, Carmen (2011): Chapter 5. Coherence. Draft of the Chapter Forthcoming in Hill, Christopher/Smith, Michael (Eds.): *International Relations and the European Union.* 2nd Ed., Oxford: Oxford University Press.

Gebhard, Carmen/Norheim-Martinsen, Per (2010): *The EU as a Comprehensive Security Provider – Theoretical and Empirical Challenges.* Paper to be presented at the UACES Conference 'Exchanging Ideas on Europe', Bruges.

Geis, Matthias (2012): Stillgestanden! *DIE ZEIT*, No. 30, 19. July 2012, p. 2.

Giegerich, Bastian (2009): *European Military Crisis Management. Connecting Ambition and Reality.* Adelphi Paper No. 397. London: International Institute for Strategic Studies.

Giegerich, Bastian/Jonas, Alexandra (2011): *"Not Simply a Cost-Saving Exercise"? Großbritanniens Neue Sicherheits- und Verteidigungspolitik nach der Strategic Defence and Security Review.* SOWI.Thema 01/2011. Strausberg: Sozialwissenschaftliches Institut der Bundeswehr.

Gill, Bates (2011): Introduction. International Security, Armaments and Disarmament. In Stockholm International Peace Research Institute: *SIPRI Yearbook 2011. Armaments, Disarmament and International Security.* Oxford: Oxford University Press, pp. 3-12.

Ginsberg, Roy H. (1999): Conceptualizing the European Union as an International Actor: Narrowing the Theoretical Capability-Expectations Gap. *Journal of Common Market Studies*, Vol. 37, No. 3, pp. 429-454.

Giorgis, Andebrhan W. (2010): Coordinating International Support for African Peace and Security Efforts: From the G8 to the EU. *The International Spectator*, Vol. 45, No. 2, pp. 69-83.

Gordon, Stuart (2006): Exploring the Civil-Military Interface and its Impact on European Strategic and Operational Personalities: 'Civilianisation' and Limiting Military Roles in Stabilisation Operations? *European Security*, Vol. 15, No. 3, pp. 339-361.

Gourevitch, Peter (1978): The Second Image Reversed: The International Sources of Domestic Politics. *International Organization*, Vol. 32, No. 4, pp. 881-912.

Gow, James (2009): The United Kingdom National Security Strategy: The Need for New Bearings in Security Policy. *The Political Quarterly*, Vol. 80, No. 1, pp. 126-133.

Grabbe, Heather (2006): *The EU's Transformative Power. Europeanization through Conditionality in Central and Eastern Europe*. Basingstoke: Palgrave Macmillan.

Green, Duncan (2009): *How Has DFID's Thinking Evolved Since 1997? Four White Papers Provide a Rough Guide*. Weblog Entry: From Poverty to Power. Entry posted on Monday, August 10th, 2009. Retrieved from http://www.oxfamblogs.org/fp2p/?p=546

Grevi, Giovanni/Helly, Damien/Keohane, Daniel (2009): *European Security and Defence Policy. The first 10 Years (1999-2009)*. Brussels: EU Institute for Security Studies.

Grieco, Joseph M. (1988): Anarchy and the Limits of Cooperation: A Realist Critique of the Newest Liberal Institutionalism. *International Organization*, Vol. 42, No. 3, pp. 485-507.

Grönberg, Göran (2010): A Swedish Approach to Comprehensiveness. In Friis, Karsten/Jarmyr, Pia (Ed.) (2008): *Comprehensive Approach. Challenges and Opportunities in Complex Crisis Management*. Oslo: Norwegian Institute of International Affairs, pp. 25-29. Retrieved from http://www.isn.ethz.ch/Digital-Library/Publications/Detail/?ots591=0c54e3b3-1e9c-be1e-2c24-a6a8c7060233&lng=en&id=57375

Grönvall, Jesper/Lundberg, Jan (2005): Crafting a Societal Security System in Sweden. In Hamilton, Daniel/Sundelius, Bengt/Grönvall, Jesper (Eds.): *Protecting the Homeland: European Approaches to Societal Security – Implications for the United States*. Washington, D.C.: Center for Transatlantic Relations, pp. 17-41.

Gross, Eva (2008): *EU and the Comprehensive Approach*. DIIS Report 2008/13. Copenhagen: Danish Institute for International Studies. Retrieved from http://www.diis.dk/graphics/ Publications/Reports%202008/R2008-13_EU_and_the_ Comprehensive _Approach.pdf

Gross, Eva (2009): *The Europeanization of National Foreign Policy. Continuity and Change in Crisis Management*. Basingstoke: Palgrave Macmillan.

Gulick, Luther (1937): Notes on the Theory of Organization. In Gulick, Luther/Urwick, Lyndall (Ed.): *Papers on the Science of Administration*. New York: Institute of Public Administration, pp. 3-45.

Hadfield-Amkhan, Amelia (2010): *British Foreign Policy, National Identity, and Neoclassical Realism*. Lanham: Rowman & Littlefield Publishers.

Haftendorn, Helga/Keohane, Robert O./Wallander, Celeste A. (1999): Introduction. In Ead. (Eds.): *Imperfect Unions. Security Institutions over Time and Space*. Oxford/New York: Oxford University Press.

Hall, Richard H. (1991): *Organizations: Structures, Processes, and Outcomes*. Englewood Cliffs, New Jersey: Prentice Hall.

Hall, Richard H./Clark, John P./Giordano, Peggy C./Johnson, Paul V./Van Roekel, Martha (1977): Patterns of Interorganisational Relationships, *Administrative Science Quarterly*, Vol. 22, No. 3, pp. 457-474.

Hall, Richard H./Quinn, Robert E. (1983): Question: Is There a Connection Between Organizational Theory and Public Policy? In Ead. (Eds.): *Organizational Theory and Public Policy*. Beverly Hills: SAGE Publications, pp. 7-20.

Hallett, Michael/Thorngren, Oke (2011): Attempting a Comprehensive Approach Definition and Its Implications for Reconceptualizing Capability Development. In Neal, Derrick J./Wells, Linton (Eds.): *Capability Development in Support of Comprehensive Approaches. Transforming International Civil-Military Interactions*. Washington D.C.: National Defense University, pp. 35-49.

Hamilton, Daniel S. (2006): Introduction: Transforming Homeland Security: A Road Map for the Transatlantic Alliance. In Brimmer, Esther (Ed.): *Transforming Homeland Security: U.S. and European Approaches*. Washington D.C.: Center for Transatlantic Relations, pp. ix-.xxxvi.

Hanssen, Måns (2010): *Civil-Military Interaction in the EU. Applying a Comprehensive Approach to CSDP Operations*. Stockholm: Swedish Defence Research Agency. Retrieved from https://www.cimicweb.org/cmo/compapp/Documents/European%20Union/FOI%20Civ-Mil%20Interaction%20in%20the%20EU%20%28M%20Hansen,%202010 %29%5B1%5D.pdf

Harnisch, Sebastian/Maull, Hanns W. (Eds.) (2001): *Germany as a Civilian Power? The Foreign Policy of the Berlin Republic*. Manchester: Manchester University Press.

Harnisch, Sebastian/Wolf, Raimund (2010): Germany. The Continuity of Change. In Kirchner, Emil J./Sperling, James (Eds.): *National Security Cultures. Patterns of Global Governance*. London/New York: Routledge, pp. 43-65.

Haugevik, Kristin M. (2007): *New Partners, New Possibilities. The Evolution of Inter-Organizational Security Cooperation in International Peace Operations*. NUPI Report Security in Practice No. 6. Oslo: Norwegian Institute of International Affairs. Retrieved from http://english.nupi.no/content/download/2801/55619/version/5/file/Haugevik-Rapport-New+partners.pdf

Hedlund, Erik/Soeters, Joseph (2010): Reflections on Swedish Peacekeepers' Self-image and Dilemmas of Peacekeeping. *International Peacekeeping*, Vol. 17, No. 3, pp. 408-414.

Helwig, Niklas/Ivan, Paul/Kostanyan, Hrant (2013): *The New EU Foreign Policy Architecture – Reviewing the First Two Years of the EEAS*. Brussels: Centre for European Policy Studies. Retrieved from www.ceps.be/ceps/dld/7711/pdf

Héritier, Adrienne (2003): New Modes of Governance in Europe: Increasing Political Capacity and Policy Effectiveness? In Börzel, Tanja/Cichowski, Rachel (Ed.): *The State of the European Union. Law, Politics, and Society*. New York: Oxford University Press, pp. 105-126.

Herolf, Gunilla (2007): Sweden and ESDP. In Brummer, Klaus (Ed.): *The North and ESDP. The Baltic States, Denmark, Finland and Sweden*. Gütersloh: Bertelsmann Stiftung, pp. 45-58.

Herrhausen, Anna (2009): *Organizing Peacebuilding. An Investigation of Interorganizational Coordination in International Post Conflict Reconstruction Efforts*. Frankfurt am Main: Peter Lang Internationaler Verlag der Wissenschaften.

Hill, Christopher (1993): The Capability-Expectations Gap, or Contextualizing Europe's International Role. *Journal of Common Market Studies*, Vol. 31, No. 3, pp. 305-328.

Hillion, Christophe/Wessel, Ramses A. (2009): Competence Distribution in EU External Relations After ECOWAS: Clarification or Continued Fuzziness? *Common Market Law Review*, Vol. 46, No. 2, pp. 551–586.

Hofmann, Herwig C. H. (2008): Mapping the European Administrative Space. *West European Politics*, Vol. 31, No. 4, pp. 662-676.

Hofmann, Herwig C./Türk, Alexander H. (2006): An Introduction to EU Administrative Governance. In Ead. (Eds.) (2006): *EU Administrative Governance*. Cheltenham: Edward Elgar Publishing Limited, pp. 1-8.

Hogan-Hern, David (2011): Relevant Experience from the UK Stabilization Unit. In Felberbauer, Ernst M./Feichtinger, Walter E./Schmidl, Erwin A. (Eds.): *International Crisis Management: Squaring the Circle*. Vienna/Geneva: Geneva Centre for Security Policy, pp. 43-54.

Hood, Christopher (2008): The Idea of Joined-Up Government: A Historical Perspective. In Bogdanor, Vernon (Ed.): *Joined-Up Government*. Oxford: The British Academy/Oxford University Press, pp. 19-42.

Hopkinson, William (2004): *Sizing and Shaping European Armed Forces. Lessons and Considerations from the Nordic Countries*. SIPRI Policy Paper No. 7, March 2004. Stockholm: Stockholm International Peace Research Institute.

Houben, Marc (2005): *International Crisis Management. The Approach of European States*. London/New York: Routledge.

Huldt, Bo (2003): Comments on the Swedish Position. In Ojanen, Hanna (Ed.): *Neutrality and Non-Alignment in Europe Today*. FIIA Report 6/2003. Helsinki: The Finnish Institute of International Affairs, pp. 46-51. Retrieved from http://www.fiia.fi/en/publication/55/

Hull, Cecilia (2011): *Focus and Convergence Through a Comprehensive Approach: But Which Among The Many?* Presented at the 16th International Command and Control Research and Technology Symposium (ICCRTS 2011). Stockholm: Swedish Defence Research Agency. Retrieved from http://www.dodccrp.org/events/16th_iccrts_2011 /papers/088.pdf

Hult, Karen M. (2003): Environmental Perspectives on Public Institutions. In Peters, B. Guy/Pierre, Jon (Eds.): *Handbook of Public Administration*. London: SAGE Publications, pp. 149-159.

Hustedt, Thurid/Tiessen, Jan (2006): *Central Government Coordination in Denmark, Germany and Sweden – An Institutional Perspective*. Forschungspapiere Regierungsorganisationen in Westeuropa. Potsdam: Universitätsverlag Potsdam.

Hynek, Nik (2011): EU Crisis Management After the Lisbon Treaty: Civil-military Coordination and the Future of the EU OHQ. *European Security*, Vol. 20, No. 1, pp. 81-102.

Independent Commission for Aid Impact (2011): *Evaluation of the Inter-departmental Conflict Pool*. London: Independent Commission for Aid Impact. Retrieved from http://icai.independent.gov.uk/wp-content/uploads/2011/11/Conflict-pool-ToRs-FINAL5.pdf

International Crisis Group (2001): *EU Crisis Response Capability. Institutions and Processes for Conflict Prevention and Management.* ICG Issues Report No. 2. Brussels: International Crisis Group. Retrieved from http://www.crisisweb.org//library/documents/report_archive/

International Crisis Group (2005): *EU Crisis Response Capability Revisited.* ICG Europe Report No. 160. Brussels: International Crisis Group. Retrieved from http://www.crisisgroup.org/~/media/Files/europe/160_eu_crisis_response_capability_revisited_edit.ashx

ISIS Europe (2012): *CSDP Map. Mission Chart.* Brussels: International Security Information Service Europe. Retrieved from http://www.csdpmap.eu/mission-chart

Jacobsson, Bengt/Lægreid, Per/Pedersen, Ove K. (2004): *Europeanization and Transnational States. Comparing Nordic Central Governments.* London/New York: Routledge.

Jäger, Thomas/Oppermann, Kai (2006): Bürokratie- und Organisationstheoretische Analysen der Sicherheitspolitik: Vom 11. September zum Irakkrieg. In Siedschlag, Alexander (Ed.): *Methoden der Sicherheitspolitischen Analyse – Eine Einführung.* Wiesbaden: VS Verlag für Sozialwissenschaften, pp. 105-134.

Jakobsen, Peter Viggo (2008): NATO's *Comprehensive Approach to Crisis Response Operations. A Work in Slow Progress.* Copenhagen: Danish Institute for International Studies. Retrieved from http://www.diis.dk/graphics/Publications/Reports%202008/Report_2008-15_NATO_Comprehensive_Approach_Crisis_Response_Operations.pdf

Jakobsen, Peter Viggo (2009): Small States, Big Influence: The Overlooked Nordic Influence on the Civilian ESDP. *Journal of Common Market Studies*, Vol. 47, No. 1, pp. 81-102.

Johannsen, Janina (2011): *The EU's Comprehensive Approach to Crisis Management. Premises, Ambitions, Limits.* Baden-Baden: Nomos.

Johansson, Karl Magnus (1999): Europeanisation and Its Limits: The Case of Sweden. *Journal of International Relations and Development*, Vol. 2, No. 2, pp. 169-186.

Johansson, Karl Magnus (2003): Sweden: Another Awkward Partner? In Wessels, Wolfgang/Maurer, Andreas/Mittag, Jürgen (Ed.): *Fifteen Into One. The European Union and its Member States.* Manchester: Manchester University Press, pp. 369-387.

Johansson, Karl Magnus/Raunio, Tapio (2010): Organizing the Core Executive for European Union Affairs. Comparing Finland and Sweden. *Public Administration*, Vol. 88, No. 3, pp. 649-664.

Jokela, Juha (2011): *Europeanization and Foreign Policy. State Identity in Finland and Britain.* London/New York: Routledge.

Jones, Alun/Clark, Julian (2010): *The Spatialities of Europeanization. Power, Governance and Territory in Europe.* London/New York: Routledge.

Jonson, Pål (2006): *The Development of the European Security and Defence Policy. An Assessment of Preferences, Bargains and Outcomes.* Stockholm: Swedish Defence Research Agency.

Juncos, Ana E. (2006): *Learning by Doing: Civil-Military Co-ordination in EU Crisis Management Policies.* Leicestershire: Loughborough University.

Juncos, Ana E. (2008): *The Lisbon Treaty and the Ongoing Problem of Co-ordination of the EU's External Action*. European Policy Brief. London: The Federal Trust for Education and Research.

Juncos, Ana E. (2011): Europeanization by Decree? The Case of Police Reform in Bosnia. *Journal of Common Market Studies*, Vol. 49, No. 2, pp. 367-389.

Juncos, Ana E./Pomorska, Karolina (2010): Secretariat, Facilitator or Policy Entrepreneur? Role Perceptions of Officials of the Council Secretariat. In Vanhoonacker, Sophie/Dijkstra, Hylke/Maurer, Heidi (Eds.): Understanding the Role of Bureaucracy in the European Security and Defence Policy. *European Integration online Papers (EIoP)*, Vol. 14, Special Issue 1. Retrieved from http://eiop.or.at/eiop/2010-007a.htm

Jupille, Joseph/Caporaso, James A. (1999): Institutionalism and the European Union: Beyond Comparative Politics and International Relations. *Annual Review of Political Science*, No. 2, pp. 429-444.

Kaldor, Mary (2007): *New and Old Wars*. Stanford: Stanford University Press.

Kamp, Karl-Heinz (2011): NATO's New Strategic Concept: An Integration of Civil and Military Approaches? In Schnaubelt, Christopher M. (Ed.): *Towards a Comprehensive Approach: Integrating Civilian and Military Concepts of Strategy*. Rome: NATO Defence College, pp. 52-69.

Kaski, Antti (2011): *The CSDP After Lisbon: Lost Opportunities or Changed Interests?* GCSP Policy Paper No.16. Geneva: Geneva Centre for Security Policy. Retrieved from http://www.gcsp.ch/Regional-Capacity-Development/Euro-Atlantic-Security/Publications/GCSP-Publications/Policy-Papers/The-CSDP-after-Lisbon-Lost-Oppor tunities-or-Changed-Interests

Kassim, Hussein (2001): Introduction: Co-ordinating National Action in Brussels. In Kassim, Hussein/Menon, Anand/Peters, B. Guy/Wright, Vincent (Eds.): *The National Co-ordination of EU Policy. The European Level*. Oxford: Oxford University Press, pp. 1-46.

Keohane, Robert O. (1988): International Institutions: Two Approaches. *International Studies Quarterly*, Vol. 32, No. 4, pp. 379-396.

Kerr, Pauline (2010): Human Security and Diplomacy. In Dunn Cavelty, Myriam/Mauer, Victor (Eds.): *The Routledge Handbook of Security Studies*. London/New York: Routledge, pp. 115-125.

Khol, Radek (2006): Civil-Military Co-ordination in EU Crisis Management. In Nowak, Agnieszka (Ed.): *Civilian Crisis Management. The EU Way*. Chaillot Paper No. 90. Paris: European Union Institute for Security Studies, pp. 123-138. Retrieved from http://www.iss. europa.eu/uploads/media/cp090.pdf

Kickert, Walter (2008): Distinctiveness in the Study of Public Management in Europe. In Id. (Ed.): *The Study of Public Management in Europe and the US. A Comparative Analysis of National Distinctiveness*. London/New York: Routledge, pp. 1-13.

King, Gary/Keohane, Robert O./Verba, Sidney (1994): *Designing Social Inquiry. Scientific Inference in Qualitative Research*. Princeton: Princeton University Press.

Kirchner, Emil J. (2010a): European Union. Moving Towards a European Security Culture. In Kirchner, Emil J./Sperling, James (Eds.): *National Security Cultures. Patterns of Global Governance*. London/New York: Routledge, pp. 102-123.

Kirchner, Emil J. (2010b): Germany's Role in European and International Security: Aims Versus Action. In Bulmer, Simon/Jeffery, Charlie/Padgett, Stephen (Eds.): *Rethinking Germany and Europe. Democracy and Diplomacy in a Semi-Sovereign State.* Basingstoke: Palgrave Macmillan, pp. 139-153.

Kiso, Jan O. (2009): The UK Response to Precarious States: Innovative Ideas on the Long Path Toward a More Interdisciplinary Policy. In Weiss, Stefani/Spanger, Hans-Joachim/Van Meurs, Wim (Eds.): *Diplomacy, Development and Defence: A Paradigm for Policy Coherence. A Comparative Analysis of International Strategies*. Gütersloh: Verlag Bertelsmann Stiftung, pp. 105-141.

Klein, Nadia (2010): *European Agents out of Control? Delegation and Agency in the Civil-Military Crisis Management of the European Union 1999-2008*. Baden-Baden: Nomos.

Klingebiel, Stephan (2006): New Interfaces between Security and Development. Introduction. In Id. (Ed.): *New Interfaces between Security and Development. Changing Concepts and Approaches*. DIE Studies No. 13. Bonn: German Development Institute, pp. 1-9.

Knight, Andy (2008): Civil-Military Cooperation and Human Security. In Ankersen, Christopher (Ed.): *Civil-Military Cooperation in Post-Conflict Operations. Emerging Theory and Practice*. London/New York: Routledge, pp. 15-30.

Knill, Christoph (2006): Introduction: Cross-National Policy Convergence: Concepts, Approaches and Explanatory Factors. In Id. (Ed.): *Cross-National Policy Convergence. Causes, Concepts and Empirical Findings*. London/New York: Routledge, pp. 1-11.

Knoepfel, Peter/Larrue, Corrine/Varone, Frédéric/Hill, Michael (2011): *Public Policy Analysis*. Bristol: The Policy Press.

Knudsen, Christian (1995): The Competence View of the Firm. What can Modern Economists Learn from Philip Selznick's Sociological Theory of Leadership? In Scott, W. Richard/Christensen, Søren (Ed.): *The Institutional Construction of Organizations. International and Longitudinal Studies*. Thousand Oaks: SAGE Publications.

Konrad-Adenauer-Stiftung (Ed.) (2009): Vernetzte Sicherheit und Entwicklung. Fachtagung des Gesprächskreises „Vernetzte Sicherheit". Berlin, 17. Juni 2009. *IM PLENUM Kompakt*. Sankt Augustin: Konrad-Adenauer-Stiftung e.V.. Retrieved from http://www.kas.de/wf/doc/kas_17738-544-1-30.pdf?091005105530

Koops, Joachim A. (2009): Introduction. In Id. (Ed.): Military Crisis Management: The Challenge of Inter-Organizationalism. *Studia Diplomatica – The Brussels Journal of International Relations*, Vol. LXII, No. 3. Brussels: Egmont – The Royal Institute for International Relations, pp. 3-6.

Korski, Daniel/Gowan, Richard (2009): *Can the EU Rebuild Failing States? A Review of Europe's Civilian Capacities*. London: European Council on Foreign Relations. Retrieved from http://ecfr.eu/page/-/ECFR18_-_Can_the_EU_rebuild_failing_States_-_a_Review_ of_Europes_Civilian_Capacities.pdf

Krause, Keith/Williams, Michael C. (Eds.) (1997): *Critical Security Studies. Concepts and Cases*. Minneapolis: University of Minnesota Press.

Kühl, Stefan/Strodtholz, Petra/Taffertshofer, Andreas (2009): Qualitative und Quantitative Methoden der Organisationsforschung – Ein Überblick. In Ead. (Eds.): *Hand-*

buch Methoden der Organisationsforschung. Quantitative und Qualitative Methoden. Wiesbaden: VS Verlag für Sozialwissenschaften/GWV Fachverlage GmbH, pp. 13-27.

Ladrech, Robert (1994): Europeanization of Domestic Politics and Institutions: The Case of France. *Journal of Common Market Studies*, Vol. 32, No. 1, pp. 69-88.

Lange, Michael A. (2010): *Sicherheit Ist Mehr, als die Summe Ziviler, Militärischer und Entwicklungspolitischer Maßnahmen.* Bundesminister de Maizière Verabschiedet Kollegjahrgang Vernetzte Sicherheit im 21. Jahrhundert. Berlin: Konrad Adenauer Stiftung. Retrieved from http://www.kas.de/wf/de/33.20079

Larsson, Torbjörn (2002): Sweden. In OECD: *Distributed Public Governance. Agencies, Authorities and Other Government Bodies.* Paris: OECD Publications, pp. 181-208. Retrieved from http://www.keepeek.com/Digital-Asset-Management/oecd/governance/distributed-public-governance_9789264177420-en

Larsson, Torbjörn/Bäck, Henry (2008): *Governing and Governance in Sweden.* Stockholm: Studentlitteratur.

Larue, T. 2006. *Agents in Brussels: Delegation and Democracy in the European Union.* Umea: Umea University/Department of Political Science.

Lawler, Peter (1997): Scandinavian Exceptionalism and European Union. *Journal of Common Market Studies*, Vol. 35, No. 4, pp. 565-594.

Lawry-White, Simon (2004): *Review of the UK Government Approach to Peacebuilding and Synthesis of Lessons Learned from UK Government Funded Peacebuilding Projects 1997-2001.* Contribution to the Joint Utstein Study of Peacebuilding. London: Department for International Development.

Lee-Ohlsson, Frederick (2008): *Sweden and the Development of the European Security and Defence Policy. A Bi-Directional Process of Europeanization.* EU Diplomacy Papers 5/2008. Bruges: College of Europe.

Lee-Ohlsson, Frederick (2009): Sweden and the Development of the European Security and Defence Policy. A Bi-Directional Process of Europeanization. *Cooperation and Conflict*, Vol. 44, No. 2, pp. 123-142.

Lehmkuhl, Dirk (2007): Some Promises and Pitfalls of Europeanization Research. In Vink, Maarten P./Graziano, Paolo (Eds.): *Europeanization: New Research Agendas.* Basingstoke: Palgrave Macmillan, pp. 337-353.

Lenschow, Andrea/Liefferink, Duncan/Veenman, Sietske (2006): When the Birds Sing. A Framework for Analysing Domestic Factors Behind Policy Convergence. In Knill, Christoph (Ed.): *Cross-National Policy Convergence. Causes, Concepts and Empirical Findings.* London/New York: Routledge, pp. 34-53.

Levin, Paul T. (2009): The Swedish Model of Public Administration: Separation of Powers – The Swedish Style. *Journal of Administration and Governance*, Vol. 4, No. 1, pp. 38-46. Retrieved from http://joaag.com/uploads/4-_4_1___LevinFinal.pdf

Liebhold, Renate/Trinczek, Rainer (2009): Experteninterviews. In Kühl, Stefan/Strodtholz, Petra/Taffertshofer, Andreas (Eds.): *Handbuch Methoden der Organisationsforschung. Quantitative und Qualitative Methoden.* Wiesbaden: VS Verlag für Sozialwissenschaften/ GWV Fachverlage GmbH, pp. 32-56.

Ling, Tom (2002): Delivering Joined-Up Government in the UK. Dimensions, Issues and Problems. *Public Administration*, Vol. 80, No. 4, pp. 615-642.

Lipson, Michael (2005): *Interorganisational Coordination in Complex Peacekeeping.* Draft Prepared for Presentation at the Annual Meeting of the International Studies Association. Honolulu. Retrieved from http://alcor.concordia.ca/~mlipson/isa_pro ceeding_58207.pdf

Lodge, Martin (2006): The Europeanization of Governance – Top Down, Bottom Up or Both? In Schuppert, Gunnar F. (Ed.): *The Europeanization of Governance.* Baden-Baden: Nomos, pp. 59-76.

Lund, Michael S. (1996): *Preventing Violent Conflicts. A Strategy for Preventive Diplomacy.* Washington D.C.: United States Institute of Peace Press.

Lynn, Laurence E. (2003): Public Management. In Peters, B. Guy/Pierre, Jon (Eds.): *Handbook of Public Administration.* London: SAGE Publications, pp. 14-24.

Macrae, Joanna/Leader, Nicholas (2000): *The Politics of Coherence: Humanitarianism and Foreign Policy in the Post-Cold War Era.* HPG Briefing No 1. London: Overseas Development Institute. Retrieved from http://www.odi.org.uk/resources/docs/ 371.pdf

Major, Claudia (2012): *Ziviles Krisenmanagement in der Europäischen Union Stand und Optionen zur Weiterentwicklung der Gemeinsamen Sicherheits- und Verteidigungspolitik.* SWP-Studie. Berlin: Stiftung Wissenschaft und Politik. Retrieved from http://www.swp-berlin.org/fileadmin/contents/products/studien/2012_S22_mjr.pdf

Major, Claudia/Bail, Martina (2011): Waiting for Soft Power: Why the EU Struggles with Civilian Crisis Management. In Gross, Eva/Hamilton, Daniel/Major, Claudia/Rieke Henning (Eds.): *Preventing Conflict, Managing Crisis. European and American Perspectives.* Washington D.C.: Center for Transatlantic Relations, pp. 15-36.

Major/Claudia/Mölling, Christian (2009): More Than Wishful Thinking? The EU, UN, NATO and the Comprehensive Approach to Military Crisis Management. In Koops, Joachim A. (Ed.): Military Crisis Management: The Challenge of Inter-Organizatio nalism. *Studia Diplomatica – The Brussels Journal of International Relations*, Vol. LXII, No. 3, Brussels: Egmont – The Royal Institute for International Relations, pp. 21-28.

Major, Claudia/Pietz, Tobias/Schöndorf, Elisabeth/Hummel, Wanda (2011): Crisis Management Toolbox – From Civilian Crisis Prevention to Peacebuilding: Principles, Actors, Instruments. In Gross, Eva/Hamilton, Daniel/Major, Claudia/Rieke Henning (Eds.): *Preventing Conflict, Managing Crisis. European and American Perspectives.* Washington D.C.: Center for Transatlantic Relations, pp. 91-137.

Major, Claudia/Schöndorf, Elisabeth (2011): *Umfassende Ansätze, Vernetzte Sicherheit. Komplexe Krisen Erfordern Effektive Koordination und Politische Führung.* SWP-Aktuell April 2011. Berlin: Stiftung Wissenschaft und Politik. Retrieved from http://www.swp-berlin.org/fileadmin/contents/products/aktuell/2011A22_mjr_snf_ks.pdf

Manners, Ian (2002): Normative Power Europe: A Contradiction in Terms? *Journal of Common Market Studies*, Vol. 40, No. 2, pp. 235-258. Retrieved from http://www.princeton.edu/~amoravcs/library/mannersnormativepower.pdf

Manners, Ian (2006): The European Union as a Normative Power. A Response to Thomas Diez. *Millenium: Journal of International Studies*, Vol. 35, No. 1, pp. 167-180.

Mannin, Michael L. (2010): *British Government and Politics. Balancing Europeanization and Independence*. Lanham: Rowman & Littlefield Publishers.

March, James G./Olsen, Johan P. (1989): *Rediscovering Institutions. The Organizational Basis of Politics*. New York: The Free Press.

March, James G./Simon, Herbert A. (1993): *Organizations*. Oxford: Blackwell.

Martens, Maria (2010): *Organized Administrative Integration. The Role of Agencies in the European Administrative System*. Oslo: Centre for European Studies/University of Oslo.

Mattelaer, Alexander (2012): *Reviewing the EU's Crisis Management Procedures*. Policy Brief No. 4. Brussels: Institute for European Studies. Retrieved from http://www.ies.be/ files/2012-04_AM_final.pdf

May, Charles A. (1973): Introduction to Strategy. In Head, Richard G./Rokke, Ervin J.: *American Defence Policy*. London/Baltimore: The Johns Hopkins University Press, pp. 40-45.

Mazey, Sonia (2001): The Swedish Permanent Representation to the EU: Melding National and Collective Interests. In Kassim, Hussein/Menon, Anand/Peters, B. Guy/Wright, Vincent (Eds.): *The National Co-ordination of EU Policy. The European Level*. Oxford: Oxford University Press, pp. 256-276.

Mearsheimer, John J. (1995): The False Promise of International Institutions. *International Security*, Vol. 19, No. 3, pp. 5-49.

Miles, Lee (2000): Sweden and Finland. In Manners, Ian/Whitman, Richard G. (Eds.): *The Foreign Policies of European Union Member States*. Manchester: Manchester University Press, pp. 181-203.

Miles, Lee (2006): Domestic Influences on Nordic Security and Defence Policy: From the Perspective of Fusion. In Bailes, Alyson J.K./Herolf, Gunilla/Sundelius, Bengt (Eds.): *The Nordic Countries and the European Security and Defence Policy*. Oxford: Oxford University Press, pp. 77-98.

Miles, Lee (2010a): Nordic Europe. In Dyson, Kenneth/Sepos, Angelos (Eds.): *Which Europe? The Politics of Differentiated Integration*. Basingstoke: Palgrave Macmillan, pp. 184-197.

Miles, Lee (2010b): The Swedish Presidency. In Copsey, Nathaniel/Haughton, Tim (Eds.): *The JCMS Annual Review of the European Union in 2009*. West Sussex: Blackwell Publishing, pp. 81-93.

Miskimmon, Alister (2007): *Germany and the Common Foreign and Security Policy of the European Union. Between Europeanisation and National Adaptation*. Basingstoke: Palgrave Macmillan.

Missiroli, Antonio (2001): Introduction. In Id. (Ed.): *Coherence for European Security Policy. Debates – Cases – Assessments*. ISS Occasional Paper 27. Paris: European Union Institute for Security Studies.

Möckli, Daniel (2010): *Strategic Trends 2010. Key Developments in Global Affairs*. Zurich: Center for Security Studies.

Möller, Almut/Rappold, Julian (2012): From Co-Founder to Wait-And-See: Germany's Policy on the European External Action Service. In European Policy Institutes Network: *Reviewing Member States' Commitment to the European External Action Service*. Working Paper No. 34, pp. 9-13. Retrieved from http://www.ceps.be/book/reviewing-member-states%E2%80%99-commitment-european-external-action-service

Möller, Ulrika/Bjereld, Ulf (2010): From Nordic Neutrals to Post-Neutral Europeans: Differences in Finnish and Swedish Policy Transformation. *Cooperation and Conflict*, Vol. 45, No. 4, pp. 363-386.

Mölling, Christian (2008): Comprehensive Approaches to International Crisis Management. *CSS Analysis in Security Policy*, Vol. 3, No. 42. Zurich: CSS ETH Zurich.

Mulford, Charles L./Rogers, David L. (1982): Definitions and Models. In Rogers, David L./Whetten, David A.: *Interorganizational Coordination: Theory, Research, and Implementation*. Ames: Iowa State University Press, pp. 9-31.

Mulgan, Geoff (2008): Joined-Up Government: Past, Present, and Future. In Bogdanor, Vernon: *Joined-Up Government*. Oxford: The British Academy/Oxford University Press, pp. 175-187.

Münkler, Herfried (2005): *The New Wars*. Cambridge: Polity Press.

Mustonen, Jari (2008): Coordination and Cooperation on Tactical and Operational Levels: Studying EU-ESDP Crisis Management Instruments in Bosnia and Herzegovina. In Henriksson, Kirsi (Ed.): *Yearbook 2008 on Civilian Crisis Management Studies*. Kuopio: Crisis Management Centre Finland, pp. 18-41. Retrieved from http://www.cmcfinland.fi/ pelastus/cmc/images.nsf/files/023BE4430570384EC225750B0038C6F8/$file/CMC_Yearbook_2008.pdf

Myrdal, Sara/Rhinard, Mark (2010): *The European Union's Solidarity Clause: Empty Letter or Effective Tool? An Analysis of Article 222 of the Treaty on the Functioning of the European Union*. UI Occassional Papers No. 2. Stockholm: Swedish Institute of International Affairs. Retrieved from http://www.societalsecurity.eu/uploads/ UIOP%202% 202010%20Myrdal%20Rhinard%20EU%20Solidarity%20Clause.pdf

Nachtwei, Winfried (2011): *Akteur oder Zuschauer? Was Vernetzte Sicherheit für den Deutschen Bundestag Bedeutet*. Erweiterte Fassung des Vortrags beim Expertengespräch „Fünf Jahre Vernetzte Sicherheit – Eine Konstruktive Zwischenbilanz" am 5.12.2011 in der Bundesakademie für Sicherheit (BAKS). Berlin.

Neal, Derrick J. (2011): The Comprehensive Approach: Lessons for its Adoption from a Change Management Perspective. In Neal, Derrick J./Wells, Linton (Eds.): *Capability Development in Support of Comprehensive Approaches. Transforming International Civil-Military Interactions*. Washington D.C.: National Defense University, pp. 5-21.

Neal, Derrick J./Wells, Linton (2011) (Eds.): *Capability Development in Support of Comprehensive Approaches. Transforming International Civil-Military Interactions*. Washington D.C.: National Defense University.

Ness, Gayl D./Brechin, Steven R. (1988): Bridging the Gap: International Organizations as Organizations. *International Organization*, Vol. 42, No. 2, pp. 245-273.

Nilsson, Claes/Hull, Cecilia/Derblom, Markus/Egnell, Robert (2008): *Contextualizing the Comprehensive Approach. The Elements of a Comprehensive Intervention*. Stockholm: Swedish Defence Research Agency.

Norheim-Martinsen, Per M. (2009): *EU Capabilities for a Comprehensive Approach: Broad Interoperability as Comparative Advantage*. FFI-rapport 2009/01300. Oslo: Norwegian Defence Research Establishment (FFI). Retrieved from http://www.ffi.no/no/Rapporter/09-01300.pdf

Norheim-Martinsen, Per M. (2010): Managing the Civil-Military Interface in the EU: Creating an Organisation Fit for Purpose. In Vanhoonacker, Sophie/Dijkstra, Hylke/Maurer, Heidi (Eds.): Understanding the Role of Bureaucracy in the European Security and Defence Policy. *European Integration online Papers (EIoP)*, Vol. 14, Special Issue 1. Retrieved from http://eiop.or.at/eiop/texte/2010-010a.htm

Noutcheva, Gergana/Tocci, Nathalie/Coppieters, Bruno/Kovziridze, Tamara/Emerson, Michael/Huysseune, Michael (2004): Europeanization and Secessionist Conflicts: Concepts and Theories. In Coppieters, Bruno/Emerson, Michael/Huysseune, Michael/Kovziridze, Tamara/Noutcheva, Gergana/Tocci, Nathalie/Vahl, Marius (Eds.) (2004): *Europeanization and Conflict Resolution. Case Studies from the European Periphery*. Gent: Academia Press, pp. 13-55.

Nowak, Agnieszka (Ed.) (2006): *Civilian Crisis Management. The EU Way*. Chaillot Paper No. 90. Paris: European Union Institute for Security Studies, pp. 123-138. Retrieved from http://www.iss.europa.eu/uploads/media/cp090.pdf

Nuttall, Simon (2005): Coherence and Consistency. In Hill, Christopher/Smith, Michael (Ed.): *International Relations and the European Union*. Oxford/New York: Oxford University Press, pp. 91-112.

Nye, Joseph S. (2004): *Soft Power: The Means to Success in World Politics*. New York: Public Affairs.

Odén, Bertil (2009): *The Swedish Policy for Global Development. Implementation and Changes*. Perspectives No. 12. Göteborg: Center for African Studies – Göteborg University.

Ojanen, Hanna (2000): *Participation and Influence: Finland, Sweden and the Post-Amsterdam Development of the CFSP*. Occasional Papers. Paris: Institute for Security Studies – Western European Union.

Ojanen, Hanna (2002): Sweden and Finland: What Difference Does It Make to Be Non-Aligned? In Græger, Nina/Larsen, Henrik/Ojanen, Hanna (Eds.): *The ESDP and the Nordic Countries. Four Variations on a Theme*. Helsinki/Berlin: The Finnish Institute of International Affairs/Institut für Europäische Politik, pp. 154-217.

Oliver, Tim/Allen, David (2006): Foreign Policy. In Bache, Ian/Jordan, Andrew (Eds.): *The Europeanization of British Politics*. Basingstoke: Palgrave Macmillan, pp. 187-200.

Olsen, Johan P. (1991): Political Science and Organization Theory. Parallel Agendas but Mutual disregard. In Windhoff-Héritier, Adrienne/Czada, Roland M. (Eds.): *Political Choice. Institutions, Rules and the Limits of Rationality*. Frankfurt a.M.: Campus/Westview, pp. 87-120.

Olsen, Johan P. (2002): *The Many Faces of Europeanization*. Arena Working Paper 02/2002. Oslo: University of Oslo. Retrieved from http://www.sv.uio.no/arena/english/research/publications/arena-publications/workingpapers/working-papers2002/02_02.xml

Olsen, Johan P. (2003): Towards a European Administrative Space? *Journal of European Public Policy*, Vol. 10, No. 4, pp. 506-531.

O'Toole, Laurence J. (2003): Interorganizational Relations in Implementation. In Peters, B. Guy/Pierre, Jon (Eds.): *Handbook of Public Administration*. London: SAGE Publications, pp. 234-244.

Overhaus, Marco (2006): Civilian Power under Stress: Germany, NATO, and the European Security and Defence Policy. In Maull, Hanns W. (Ed.): *Germany's Uncertain Power. Foreign Policy of the Berlin Republic*. Basingstoke: Palgrave Macmillan, pp. 66-78.

Page, Edward C. (2008): Joined-Up Government and the Civil Service. In Bogdanor, Vernon (Ed.): *Joined-Up Government*. Oxford: The British Academy/Oxford University Press, pp. 139-155.

Paris, Roland (2009): Understanding the "Coordination Problem" in Postwar Statebuilding. In Paris, Roland/Sisk, Timothy (Eds.): *The Dilemmas of Statebuilding. Confronting the Contradictions of Postwar Peace Operations*. London/New York: Routledge, pp. 53-78.

Paris, Roland/Sisk, Timothy D. (Eds.) (2007): *Managing Contradictions: The Inherent Dilemmas of Postwar Statebuilding*. New York: International Peace Academy.

Patrick, Stewart M. (2009): The US Response to Precarious States: Tentative Progress and Remaining Obstacles to Coherence. In Weiss, Stefani/Spanger, Hans-Joachim/Van Meurs, Wim (Eds.): *Diplomacy, Development and Defence: A Paradigm for Policy Coherence. A Comparative Analysis of International Strategies*. Gütersloh: Verlag Bertelsmann Stiftung, pp. 54-104.

Patrick, Stewart M./Brown, Kaysie (2007a): *Greater Than the Sum of Its Parts? Assessing "Whole of Government" Approaches to Fragile States*. New York: International Peace Academy.

Patrick, Stewart M./Brown, Kaysie (2007b): *Greater Than the Sum of Its Parts? Assessing "Whole of Government" Approaches to Fragile States*. CGD Brief. Washington D.C.: Center for Global Development. Retrieved from http://www.cgdev.org/doc/ weakstates/Fragile_States.pdf

Penttinen, Elina (2010): Enhancement of Expertise in Civilian Crisis Management: Positivity as a Key Towards Personal Dedication for a Comprehensive Approach. In Henriksson, Kirsi (Ed.): *Crisis Management Centre Finland Yearbook 2010 on Peacebuilding and Civilian Crisis Management Studies*. CMC Finland Crisis Management Studies, Vol. 3, No. 6. Helsinki: Crisis Management Centre, pp. 46-59. Retrieved from http://www.cmcfinland. fi/pelastus/cmc/images.nsf/files/2A0BB51A8 D901CA0C22577D6003A4912/$file/CMC_Yearbook_2010_penttinen.pdf

Perlo-Freeman, Sam/Ismail, Olawale/Kelly, Noel/Sköns, Elisabeth/Solmirano, Carina (2011): Appendix 4A. Military Expenditure Data, 2001-10. In Stockholm International Peace Research Institute: *SIPRI Yearbook 2011. Armaments, Disarmament and International Security*. Oxford: Oxford University Press, pp. 181-222.

Perrow, Charles (1986): *Complex Organizations. A Critical Essay*. New York: McGraw-Hill.

Peter, Mateja (2012): *Double-Hatting in EU External Engagements. EU Special Representatives and the Question of Coherence Post-Lisbon*. SWP Comments 46. Retrieved from http://www.swp-berlin.org/fileadmin/contents/products/comments/2012 C46_ptr.pdf

Peters, B. Guy (1998a): *Globalization, Institutions and Governance*. Jean Monnet Chair Papers 51. San Domenico: The Robert Schumann Centre at the European University Institute.

Peters, B. Guy (1998b): *Managing Horizontal Government. The Politics of Coordination.* Research Paper 21. Ottawa: Canadian Centre for Management Development.

Peters, B. Guy (2005): *Institutional Theory in Political Science. The 'New Institutionalism'.* London: Continuum.

Peters, B. Guy/Pierre, Jon (2003): Introduction: The Role of Public Administration in Governing. In Ead. (Eds.): *Handbook of Public Administration.* London: SAGE Publications, pp. 1-9.

Plattform Zivile Konfliktbearbeitung/Forum Menschenrechte (2010): *Stillschweigender Abschied vom Aktionsplan Zivile Krisenprävention? Stellungnahme zum 3. Umsetzungs-bericht „Zivile Krisenprävention, Konfliktbewältigung und Friedenskonsolidierung" der Bundesregierung vom 23. Juni 2010 zum Aktionsplan „Zivile Krisenprävention, Konflikt-lösung und Friedenskonsolidierung".* Köln/Berlin: Plattform Zivile Konflikt-bearbeitung/Forum Menschenrechte.

Plattform Zivile Konfliktbearbeitung (2012): *Zivile Krisenprävention – Parlamentarische Aufgabe und Verantwortung. Stellungnahme der Plattform Zivile Konfliktbearbeitung e.V. zur Zwischenbilanz des Bundestags-Unterausschusses „Zivile Krisenpräsention und Vernetzte Sicherheit".* Köln/Berlin: Sprecherrat der Plattform Zivile Konfliktbearbeitung.

Porter, Patrick (2010): Why Britain Doesn't Do Grand Strategy. *The RUSI Journal*, Vol. 155, No. 4, pp. 6-12.

Powell, Walter W. (1990): Neither Market nor Hierarchy: Network Forms of Organization. In *Research in Organizational Behaviour*, Vol. 12, pp. 295-336. Retrieved from http://www. stanford.edu/~ woodyp/papers/powell_neither.pdf

Powell, Walter W./DiMaggio, Paul J. (1991): Introduction. In Ead. (Ed.): *The New Institutionalism in Organizational Analysis.* Chicago/London: The University of Chicago Press.

Pradetto, August (2006): The Polity of German Foreign Policy: Changes Since Unification. In Maull, Hanns W. (Ed.): *Germany's Uncertain Power. Foreign Policy of the Berlin Republic.* Basingstoke: Palgrave Macmillan, pp. 15-28.

Pradetto, August (2011): Zivil-Militärische Zusammenarbeit und Comprehensive Approach im Kontext Post-Bipolarer Weltordnungspolitik. *Studien zur Internationalen Politik*, Heft 2/2011. Hamburg: Helmut-Schmidt-Universität.

Pressman, Jeffrey L./Wildavsky, Aaron B. (1973): *Implementation – How Great Expectations in Washington are Dashed in Oakland; or, Why It's Amazing that Federal Programmes Work at All, This Being a Saga of the Economic Development Administration as Told by Two Sympathetic Observers Who Seek to Build Morals on a Foundation of Ruined Hopes.* Berkeley/Los Angeles: University of California Press.

Quille, Gerrard (2010): The European External Action Service and the Common Security and Defence Policy (CSDP). In Greco, Ettore/ Pirozzi, Nicoletta/ Silvestri, Stefano (Eds.): *EU Crisis Management: Institutions and Capabilities in the Making.* Rome: Istituto Affari Internazionali, pp. 55-78.

Radaelli, Claudio M. (2003): The Europeanization of Public Policy. In Featherstone, Kevin/Radaelli, Claudio M. (Eds.): *The Politics of Europeanization.* Oxford: Oxford University Press, pp. 27-56.

Radin, Beryl A. (2003): The Instruments of Intergovernmental Management. In Peters, Guy/Pierre, Jon (Eds.): *Handbook of Public Administration*. London: SAGE Publications, pp. 606-618.

Rainey, Hal G. (2003): Public Management: Old and New. Introduction. In Peters, Guy/Pierre, Jon (Eds.): *Handbook of Public Administration*. London: SAGE Publications, pp. 11-13.

Ramsbotham, Oliver (2010): *Transforming Violent Conflict. Radical Disagreement, Dialogue and Survival*. London/New York: Routledge.

Ramsbotham, Oliver/Woodhouse, Tom/Miall, Hugh (2011): *Contemporary Conflict Resolution. The Prevention, Management and Transformation of Deadly Conflicts*. 3rd Ed., Cambridge/Malden: Polity Press.

Ranade, Wendy/Hudson, Bob (2003): Conceptual Issues in Inter-Agency Collaboration. *Local Government Studies*, Vol. 29, No. 3, pp. 32-50.

Rees, Wyn (2011): Britain and the Wider World. *Defence and Security Analysis*, Vol. 27, No. 1, pp. 31-41.

Reichard, Christoph (2008): The Study of Public Management in Germany. Poorly Institutionalized and Fragmented. In Kickert, Walter (Ed.): *The Study of Public Management in Europe and the US. A Comparative Analysis of National Distinctiveness*. London/New York: Routledge, pp. 42-69.

Rettman, Andrew (2011): *Staff Leaving EU Diplomatic Service Amid Bad Working Conditions*. Brussels: EUobserver. Retrieved from http://euobserver.com/18/113777

Rhodes, R.A.W. (1995): From Prime Ministerial Power to Core Executive. In Rhodes, Rod/Dunleavy, Patrick (Eds.): *Prime Minister, Cabinet and Core Executive*. London: MacMillan Press, pp. 11-37.

Rhodes, R.A.W. (1996): The New Governance. Governing without Government. *Political Studies*, No. XLIV, pp. 652-667.

Rhodes, R.A.W. (2007): Understanding Governance: Ten Years On. *Organization Studies*, Vol. 28, No. 8, pp. 1243-1264.

Rieker, Pernille (2002): *From Territorial Defence to Comprehensive Security? European Integration and the Changing Norwegian and Swedish Security Identities*. [626] Paper March 2002. Oslo: Norwegian Institute for International Affairs. Retrieved from http://www.isn.ethz.ch/Digital-Library/Publications/Detail/?ots591=0c54e3b3-1e9c-be1e-2c24-a6a8c7060233&lng=en&id=27374

Rieker, Pernille (2004): Europeanization of Nordic Security. The European Union and the Changing Security Identities of the Nordic States. *Cooperation and Conflict*, Vol. 39, No. 4, pp. 369-392.

Rieker, Pernille (2006a): From Common Defence to Comprehensive Security: Towards the Europeanization of French Foreign and Security Policy? *Security Dialogue*, Vol. 37, No. 4, December 2006, pp. 509-528.

Rieker, Pernille (2006b): *Europeanization of National Security Identity. The EU and the Changing Security Identities of the Nordic States*. London/New York: Routledge.

Rieker, Pernille (2006c): The Nordic Countries and EU Security Policy: Convergent or Divergent Agendas? In Bailes, Alyson J.K./Herolf, Gunilla/Sundelius, Bengt (Eds.): *The Nordic Countries and the European Security and Defence Policy*. Oxford: Oxford University Press, pp. 301-314.

Rieker, Pernille (2007): *The EU as a Security Actor: The Development of Political and Administrative Capabilities*. NUPI Working Paper 725. Oslo: Norwegian Institute of International Affairs.

Rintakoski, Kristiina/Autti, Mikko (Eds.) (2008): *Comprehensive Approach. Trends, Challenges and Possibilities for Cooperation in Crisis Prevention and Management*. Seminar Publication from the Comprehensive Approach Seminar (17 June 2008). Helsinki: Crisis Management Initiative. Retrieved from http://reliefweb.int/sites/reliefweb.int/files/resources/3185D85A55E647638525750B0076D291-comprehensive-approach.pdf

Risse, Thomas/Cowles, Maria G./Caporaso, James (2001): Europeanization and Domestic Change: Introduction. In Cowles, Maria G./Caporaso, James/Risse, Thomas (Eds.) (2001): *Transforming Europe. Europeanization and Domestic Change*. Ithaca: Cornell University Press. pp. 1-20.

Rockberger, Nicolaus (2005): *Scandinavia and ESDP. Are the Nordic States Holding Back?* Strategische Analysen. Vienna: Büro für Sicherheitspolitik des österreichischen Bundesministeriums für Landesverteidigung.

Rotte, Ralph/Schwarz, Christoph (2011): Strategiefindung unter Reformdruck. Die Neue National Security Strategy und Strategic Defence and Security Review des Vereinigten Königreichs. *Europäische Sicherheit*, Vol 60., No. 2, pp. 57-60.

Rummel, Reinhardt (2006): *Deutscher Einfluss auf den Ausbau ziviler Krisenintervention in der EU*. Berlin: Stiftung Wissenschaft und Politik. Retrieved from http://www.swp-berlin.org/fileadmin/contents/products/arbeitspapiere/Rummel_AFK_ks.pdf

Rynning, Sten (2011): Of Sirens and Deceptive Virtue. A Critical Look at NATO's Comprehensive Approach. In Biscop, Sven (Ed.): *Studia Diplomatica – The Brussels Journal of International Relations*, Vol. LXIV, No. 2, Brussels: Egmont – The Royal Institute for International Relations, pp. 37-55.

Scharpf, Fritz W. (1976): *Does Organization Matter? Task Structure and Interaction in the Ministerial Bureaucracy*. Berlin: International Institute of Management.

Scharpf, Fritz W. (1993): Coordination in Hierarchies and Networks. In Id. (Ed.): *Games in Hierarchies and Networks. Analytical and Empirical Approaches to the Study of Governance Institutions*. Frankfurt am Main: Campus, pp. 125-165.

Schnaubelt, Christopher M. (2009): Introduction. Operationalizing a Comprehensive Approach in Semi-Permissive Environments. In Id. (Ed.): *Operationalizing a Comprehensive Approach in Semi-Permissive Environments*. Rome: NATO Defence College. Retrieved from http://www.isn.ethz.ch/Digital-Library/Publications/Detail/?ots591=0C54E3B3-1E9C-BE1E-2C24-A6A8C7060233&lng=en&id=102399

Schröder, Ursula C. (2006): *Coping with Complexity. An Organizational Perspective on European Security Governance*. EUI Working Papers. SPS No. 2006/9. Frankfurt: Hessische Stiftung Friedens- und Konfliktforschung. Retrieved from http://www.efsps.eu/ docs/300SCHROEDER_SPS_EUI_WorkingPaper2006.pdf

Scott, Richard W. (2001): *Institutions and Organizations*. 2nd Ed., London: SAGE Publications.

Self, Robert (2010): *British Foreign and Defence Policy Since 1945. Challenges and Dilemmas in a Changing World*. Basingstoke: Palgrave Macmillan.

Seybolt, Taylor B. (2000): *Systemic Network Analysis of Refugee Relief: A Business-like Approach*. Stockholm: Stockholm International Peace Research Institute. Retrieved from http://www.reliefweb.int/rw/lib.nsf/db900sid/VONN5E6CVB/$file/Refugee %20Relief.pdf?openelement

Shafritz, Jay M./Ott, J. Steven (Eds.) (1992): *Classics of Organization Theory*. Pacific Grove: Brooks/Cole Publishing Company.

Siedschlag, Alexander/Eder, Franz (2006): Akteure und Zusammenspiel im EU-Krisenmanagement. In Feichtinger, Walter/Gebhard, Carmen (Eds.): *EU als Krisenmanager. Herausforderungen – Akteure – Instrumente*. Vienna: Landesverteidigungsakademie (LVAk)/Institut für Friedenssicherung und Konfliktmanagement (IFK), pp. 61-89. Retrieved from http://www.bmlv.gv.at/pdf_pool/publikationen/eu_km _akteure_im_eu_km_siedschlag_f_eder.pdf

SIPRI Multilateral Peace Operation Database. Stockholm: Stockholm International Peace Research Institute. Retrieved from http://www.sipri.org/databases/pko

Smith, Andy (2003): Multi-level Governance: What It Is and How It Can Be Studied. In Peters, Guy/Pierre, Jon (Eds.): *Handbook of Public Administration*. London: SAGE Publications, pp. 618-628.

Smith, Dan (2004): *Towards a Strategic Framework for Peacebuilding: Getting Their Act Together*. Overview report of the Joint Utstein Study of Peacebuilding. Oslo: International Peace Research Institute. Retrieved from http://www.regjeringen.no/ upload/kilde/ud/ rap/2004/0044/ddd/pdfv/210673-rapp104.pdf

Smith, Steve (2005): The Contested Concept of Security. In Booth, Ken (Ed.): *Critical Security Studies and World Politics*. Boulder/London: Lynne Rienner Publishers, pp. 27-62.

Smith, Martin A. (2010): United Kingdom. How Much Continuity? How Much Change? In Kirchner, Emil J./Sperling, James (Eds.): *National Security Cultures. Patterns of Global Governance*. London/New York: Routledge, pp. 85-102.

Soder, Kirsten (2010): *EU Military Crisis Management: An Assessment of Member States' Contributions and Positions*. Draft for the COST Action Meeting on 10 June 2010. Retrieved from http://www.peacekeeping-cost-is0805.eu/siteweb/images/ ACTIVITIES/Publications/100331%20-%20+Article+Soder.pdf

Sperling, James (2010): National Security Cultures, Technologies of Public Goods Supply and Security Governance. In Kirchner, Emil J./Sperling, James (Eds.): *National Security Cultures. Patterns of Global Governance*. London/New York: Routledge, pp. 1-18.

Stepputat, Finn (2009a): *Integrated National Approaches to International Operations. The Cases of Denmark, UK, and the Netherlands*. DIIS Report 2009:14. Copenhagen: Danish Institute for International Studies. Retrieved from http://www.diis. dk/graphics/Publications/Reports2009/DIIS_Report_2009_14_Integrated_National_ Approaches_International_Operations_web.pdf

Stepputat, Finn (2009b): *Synthesis Report. Civil-Military Relations in International Organisations. A Danish Perspective*. DIIS Report 2009:16. Copenhagen: Danish Institute for International Studies. Retrieved from http://www.diis.dk/graphics/publica tions/reports 2009/diis_report_2009_16_synthesis_report_web.pdf

Stewart, Emma J. (2006): *The European Union and Conflict Prevention. Policy Evolution and Outcome*. Berlin: LIT Verlag.

Strömvik. Maria (2006): Starting to 'Think Big': the Nordic Countries and EU Peace-Building. In Bailes, Alyson J.K./Herolf, Gunilla/Sundelius, Bengt (Eds.): *The Nordic Countries and the European Security and Defence Policy*. Oxford: Oxford University Press, pp. 199-214.

Suhrke, Astri/Samset, Ingrid (2007): What's in a Figure? Estimating Recurrence of Civil War. *International Peacekeeping*, Vol. 14, No. 2, pp. 195-203.

Sundberg, Anna/Nilsson, Claes (2009): *Swedish Presidency Ambitions and ESDP*. ARI 156/2009. Madrid: Real Instituto Elcano (ARI). Retrieved from http://www.realinstitutoelcano.org/wps/wcm/connect/1059ae004054db949af39f5e01ac4adf/ARI156-2009_Sundberg-Nilsson_Swedish_Presidency_ESDP.pdf?MOD=AJPERES&CACHEID= 1059ae004054db949af39f5e01ac4adf

Tannous, Isabelle (2008): EU Institutions' and Member States' Approaches to Promoting Policy Coherence of Development and Security. In Brockmann, Kathrin/Hauck, Hans B./Reigeluth, Stuart (Eds.): *From Conflict to Regional Stability. Linking Security and Development*. Berlin: German Council on Foreign Relations.

Tardy, Thierry (2011): Peace-Operations: The Fragile Consensus. In Stockholm International Peace Research Institute: *SIPRI Yearbook 2011. Armaments, Disarmament and International Security*. Oxford: Oxford University Press, pp. 87-109.

Teuten, Richard (2010): *Civilian-Military Collaboration. The Stabilisation Unit Coming Of Age?* London: Lecture Delivered at the Royal United Services Institute (RUSI). Retrieved from http://www.rusi.org/analysis/commentary/pauls@rusi.org/www.rusi.org/research/hsr/ref:C4B715FF659286/

Teuten, Richard/Korski, Daniel (2010): *Preparing for Peace. Britain's Contribution and Capabilities*. Whitehall Paper 74. Royal United Services Institute for Defence and Security Studies. Abingdon: Taylor & Francis.

The Economist (2012): Germany's Army. No Shooting Please, We're German. Germans Still Have a Uniquely Complicated Relationship With Their Soldiers. *The Economist*, October 13, 2012. Retrieved from http://www.economist.com/node/21564617?fsrc=rss|eur

Thiele, Ralph (2012): Zur Bedeutung der Vernetzten Sicherheit – Eine Militärische Perspektive. *S + F. Sicherheit und Frieden. Security and Peace*, Vol. 30, No. 2, pp. 81-87.

Thoenig, Jean-Claude (2003): Institutional Theories and Public Institutions: Traditions and Appropriateness. In Peters, Guy/Pierre, Jon (Eds.): *Handbook of Public Administration*. London: SAGE Publications, pp. 127-137.

Thompson, Grahame F. (2003): *Between Hierarchies and Markets. The Logics and Limits of Network Forms of Organization*. New York: Oxford University Press.

Tolksdorf, Dominik (2012): *The Role of EU Special Representatives in the Post-Lisbon Foreign Policy System: A Renaissance?* IES Policy-Brief. Brussels: Institute for European Studies.

Tonra, Ben (1999): The Europeanization of Irish Foreign Policy. *Irish Studies in International Affairs*, Vol. 10, pp. 149-165.

Tonra, Ben (2001b): *The Europeanization of National Foreign Policy. Dutch, Danish and Irish Foreign Policy in the European Union.* Aldershot: Ashgate.

Toonen, Theo A.J. (2003): Administrative Reform – Introduction. In Peters, Guy/Pierre, Jon (Eds.): *Handbook of Public Administration.* London: SAGE Publications, pp. 465/466.

Torreblanca, José I. (2001): *Ideas, Preferences and Institutions: Explaining the Europeanization of Spanish Foreign Policy.* Arena Working Paper 26/2001, University of Oslo. Retrieved from http://www.sv.uio.no/arena/english/research/publications/arenapublications/workingpapers/working-papers2001/wp01_26.htm

Trinity College Dublin (2011): *Policy Coherence – Exploring Links between EU Agricultural Policy and World Poverty.* Dublin: Institute for International Integration Studies. Retrieved from http://www.tcd.ie/iiis/policycoherence/about/

Trondal, Jarle (2009): Administrative Fusion: Less than a European 'Mega-Administration'. *European Integration*, Vol. 31, No. 2, pp. 237-260.

Trondal, Jarle/Marcussen, Martin/Torbjörn, Larsson/Veggeland, Frode (2010): *Unpacking International Organisations. The Dynamics of Compound Bureaucracies.* Manchester: Manchester University Press.

Ulrich, Bernd (2011): Machtwechsel. Schneller, Internationaler, Gefährlicher – Die Dauerkrise Sortiert die Gewichte in Deutschland Neu. Ein Überblick über Gewinner und Verlierer. *DIE ZEIT*, No. 38, 15. September 2011, p. 4.

Ulrich, Bernd (2012): Reden, Reden, Reden. *DIE ZEIT*, No. 30, 19. July 2012, p. 1.

Unger, Barbara (2009): Germany's Response to Precarious States. In Weiss, Stefani/Spanger, Hans-Joachim/Van Meurs, Wim (Eds.): *Diplomacy, Development and Defence: A Paradigm for Policy Coherence. A Comparative Analysis of International Strategies.* Gütersloh: Verlag Bertelsmann Stiftung, pp. 142-184.

Van de Goor, Luc/Major, Claudia (2012): *How to Make the Comprehensive Approach Work. Preparation at Home is Key to Effective Crisis Management in the Field.* CRU Policy Brief. Clingendael Conflict Research Unit. The Hague: Netherlands Institute of International Relations Clingendael. Retrieved from http://www.swp-berlin.org/fileadmin /contents/products/fachpublikationen/CRUPolicyBrief_ks.pdf

Vanhoonacker, Sophie/Dijkstra, Hylke/Maurer, Heidi (2010): Understanding the Role of Bureaucracy in the European Security and Defence Policy: The State of the Art. In Ead. (Eds.). Understanding the Role of Bureaucracy in the European Security and Defence Policy. *European Integration online Papers (EIoP)*, Vol. 14, Special Issue 1. Retrieved from http://eiop.or.at/eiop/texte/2010-004a.htm

Van Vooren, Bart (2011): The European External Action Service: Avoiding Past Disputes in the Security-Development Nexus? In Larik, Joris/Moraru, Madalina (Eds.): *Ever-closer in Brussels – Ever Closer in the World? EU External Action after the Lisbon Treaty.* EUI Working Papers. Florence: European University Institute, pp. 19-28.

Väyrynen, Tarja (2006): 'The Higher Cause of Peace': What Could and Should the Nordic Countries Contribute to the Development of Conflict Mediation in the EU Context? In Bailes, Alyson J.K./Herolf, Gunilla/Sundelius, Bengt (Eds.): *The Nordic Countries and the European Security and Defence Policy.* Oxford: Oxford University Press, pp. 217-233.

Verhoest, Koen/Bouckaert, Geert/Peters, B. Guy (2007): Janus-Faced Reorganization: Specialization and Coordination in Four OECD Countries in the Period 1980-2005. *International Review of Administrative Sciences*, Vol. 73, No. 3, pp. 325-348.

Vink, Maarten P./Graziano, Paolo (2007): Challenges of a New Research Agenda. In Ead. (Eds.): *Europeanization: New Research Agendas*. Basingstoke: Palgrave Macmillan, pp. 3-20.

Voget, Bernhard G. (2008): Civil-Military Cooperation of the German Armed Forces. Theoretical Approach and Contemporary Practice in Kosovo. In Ankersen, Christopher (Ed.): *Civil-Military Cooperation in Post-Conflict Operations. Emerging Theory and Practice*. London/New York: Routledge, pp. 143-172.

Vogt, Henri (2009): Zwischen Pragmatismus und Moralismus: Überlegungen zum Finnischen und Skandinavischen Internationalismus. *Internationale Politik und Gesellschaft*, No. 3, pp. 139-157. Retrieved from http://library.fes.de/pdf-files/ipg/ipg-2009-3/10_a_vogt_d.pdf

Wæver, Ole (1995): Securitization and Desecuritization. In Lipschutz, Ronnie D. (Ed.): *On Security*. New York: Columbia University Press, pp. 46-86.

Wall, Rob/West, Paul (2002): United Kingdom. In OECD: *Distributed Public Governance. Agencies, Authorities and Other Government Bodies*. Paris: OECD Publications, pp. 209-239. Retrieved from http://www.keepeek.com/Digital-Asset-Management/oecd/governance/distributed-public-governance_9789264177420-en

Waltz, Kenneth N. (1959): *Man, the State and War. A Theoretical Analysis*. New York: Columbia University Press.

Waltz, Kenneth N. (1979): *Theory of International Politics*. New York: McGraw-Hill.

Weber, Max (1946): Bureaucracy. In Shafritz, Jay M./Ott, J. Steven (Eds.) (1992): *Classics of Organization Theory*. Pacific Grove: Brooks/Cole Publishing Company, pp. 81-86.

Weber, Max (1966): *The Theory of Social and Economic Organization*. New York: The Free Press.

Wedin, Lars (2005): The Concept of Crisis Management. In Koivula, Tommi/Tammiletho, Tuomas: *Crisis Management: A New Form of Nation-Building?* Helsinki: National Defence College, pp. 11-25.

Wedin, Lars (2006): The Impact of EU Capability Targets and Operational Demands on Defence Concepts and Planning. The Case of Sweden. In Bailes, Alyson J.K./Herolf, Gunilla/Sundelius, Bengt (Eds.): *The Nordic Countries and the European Security and Defence Policy*. Oxford: Oxford University Press, pp. 141-149.

Wehmeier, Sally (Ed.) (2000): *Oxford Advanced Learner's Dictionary of Current English*. 6th Ed., Oxford: Oxford University Press.

Weiland, Severin (2012): Westerwelle bekommt 95 Millionen von Niebel. *SPIEGEL ONLINE*, 24. Mai 2012. Retrieved from http://www.spiegel.de/politik/deutschland/entwicklungshilfe-westerwelles-ministerium-erhaelt-95-millionen-von-niebel-a-834892.html

Weiss, Stefani/Spanger, Hans-Joachim/Van Meurs, Wim (2009): Precarious States Strategies. Toward a Culture of Coherence. In Ead. (Eds.): *Diplomacy, Development and Defence: A Paradigm for Policy Coherence. A Comparative Analysis of International Strategies*. Gütersloh: Verlag Bertelsmann Stiftung, pp. 9-53.

Weller, Christoph (2007): Zivile Krisenprävention und Konfliktbearbeitung: Politische Herausforderungen und der Aktionsplan der Bundesregierung. To be published in Grasse, Renate/Nolte, Wilhelm/Schlotter, Peter (Eds.): *BerlinerFriedenspolitik? Militärische Transformation –Zivile Impulse – Europäische Einbindung*. Baden-Baden: Nomos. Retrieved from http://www.inef.uni-duisburg.de/page/documents/Weller_-_Zivile_ Krisenpraevention.pdf

Wendling, Cécile (2010): *The Comprehensive Approach to Civil-Military Crisis Management. A Critical Analysis and Perspective*. IRSEM Report. Paris: Institut de Recherche Stratégique de l'École Militaire. Retrieved from http://www.humansecurity gateway.com/documents/IRSEM_TheComprehensiveApproachtoCivilMilitaryCri sisManagement.pdf

Wenger, Andreas/Mauer, Viktor/Bruno, Stefano/Callsen, Cristiane/Trachsler, Daniel (2006): *Zivile Friedensförderung als Tätigkeitsfeld der Aussenpolitik. Eine vergleichende Studie anhand von fünf Ländern*. Zürich: Center for Security Studies. Retrieved from http://www.css.ethz.ch/publications/pdfs/Zivile-Friedensfoerderung.pdf

Wildavsky, Aaron (1987): Choosing Preferences by Constructing Institutions: A Cultural Theory of Preference Formation. *The American Political Science Review*, Vol. 81, No. 1, pp. 3-22.

Wilke, Tobias M. (2007): *German Strategic Culture Revisited. Linking the Past to Contemporary German Strategic Choice*. Münster: LIT Verlag.

Windhoff-Héritier, Adrienne/Czada, Roland M. (1991): Introduction. In Ead. (Eds.): *Political Choice. Institutions, Rules and the Limits of Rationality*. Frankfurt a.M.: Campus/Westview, pp. 9-23.

Wittkowsky, Andreas (2012): *Vernetzte Sicherheit: Begriff, Einordnung und Umsetzung in der Konfliktbearbeitung. Thesenpapier zur Anhörung des Unterausschusses „Zivile Krisenprävention und vernetzte Sicherheit"*. ZIF Analyse/AG Vernetzte Sicherheit. Berlin: Zentrum für International Friedenseinsätze. Retrieved from http:// www.zif-berlin.org/fileadmin/uploads/analyse/dokumente/veroeffentlichungen/ZIF _Thesenpapier_UA_ZKvS-2012.pdf

Wittkowsky, Andreas/Hummel, Wanda/Meierjohann, Jens P./Pietz, Tobias (2011): *Vernetztes Handeln auf dem Prüfstand: Einschätzungen aus den Deutschen Ressorts*. Berlin: Zentrum für Internationale Friedenseinsätze.

Wollmann, Hellmut (2003): Coordination in the Intergovernmental Setting. In Peters, B. Guy/Pierre, Jon (Eds.): *Handbook of Public Administration*. London: SAGE Publications, pp. 594-606.

Wong, Reuben (2005): The Europeanization of Foreign Policy. In Hill, Christopher/Smith, Michael (Eds.): *International Relations and the European Union*. Oxford: Oxford University Press, pp. 134-153.

Wong, Reuben (2007): Foreign Policy. In Vink, Maarten P./Graziano, Paolo (Eds.): *Europeanization: New Research Agendas*. Basingstoke: Palgrave Macmillan, pp. 321-334.

Wong, Reuben/Hill, Christopher (Eds.) (2012a): *National and European Foreign Policy*. London/New York: Routledge.

Wong, Reuben/Hill, Christopher (2012b.): Introduction. In Ead. (Eds.): *National and European Foreign Policy*. London/New York: Routledge. pp. 1-18.

Wunderlich, Carmen (2010): *Moving Beyond Neutrality: Sweden's Changing Attitude Towards the Military Use of Force*. Paper presented at the 7th Pan-European International Relations Conference of the Standing Group on International Relations (SGIR) of the ECPR, September 9-11, 2010. Stockholm.

Yates, Jacqueline (2000): Sweden. In Chandler, J.A. (Ed.): *Comparative Public Administration*. London/New York: Routledge, pp. 148-172.

Zürn, Michael/Herrhausen, Anna (2008): Post-Conflict Peacebuilding: The Roles of Ownership and Coordination. In Rittberger, Volker/Fischer, Martina (Eds.): *Strategies for Peace. Contributions of International Organizations, States, and Non-State Actors*. Opladen: Verlag Barbara Budrich, pp. 271-287.

6, Perri (2008): Joined-Up Government in the West Beyond Britain: A Provisional Assessment. In Bogdanor, Vernon (Ed.): *Joined-Up Government*. Oxford: The British Academy/Oxford University Press, pp. 43-106.

MIX
Papier aus verantwortungsvollen Quellen
Paper from responsible sources
FSC® C105338

If you have any concerns about our products,
you can contact us on
ProductSafety@springernature.com

In case Publisher is established outside the EU,
the EU authorized representative is:
Springer Nature Customer Service Center GmbH
Europaplatz 3, 69115 Heidelberg, Germany

Printed by Libri Plureos GmbH
in Hamburg, Germany